CZECH
REPUBLIC

UNITED
KINGDOM

RUSSIA

GERMANY

BELGIUM

HUNGARY

FRANCE

UKRAINE

SPAIN

GEORGIA

UZBEKISTAN

PORTUGAL

CROATIA

ROMANIA

GIBRALTAR
(UK)

ITALY

GREECE

IRAQ

CHINA

ALGERIA

ISRAEL

MOROCCO

CHAD

ERITREA

ETHIOPIA

KENYA

TANZANIA

INDONESIA

ZAMBIA

SOUTH AFRICA

	H. floresiensis	Germany	H. heidelbergensis		H. sapiens neanderthalensis	
Iraq	H. sapiens neanderthalensis		H. sapiens neanderthalensis		H. sapiens sapiens (Cro-Magnon)	
Israel	H. sapiens neanderthalensis		H. sapiens sapiens (Cro-Magnon)	United Kingdom	H. heidelbergensis	
	H. sapiens sapiens (Cro-Magnon)	Gibraltar (UK)	H. sapiens neanderthalensis		H. sapiens neanderthalensis	
Russia	H. sapiens neanderthalensis	Greece	H. heidelbergensis/H. sapiens		H. sapiens sapiens (Cro-Magnon)	
Uzbekistan	H. sapiens neanderthalensis	Hungary	H. sapiens neanderthalensis	Ukraine	H. sapiens neanderthalensis	
EUROPE			H. sapiens sapiens (Cro-Magnon)	NORTH AMERICA		
Belgium	H. sapiens neanderthalensis	Italy	H. cepranensis	Canada	H. sapiens sapiens (Paleo-Indian)	
Croatia	H. sapiens neanderthalensis		H. sapiens neanderthalensis	Mexico	H. sapiens sapiens (Paleo-Indian)	
Czech Republic	H. sapiens neanderthalensis		H. sapiens sapiens (Cro-Magnon)	United States	H. sapiens sapiens (Paleo-Indian)	
France	H. heidelbergensis	Portugal	H. sapiens neanderthalensis	SOUTH AMERICA		
	H. sapiens neanderthalensis	Romania	H. sapiens neanderthalensis	Brazil	H. sapiens sapiens (Paleo-Indian)	
	H. sapiens sapiens (Cro-Magnon)		H. sapiens sapiens (Cro-Magnon)	Chile	H. sapiens sapiens (Paleo-Indian)	
		Spain	H. heidelbergensis	Peru	H. sapiens sapiens (Paleo-Indian)	
			H. antecessor			

INTRODUCTION TO
PHYSICAL ANTHROPOLOGY

John Steckley

OXFORD
UNIVERSITY PRESS

8 Sampson Mews, Suite 204, Don Mills, Ontario M3C 0H5
www.oupcanada.com

Oxford University Press is a department of the University of Oxford.
It furthers the University's objective of excellence in research, scholarship,
and education by publishing worldwide in

Oxford New York

Auckland Cape Town Dar es Salaam Hong Kong Karachi
Kuala Lumpur Madrid Melbourne Mexico City Nairobi
New Delhi Shanghai Taipei Toronto

With offices in

Argentina Austria Brazil Chile Czech Republic France Greece
Guatemala Hungary Italy Japan Poland Portugal Singapore
South Korea Switzerland Thailand Turkey Ukraine Vietnam

Oxford is a trade mark of Oxford University Press
in the UK and in certain other countries

Published in Canada by Oxford University Press

Copyright © Oxford University Press Canada 2011

The moral rights of the author have been asserted

Database right Oxford University Press (maker)

First Published 2011

Library and Archives Canada Cataloguing in Publication

Steckley, John, 1949–
Introduction to physical anthropology / John Steckley.

Includes bibliographical references and index.
ISBN 978–0–19–543215–2

1. Physical anthropology—Textbooks. 2. Physical anthropology—Canada—Textbooks. I. Title

GN60.S74 2011   599.9   C2010-906637-5

Cover credits: (top) Archaeologist excavating inside a circular hearth structure on a pyramid in the sacred city of
Caral, Peru: Pasquale Sorrentino/Science Photo Library; (bottom) Cave paintings from the Great Hall of the Bulls, in the
Lascaux caves in France: Pascal Goetgheluck/Science Photo Library; (inset, top) Gibbon: toos/iStockphoto.com;
(inset, middle) Skeleton: R.V. Bulck/iStockphoto.com; (inset, bottom) Small ground finch, one of Darwin's finches,
from Santa Cruz Island, Galapagos Islands: Gerald & Buff Corsi/Visuals Unlimited, Inc.

This book is printed on permanent (acid-free) paper ∞ which contains a minimum of 10% post-consumer waste.

Printed and bound in the United States of America

1 2 3 4 — 14 13 12 11

Contents Overview

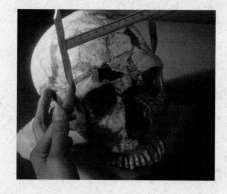

Contents

1 Introduction   2

2 Tools of the Trade: How Old is That?   26

3 | Evolution 50

4 | Evolutionary Pathways 76

5 | Primates: Taxonomy and Behaviour   102

6 | Hominins before *Homo*   130

7 | Early *Homo*   154

8 | Transition to Modern *Homo sapiens*   178

continued

8 | *continued*

9 | Neandertal 202

10 | Human Variation 226

11 | Recent History: Ethics, Migrations, and the Physical Costs of Agriculture 250

12 | Forensic Anthropology   276

13 | Conclusions   304

Preface

Why Write a Physical Anthropology Textbook

You have to be dissatisfied with what is already out there to write a physical anthropology textbook. Very dissatisfied. It's certainly not for the elusive fame and fortune. Here is why I wrote this book.

First, I wanted to provide a book that is written by a Canadian (me) for Canadian students. I think that's important. Canadianized versions of textbooks originally written for non-Canadians (usually Americans) lack a few things, and not just references to hockey, beer, and doughnuts. While anthropology is an international discipline and the main finds reported in this and other textbooks are essentially the same, there are just some aspects that resonate more with Canadian students than with others. Including references to places, schools, and individuals that are Canadian—and ensuring that these are relevant and not just token—is an important part of writing a textbook for Canadian readers. Further, the relationship between Canadian anthropologists and Canadian Aboriginal people—a major cultural group studied by those anthropologists, both for the better and for the worse—is as unique as it is significant and merits serious treatment in a textbook written for a Canadian audience.

Another reason for writing this textbook is that most anthropology texts profess greater certainty and consensus about research findings than I believe they should. Species are presented as if everyone in the field believes they exist, when many of us do not (look up *Kenyanthropus platyops* in Chapter 6 for a good example of that). Science, tentative by its very nature, is more comfortable with expressions like 'appears to be' and 'is consistent with' than with terms like 'definitely' or 'conclusively'. That's why a number of physical anthropologists and other scientists do not do well as expert witnesses in court, where an artificial legal certainty is required—a certainty that honest scientists are reluctant to provide. Textbooks in many areas, including physical anthropology, tend to read as though the authors are absolutely certain about everything, which they are not. The same authors as anthropologists wouldn't present that kind of certainty in an academic journal article, or a conference presentation, or even a classroom lecture. Why, then, should they be so certain in a textbook for their students?

Both of these views reflect what I might call the *subjectivity* of science. Scientists are human, and committed though they may be to 'letting the data speak for themselves', much of what they write is not the data speaking but the scientist speaking. That's a good thing, I believe: a textbook should be as much a narrative that engages students as a reference work. Stories sell important ideas; dullness does not attract bright, young minds.

Textbooks should be human, too. Contemporary literature in the social sciences and the humanities informs us, rightly, that the author of a text should not pretend to be an anonymous, machine-like 'expert', but a human with views, some right, some wrong. The eminent Canadian philosopher Red Green said that the three words that men (Canadian men, specifically) never like to say are *I don't know*. Textbook writers should be able to say that, too. As well as being aware of his or her own limitations, the textbook writer should be able to recognize the lighter side of the material covered. Humans have a sense of humour—so should your textbook.

A further reason for writing this textbook is that while anthropology tends to focus on the past, so much is happening in the field right now (see the last chapter for 'New Thoughts'). Anthropologists involved in teaching have to keep up with the physical anthropology 'news' or they will very much look behind the times. Physical anthropology books have to be continually rewritten and brought up to date. In April 2010, just as I was putting the finishing touches on this text, a major find was announced to the world: the discovery of *Australopithecus sediba*. I'm glad that I was able to include it in this textbook. I live in fear that as soon as this work is published, a team of anthropologists will announce another exciting discovery. Slow it down, guys—keep it for the second edition.

Finally, a personal objective for me in writing this textbook was to learn. Instructors know well that one of the best ways of learning a subject is to teach it. Writing a textbook is an even more intense form of learning—at least it was for me. I know that of all my aims in writing this textbook, I am guaranteed at least one success: I learned a lot.

What to Expect: An Overview

1 | INTRODUCTION

I address the typical student questions *What is anthropology?* and *What is physical anthropology?* (but not 'Is it like psychology?'). I go about it by looking at what the discipline (anthropology) and the sub-discipline (physical anthropology) do to show us why the sasquatch legend is b.s., and what ever happened to the St Lawrence Iroquoians (the people who named our country 'Canada'), and how the story of Piltdown Man—perhaps the biggest scientific fraud of the twentieth century—tells us a little bit about how overzealous scientists (physical anthropologists among them) can sometimes be led astray.

2 | TOOLS OF THE TRADE

So what does an anthropologist have to do to get a date? Dig up an old friend? This chapter looks at *relative dating*—how we know that something is older than something else—and *absolute dating*, highlighting the Canadian site of Crawford Lake in southern Ontario. We'll also see how assigning a date to a piece of volcanic rock can cause a huge argument that a professor teaching at a university in Atlantic Canada helped resolve with his knowledge of old pigs.

3 | EVOLUTION

Here we tackle the most controversial concept in science, and illustrate how religion and science do not have to clash. Readers will learn what is and isn't evolution, why we are all X-Men, and how to convince a doubter that evolution is true. Finally we will learn that moths can lie.

4 | EVOLUTIONARY PATHWAYS

Time is deep, continents move, and species can be extremely difficult to define. Taxonomy is changing fast and in ways that force us to relearn (or at least rename) our animal kingdom. Birds might not be birds, and there is now no pure class called reptiles. What does it take the join the primate club? And how is a cow like a primate? Finally, we'll learn how a fossil specimen can go from potential human ancestor to distant ape in just 20 years.

5 | PRIMATES

Primate taxonomy is changing, too. There is no clearly defined group we can call monkeys, although we know this: if it doesn't have a tail, it's *not* a monkey. Zoo visitors please take note. Gibbons don't have tails, so they are not monkeys. And Celebes apes, which have very tiny tails, *are* monkeys. Probably most interesting in this chapter are the many ways in which primates other than humans can have culture. Finally, we'll dispel the myth—spread widely on the Net—of the hundredth monkey.

6 | HOMININS BEFORE *HOMO*

In this chapter students will learn that they will have more species names to memorize than students 20 years ago, 10 years ago, or even last year. Significant finds are being reported every few years. The emphasis is on the fact that there is no agreement—no agreement on how many pre-*Homo* species there are, and no agreement on what the pathway to *Homo* (our genus) is. This is a source of much argument. Certainty (unlike a gibbon) hangs on wobbly branches. Even our path to understanding how we came to be bipedal (walking on two limbs rather than the more usual four) has had obstacles placed in its way.

7 | EARLY *HOMO*

This chapter offers more of the same—more species claimed, more lack of certainty about where they all fit—only this time it's closer to home, closer to us. Arguments abound. Some interesting fossil specimens are discussed here, including our first really early European and a small and comparatively recent Indonesian nicknamed 'the Hobbit' (because of her height; no ring, unfortunately). Sadly, our descent from giants rings false.

8 | TRANSITION TO MODERN *HOMO SAPIENS*

Modern humans, where did they (i.e. you) come from? Did they all come from Africa, or did some originate elsewhere as well? Is there a hybrid theory that will keep most physical anthropologists happy? Physical anthropologists are following the molecular trails across the continents—of mtDNA for the girls and Y-chromosome for the boys—from Africa to India, to Australia, winding up in Europe. But what do we do with the early moderns of Asia? How did they get there? That isn't settled—yet. In this chapter, we'll see that there are no tools like old tools, and they're put to work in four different diagnostic methods. We conclude by wondering aloud whether there has been too much stress on the importance of Europe, the ancestral land of most physical anthropologists, in charting the evolutionary journey of *Homo sapiens*.

9 | NEANDERTAL

Neandertal is a pop-cultural bad boy, made famous in books and movies. It's also a good shot taken at that big, awkward, not-so-bright opponent you run into at your Thursday-night rec league hockey game. But is Neandertal in or out of our gene pool? Or, as one Canadian suggests, is he just dipping his toe in the cold water of humanity? In this chapter we'll look at the changing images of the big guy with the brow ridges and the bun in the back, from Boule's bent-over freak to Blanc's spiritual cannibal, to the compassionate flower child of the 1960s. And we'll address the question of why and how Neandertal disappeared. Did they, like the bulkier, fuel-inefficient cars of recent times, simply become hybrids?

10 | HUMAN VARIATION

What is 'race'? Can it be used scientifically to classify human beings into different groups? These are loaded questions. The most famous classifier, skin colour, has been used for a long time, but we know it isn't reliable (quite apart from the fact that 'redskins' do not have red skin, and 'white folks ain't white'). How do you feel about using ear wax? Bones, the favourite object of study for most physical anthropologists, have been used by forensic anthropologists to classify people racially—how does *that* work? And certain geographic populations are more prone or resistant to particular diseases and disorders than others are. How do we work that into the race picture?

11 | RECENT HISTORY

In this chapter we'll look at the rights of science, typically embodied in 'outsider scientists' (i.e. scientists merely visiting from outside of the community their research findings will affect). We'll also look at the rights of the dead of the last 15,000 years and their descendants. We will see how agriculture makes the body pay for food in ways that hunting and gathering does not, and then we'll consider the peopling of the Americas—once a simple picture of a single group crossing a land bridge, going through an Albertan corridor, and being Clovis First, the story is now more complicated, with questions about timing and the possibility that initial immigration occurred in different waves and by different means (think boats) to various points along the BC coast. Finally, we examine just one in a long series of failed attempts by anthropologists of European origin to demonstrate that Europeans have been in the Americas for longer than we think. Give it up already!

12 | FORENSIC ANTHROPOLOGY

The TV crime series *CSI* (which, we will see, distorts the picture), physical anthropologists turning their talented hands to writing murder mysteries, and the opening-up of new positions in forensic anthropology across Canada all show and contribute to the popularity of this form of physical anthropology. And then there is that fascinating anthropological research centre known as 'The Body Farm'—it grows research, not bodies. How do forensic anthropologists determine how old skeletal (or partially skeletalized) remains are? How do they sex the skeleton (i.e. determine its sex)? How do they establish the controversial 'race' of the skeleton? And how do they engage in human rights work across the world? This chapter digs deep into the questions of forensic science and its applications in anthropology.

13 | CONCLUSIONS

This is a chapter of new thoughts—about the species concept, genetic or molecular dating (blame the penguins), race and the racial face of physical anthropologists, bipedalism, human evolution, and the Aboriginal genetic input into the peoples of the Caribbean. It ends with ten great books that you might want to read over the summer.

Acknowledgements

I have many people, human and non-human, to acknowledge for their assistance. First, I must thank the people at Oxford University Press, who know that I am eccentric, and yet put faith in me and treat me with respect. Among them are Jodi Lewchuk, Nancy Reilly, Kathryn West, Eric Sinkins (who fortunately has a sense of humour that is similar to my own), and Steven Hall, as well as several others whom I did not get to know because they were involved behind the scenes in the production of this book. I owe special thanks to David Stover, the president of Oxford University Press Canada, who asked me three times before I agreed to write an anthropology textbook for his company. I'd also like to thank the following reviewers, and those who remained anonymous, whose thoughtful comments and suggestions helped give shape to this book:

- **Julie Cormack**, Mount Royal University
- **Elizabeth Finnis**, University of Guelph
- **Moira McLaughlin**, St Thomas University
- **Carol MacLeod**, Langara College
- **Farid Rahemtulla**, University of Northern British Columbia
- **Anna Lucy Robinson**, Lambton College
- **Jocelyn S. Williams**, Trent University

You need colleagues before you write a textbook. Two in particular shone through for me. Humber College's Mike Badyk I must thank for constantly sending me material that improved the information base of this book (he is still sending me stuff on Neandertal). Bryan Cummins was there to support all of my efforts, especially in the Toby Jug, the pub home of many good ideas. With his broad knowledge, Bryan was the perfect person to bounce anthropological ideas off of when I wasn't sure.

Closer to home (actually at home), there is Angelika (Angie), my wife, who had to live with half a husband in the summer of 2009 when my mind was totally immersed in this book. Without her support, nothing would ever be written. Then, there are the dogs. Egwene, the beautiful border collie whom we lost in summer of 2010; she taught me so much about life and how to live. Wiikwaas and Cosmo, the 'Biscuit Brothers', helped me with their demands for walking when my brain was too crowded with ideas. Finally, there are the parrots: Quigley, Stanee, Louie, Lime, Finn, Meesha, and the newcomers, Tika and Sammy. Never has a book that has not been specifically about parrots had so many references to parrots. You have taught me that birds are people too, and have opened me up to a world I never knew existed.

From the Publisher

Oxford University Press is delighted to introduce this brand-new text, designed to provide students with a fresh take on physical anthropology. Drawing on Canadian content in an international field, author John Steckley spins a narrative that sets the foundation for introductory courses while adding a contemporary flavour throughout. Profiles of Canadian anthropologists, snapshots of current debates, a stand-alone chapter on the increasingly popular world of forensics, and a concluding chapter dedicated to new trends in the discipline together make for the most current and relevant textbook on the market. We hope that as you flip through the following pages, you will get a sense of the text's accessible content, engaging features, and unique approach, all combining to hook students on a fascinating and ever-evolving field of study.

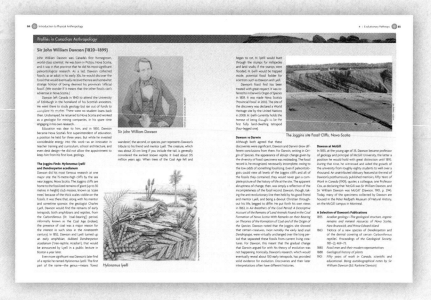

A Canadian textbook for Canadian students. Canadian from cover to cover, *Introduction to Physical Anthropology* draws on Canadian profiles, examples, and research throughout. A focus on the relationship between Canadian anthropologists and Canadian Aboriginal people provides relevant and interesting cases for student study.

Forward-looking coverage. Stand-alone chapters on forensic anthropology and on ethics, migrations, and the physical costs of agriculture combine with up-to-the-minute findings, current debates and controversies, and new media coverage to reflect emerging trends in the discipline.

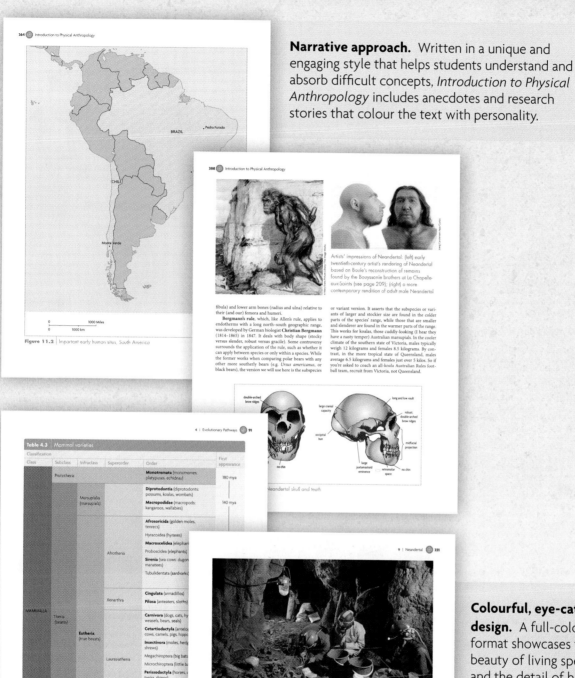

Narrative approach. Written in a unique and engaging style that helps students understand and absorb difficult concepts, *Introduction to Physical Anthropology* includes anecdotes and research stories that colour the text with personality.

Colourful, eye-catching design. A full-colour format showcases the beauty of living species and the detail of historic finds to seize the student reader's attention. Figures and maps are clear, bright, and easy to navigate.

Student-centred pedagogy. Boxed features include topical case studies, Canadian biographies, and stories of 'Science Gone Wrong'. Teaching and test tools include a full set of review and critical thought questions, key terms, and typical student questions that tackle the issues your students are most likely to raise in the classroom.

Complete ancillary package. Online supplemental material includes an instructor's manual, PowerPoint slides, a test generator, and a student study guide—a full suite of additional teaching and learning tools for instructors and students.

COMPANION WEBSITE

John Steckley

Introduction to Physical Anthropology
ISBN 13: 9780195432152

Inspection copy request

Ordering information

Contact & Comments

About the Book

A new Canadian entry on the market, *Introduction to Physical Anthropology* offers a fresh take on the fundamentals of the discipline. Author John Steckley guides students through the field step by step, covering foundational aspects such as archaeological tools and dating methods, evolution and genetics, and taxonomy in the early chapters before moving on to more complex topics. After tracing the transition from hominids to modern Homo sapiens, students are introduced to Neandertals; human variation; ethics and migration; and forensic anthropology. Combining scientific discoveries with personal anecdotes and case studies on controversies, students are shown the human side of scientists and the subjective side of science.

Instructor Resources

You need a password to access these resources. Please contact your local Sales and Editorial Representative for more information.

Student Resources

Tables, Figures, and Maps

Tables

Figures and Maps

Boxed Features

Profiles in Canadian Anthropology

Case Studies Controversies

Science Gone Wrong

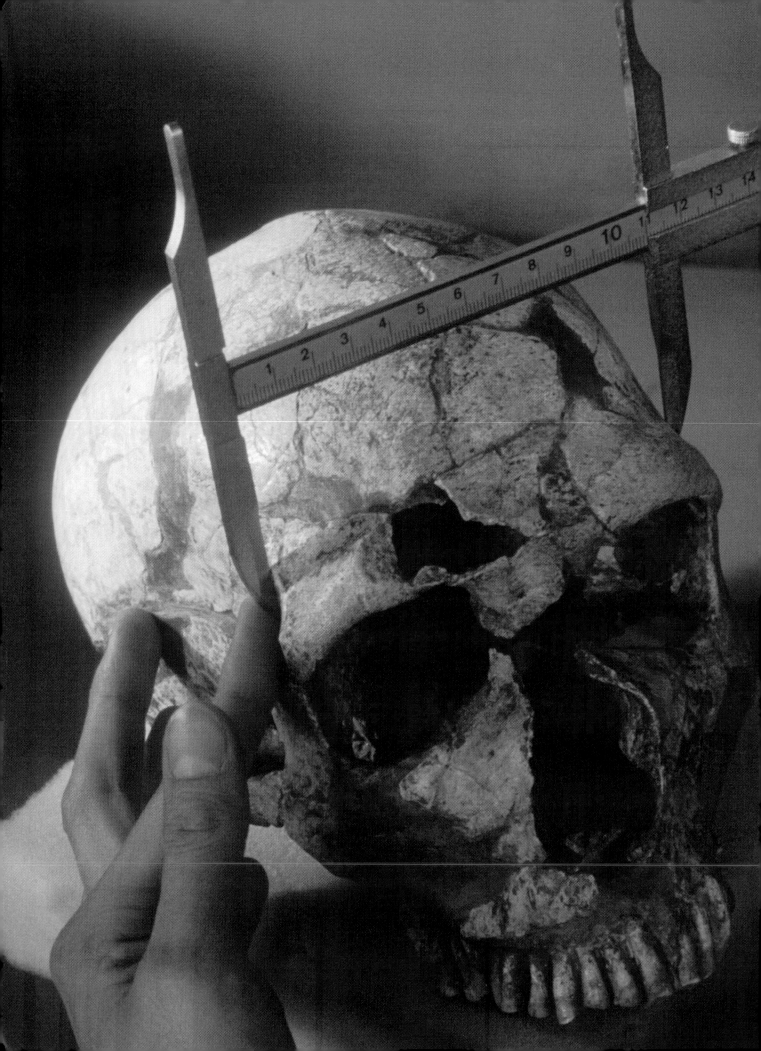

Introduction

Measuring the skull of a Neandertal
© Pierre Vauthey/Corbis Sygma

Introduction: Physical Anthropology Faces Bigfoot

You see it in the distance, and can barely believe your eyes: a somewhat blurry, definitely furry figure, large in size, that's looking back at the person holding the camera. You can imagine that it might be trying to decide whether to take a few long strides back and attack the much smaller human invading its privacy.

The brief footage—under a minute long—was shot in 1967 by Roger Patterson. It is considered by many self-described sasquatch investigators to be the best evidence yet for their elusive quarry. The Discovery Channel presenters, with their authoritative voices, will inform you that scientists have been unable to prove that the film is a fake. More to the point is that no one can prove it is real.

What can a physical anthropologist add to the discussion? Physical anthropologists work primarily with bones. The technical name for the study of bones is **osteology** (lit. 'bone study'). Are there any sasquatch bones to investigate? If there has been a successful breeding population (of, say, a few hundred at a time, each living maybe 50 years) over the more than 200 years that Europeans have claimed to have seen them, then there should be hundreds of remains. If you add to that the hundreds, maybe thousands, of years more that the Salish people of British Columbia have spoken of *se'sxac* ('wild men'), there should be bones of thousands of individuals. But no bones have been found. Believers ask skeptical physical anthropologists to accept that, unlike other mammals, fish, reptiles, and birds native to the area, the sasquatch have hidden the bones of their dead well. That seems unlikely—humans hide their bones, and yet anthropologists commonly discover human remains to study.

How about teeth? Physical anthropologists have been known to discover some amazing things about a species by just looking at a few teeth. Canadian anatomist **Davidson Black** (1884–1934) was able to declare *Sinanthropus pekinensis* (**Peking Man**) a unique species based on his analysis of just three teeth. However, no sasquatch teeth have been found. Perhaps some hair has been discovered, caught in branches near that famous sighting and providing valuable evidence for forensic analysis. As it turns out, no. Poop, maybe? Nope, no dice—not even the relatively sanitary fossilized poop called 'copralites' that some archaeologists and physical anthropologists use to determine diet and DNA. In fact, no physical evidence exists for a physical anthropologist to study in order to prove the existence of the creature.

The so-called evidence of Bigfoot is all easily faked: film, footprints, and lots of stories. The first published reference to sasquatch 'footprints' was in an account written by the normally reliable explorer David Thompson in 1811 (around the time that fossilized dinosaurs and frozen mammoths were being found and reported on). Shortly thereafter came the first eyewitness accounts. Since around the 1840s there have been hundreds of alleged 'sightings'.

What is it that people claim they have 'seen'? Here's a summary:

A hairy, foul-smelling creature having the following physical characteristics:

- height: anywhere from 6 feet to 15 feet
- weight: between several hundred and over a thousand pounds
- footprint size: up to 24 inches long and 8 inches wide
- sound: either completely silent or else making a high-pitched whistle.

One famous sasquatch sighting occurred in 1884 in the Fraser canyon of BC. 'Jacko', as the creature was nicknamed, was described as being about 1.35 metres (4′ 6″) and weighing between 54 and 68 kilograms (about 120–150 lbs), with strong hands and a distinctive method of locomotion (specifically, 'knuckle walking'). These relatively consistent details suggest that Jacko was really an escaped chimpanzee.

As for the sparse and widely varying details from the body of sasquatch sightings, there isn't a lot a physical anthropologist can say other than to declare that there is absolutely no solid evidence for the existence of sasquatch (i.e. Bigfoot) or Yeti (a.k.a. the Abominable Snowman). On the other hand, when presented with even the barest

Case Study Controversies

The Meaning of Sasquatch Legends from an Anthropological Point of View

The four branches of anthropology can help us address (okay, debunk) reported sasquatch sightings. But anthropology can also help us understand the prevalence of sasquatch legends that have existed in Western culture for centuries.

Sociocultural anthropologists can tell us that the word 'sasquatch' comes from se'sxac, a Salish word meaning 'wild men'; the se'sxac figure prominently in traditional stories of the Salish people of British Columbia. Similar figures are common throughout Aboriginal Canada. Legends of the Kwakiutl people feature a female cannibal called

Frame from Roger Patterson's film footage of a sasquatch, taken in October 1967 at Bluff Creek, California

Dsonoqua, who is often portrayed on totem poles. On the opposite coast of Canada, the Mi'kmaq of the early 1600s told stories of a female cannibal giant known as Gougou, who was said to live by Gaspé and who, according to their legends, terrified the men travelling with French explorer and trader Samuel de Champlain as they sailed past the Gaspé peninsula. Linguistic anthropologists can tell us that related terms and concepts include kukwes ('little Gougou'), the name for a cannibal giant connected with earthquakes, and kukwech (also 'little Gougou'), a name for owls.

The figure of a wild and hairy creature living just beyond society is common in the traditional stories not just of the Aboriginal people of Canada but of people all over the world, from northern California ('Bigfoot') to the Himalayas (the 'Yeti', or 'Abominable Snowman'). Around 3,000 years ago the people of ancient Mesopotamia (in the area of modern-day Iraq and Iran) told a story of Enkidu, a wild man created by the god Anu. Enkidu lived with animals, and was 'tamed' by the hero Gilgamesh.

There is no archaeological evidence for any of these creatures, no physical evidence for physical anthropologists to study and make pronouncements on. What, then, can anthropologists say about such stories, which we hear around the world?

Storytelling is very important to humans the world over. It is universal. All societies have stories that are alternately scary, funny, and instructive. Not all stories have to be true. We often don't let 'fact' get in the way of a good story. We are like our ancestors in many ways, including the love of tales starring creatures (sasquatch, vampires, werewolves, bogeymen, Frankenstein's monster) that may not exist but that nevertheless scare and amuse us and teach us lessons.

What Do You Think?

1. What are the weaknesses of the evidence that has been given for the existence of sasquatch and its 'cousins'?
2. Why are there so many stories about the sasquatch and sasquatch-like creatures?

physical evidence—bones, teeth, hair, and so on—the physical anthropologist can make some amazing discoveries. This is a message I hope you will take from this book.

The Four Branches of Anthropology

The term **anthropology** (lit. 'study of man'—a meaning applicable to the early anthropologists, almost all of them men focusing almost all of their attention on other men) involves studying humans in all times, situations, and places. For anthropologists, such breadth and diversity are strengths of the discipline. There are four principal branches of anthropology:

- sociocultural anthropology
- physical anthropology
- linguistic anthropology
- archaeology.

Sociocultural anthropology (sometimes called simply **cultural anthropology**) studies the cultures of people. The main research techniques of the sociocultural anthropologist include doing fieldwork with people, living with them for significant periods of time, and making generalizations based on cross-cultural analyses. For example, sociocultural anthropologists have noted that in societies that are heavily engaged in the type of relatively

small-scale agriculture known as horticulture (sometimes called 'slash-and-burn agriculture'), women tend to be the farmers, and their work is highly respected. The Huron of the seventeenth century, who were heavily involved in the cultivation of corn, beans, and squash, are a model of this kind of society.

Physical anthropology (or **biological anthropology**) looks at bones, teeth (and sometimes the rest of the body), diseases, and our **primate** relatives. It classifies species and subspecies, considers how the lives and lifestyles of people made their bodies the way that they are, and, in recent times, is very involved in investigating the sometimes mysterious circumstances surrounding the deaths of individuals.

Linguistic anthropology (or **anthropological linguistics**) explores languages, their relationship to cultures, and how even fragments of language can be used as important sources of historical information. For example, we know that the Apache and the Navaho of the American Southwest used to live in northwestern Canada, based on the fact their languages are closely related to those of Aboriginal people such as the Sarcee living in the Calgary area.

Archaeology digs into the past literally, like a detective trying to find 'what happened here' and 'who was here'. The trowel (a small, triangular digging implement) is the archaeologist's main tool of investigation. One of the gifts that this branch of anthropology gives us is a fresh understanding of people who did not write (e.g. Canada's Aboriginal people), but who were written about negatively by those who did (e.g. the European newcomers).

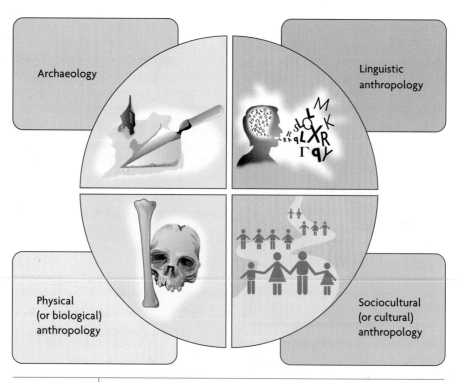

Figure 1.1 | The four branches of anthropology

A Case Study: Using the Four Branches to Solve a Historical Mystery

One way to illustrate the similarities, differences, and information sharing among the four branches of anthropology is to apply them to a common problem. Here is one that is a great historical mystery in Canada: what happened to the **St Lawrence Iroquoians**?

The St Lawrence Iroquoians were a people that French explorer Jacques Cartier encountered along the St Lawrence River in the 1530s. The interactions between the French and Iroquoians were very unequal, establishing the you-give-and-we-take pattern that would characterize the one-sided (favouring Europeans) relationship for centuries to follow. The St Lawrence Iroquoians saved Cartier's men from death by scurvy by showing him how to boil the inner bark of an evergreen tree to obtain the vitamin C needed to survive; in return for this hospitality, Cartier kidnapped the sons of Chief Donnacona and brought them to France on his first return home (an offence he committed again on his second visit to the 'New World'). Among the short list of Iroquoian words that Cartier recorded was the word 'Canada', which means 'village' (although the vast majority of Canadians don't know that).

When the French returned in full force to Canada during the early 1600s, the St Lawrence Iroquoians, as far as the French could see, had disappeared. What had happened to them? We can gain important clues from the four branches of anthropology.

We'll look to physical anthropology for our first clue, which will also contribute others. One aspect of the biological human that physical anthropologists study is the *social patterning of diseases*—in other words, who is likely or unlikely to catch and die from certain diseases. In this case, physical anthropologists tell us that when visiting Europeans first make contact with Aboriginal people, the former transfer diseases to the latter, with a significant negative impact on the size of the Aboriginal population. It's quite likely that the population of the St Lawrence Iroquoians was greatly reduced by diseases such as flu and measles spread from contact with Jacques Cartier and his men. This could have devastated Aboriginal communities and tribal units. To confirm this hypothesis, a physical anthropologist could look at the teeth of St Lawrence Iroquoians for signs of **hypoplasia**: when your so-called permanent teeth grow in, the growth of the enamel covering the teeth is affected by elements that cause nutritional stress, particularly disease and periodic lack of nutritious food. Hypoplasia records the times of nutritional stress in striations and grooves in the teeth.

Using archaeology we can access clues about the demise of the St Lawrence Iroquoians from their pottery. Like other Iroquoians (people who speak or spoke languages in the Iroquoian family of languages), they made pottery, largely for cooking corn soup. Every nation's pottery features distinctive decoration and finishes, particularly around the rim. This is the sort of detail that archaeologists can make much out of—a PhD dissertation at the very least.

Among the characteristic elements of St Lawrence Iroquoian pottery is a distinctive rim pattern that features a repeated image of what looks like a tiny cob of corn. These pots have been discovered not just in known St Lawrence Iroquoian communities but at other sites also. A small percentage (1.2 and 3.4 per cent respectively) of pottery found in two east Huron sites along the Trent River system in eastern Ontario was St Lawrence Iroquoian pottery; these communities were inhabited between roughly 1450 and 1530. A site in northwestern Toronto dating to around 1450 has almost 10 per cent of its pottery as St Lawrence Iroquoian. The percentage of recovered pottery dating from about the time of St Lawrence Iroquoian contact with Cartier to the return of the French increases on sites along the Trent system, reaching a high of 15.6 per cent in one community. South of Ontario and the St Lawrence, in what is now upstate New York, pottery from this period and throughout the sixteenth century has been discovered on sites occupied by the Mohawk, Oneida, and Onondaga (the three eastern tribes of the Iroquois). Farther north and east, in Vermont and Maine, the pottery shows up on what appear to be Abenaki sites.

Sociocultural anthropology helps us out here with some clues related to gender roles. We know from historical sources that women were potters, and that pottery was not an item of trade—after all, why trade for what you have in abundance? We can say, then, that where St Lawrence Iroquoian pottery was, so were St Lawrence Iroquoian women. What neither sociocultural anthropology nor archaeology can tell us definitively right now is under what circumstances the women arrived at sites where their pottery has been found. Were they captured as prisoners? Did they travel to other people's lands through marriage exchanges with trading partners? Were they part of a general St Lawrence Iroquoian merger with neighbouring peoples, as disease and potential warfare with others threatened their tribal or national structure? And were they accompanied by any St Lawrence Iroquoian men, or did they travel this path to other nations alone?

We could turn again to physical anthropology for help in answering these questions. Physical anthropology could address the situation by determining whether a significant proportion of the women in a community or area came from 'away'. April DeLaurier and Michael Spence performed a study of this kind with **cranial** (i.e. skull) analysis of women buried in the Moatfield ossuary in northeastern Toronto, just a little south of the 401 and west of Don Mills Road. They were able to conclude, based on a few cranial features, that 'At some point, or repeatedly over some time, a substantial proportion of the Moatfield females may have married into the community from some more distant, and genetically different, population' (DeLaurier & Spence, 2003, p. 291). Now, while this does not directly help us with the St Lawrence Iroquoian story, since the Moatfield site is too early (AD 1280–1320), it does show the methods at the disposal of a physical anthropologist attempting to answer questions about the disappearance of the St Lawrence Iroquoians.

Linguistic anthropology helps to fill in some gaps in the story. We have a short list of St Lawrence Iroquoian words compiled by Jacques Cartier, in addition to some others found in a Huron dictionary published in 1632. Some of the St. Lawrence Iroquoian words found in the early Huron dictionary relate to trade, such as their words for glass beads (based on the word for 'eyeball') and awls (sharp metal tools used to etch marks and bore holes in wood, bone, or antler). Some others relate to animals then found abundantly in the St Lawrence River, including eels and whales. And many of the Huron dictionary entries are recorded

in structurally simplified forms that are typical of a trade language or pidgin (a simplified form of language used by speakers of two different languages when they want to communicate). From sociocultural anthropology we know that Iroquoian men were involved in the trading of goods, so it seems reasonable to speculate that St Lawrence Iroquoian men were living with the Huron, acting as middlemen in the trade between the Huron and the French. Linguistic evidence suggesting that the St Lawrence Iroquoians may have joined the Huron includes a place name from about this time, associated with St Lawrence Iroquoian territory and meaning 'they join', and a people's name that has not been connected with any particular people in the area, which may have been used to refer to a joint St Lawrence Iroquoian–Huron community.

In sum, then, we can say that the nation of St Lawrence Iroquoians may have disappeared by the early seventeenth century, owing largely to the introduction of foreign

Profiles in Canadian Anthropology

James F. Pendergast (1921–2000): Scientific Passion

James (Jim) F. Pendergast had the one characteristic an anthropologist (or any scientist, really) cannot succeed without: a passion to learn. He is best known in archaeological circles for what one respected colleague described as 'his almost single-handed resurrection of the St Lawrence Iroquois from an undeserved obscurity' (Wright & Pilon, 2004, p. 2). His career is full of examples of the passion that drove him. In 1940, at the age of 19, he joined the Canadian army as it went off to World War II. He entered the war as a private, and left it a lieutenant (pronounced the Canadian way: leff–tenant). After the war he remained in the army, where he continued to rise through the ranks. But it was archaeology that would take up an increasing portion of his energy.

Pendergast's specialty was military intelligence, and one of the requirements for his promotion to major was to demonstrate his ability to investigate and intellectually dissect a historic battle. Rather than choosing the American case studies that were readily available at the time, he chose to study something closer to home: the Battle of Crysler's Farm, fought during the War of 1812 along the St Lawrence River, near his birthplace of Cornwall, Ontario. In 1947–8, he investigated the site of the battle, where he uncovered military artifacts, including objects he located with a metal detector. But what interested him most was not the metal items, or even the military ones. What piqued his interest were much older artifacts: **sherds** of pottery.

Pendergast took his findings north to Ottawa to have the pottery identified by Douglas Leechman of the National Museum. He was told that it was Iroquoian, but possibly not related to the Iroquois (who lived in New York), being similar to material from a site 40 miles south of Ottawa, the Roebuck site, on which the Museum had published a report over 10 years earlier. It was then that Pendergast established two lifelong relationships: with staff of the National Museum and with the St Lawrence Iroquoians.

Pendergast studied archaeology under the legendary Richard 'Scotty' MacNeish, who trained him in the field like a drill sergeant, a method that suited Pendergast perfectly. Soon he was excavating sites and analyzing artifacts on his own. He didn't let being stationed outside of Canada stop his unpaid passion. He had pottery rim sherds from the Roebuck site added to the cargo of military aircraft flying supplies in to the base in the Netherlands. His own comments on what he chose to study tell us much about him:

Very early in the game I asked myself, . . . what should I concentrate on if I am interested in archaeology? Should I concentrate on antiquity alone, because carbon 14 . . . was coming into use and age was becoming important? Status lay in the oldest, not necessarily the biggest or best. I

diseases. Smaller groups or individuals of the St Lawrence Iroquoians may have joined up with the Huron to the west and, to a lesser extent, with the more eastern three nations of the Iroquois and with the Abenaki. There, they brought their skills and knowledge to their new compatriots, particularly in the areas of trade and pottery making, and have descendants among those peoples today.

Branches of Physical Anthropology

We have set physical anthropology within the broader context of anthropology. Now it's time to break physical anthropology up into its own branches of study. A word of warning to start: no two physical (or biological or evolutionary) anthropology textbooks divvy the discipline up exactly the same way. Even the name itself is a source of disagreement.

The discipline was originally called physical anthropology because it dealt with the physical aspects of human beings—their bones and teeth, in particular—and also, to a certain extent, the physical objects that surrounded them, such as their tools. Many of those involved in the discipline today prefer the term 'biological anthropology', since those in the field deal mainly with humans as biological entities. I've stuck with physical anthropology because of the close connection between the biological entities studied and their physical environment (including material culture, climate, and geographic features). The biological is not in isolation.

It's also important to know that no discipline stands alone. Even though there are separate departments in universities dedicated to offering courses named for their particular disciplines, physical anthropology often involves research done in a variety of areas, and physical anthropologists can often be called experts in disciplines outside the field of physical anthropology.

(also) thought, why should I become interested in Aztec or Egyptian archaeology? I'll never go to Mexico and I'll never go to Egypt. Why don't I just find out what Roebuck is all about? At the time we had a date on but one village and I was fairly certain that we couldn't have (just) one of anything. There must be other villages to be investigated. So, I made a deliberate choice to keep my archaeological interests close to home, where I could work on weekends. That seemed to be more practical, as opposed to getting involved in the glamour of the biggest, oldest and so on. (as cited in Dyck, 2000, p. 10)

Pendergast retired from the army in 1972 and was immediately hired on as an assistant director at the National Museum. He knew that archaeological findings needed to be published to be known, but the avenues for publishing archaeology news in Canada were few. So he developed the *Mercury Series*, a set of stand-alone, book-length publications designed to spread the word on current findings of interest to archaeologists, anthropologists, ethnologists, and museum curators. While he officially retired from the Museum in 1978, he never retired from archaeology—this list of selected publications, most of them from after 1978, is evidence of that (and of his single-mindedness).

James Pendergast: Selected Publications

1973   *The Roebuck Prehistoric Village Site Rim Sherds—An Attribute Analysis*

1978   (with Bruce Trigger) 'Saint Lawrence Iroquoians'

1981   *The Glenbrook Village Site—A Late St Lawrence Iroquoian Component in Glengarry County, Ontario*

1985   'Huron—Saint Lawrence Iroquois Relations in the Terminal Prehistoric Period'

1990   'Emerging Saint Lawrence Iroquoian Settlement Patterns'

1991   'The Saint Lawrence Iroquoians: Their Past, Present and Immediate Future'

1993a   'Some Comments on Calibrated Radiocarbon Dates for Saint Lawrence Iroquoian Sites'

1993b   'More on When and Why the St Lawrence Iroquoians Disappeared', in *Essays in Saint Lawrence Iroquoian Archeaology*

1996   'Problem Orientation for St Lawrence Iroquoian Archaeological Research'

1999   'The Ottawa River Algonquin Bands in a St Lawrence Iroquoian Context'

Postscript

I enjoyed writing about Jim, as he had an influence on my own career. He was supportive of my research and encouraged me to pursue my seemingly obscure work with the Huron language.

Human Evolution

Humans, like all other living entities, started off as something else. Before we were **Homo sapiens** (lit. 'Man thinking'), we were *Homo something-else* (such as the perpetually funny **Homo erectus**), and even *Something-Else something-else* (e.g. **Australopithecus afarensis**). Confused? Then it's time for a quick lesson in *binomials*, the two-part Latin names used to classify living things.

Two-part names begin with the **genus**, which is capitalized and italicized and which may be abbreviated to its first letter after the first reference. One genus can be shared by several related species. An example is the *Marmota* genus, which gives us *Marmota monax* (woodchucks), *M. flaviventus* (the yellow-bellied marmot), *M. vancouverensis* (the Vancouver Island marmot), and *M. caligata* (the hoary marmot, from whose whistling call BC's Whistler Mountain got its name). The **species** name is not capitalized, though it is italicized.

Genus and species are the lowest two categories in a hierarchy of groups to which all organisms belong. The usual hierarchy (but see the discussion of taxonomy in Chapter 4) is as follows (in descending order):

kingdom
phylum
class
order
family
genus
species.

Fluffy the housecat's description beyond being a mammal is Carnivora (order), Felidae (family), *Felis* (genus), *catus* (species).

Human evolution entails the adaptation or change of our ancestors and some of their 'cousins' across millions of years. The time period is important to keep in mind, as sometimes when you read about changes, it seems as though they happened practically overnight—like one day *Homo erectus* looks in the mirror and sees *Homo sapiens*. A hard concept to grasp in studying evolution is the massive time period required for a species to evolve into something different.

Genetics

The field most directly connected with evolution is **genetics**, as it deals specifically with the mechanics and the signs of evolutionary change. Genetics is a discipline unto itself, technically separate from physical anthropology. However, physical anthropologists are usually well educated on the subject, and they sometimes rely heavily on the conclusions of geneticists, particularly when it comes to talking about how closely related living primates are.

All is not peaceful between geneticists and physical anthropologists, though. There have been some clashes between members of the two disciplines dating back to the 1960s, particularly concerning the dating of the split between chimpanzees and humans: geneticists believe that humans and chimps parted company about five to seven million years ago (the generally accepted figure now), while physical anthropologists put the date anywhere from 15 to 30 million years ago. The two fields have had to learn to respect each other's research methodologies. There will be more on this little squabble later.

Paleoanthropology

Paleoanthropology (lit. 'old anthropology') involves the study of fossil evidence for signs of human evolution. The holy grail of paleoanthropologists is the illusive and reputation-making **missing link** between humans and other primates—the common ancestor, perhaps, of chimpanzees, bonobos, and humans. It's 'dirt-and-bones' anthropology, mixing physical anthropology with archaeology. This made paleoanthropologists the romantic heroes of physical anthropology, at least until the forensic anthropologists became so popular. The list of famous ones includes the renowned **Leakeys**—**Louis** (1903–1972), **Mary** (1913–1996), **Richard** (b. 1944), and **Meave** (b. 1942)—as well as **Donald Johanson** (b. 1943), the discoverer of 'Lucy', the skeleton of a female *Australopithecus afarensis*. The following passage gives a good sense of both the romance and the real-life hard slog of the occupation, thanks to the artful writing of

Famous paleoanthropologist Donald Johanson

Famous paleoanthropologist Mary Leakey

Richard Leakey. The year was 1984; the location was the very-fruitful-in-terms-of-bones (there being no term 'boneful') east shore of Lake Turkana, in Kenya:

> On August 23rd, **Kamoya Kimeu**, my oldest friend and colleague, spotted a small fragment of an ancient **cranium** lying among pebbles on a slope near a narrow gully that had been sculpted by a seasonal stream. Carefully we began to search for further fragments of the skull and soon found more than we dared hope for. During the five seasons of excavation that followed this find, amounting to more than seven months in the field, our team moved fifteen hundred tons of sediment in the massive search. We uncovered what eventually turned out to be virtually the entire skeleton of an individual who had died at the edge of the ancient lake, more than 1.5 million years ago. Dubbed by us the **Turkana boy,** he was barely nine years old when he died; the cause of his death remains a mystery. (Leakey, 1994, p. xiv)

Anthropometry

A major traditional aspect of a paleoanthropologist's job is **anthropometry**, the measurement of the human body, particularly the human skeleton. During the nineteenth century, the most favoured aspect of this field by far was **craniometry**, the measurement of the cranium, or skull. The main purpose to which craniometry was set was to compare 'races', typically so that white scientists could attempt to demonstrate, quantifiably, the superiority of their race to others, particularly the blacks and 'Indian' races. To put it mildly, this kind of thinking was wrong-headed.

A classic case of using craniometry to justify racial prejudice is described in the writings of **Samuel George Morton** (1799–1851), a Philadelphia doctor/scientist who collected more than 1,000 skulls. He was a supporter of **polygeny**, the idea that different races were created separately by God, each with a different measure of natural intelligence. He believed that races could be objectively ranked for intelligence based on the physical capacity of the brain. He would determine this by filling the skull with small objects (first, white mustard seeds; later, lead shot pellets an eighth of an inch in diameter) and measuring the amount in cubic inches. In 1849 he published his epic racializing work 'Observations on the Size of the Brain in the Various Races and Families of Man'. While Morton measured his skulls very carefully, and he doubtless believed in the objectivity of his work, anthropologist John S. Michael's 1988 comments deliver a good assessment of the value of the work. Michael remeasured the skulls that Morton had used and concluded:

> Morton's tables contain miscalculations and omissions of data, but his 1849 data are reasonably accurate and there is no clear evidence that he doctored these tables for any reason. His tables are nevertheless scientifically unsound because his so-called samples were never adequately defined. His failure to define 'race' makes his work statistically meaningless. (Michael, 1988, p. 354)

What is especially important about this last point is that not only did Morton fail to define 'race' as it applies to

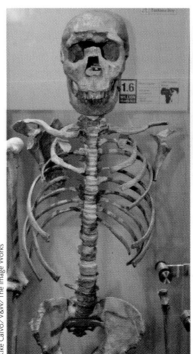
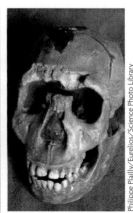

The skeleton (left) and skull (right) of Turkana boy.

Walker/AnthroPhoto

Richard Leakey working with Kamoya Kimeu in Lake Turkana, Kenya, 1984

nature of the human body in all its variety. An early figure in Canada whose work served those ends was **John Charles Boileau Grant** (1886–1973), whose books on anatomy, especially his famous *Grant's Atlas of Anatomy* (first published in 1943 and, currently [2009], in its twelfth edition), rank with *Gray's Anatomy* as classic works in the discipline. In a 1992 tribute, the South African paleoanthropologist Phillip Tobias identified Grant as 'one of the most outstanding teachers of anatomy in the 20th century' (Tobias, 1992).

Grant came to Canada from Edinburgh, Scotland, to teach anatomy at the University of Manitoba and, later, the University of Toronto. He served the young discipline of physical anthropology well. The human osteology collection of the University of Toronto's Department of Anthropology grew out of what was known as the 'Grant Collection', and anatomical specimens

humans in a scientifically meaningful way, but biological race among humans has never been defined in a scientifically meaningful way.

It would be wrong and short-sighted, however, to see anthropometry as merely serving a racist agenda. In Canada, it has proven useful in medical and physical anthropology departments to teach students about the

that he dissected are still found in the Grant's Museum at U of T's Medical Sciences Building. His anthropometry studies were published by the National Museum of Canada in 1929 (*Anthropometry of the Cree and Saulteaux in Northeastern Manitoba*) and 1931 (*Anthropometry of the Chipewyan and Cree Indians of the Neighbourhood of Lake Athapaska*).

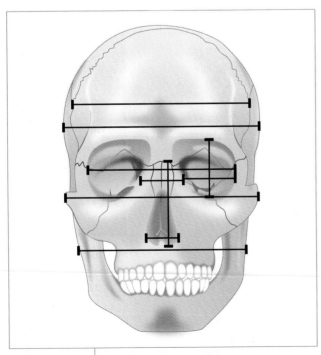

Figure 1.2 | Location of cranial measurements, front view

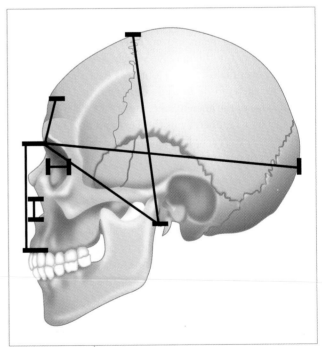

Figure 1.3 | Location of cranial measurements, lateral view

Anthropometry continues to be an important avenue of anthropological investigation today, and craniometry in particular has a number of uses. We'll take a closer look at craniometry in Chapter 10, in connection to the cephalic index, and in Chapter 11, where we'll see how it has provided important evidence suggesting that the earliest peoples to inhabit the American continents were different in several features from those who followed.

Medical Anthropology

When we present ourselves as medical anthropologists, some of us have been asked:

- *do you study dead people?*
- *do you study bones?*
- *what is a medical anthropologist?*
- *what is an anthropologist?*
- *and ...what?!* (Sargent, 2010)

Medical anthropology is the study of health, illness, and healing from a cultural and/or cross-cultural perspective. Like the study of genetics, it can be seen as a field in itself, but one that other physical anthropologists can both learn from and play an important part in. It has ties to sociocultural anthropology in that medical anthropologists study the various diagnostic, healing, and curing practices used by a variety of people in a variety of cultures. From a purely physical anthropological perspective, medical anthropology can be applied to examining the effects of living a particular lifestyle, either currently or in the past. Hunters and gatherers, for example, tend to have fewer dental problems and fewer diseases, and tend to be taller than people who live exclusively by farming, especially by *monoculture*, or 'one-crop' farming. When the hunting-and-gathering Aboriginal peoples of Canada first met the farming peoples from France, they must have wondered, 'Who *are* these busy short people?' When the English encountered the now extinct Beothuk of Newfoundland, they thought of them as giants.

Medical anthropologists study the ways that certain geographic populations tend to be more affected than others by particular diseases or disorders. For example, certain Native populations, such as the Ojibwa or Anishinabe people of northwestern Ontario, have something called the **thrifty gene**, which means that they are genetically predisposed to maintaining a fairly high amount of blood sugar. This makes them particularly susceptible to **diabetes type II**.

A recent practice in medical anthropology—an approach shared with researchers in kinesiology, anatomy, and nutritional studies—involves the examination of **secular trends** to identify long-term health changes. In physical anthropology specifically, the study of secular trends is used to track such changes as increased adult height, earlier onset of puberty (particularly in girls), and increased body mass or obesity in children (see, for example, Tremblay and Willms, 2000).

Forensic Anthropology

Forensic anthropology applies the methodologies and knowledges of physical anthropology, especially those related to bones and teeth, for legal purposes, such as identifying the victim of a fatal crime or accident and pinpointing the cause of death. Thanks to all of the shows in the *CSI* franchise and the TV hit *Bones* (produced by a real-life forensic anthropologist, Kathy Reichs), it is currently physical anthropology's 'megastar', the most popular branch of the discipline (though just wait until a paleoanthropologist turns up the next 'missing link').

Forensic work can be very localized: in and around any major urban centre, a forensic anthropologist can keep pretty busy identifying skeletons found on construction sites, rural roads, or downtown alleyways. Increasingly, though, forensic anthropologists are performing their work on the world stage. For years, in class, I have assigned **Clea Koff's** *The Bone Woman: A Forensic Anthropologist's Search for Truth in Rwanda, Bosnia, Croatia, and Kosovo* (2005), a true account of a young physical anthropologist's work identifying bodies and causes of death (typically by execution) in areas of mass killing, such as Rwanda, Kosovo, and Bosnia. Her work has helped families find their loved ones while aiding international prosecutors in identifying the guilty parties. Koff's story is one that has caused some students to seriously reconsider their non-anthropology major.

Primatology

Primatologists study primates. We are primates. The word comes from the Latin *primus*, meaning 'first', so in a way, identifying ourselves as primates is like saying, 'We're number one'. But we are not alone—there are others like us. Primate is a large order of the mammal class—equivalent in rank to the order Carnivora (carnivores), which includes Fluffy (as we noted above), dogs, skunks, meerkats, bears, and pinnipeds ('fin feet', including seals). Some primates—the wet-nosed, claw-handed lemurs of Madagascar, for instance—are more distantly related to us than are others. On the other hand, some primates are very close, such as the Old World monkeys, who share with us a dental pattern (4×2 incisors, 1 canine, 2 premolars, and—if you've decided to keep your wisdom teeth—3 molars) and visual ability (the so-called 'full colour vision' possessed by very few mammals). Chimpanzees and bonobos are primates with no closer relative than us.

Here is an important point to clear up, since it is a source of misinformation and misunderstanding: other *living* primates are NOT OUR ANCESTORS, although we may *share* ancestors, some living more recently (five to seven million years ago) than others. Darwin did not say that we descended from the chimps or gorillas we see at the zoo—no credible physical anthropologist has ever said that. All primates living today have changed over the history of the order.

Why study *other* primates? There are several good reasons. Anthropologists may study other primates to look for signs of shared biological heritage. If you see a 'human' characteristic in another primate, it suggests that the characteristic has something to do with a shared ancestry. Tool use was once thought by anthropologists to be a characteristic unique to humans. However, primatologists such as the famous **Jane Goodall** (b. 1934) have noted that chimps can make and use tools. Some have been observed to strip down branches and dip them into termite hills as a ramp for unwitting termites, who are then eaten by the tool-using chimps. Others use separate hammer and anvil stones for cracking open particularly stubborn nuts. Perhaps humans are not as unique as we once thought.

One of the big philosophical issues involved in **primatology**, as well as in other branches of physical anthropology, is the degree to which we can say that other primates think the way we do. There has been a long tradition in the West of throwing the accusation of **anthropomorphism** at anyone who points to parallels in animal and human behaviour and emotions. This stems in part from the Christian tradition of assigning souls to humans but not to other animals, a distinction that Eastern religions do not make. I wonder whether the West's distancing itself from other primates might have something to do with Europe's lack of prolonged contact with other primates. Over the past few thousand years, the primate living closest to Europe has been the Barbary macaque, native to North Africa and Gibraltar. By contrast, in Indonesia, home to the orangutan, there is a tradition of stories representing the big orange ape as a human who did not speak so that it would not be put to work. The field of physical anthropology is in debt to primatologist **Frans de Waal** (b. 1948) for coining the term **anthropodenial**, which he defined as 'a blindness to the humanlike characteristics of other animals, or the animal-like characteristics of ourselves' (de Waal, 1997, p. 51). Here is the example he used:

> When guests arrive at the Yerkes Regional Primate Research Center in Georgia, where I work, they usually pay a visit to the chimpanzees. And often, when she sees them approaching the compound, an adult female chimpanzee named Georgia will hurry to the spigot to collect a mouthful of water. She'll then casually mingle with the rest of the colony behind the mesh fence, and not even the sharpest observer will notice anything unusual. If necessary, Georgia will wait minutes, with her lips closed, until the visitors come near. Then there will be shrieks, laughs, jumps—and sometimes falls—when she suddenly sprays them. . . .

> I found myself in a similar situation with Georgia; she had taken a drink from the spigot and was sneaking up to me. I looked her straight in the eye and pointed my finger at her, warning in Dutch, 'I have seen you!' She immediately stepped back, let some of the water dribble from her mouth, and swallowed the rest. I certainly do not wish to claim that she understands Dutch, but she must have sensed that I knew what she was up to, and that I was not going to be an easy target. (de Waal, 1997, p. 50)

He should try living with parrots.

Science

One of the more unforgivable things that our education system in Canada is guilty of is turning many students off science. To me, there is nothing more exciting than science. It is about curiosity and the joy of discovery, two principal human traits. Children especially are naturally curious about science. A long time ago, when my career was heading in several directions (none of them up), I worked for the Ontario Ministry of Natural Resources, Wildlife Branch, where I was responsible for teaching primary school teachers some of the interactive and genuinely fun activities used to teach students about wildlife conservation. I also got to work with kids myself. We once did a squirrel count of the infamous eastern grey squirrels living in Queen's Park, where the provincial parliament of Ontario sits (and sometimes does little else). The point was to show young students how government wildlife biologists keep a count of game animals like deer and moose so they know how many hunting licences to issue without threatening the population of the animal. The kids thoroughly enjoyed the experience, mostly because they are curious and like to discover things. I hate that we often kill that curiosity with schooling.

Physical anthropology is a science. The curiosity of physical anthropologists is what drives them to weird (and sometimes disgusting—wait till you read about the Body Farm) as well as wonderful places. The thrill of scientific discovery never gets old. To use a Canadian analogy, it is like a shootout at the end of a hockey game. Or the first time a parrot steps up onto your finger (okay, so not everyone will get that one). Sure, a lot of hard and sometimes boring work comes both before and after the discovery, but the thrill of science is a major reason why people get involved with it, and with physical anthropology especially.

The Scientific Method

The **scientific method** is simple and fundamental. Its principles have been stated many times, and they will be presented here because they need to be remembered. In

Frauds, Myths, and Mysteries: Science and Pseudoscience in Archaeology, an important and very readable book of what is and isn't science, Kenneth L. Feder explains that the scientific method typically includes the following principles:

1. Observe.

2. Induce general hypotheses or possible explanations for what we have observed.

3. Deduce specific things that must also be true if our hypothesis is true.

4. Test the hypothesis by checking out the deduced implications. (Feder, 1990, p. 20)

In a textbook on anthropology, the following needs to be added. The scientific method, minus the Western name 'science', has been applied through the centuries and all over the world by peoples developing medicines and other knowledges that were useful to them. Some of these knowledges are useful to us today, especially in medicine. 'Science', so called, may have begun in the West in the seventeenth century, but the scientific method did not. Read Canadian anthropologist **Wade Davis**'s *Wayfinders* (2009) for a cogent discussion of this issue.

Fact, Theory, and Hypothesis

One of the most important distinctions in science is the one between fact and theory. Bear it in mind when, in the next chapter, we talk about the theory (or, more accurately, theories) of evolution. A **fact** is an observation that is true as far as we can determine truth to be. We use the term 'observation' here, although it can be something smelled, heard, or touched as well. Of course, sometimes what we

think of at first as facts are later overturned by clearer observation, either physical (my stronger microscope enables me to see that the cell is not the smallest unit) or intellectual. Galileo was the first human to see the rings of the planet Saturn, but what his mind 'saw' was a double planet, because the idea of rings was not in his head. This is similar to the situation in which, because I have dogs, cats, and birds as pets, I sometimes perceive dead dogs, cats, and birds on the road, only to learn, on closer inspection, that what my brain saw was wrong, that they were really cardboard boxes, old shoes, and fast food containers. In the realm of physical anthropology, there is a notable range in the brain size of fossil specimens that have been identified and labelled as ***Homo habilis***, notably the skulls known as **KNM–ER 1813** (Kenya National Museum–East Rudolf; fossil number 1913), with an estimated cranial capacity (i.e. brain size) of 510 cubic centimetres, and the one known as **KNM–ER 1470**, cranial capacity 752 cc. Close observation of the evidence leads us to question the 'fact' that skulls with such different brain sizes are really specimens of the same species. Of course, the facts that physical anthropologists throw their theories at can change, particularly when they involve either dates assigned to sites and their fossils or skulls reconstructed from parts. As we'll see in Chapter 7, further study of these two skulls has led anthropologists to reconsider the cranial measurement of KNM–ER 1470, which has been 'downsized' to 526 cc.

A **theory** is an explanation. It is usually much easier to observe something than to explain it. I have observed that Quaker parrots are more aggressive than conures are. Theorizing about the different behaviours of Quaker parrots and conures may contain reference to their different social environments. Quaker parrots live in huge

Two skulls of *Homo habilis*: (left) KNM–ER 1813; (right) KNM–ER 1470

nests that are like avian apartment buildings, and that are home to a number of fiercely territorial pairs. They have to be aggressive. The smaller green-cheeked conures live in smaller nests, and are superior fliers, better suited to flight than to fight.

Two different theories have been advanced to account for the wide difference in brain size between KNM–ER 1813 and KNM–ER 1470. One states that the difference lies in **sexual dimorphism** within the species *Homo habilis*, having to do essentially with size differences between males and females—the males have the largest brains in the sample (a function of their greater body size overall; this has nothing to do with intelligence). A number of contemporary mammals show sexual dimorphism. The gorilla comes to mind. Adult male gorillas range in height from 165 cm to 175 cm (5′ 5″–5′ 9″) and in weight from 140 kg to 204.5 kg (310–450 lb). Adult females are often half the size of a silverback, averaging about 140 cm (4′ 7″) in height and 100 kg (220 lb) in weight. Sexual dimorphism, then, could account for the difference in cranial capacity of the two *Homo* skulls. But the other theory posits that the difference reflects a species-level distinction, in other words, that the two skulls represent different species: *Homo habilis* (KNM–ER 1813) and *Homo rudolfensis* (KNM–ER 1470).

A **hypothesis** differs from both fact and theory. It sets up a study in which proof of a theory comes from observing a particular pattern predicted by the theory. If I were to set up a nest of the kind Quaker parrots build and put mated pairs of conures in them, and engaged in this study over 50 years or so (parrots live long), my hypothesis would be that the conures would become more aggressive. The thing about a hypothesis is that you can be wrong and still learn something. If the conures didn't become more aggressive, I might be forced to take an alternative theory or even question my so-called facts. A hypothesis coming from the sexual dimorphism theory would be that large–cranial capacity males and smaller–cranial capacity females would eventually be found on the same site, showing that they lived together as one species. If the hypothesis were wrong, then only similar-sized males and females would be found at common sites.

Objectivity and Subjectivity: Lumpers and Splitters

As I was writing this chapter, I was reading Richard Leakey's *The Origin of Mankind: Unearthing our Family Tree* (1994). While the book was both entertaining and enlightening, it demonstrated distinct biases—not stated, but presented as objective science, which irritated me. We'll come back to Richard Leakey and his book in a moment. First, though, we'll look at the notion of 'objective science'.

Traditionally, teachers of science taught students that science was **objective**, free from bias and prejudice, and

that everyone who performs a particular experiment will produce the same findings because, to quote *The X-Files*, 'the Truth is out there'. But this notion (like the show) contained a good dose of fiction—the idea that humans can be objective. We can't be totally objective, and that is the case especially when we are talking about other humans, past or present. Scientists have emotional investments in what they have said in the past. They defend their theories, sometimes beyond good intellectual sense, because they take criticism of their ideas as attacks on themselves personally (which they sometimes are, of course). British science writer **Roger Lewin** puts it differently, but well, in the following comments:

> In fact, a completely unbiased, unprejudiced exploration of nature is a methodological impossibility, as biologist and philosopher of science Sir Peter Medawar is fond of pointing out. Without a set of expectations to act as a guide, such a search would be a chaotic and largely unprofitable enterprise. Moreover, the way in which scientists typically report their findings, in formal papers submitted to learned journals, is, he says, 'notorious for misrepresenting the process of thought that led to whatever discoveries they describe' [Medawar, 1984]. Preconceptions are rarely acknowledged, because this, after all, would be 'unscientific'. And yet preconceptions are an individual scientist's guide to how to view the world with a degree of order that allows structured questions to be asked. (Lewin, 1987, pp. 18–19)

And while all science is coloured by hues of **subjectivity**, physical anthropology—especially in its search for the 'missing link'—may well be the subjectivity champ. This is essentially the thesis that Lewin, a scientist and prolific writer and editor (though not physical anthropologist), puts forward in his bestselling book *Bones of Contention: Controversies in the Search for Human Origins*, which focuses on some of the more dramatic confrontations in the field (and from which the comments quoted above are taken).

In the introductory chapter of *Bones of Contention*, Lewin tackles anthropology's subjectivity in terms of the very human psychology involved in the quest for fame available to discoverers of the 'great finds' in palaeoanthropology. Make a significant find, as all four Leakeys and their archrival Don Johanson have, and you can become famous almost instantly. Being merely an important thinker in the discipline, such as **Wilfrid Le Gros Clark** (1895–1971), who helped debunk the **Piltdown Man** hoax (see below), or an influential teacher will earn you little more than relative obscurity. If you weigh the options, it's easy to see the allure of anthropological discovery and the fame and glory it brings. Unfortunately, the pull of fame and glory has sometimes led anthropologists to embellish or altogether falsify their claims.

This subjectivity issue was brought to light by an American physical anthropologist with the interesting name of **Earnest Hooton** (1887–1954) in what is perhaps the best-named book in the discipline—*Apes, Men and Morons*, published in 1937 (Hooton bought the Piltdown Man hoax right to the end, so who's the moron after all?). In this book he discussed the opposing impulses felt by 'missing link' anthropologists, who spend lifetimes searching for evidence of primitive ancestors who are similar to yet separate from *Homo sapiens*. In Lewin's well-crafted paraphrase of Hooton,

> on the one hand you exaggerate differences between your fossil and modern humans, thus getting for yourself a nice, ancient, discrete ancestor. And on the other, you overlook the differences and exaggerate the similarities, thus setting your fossil on the threshold of the noble *Homo sapiens*. (Lewin, 1987, p. 26)

While Hooton was writing in the first heyday of the 'species-naming frenzy', it is well to consider the principles involved, as a similar frenzy occurred in the 1990s and early twenty-first century, continuing to the time of the writing of this book.

Lumpers and Splitters: Declaring Which Side You Are On

How can we do science when we are not completely objective? In part, we declare our biases up front, so that colleagues and other readers know where we might wander off the path of good intellectual sense. A key area in the biological sciences where declaring biases is quite necessary but too rarely done is in the division between lumpers and splitters—a division that separates anthropologists like the rift valley that is Olduvai Gorge separates East and Central Africa. Before I define the two terms, I will take my own advice and make a declaration of bias: I am a lumper. Now I can define the terms.

Lumpers are those who, in categorizing biological entities, prefer to think that within a species or genus, there is a significant range, such as one finds with humans or black bears (*Ursus americanus*). The latter come in black, cinnamon-brown, white, and a kind of blue, but they're all of the same species (the lumper says). In interpreting the evolutionary line of humans, lumpers tend to see more similarities among humans past and present. They argue that ancestral-to-human species, and their cousins, like gorillas, would exhibit considerable sexual dimorphism, and that they would, like contemporary humans, demonstrate variety in other ways.

Splitters are those who see in smaller differences, greater distinctions. They believe that in the large time periods represented by early fossil finds, there is ample opportunity for a lot of *speciation* (variation in species). And they are right in that regard, though they are probably not more right, or more wrong, generally, than lumpers—they are more right in certain circumstances, less right in others. Splitters tend to speak as though there is a greater number of hominids than lumpers accept. I'm sure they're right, to an extent, that more hominid species have existed than we're aware of, although the evidence is still not complete enough to make a definitive assertion. Traditional physical anthropology textbooks (i.e. those not like this one) have a tendency to present a splitter worldview. They present different hominid species as if it were clearly established that every proposed species actually existed. The fossils exist, but the currently theorized labels might not. Addressing this point is one reason why I wanted to write this textbook.

Richard Leakey is a splitter, like his father, Louis; however, he did not declare that bias in *The Origin of Mankind* (though it's pretty obvious to any anthropologist in the title of Chapter 2, 'A Crowded Family'). Leakey also does not state that he and Donald Johanson, whom he typically presents as someone who merely theorizes incorrectly, 'really don't like each other' (think Oilers and Flames, or Ti-Cats and Argos). They have a history together that should be mentioned (see Lewin, 1987). I don't mind Leakey's being a splitter. I do mind that although he presents two sides, he doesn't declare his bias and acknowledge that it could be right or wrong. He doesn't inform the reader that he is 'stacking the deck' in his favour when he presents his information.

I want to stress that I am not using Richard Leakey as an example here because he holds a view that's different to mine. In fact, I agree with and am greatly informed by much of what he writes in this book and elsewhere. It's just that the example presented itself at the right time. And—I admit it—it's fun to make examples of famous people.

Marion Kaplan/GetStock

Famous paleoanthropologist Richard Leakey

Science Gone Wrong

The Piltdown Man Story

The year 1912 was a good year *not* to believe what you read in the newspaper. On its maiden voyage, before its tragic encounter with an iceberg on 15 April, the *Titanic* was declared unsinkable. On 13 August, Icelandic-Canadian Vilhjalmur Stefansson, explorer, public speaker, and academic conman in need of funding for his research, asserted that he had discovered a mixed race of blue-eyed Viking–Inuit 'Blonde Eskimos' in the centre of the Canadian Arctic. And on 18 December, leading British anthropologist (later knighted 'Sir') **Arthur Smith Woodward** (1864–1944) and aristocratic lawyer and gentleman archaeologist/geologist **Charles Dawson** (1864–1916) proclaimed to an excited crowd at the Geological Society at Burlington House in London, England, that they had discovered the missing link between ape and human: ***Eoanthropus dawsoni***, or 'The Dawn Man of Piltdown' (i.e. Piltdown Man). It would turn out to be the academic forgery of the century. How did it happen?

We'll begin with Charles Dawson's own account:

> I was walking along a farm-road close to Piltdown Common..., when I noticed that the road had been mended with some peculiar brown flints not usual in the district. On inquiry I was astonished to learn that they were dug from a gravel-bed on the farm, and shortly afterwards I visited the place, where two labourers were at work digging the gravel for small repairs to the roads.... I asked the workmen if they had found bones or other fossils there.... I urged them to preserve anything that they might find. Upon one of my subsequent visits to the pit, one of the men handed to me a small portion of an unusually thick human parietal bone. I immediately made a search, but could find nothing more, nor had the men noticed anything else.... [A]lthough I made many subsequent searches, I could not hear of any further find nor discover anything—in fact, the bed seemed to be quite unfossiliferous. It was not until some years later, in the autumn of 1911, on a visit to the spot, that I picked up, among the rain-washed spoil-heaps of the gravel pit, another and larger piece. (as cited in Weiner, 1980)

Joseph Sydney Weiner (1915–1982) provides the definitive account of what happened next in *The Piltdown Forgery*, first published in 1955. Weiner was one of the three researchers who conclusively declared that Piltdown Man

had been faked. Although a large number of books and articles have been written on the subject, I still find *The Piltdown Forgery* to be the most insightful and informative.

The **parietal** (one of two bones—left and right—on top of your head) had come into Dawson's possession by at least 1908. Curiously he would tell close friends about it but did not tell Woodward, whom he knew well, until February 1912, when he declared in a letter: 'I have a portion of a skull which will rival *Heidelbergensis* in solidity' (as cited in Weiner, 1980, p. 137). He was referring to the famous **Heidelberg jaw** (discussed in full in Chapter 8).

The cranium was eventually put together from nine fossilized pieces, five of which, supposedly, had been found before Dawson wrote to Woodward. Woodward himself would discover at least one piece. There was a story, unsubstantiated, that the workers had smashed what they referred to as a 'coconut' with their shovels, and that that was the skull. It adds flavour to the story, but is probably not true.

The skull was greeted with almost universal approval in academic circles; not so the Piltdown **mandible** (lower jaw bone), containing two **molars** (back teeth). There were several problems with the mandible. One of them relates to something archaeologists and physical anthropologists today are very aware of: **wear patterns**. Scientists look at tools or teeth to gain a sense of how they were typically used, and what they were used on, based on their patterns of wear. The molars were worn down almost flat, more characteristic in shape of human teeth than of ape teeth, which made them perfectly acceptable to some (see page 22 for a discussion of early twentieth-century pithecophobia and attitudes towards apes). One difficulty was recognized by a dentist, Dr W. Courtney Lyne, who noted that the teeth themselves were very immature, inconsistent with having a lot of wear—you'd have to have been eating gravel to have incurred that kind of wear in teeth so young.

Another tooth, a **canine**, was also problematic. Weiner pointed out the strategic timing and nature of its discovery (by French Jesuit anthropologist **Teilhard de Chardin**, 1881–1955) on 30 August 1913. It was just what was needed to convince some of the remaining skeptics. In Weiner's words,

> The eye-tooth [canine] was just what they had hoped for and closely fulfilled Smith Woodward's prediction of its shape, size, and above all of the nature of its wear. (Weiner, 1955, p. 9)

At the museum where Piltdown Man was held, Woodward and company had already made a plaster model that looked just like the newly discovered canine. Also suspicious was the fact that two parts of the mandible—two parts that would have clearly identified the jaw as being either human or ape—were conspicuously broken off of the jaw: the chin and an articulator condyle (a knobby protuberance).

Supposed tools made of flint were found 'with' (exact locations of objects found at Piltdown were rarely stated, or not stated clearly) Piltdown Man. Certain ones, called *paleoliths* ('old stones'), definitely looked like **artifacts**—human-made implements. But from the beginning, skeptics noted their uncharacteristically poor workmanship and their unusual shaping. Even dodgier were the *eoliths* ('dawn stones'), which looked more like **geofacts** (shaped by nature rather than by humans).

Dating human fossils was difficult then, as the means available were rough at best. Human fossils had to be compared with others found at the same geological level, or *stratum*. The age that Dawson and Woodward arrived at for Piltdown Man was 500,000 years (very much a ballpark figure). The determining factor here came from animal (or **faunal**) fossils, which included 18 fossilized bones from the remains of horses, rhinos, hippos, deer, elephants, mastodons, and beavers, most of them found, supposedly, at the same level. That the bones were all the same colour helped persuade the critics that they were, indeed, from the same period.

Not every scholar accepted the claims put forward by Dawson and Woodward. The duo had their fair share of critics, particularly from outside the UK. Notable among the skeptics was the very influential French paleontologist **Marcellin Boule** (1861–1942), whom we'll meet again in Chapter 9, and the American Fairfield Osborn. Convenient, then, was Dawson's sudden discovery in 1915 of two more skull pieces, fragments of a frontal and an **occipital** bone (i.e. front and back bones), in Sheffield Park, about two miles away from Piltdown. The first was announced by Dawson in a letter to Woodward dated 20 January: 'I believe we are in luck again. . . . I have got a fragment of the left side of a frontal bone with a portion of the orbit and root of nose.' Later, in July 1915, he reported: 'I have got a new molar (*Eoanthropus*) with the new series.' The

John Reader/Science Photo Library

Composite photograph of Charles Dawson (left) and Arthur Smith Woodward (right) during excavation work and the lower right canine ascribed to Piltdown Man

The Piltdown Man Story *continued*

frontal bone was especially significant in convincing the disbelievers (including Boule and Osborn, who shifted sides), as Weiner notes:

> When the experts in 1912 decided that *Eoanthropus* must have possessed a modern forehead and that it gained a special evolutionary significance from this very fact, the appropriate piece, with its 'angelic' forehead, as Dawson put it, was forthcoming at Piltdown II. (Weiner, 1980, pp. 115–16)

And to complete the set, this 'find' was followed by the 'discovery' of a human molar and the tooth of a rhinoceros, the latter of the appropriate period for dating purposes.

The announcement of Piltdown II was Dawson's last hurrah. At the end of 1915 he fell ill, dying the next year. But his brainchild, Piltdown Man, did not die with him. Over the next few decades, with the discovery of the first *Australopithecus* ('southern ape') by Raymond Dart in 1924, and more *Homo erectus* ('man standing', remember) in Indonesia, Piltdown Man became one of two routes followed by ancestors, a highly specialized difference, almost irrelevant.

Then came the trio of researchers—Wilfred Le Gros Clark, **Alex Oakley**, and Jonathan Weiner—who, armed with new techniques of a growing science, identified Piltdown Man once and for all as a forgery. One of the techniques they relied on was **fluorine analysis**, which was used to test how much fluorine had seeped into each fossil during its time in the ground. It is a **relative dating** technique, a way of saying whether something is older, younger, or the same age as something else. Oakley had already tested the Piltdown fossils in 1949, finding that they seemed to have had a lot less fluorine than they should have for the considerable (500,000 years) age claimed. For their 1953 paper, the three researchers did a full-bore attack on the material from both sites. They performed a series of tests, most famously a fluorine test and a nitrogen test. (In nitrogen analysis, another relative dating technique, the amount of nitrogen *decreases* over time.) The results are summarized in Table 1.1.

The results show that at both sites, the skull was older than the jaw and teeth. Moreover, the molar found at Piltdown I was older than the canine, and at Piltdown II, the occipital bone was older than the frontal bone.

Other things were revealed in the course of the investigation. The thickness of the Piltdown II frontal bone was very similar to that of the unusually thick cranial bones uncovered at Piltdown I, but not with that of the more 'normal' occipital from Piltdown II. Weiner suggested that it had come from the same skull, that the pieces had been 'planted' by the forger at different times at the two locations. Further, it was clear upon further scrutiny that the teeth had been worn down long after they were fossilized. The club-like worked elephant bone was unique in shape, suggesting that it, too, had been recently manufactured, along with the flints, the likes of which were never found elsewhere in the area. And the faunal remains were not local, but bore a striking resemblance to fossils found in Mediterranean islands such as Malta and in North Africa (Tunisia, specifically). Finally, it was found that the match in colour among the remains was the result—at least in part—of artificial staining of some kind, clearly designed to make the remains look like part of a matching set. That a forgery took place is undeniable.

The mystery remains as to who committed the forgery. Over the years the lights of suspicion have been made to shine in the eyes of several suspects: Dawson, Woodsworth, de Chardin, an assistant named Martin Hinton (carrying out an elaborate practical joke), and even Sir Arthur Conan Doyle, author of the Sherlock Holmes books and *The Lost World*. No clear culprit has been identified. If I were to bet, I would place my money on Dawson himself. More than any other suspect, he had opportunity, as well as motives. Dawson had enemies, and people that he badly wanted to impress. Though he had achieved some recognition in geological and archaeological circles, he was in many ways standing on the margins of scientific respectability. He had a demonstrated passion for finding 'missing links' or other kinds, including *Plagiaulax*, supposedly transitional between reptile and mammal; a type of fish midway between carp and goldfish; a horseshoe that was part Roman 'hippo-sandal' and part modern horseshoe; and even a boat that he claimed was transitional between a coracle and a canoe. Clearly Dawson longed for the kind of recognition attained by successful anthropologists. He could be innocent—but then, he was a lawyer.

Site	Fossil type	Amount of fluorine/ nitrogen detected (%)	Relative age
Table 1.1 Results of relative dating tests performed by Clark, Oakley, and Weiner on Piltdown Man I and Piltdown Man II			
Fluorine test			
Piltdown I	cranial	0.1	older
	jaw and teeth	0.03	younger
Piltdown II	cranial	0.1	older
	molar	0.01	younger
Nitrogen test			
Piltdown I	cranial	1.4	oldest
	jaw	3.9	
	molar	4.3	
	canine	5.1	youngest
Piltdown II	occipital	0.6	oldest
	frontal	1.1	
	molar	4.2	youngest

Adapted from Weiner, 1980, pp. 37–9.

Piltdown hoax: Assessing the skull (portrait by John Cooke, 1915)

How Do You Feel about Apes?

A similar philosophical difference that produces bias has to do with how a physical anthropologist feels about the closeness of humans to apes—in body shape, tool use, culture, and communication skills)—and with that the closeness of apes to possible human ancestors in a fossil find. If an anthropologist has some discomfort with the closeness—a form of anthropodenial that Lewin calls **pithecophobia** ('irrational fear of apes')—then any find needs to resemble a human more than an ape for that scientist to consider the find an ancestor. Historically, scientists have situated the divide between human and ape very deep in time. The acceptance of Piltdown Man (mostly human) and the rejection of the Taung Baby (more apelike) from 1912 to the early 1950s were influenced by feelings of pithecophobia prevalent at the time. Of course, a healthy dose of pithecophobia aided the eventual rejection of the South Asian fossil find *Ramapithecus* ('Rama ape') as a human ancestor in the late 1970s. And there is a whiff of pithecophobia in much of the automatic opposition to theories of Neandertal as ancestor or at least gene pool contributor, as we will see. For the record, my bias is to see a link between humans and other animals, a feature of the influence of Buddhism and Aboriginal beliefs on my thinking, as well, perhaps, as the fact I have 11 pets.

So what should the beginning student of anthropology take from all of this? Part of being an 'expert' or professional in a scientific field is having the ability to detect subjectivity disguised as objectivity. Part of being a good teacher in such a field is trying to develop in students a critical ability to detect bias—in classmates, 'experts', textbook writers, even the instructor her- or himself. No one presents 'just the facts'; it isn't possible.

Author's Message

What can we learn from stories of academic bias and even forgery that will serve us in our critical investigations of works of physical anthropology and other scientific disciplines? We need to ask ourselves what could motivate the scholars who concoct or accept evidence that is clearly inauthentic. Specifically, what do we make of those who backed the idea of Piltdown Man as the elusive 'missing link'? Clearly the Piltdown Man supporters wanted to believe that the missing link had developed the large brain first, and other 'human' features later; this was the

kind of ancestor they could live with. Darwin had already shut the door to the possibility that humans had heavenly ancestors; Piltdown Man gave hope that modern humans were descended of something more human than apelike. The Neandertal, which had been put forward as a possible ancestor (and is still recognized as such by some people today), was a more brutish, more primitive, less appealing alternative. In science it is sometimes said that you can have the findings you want if you look (or turn your analytical eyes away) long enough. Be careful what you wish for: it might be a distortion of what you are actually seeing.

Alongside the motivation of personal glory and species pride is national pride—specifically, in the case of Piltdown Man, British nationalism. At the time of Dawson's supposed finds, other Europeans—French, German, Belgian—were unearthing their own candidates for the missing link. These were the same people competing with Britain for pieces of the then carved-up colonial pie of Africa, and the same ones vying for military supremacy (an early arms race that was laying the groundwork for the First World War). Here, in Piltdown Man, was a proper English ancestor that one could be proud of.

The moral of the story is that physical anthropologists, like other scientists, are people with nationalities and genders, among other social traits, and they can think with the social location of their bodies as much as they do with their scientific minds. Scientists can compete with all the passion and prejudice of athletes, can deploy their imagination with all the creativity of an artist, and can strive to win like lawyers, the ones who are willing to get by on a technicality. They are not machines.

Summary

Physical anthropology is a branch of the larger discipline, or field, of anthropology, which studies human beings over time, place, and situation. What makes physical anthropology unique is its focus on physical evidence, especially bones and teeth (whether fossilized or less than a week old). But physical anthropology does not stand alone: it draws from the other branches, as well as from genetics, in its study of human evolution. Physical anthropology is a science, and like any other science—and probably more than some—it involves the personalities of its practitioners, and this is what gives the field a large measure of its charm and makes it so fascinating to its various audiences.

Typical Student Questions

▶ *How do we know that sasquatch does not exist?*

From a physical anthropological perspective, we can say that sasquatch does not exist—and never has—because there are no physical samples (bones, teeth, hair, excrement) that can be connected with the creature.

▶ *Didn't Darwin say that we descended from chimps?*

No, he said that we share a common ancestor. This is still the position taken by physical anthropologists. Think of chimps as our cousins, not our great, great, great . . . grandparents.

▶ *Doesn't science have to be purely objective?*

No: all science, like any human activity, necessarily mixes objectivity and subjectivity.

Review Questions

1. What evidence is lacking that might prove the existence of sasquatch?
2. What are the four branches of anthropology, and what do they involve?
3. Who were the St Lawrence Iroquoians, and how can the four branches of anthropology be used to discover what happened to them?
4. What are the main areas of study of physical anthropology?
5. Distinguish between objectivity and subjectivity in science.
6. What is the difference between 'lumpers' and 'splitters'?
7. Briefly explain the Piltdown Man forgery. What might have motivated someone to carry out such an elaborate hoax? Why were researchers so eager to believe that Piltdown Man was a genuine find?
8. What methods were used to prove that Piltdown Man was a forgery?

Discussion Questions

1. Discuss why pure objectivity is considered impossible when humans are studying other humans.
2. Why would Aboriginal people today be hesitant to have the skulls of their ancestors measured? If you were a physical anthropologist involved with anthropometry, how would you deal with that resistance?
3. What can be learned from the Piltdown Man story?

Key Species and Specimens

*Australopithecus
    afarensis*
Eoanthropus dawsoni
    (= Piltdown Man)

Heidelberg jaw
Homo erectus
Homo habilis
Homo sapiens

KNM–ER 1470
KNM–ER 1813
Peking Man
Piltdown Man

Ramapithecus
Sinanthropus pekinensis
    (= Peking Man)
Turkana Boy

Key Terms

anthropodenial
anthropological
    linguistics (= linguistic
    anthropology)
anthropology
anthropometry
anthropomorphism
archaeology
artifacts
biological anthropology
    (= physical anthropology)
canine
cranial
craniometry

cranium (plural: crania)
cultural anthropology
diabetes type II
fact
faunal
fluorine analysis
forensic anthropology
genetics
genus
geofacts
hypoplasia
hypothesis
linguistic anthropology
lumper

mandible
medical anthropology
missing link
molar
objective, objectivity
occipital
osteology
paleoanthropology
parietal
pithecophobia
polygeny
physical anthropology
primate
primatology

relative dating
scientific method
secular trends
sexual dimorphism
sherd
sociocultural
    anthropology
species
splitter
St Lawrence Iroquoians
subjective, subjectivity
theory
thrifty gene
wear patterns

Key Individuals

Davidson Black
Marcellin Boule
Wilfrid Le Gros Clark
Wade Davis
Charles Dawson (that's
    Dawson, not Darwin)
Teilhard de Chardin

Frans de Waal
Jane Goodall
John Charles Boileau
    Grant
Earnest Hooton
Donald Johanson
Kamoya Kimeu

Clea Koff
Louis Leakey (father of
    Richard, husband
    of Mary, and father-
    in-law of Meave)
Meave Leakey
Mary Leakey

Richard Leakey
Roger Lewin
Samuel George Morton
Alex Oakley
Joseph Sydney Weiner
Arthur Smith Woodward

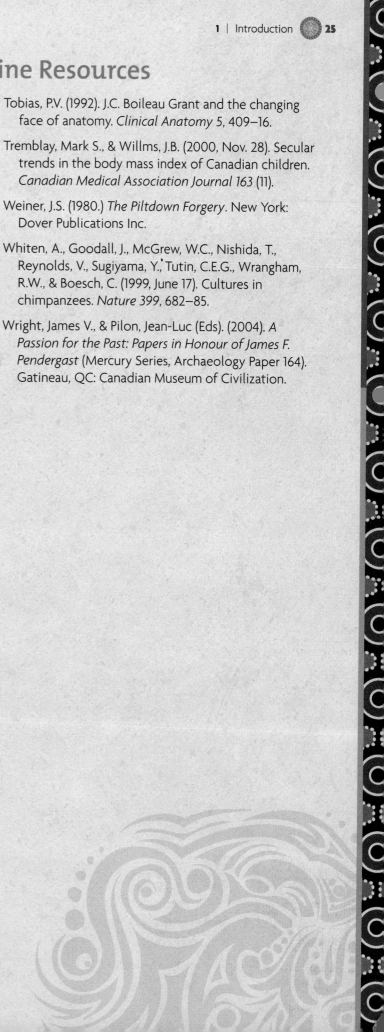

Recommended Print and Online Resources

DeLaurier, April, & Spence, Michael W. (2003). Cranial genetic markers: Implications for Postmarital residence patterns. In Ronald F. Williamson & Susan Pfeiffer (Eds), *Bones of the Ancestors: The Archaeology and Osteobiography of the Moatfield Ossuary* (Mercury Series, Archaeology Paper 163). Gatineau, QC: Canadian Museum of Civilization.

de Waal, Frans. (1997). Are we in anthropodenial? *Discover 18* (7), 50–3.

Koff, Clea. (2005). *The Bone Woman: A Forensic Anthropologist's Search for Truth in Rwanda, Bosnia, Croatia, and Kosovo*. Toronto: Random House.

Leakey, Richard. (1994). *The Origin of Humankind*. New York: Basic Books.

Michael, John. (1988, April). A new look at Morton's craniological research. *Current Anthropology 29* (2), 349–54.

Tobias, P.V. (1992). J.C. Boileau Grant and the changing face of anatomy. *Clinical Anatomy 5*, 409–16.

Tremblay, Mark S., & Willms, J.B. (2000, Nov. 28). Secular trends in the body mass index of Canadian children. *Canadian Medical Association Journal 163* (11).

Weiner, J.S. (1980.) *The Piltdown Forgery*. New York: Dover Publications Inc.

Whiten, A., Goodall, J., McGrew, W.C., Nishida, T., Reynolds, V., Sugiyama, Y., Tutin, C.E.G., Wrangham, R.W., & Boesch, C. (1999, June 17). Cultures in chimpanzees. *Nature 399*, 682–85.

Wright, James V., & Pilon, Jean-Luc (Eds). (2004). *A Passion for the Past: Papers in Honour of James F. Pendergast* (Mercury Series, Archaeology Paper 164). Gatineau, QC: Canadian Museum of Civilization.

Tools of the Trade
How Old Is That?

◀ Archaeologists search for hominin remains
at the Gran Dolina dig site, Spain
Javier Trueba/Madrid Scientific Films/
Photo Researchers, Inc.

Introduction

There are basically two ways to talk about how old an archaeological artifact or site is: relative dating, and absolute dating. With **relative dating** you compare the artifact or site with another, or others, in an attempt to discover which is older and which is younger. Remember from the previous chapter how fluorine analysis was used on Piltdown Man to show that the cranium was older than the jaw and the teeth. The more precise, or at least countable, technique is **absolute dating**, though it usually involves establishing a range of dates rather than one specific date. It is a lot like assigning an age of death to a skeleton: an anthropologist's measurement will yield a range of ages, but not a precise year (as we'll see in Chapter 12, on forensic anthropology). Some situations call for an approach that combines the two dating methods. In these cases, absolute dating techniques are used to determine figures above (i.e. younger than) and below (i.e. older than) the artifact in question, thus establishing an upper and lower range for its age.

Relative Dating

Stratigraphy

When I was 'between marriages' I had an almost sure-fire way to tell, by the end of the week, what shirt I had worn on which day. My Monday shirt was lying on the floor by the bed; on top of it was the Tuesday shirt, then the Wednesday shirt, and so on, with Saturday's shirt lying on the top of the pile. It wasn't foolproof, of course—I might have disturbed the shirts looking for socks or something else—but it was fairly reliable. Then there was my method

of determining what I had eaten on Monday just by looking at the sink, but you don't need to read about that. The point is that studying things in layers is a useful dating

A typical archaeological dig site

© James L. Amos/CORBIS

© Othon Alexandrakis/iStockphoto

The badlands near Drumheller, Alberta—site of many dinosaur discoveries—showing distinct stratigraphic layers in the hillsides

technique. I may call it the bachelor laundry method, but it's actually called stratigraphy.

Stratigraphy, the study of the order and relative position of *strata*, or layers, was a method of analysis first used by early geologists James Hutton, William Smith, and Charles Lyell to study different levels of rock and how they related to the geological timescale. Technically they were engaged in *stratigraphic observation*—that is, observing that different strata exist, in geological cases as formed by nature.

Different from stratigraphic observation is *stratigraphic excavation*, which involves excavating (or digging up) a site in order to identify the different strata. It's delicate work that demands the utmost care not to mix the strata; any sign of negligence in this regard will have you sentenced to daylong wheelbarrow duty by the site director. Once on vacation in Edinburgh, Scotland, I talked my way onto a dig of a Roman fort that was on a hill just outside of town. We had to be very, very careful to keep artifacts recovered from the stratum of the Roman fort separate from anything that might have belonged to strata representing earlier Celtic occupations on the same site. In the Canadian context, an especially important reason

to be absolutely sure that you have objects from each occupational stratum clearly separated from the others is the matter of pre-contact (i.e. before European explorers and missionaries arrived) versus post-contact. A single artifact, such as a glass bead or an object of European-manufactured metal, can tell you whether or not a site is pre-contact. Misidentify the stratum the artifact is from, and you could seriously misdate the site.

The science of stratigraphy dates back to the 1600s. The seventeenth century was a time of scientific innovation, a point that tends to be forgotten when we look at the advances since then. One of the century's leading scientific figures was the Danish scholar and eventual archbishop (an unusual destination for a man of science) **Nicolas Steno** (1638–1686), who specialized in anatomy and geology. He was not the first to claim that fossils were once living things (although he strengthened that position), nor was he the first to suggest that lower strata were older than the strata above them (that observation was made by the amazing eleventh-century Persian thinker Ibn Sina, more commonly known by his Latinized name Avicenna); however, he was the first to set down the basic principles of stratigraphy.

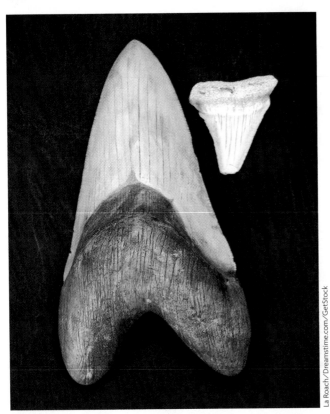

La Roach/Dreamstime.com/GetStock

The fossilized tooth of a stenoshark (*Carcharocles megalodon*) compared with that of a shark (right).

In October 1666, two Italian fishermen caught a huge shark. The Grand Duke of Tuscany had the head of the beast sent to Steno. Steno noted that its teeth looked remarkably like some strange stones that others had called *glossopetrae* (lit. 'tongue stones'). Following the 1616 writings of Italian scientist Fabio Colonna, he realized that the stones were fossilized shark's teeth. He wondered, though, how it was that a solid could form inside another solid—a problem that had troubled him also with crystals and veins of rock. His theory was that the surrounding solid stone was fluid at the time the other solid object entered it. This led him to hypothesize about the nature of different strata, or layers, of rock. In 1669 he published his thoughts in *Dissertation prodromus*, which outlined four of the basic principles of modern stratigraphic analysis.

The first was the **law of superposition**, which states that upper strata formed after those that are lower. Different strata didn't form at the same time: the lower the stratum, the older the rock. (This sounds so logical that it's hard to believe how controversial such views were in the seventeenth century, when many Europeans still believed that God had created the whole earth in a single day!) The second was the **principle of original horizontality**, which essentially states that rock typically forms in a horizontal position. Think of how fluids (like water in a pool or beer

in a glass) tend to being horizontal at their top (it was summer when I wrote this). Since a stratum of solid rock forms from a molten state or from the accumulation of sediment, any other position it takes (i.e. perpendicular to the horizon or leaning towards the horizon) must have resulted from a disturbance to its original horizontal position. Third, Steno articulated the **principle of lateral continuity**, which stated that strata originally extended unbroken over a great distance, only to have their lateral continuity broken by intruding (i.e. upward-thrusting) rocks. Steno's fourth principle is the **principle of cross-cutting relationships**, which claims essentially that any rock that cuts through (or intrudes) a given stratum was formed after that stratum was.

Steno's four principles were meant to account for geological phenomena, but archaeologists working on sites where there have been successions of people living in the area use stratigraphy in their digging. Each stratum they uncover is a cultural–chronological one, showing what people had and were doing at a given period of time. It is not exactly a snapshot of one precise year, as Ontario archaeologists would tell you when they find the postmoulds (decayed remnants of poles) of one house crossing those of another, but it is still a very useful tool. By identifying everything belonging to a given stratum (poetically called a 'living floor') you learn the context of each artifact it contains, without which the artifact has only a very partial story. One of the misconceptions perpetuated by Hollywood portrayals of archaeology such as the Indiana Jones movies and TV's *Relic Hunter* is that the work is all about finding the great single artifact. On the contrary, real-dirt archaeology is about reading any artifact in the context of its contemporary surroundings. Think of an artifact as a page in a book: by itself it tells something of a story, but you need the rest of the book to get to the whole story.

Seriation

Another relative dating technique is **seriation**. It involves putting particular artifacts (e.g. pottery, beads, arrowheads) from a number of sites within a given culture area into chronological order, usually in a series of older-to-younger patterns; a standard one is A (oldest) – B (younger) – C (youngest). The relative age of the artifacts is ascertained by differences in the styles they show and where they occur among the various strata. Say you found the rims of an Iroquoian pot that is decorated with back-slash lines—let's call that type A. Quite often you will find these type-A pots at sites where pottery decorated with dotted rims (call that pattern type B) occurs in the stratum above. We know that type A is older than type B, since it represents the lower stratum. Elsewhere within the Iroquoian culture area you may find pots with another

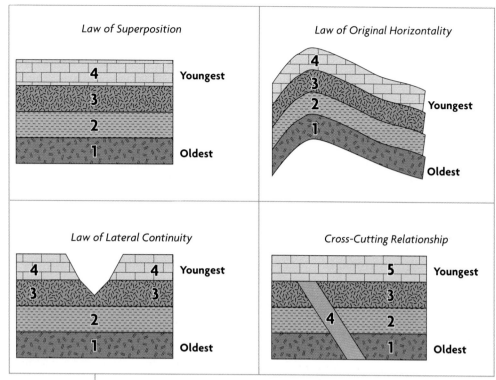

Figure 2.1 | Steno's four principles of stratigraphy

decorated pattern occuring above the type-B stratum: we will call this pattern type C. You know, then, that C is the youngest, followed by B (younger), and A (oldest). Seriation is particularly useful when you can assign an absolute date to one or more of the stages. If you were to ascertain that the the type-B pottery dates to the year 1460, then you could say that the type-A pottery (and anything else found in the same stratum) is more than 550 years old, and that the type-C pottery is less than 550 years old.

Table 2.1 shows the various kinds of music recordings and the numbers of each found at a series of hypothetical junkyards. The types of album are presented in chronological order. The first three are vinyl records, the second two tapes.

WHAT DO YOU THINK?

1. Which of these junkyards was open for business first?

2. Which of these junkyards opened for business last?

The answers are at the end of the chapter.

Table 2.1	Seriation: Step by step through the junkyards					
	Type of recording					
Junkyard	78 rpm	33 1/3 rpm	45 rpm	8 track	Cassette	CD
A	10	20	30	40	—	—
B	50	20	5	20	5	—
C	20	40	25	15	—	—
D	35	50	15	—	—	—
E	10	20	30	20	10	10
F	—	10	40	50	—	—

Source: Hirst, n.d.(b).

Absolute Dating

A Warning about Websites and Absolute Dating

Before we embark on any discussion of the various means of putting an absolute date (more accurately, an absolute range) on materials and sites, I must issue a disclaimer concerning creationists and their websites. Name a dating method—any dating method—and there will be mention of it on websites promoting or espousing creationism (the belief that the universe and living things are products of divine creation, as set out in the Bible, rather than natural processes such as evolution). They'll mention all the disagreements that surround scientific dating techniques; they may note recent improvements to these techniques, but only in the context of how they prove that earlier attempts at dating were flawed, and how there is no consistency or consensus. It's like describing a person only in terms of character defects, which, if they've been improved, only point to how bad the person was a long time ago. Of course, none of these websites talk about scientific evidence for their own position.

This is as a cautionary statement for students looking on the Web for more information about the various methods. Be suspicious of sites that seem to just dismiss the various methods, that are merely critical without offering up any scientific alternatives. I like the Internet for fast information that can lead me to articles or books; unfortunately, any fool can have a website, but only a scientist can write a refereed journal article, and generally only well-informed people write books of non-fiction.

Writing

The earliest method used to determine absolute dates is the written record. Often the term **history** is used to mean written history, while **prehistory** is used to refer to non-written history, although this is misleading and reflects a bit of prejudice against peoples who don't write. The

---

Profiles in Canadian Anthropology

A Canadian Example of Paleography

The following is an entry taken from a French–Huron dictionary written down by an anonymous Jesuit missionary during the seventeenth century:

Agasser ,aato,ati p. θa f t n Sonaato,aθa, il nous agasse

Here, *Agasser* meaning 'to bother', is used as the French translation of the Huron, *aato,ati,*; the last part of the entry provides a specific use or example of the verb.

Two aspects of this fragment indicate that the dictionary was recorded sometime between the mid-1640s and the mid-1650s. First is the example *Sonaato,aθa,*, which means 'He bothers us'. The pronominal prefix *sona-* means 'he . . . us', and in this way represents both the subject ('he') and object ('us') of the verb element, 'bothers'. This is the part of the word that suggests the date. From the mid-1630s up until the mid-1640s, the Jesuits, in annual reports known as the Jesuit Relations, wrote in a dialect of Huron that had this form. In their documents written in the second half of the seventeenth century, the same word would have been written *Son,8ato,aθa*, with the internal comma recording a sound like the English letter *y*, and the letter 8 recording a sound like *w*. The sona- spelling places this fragment in the first half of the seventeenth century.

The second clue is the spelling of the French infinitive *agasser*. Later in the century the word began to be spelled *agacer* (as it is today).

Archive Séminaire de Québec, MS. 67.

time period in which humans have written is relatively short: in rough figures, we can date the earliest writings to 5500 BP (before present), in Mesopotamia (present-day Iran and Iraq), for the system of writing that eventually became the alphabet we use today in English; 5250 BP for Egyptian hieroglyphics; 3500 BP for Chinese character writing; and 2600 BP for writing in Mesoamerica (Mexico and Central America).

Paleography involves the study of old written materials. An experienced paleographer can discern a date from such elements as the dialect in which a text is written, the script used (e.g. alphabet, syllabic, or hieroglyphic), the nature of the knowledge displayed, and references to specific individuals. The box on page 32 features a Canadian example of paleography.

Dendrochronology

Remember when you were told that every year, a tree grows a new 'ring', and that you could tell the age of a chopped-down tree by how many rings it has? If you ever suspected your parents were just making it up, rest assured: it's true, and in fact, it's the basis of a form of absolute dating called **dendrochronology**, 'the highly specialized science of assigning accurate calendar dates to annual growth rings in trees' (Nash, 2000, p. 60). It should be pointed out that this does not apply to all trees, but only to those that live in a climate in which there is a distinct change between growing and non-growing seasons. Growth has to stop and resume before a ring is formed; with continous growth there are no marks. So we can apply the principles of dendochronology in areas with winters of growth stoppage and springs of growth beginnings, or areas in which there is a growth-filled rainy season followed by a growth-stopped dry season.

Dendochronology, though, isn't quite as simple as just counting a year for each ring. For one thing, there are some environments where growth stops and starts more than once a year, and this must be factored into the counting. Second, growing seasons aren't all the same: they can differ in terms of how much growth actually happens. More favourable conditions (e.g. more sunlight and/or more rain) will produce wider rings; poorer conditions (e.g. drought) will produce thinner rings. This is useful when comparing two trees or wooden products that are from the same local area: the rings for items of the same age should match. Differences in the spacing of the rings could suggest that two articles came from different climate zones or they could point to local variations in the same climate zone.

A third fundamentally important concept is overlap (or, more properly, overlapping rings). Typically, for any area in which dendrochronology is used to a significant extent, there is a chart or timeline that extends farther back in time than the age of the oldest tree. How can that be? It has to do with overlap. Growing seasons differ from year to year: for instance, a particularly long spring or summer will produce a ring that may be wider than the previous year's ring, so the pattern of rings produced over a succession of years is unique. Knowing this, you can match two trees or wooden products and identify the extent of their overlap—the period in which they were alive at the same time. Ideally, for this kind of analysis, you'd be looking at an overlap of at least 50 years. Here's an example. Let's say a particular species of tree lives only 1,000 years. You have one fully mature tree on its last legs (or roots) that was cut down in 2005 and another tree of identical age (in growth or development years) that was felled in the same area by, say, someone who travelled with explorer Jacques Cartier to the St Lawrence in 1535 (it may have been cut down to make a basic dwelling). As Table 2.2 shows, your two trees would theoretically have an overlap of 530 years (the first tree was 'born' in 1005, the latter in 535; both were alive for a 530-year period from 1005 to 1535). With this knowledge, you could develop a continuous chronology that now goes back 1470 years. (Warning: this chapter may contain math.)

Table 2.2	Developing a deep chronology based on the hypothetical overlap of growth rings	
	Growth rings begin	Growth rings stop
Tree 'A'	1005	2005
Tree 'B'	535	1535

Master chronology			
535	1005	1535	2005
|	|	|	|
Tree 'A'	n w w n n w w n n w w n n w n n w n w n n w w n w n w n w w w		
Tree 'B'	w w w w w n n w n w n w n n n n w w n n w w n n w w n n w n		

Notes: n = narrow rings; w = wide rings; orange = matching rings in overlap

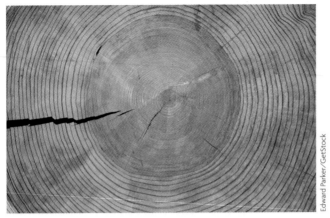

Tree rings, the focus of dendrochronology

The wood-wise scientist who first discovered this method was actually an astronomer, **Andrew Ellicott Douglass** (1867–1962), who developed the methodology and local chronologies for the American Southwest over the first three decades of the twentieth century. Just how useful his pioneering dating methods were can be seen in the differences between educated guesses of time and his dendrochronologically determined dates. Sites such as Pecos Pueblo, guessed at being 1,000–1,500 years old, Chaco Canyon and Mesa Verde, estimated at 2,000 years old, and the Basketmaker sites, presumed to be 3,000–4,000 years old, all had their ages cut in half by Douglass's calculations (Nash, 2000, p. 62–3).

Varves

Varves are the layers of sediment or of sedimentary rock deposited each year, typically by 'retreating' glaciers as they melt and in glacial lakes with little or no **bioturbation** (disturbances in the sediments caused by small life forms). Varves were first studied for dating purposes by Swedish researchers in the nineteenth century. As a stratigraphic absolute dating technique, varve dating is restricted to a relatively short period. In Sweden, for example, they have a varvechronology of just about 13,500 years.

Radiocarbon Dating: Carbon-14

Radiocarbon dating and the related methods of assigning dates to material are based on the principle known as **radioactive decay**. According to this principle, radioactive **isotopes** (different forms of the same element) are made unstable thanks to a particular kind of molecular event (in the case of carbon-14, we're looking at cosmic ray bombardment—and you thought you were safe!). These unstable elements, over time, change back to more stable forms at a regular pace like the ticking of a clock. The pace, or rate, of this process is the element's **half-life**, which is the amount of time it takes for half of the remaining unstable isotope to change back into a more stable isotope. Here's an example. Say a particular element—we'll

Settlements of Chaco Canyon, New Mexico

Clay varves

the concept of radioactive decay and the related principle that radioactive decay takes place at a fixed rate calculable in terms of a half-life were discoveries made in 1908 by New Zealand-born, Nobel Prize–winning physicist/chemist **Ernest Rutherford** (1871–1937), who at the time was chair of physics at McGill University.

Carbon is an element that exists in all living things. In fact, all living things, through life, accumulate and give off carbon. This makes carbon a good element to use to establish dates for relics of living things, like fossils, bones, and teeth. There are three naturally occurring isotopes for the element carbon: carbon-12 (12C), carbon-13 (13C), and carbon-14 (14C). The first two are stable (and therefore not so useful for dating); the last one, though, is unstable, and therefore very useful for dating. The most commonly occurring of the three isotopes by far is 12C, which accounts for 98.89 per cent of all carbon found on earth. 13C makes up about 1.11 per cent of the earth's carbon, leaving 14C in last place, accounting for less than one-trillionth of our carbon. No wonder it took until 1940 to discover the isotope, an achievement Toronto-born physicist Martin David Kamen shared with his lab partner, Sam Ruben. You have a better chance of winning the lottery than of being a carbon-14 isotope. As 14C decays, it emits an electron, turning it back into nitrogen-14.

The American physical chemist **Willard F. Libby** (1908–1980), who pioneered the technique of radiocarbon dating in 1949, worked out a half-life of 5,568 ± 30 years for 14C. This was the figure used until about 1970, when researchers cross-checking radiocarbon dates with dendrochronology dates noted discrepancies and recognized the need to make a correction, or **calibration**, to account for fluctuations in the amount of carbon in existence at certain times. The more accurate figure of 5,730 ± 40 years is sometimes known as the 'Cambridge half-life'. Table 2.4 shows what percentage of the unstable carbon-14 would be changed to the stable isotope nitrogen-14 over the course of 22,920 years.

call it unstable isotope 'A'—had a half-life of 1,000 years (it would be great for the math-challenged among us if half-lives were nice round numbers, but that's never the case outside of made-up examples like this). Table 2.3 below shows how much of the unstable isotope would reach a stable state (stable isotope 'B') in 4,000 years.

When I do this as a class exercise, and I ask what percentage of the two isotopes remain after the first thousand years, students uniformly reply with the correct answer: 50 per cent for each. However, when I ask what remains of each after 2,000 years, there are always a few students who say 0 per cent for isotope 'A' and 100 per cent for isotope 'B', which is not correct. Since the scale considers the time it takes for half the remaining unstable isotope to change, we're always dividing the remaining percentage by 2, so that percentage can never (mathematically speaking) reach 0.

The main idea here is that an isotope's half-life is like a natural clock that changes or 'ticks' at a uniform rate as the isotope decays. Knowing an isotope's half-life enables us to date items made up in part of that element, which brings us to carbon-14. First, though, a bit of Canadian content:

Table 2.3	The half-life principle illustrated	
	Percentage of total sample remaining (%)	
Time period elapsed (years)	Unstable isotope 'A'	Stable isotope 'B'
0	100	0
1,000	50	50
2,000	25	75
3,000	12.5	87.5
4,000	6.25	93.75

Table 2.4	Radiocarbon dating: From carbon-14 to nitrogen-14	
	Percentage of total sample remaining (%)	
Time period elapsed (years)	Unstable isotope 14C	Stable isotope 14N
0	100	0
5,730	50	50
11,460	25	75
17,190	12.5	87.5
22,920	6.25	93.75

Joel Arem./Photo Researchers, Inc.

How reliable is carbon-14 dating? In R.E. Taylor's words, radiocarbon dating is good for making what would be called a conservative estimate:

The C14 time scale currently extends from about 300 years to between 40,000–60,000 years. The limitations on the young end of the C14 time scale are a consequence of three factors: significant variability in C14 production rates because of seventeenth-century changes in solar magnetic fields, the effect of the combustion of large quantities of fossil fuels beginning in the late nineteenth century, and the production of artificial C14 ('bomb' C14) as a result of the detonation of

Case Study Controversies

Crawford Lake: Varves, Corn Pollen, and Goose Poo

If you drive west on Highway 401 from Toronto to the Niagara escarpment, just past Milton you will see signs for the Crawford Lake Conservation Area. Visitors to the site will see not only a beautiful view from the escarpment, where vultures soar, but also a reconstructed Iroquoian (*not* 'Iroquois') village, the first pre-contact community in the Eastern Woodland area to be accurately dated. How was it done? Using varves.

Crawford Lake is very unusual in that it is only partly circulating (or **meromictic**). In the upper 15 metres, there is a little circulation; below that depth (the lake is about 22.5 metres deep), there is none. This means that there is a renewal of dissolved oxygen in the water near the lake's surface, but not near the bottom, where, as a result, toxic sulphur compounds are created. These compounds are inhospitable to detrivores such as insect larvae that eat and thus disturb the sediments, or detritus, on the bottom of a lake. Thus, the sediment that falls to the bottom of Lake Crawford is typically well preserved in annual layers.

Every year, two varves about a millimetre thick are laid down at the bottom of the lake. In June, the varve is a layer of white calcium carbonate; in October, the varve consists of black organic matter that produces a distinct pattern not unlike a tree's annual growth ring. In 1971, an analysis of the sediments of the lake suggested that there had been corn pollen between 1434 and 1459. In its initial, 'wild' state, corn was very much like any other grass growing in fields: it would easily reproduce naturally. Even today it looks so much like grass in its early growth state that I once inadvertently weeded some out of my garden. Today, after centuries of domestication, corn and humans are co-dependent: corn, essentially, cannot reproduce without us. The kernels do not detach easily from the ears on their own, so the seeds cannot be spread. This is a very longwinded way of saying that where there is corn pollen, there is human occupation. The discovery of corn pollen in the varves of Crawford Lake in 1971 led anthropologists to search for and find, within two years, the site of the Iroquoian village.

In 2001, researchers took another look at the varves. This is done by probing the varves with a cold metal pole to which the sediments stick (it's a little like being goosed by a spoon that has been in the freezer). They checked their dating with accelerator mass spectrometry carbon-14 dating (explained on page 37) and found that from about the year 1300, 100 varves were missing. This effectively moved the dates that had previously been assigned to the corn pollen back by a century, to the period 1268–1468. The new dating made the nearby archaeological site a fourteenth-century village rather than a fifteenth-century one.

In addition to the corn pollen, researchers found the pollen of two other crops grown by Iroquoians of that time: sunflower and squash. They also found nodules that turned out to be Canada goose poo (as determined by DNA analysis), which, when analyzed, showed that the geese had been eating corn in the fields. Jock McAndrews, a palynologist (one who studies pollen) with the Department of Botany at the University of Toronto, from whose 2005 article much of this material was gathered, reckons that prior to the coming of the Iroquoians and the geese the lake was *dimictic* (circulating), but that the presence of both helped the lake become meromictic. His conclusion runs as follows:

[A]t the beginning of the millennium, a fully circulating Crawford Lake was surrounded by virgin forest of beech and maple. About 1280, Iroquoians moved into the area, cleared forest, burned the clearings and grew corn, sunflower, squash and bean. During migration, Canada geese fed in the fields and roosted on the lake where they cast pellets. At first the pellets dispersed in the water to release nutrients and these nutrients caused algae, particularly diatoms, to bloom. On sinking to the bottom, the algal organic matter decayed to deplete the bottom water of oxygen, exclude bioturbating bottom fauna and allowing the varves and some subsequent pellets to persist undisturbed. (McAndrews & Turton, 2005)

And we think that our parrot poo is bad.

nuclear and thermo-nuclear devices in the atmosphere, particularly during the period between 1955 and 1963. (Taylor, 2000, p. 87-8)

In terms of Canadian history, this means that radiocarbon dating would be useful for the entire pre-contact history of Aboriginal life before the arrival of Europeans (i.e. more than 90 per cent of our human history, from roughly 12,000 years ago minimum), including the Vikings' visits to L'Anse aux Meadows in northern Newfoundland, Jacques Cartier's voyages of the 1530s, and the early history of New France up until the end of the seventeenth century.

There are ways of making radiocarbon dating extend deeper into time. One of these is **accelerator mass spectrometry** (AMS), a method pioneered in labs at a number of universities, including the University of Toronto, McMaster University, and Simon Fraser University. AMS enables us not just to date much older artifacts—as old as 70,000 or even 100,000 years—but to produce accurate dates from much smaller samples, which is particularly useful when you don't want to destroy the item you're trying to assign a date to.

The Dead Sea Scrolls: An Exercise in Radiocarbon Dating

One of the greatest archaeological finds of the twentieth century, the Dead Sea scrolls were discovered in pottery jars in a cave near the ancient village of Qumran, Israel, between 1947 and 1956. These were manuscripts in Hebrew and Aramaic, written on parchment (made from goat skin) and papyrus (a reed) and covering matters of religion, commerce, and astronomy. For Bible scholars, Middle Eastern historians, and theologians, this was a sensational discovery.

The scrolls have been analyzed using a number of methods. Paleography, drawing on the old and well-established tradition of Bible scholarship, has proven useful here in a number of ways, including dating the texts. DNA analysis, especially of the goatskin parchment, has also yielded insight in the study of these scrolls. We can find out such things as which parchments might have come from the same or related animals, or from animals connected to goats living in the area today. This information can be used to learn whether the scrolls were written in the area in which they were found or much farther away.

Because both skin and reed came from living entities, radiocarbon dating is particularly useful here. We will look at one study in particular, published in 1995 by Jull and colleagues. They analyzed 18 texts and 2 linen fragments (such as those that were used to bind the scrolls). In most cases, in order to be relatively sure of their dating, they took multiple measures of sections of each text or linen fragment. Included in their study were three texts whose exact dates were already known; the previously determined dates were not revealed to the scientists working on the dating. Another of the pieces had already been carbon-dated by another laboratory.

Partial results of Jull et al.'s analysis are shown in Table 2.5. You should know that there were two calibrations for each year range ('calibration' being a term for the magical, mystical, highly technical alterations done to try to produce the most reasonable, precise date for a work, taking into account the specific circumstances in which it was created). Also, multiple runs were done with each piece. Remember when looking at this table that the 'current date' is 1950. There were 18 materials dated; Table 2.5 presents the results for the three in which there was a precise date already known (DSS-25, DSS-52, and DSS-53) and the one that was previously dated by another lab (DSS-50).

You can see, in Table 2.5, that for DSS-25 and DSS-53 the ranges of the radiocarbon dating include the precise date known, but that such is not the case for DSS-52, although one of the calibrations is only nine years off—generally speaking, not bad. With the one example in which the two labs are compared (DSS-50), they are very close.

Table 2.5	Selected results from Jull et al.'s radiocarbon dating of scrolls and linen fragments from the Judean desert		
Text number	Radiocarbon age	Year range	Paleographic date
DSS-25	1,799 ± 57	AD 130–321 AD 80–380	AD 130
DSS-52	1,758 ± 36	AD 231–332 AD 144–370	AD 135
DSS-53	1,827 ± 36	AD 126–234 AD 86–314	AD 128
DSS-50	2,141 ± 32	335–122 BC 250–103 BC	150–125 BC

Stonehenge: Radiocarbon Dating Rewriting History

Stonehenge, the vast circle and semi-circle of huge stones (*megaliths*) in southern England, is one of the archaeological sites that has attracted the most and the longest run of interest in the world. It is composed of a circle of sarsen (a form of sandstone) stones enclosing a circle of smaller bluestones; within this inner circle is a horseshoe arrangement of five trilithons (tri- 'three' and –lith- 'stone'), each composed of two upright stones supporting a third stone across the top. The origins and purpose of Stonehenge have been sources of a great deal of speculation. In the seventeenth century, the antiquarian (a term used to refer to people interested in old artifacts) John Aubrey declared that it was made and used by the Celtic priests well-known as Druids. His idea caught on, and remains the most common popular belief today. Other theories, which developed during the nineteenth and early twentieth centuries, speculated that the ancient Egyptians or Greeks, or other more southerly civilizations, were the ones who built and used the site. Radiocarbon dating has been used to prove this is not the case.

The most recent radiocarbon date for the construction of the ring of bluestones was determined in 2008 by Tim Darvill and Geoff Wainwright, based on analysis of around 100 pieces of organic material at the sockets or holes where the first bluestones once stood. They date this phase of construction to roughly 4,300 years ago, plus or minus 100 years. That's certainly well before the Celtic people came to Britain (about 2,300 years ago), and even earlier than the prime time of the ancient Greeks and Egyptians. Quite a feat when you consider that mineral analysis of the bluestones (which weighed up to around four tons each) has revealed that they came from some 240 miles away, in the Preseli Mountains in Wales.

Darvill and Wainwright's date is especially interesting as it situates this part of the monument around the radiocarbon-determined time of the 'Stonehenge Archer', suggesting a connection between the bluestones and the man. In 1978, Richard Atkinson and John Evans unearthed a skeleton in the area close to where the material used to date Stonehenge was found. Physical anthropological research revealed that he was young (25–30), 1.75 metres (5′ 10″) tall, probably local (unlike the 'Amesbury Archer', another skeleton found three miles away, who may have been from Switzerland), and strong, with back problems probably connected with heavy lifting. Maybe he had helped erect the bluestones. Fragments of bluestone were found around him. He has been nicknamed the Stonehenge Archer because he was wearing on his left

Stonehenge

arm what appears to be a stone wrist guard—if you have ever engaged in archery, you know how the bow string can sting. Arrows brought about his death, as the tips of at least three arrowheads are embedded in his skeleton. Was he sacrificed or killed in defence of the stones? Further research might tell; however, here is one small weakness of radiocarbon dating: the dates for both the bluestones and the Stonehenge Archer have a range of a few hundred years. Putting them together at the same time might be difficult, but anthropologists will have a good time trying.

How to Test a New Form of Dating: Consilience

So, you've developed this brilliant new way of dating archaeological material. How do you test its accuracy? In the historical sciences (archaeology, geology, physical anthropology, evolutionary biology, history), we have little access to experiments as proof, to the kinds of test that will allow us to say, 'If I find something and you can reproduce my experiment and find the same thing, then my findings are confirmed.' **Consilience** refers to agreement among a variety of different proofs that all point to the same conclusions, and it is a much valued test in the historical sciences. A small example of this involves a map of the Great Lakes that was found by historical geographer Conrad Heidenreich and labelled something like 'Old Indian Map' in a British museum. It was made of deer hide, and the view of the Great Lakes was greatly distorted. Heidenreich

demonstrated, based on the knowledge shown by the European cartographer, that the map must have been made in the early 1640s. Most of the map had Huron writing on it: in a separate paper, I demonstrated that, based on the dialect of Huron used, the map must have been from 1641–2. Two different paleographic tests producing nearly identical results means that we managed to achieve consilience.

In December 1949, J.R. Arnold and Willard F. Libby published an article in which they demonstrated how they had successfully tested radiocarbon dating on six samples whose ages were already known (Arnold & Libby, 1949). The first material tested was a piece of Douglas fir that had an inner-ring date of AD 530 and a cutting-down date of AD 623, as determined using dendrochronology. The radiocarbon date put its age at 1,100 years ± 150 from 1950, for a date of AD 750 (or a range of AD 600–900, when the '±150' is factored in).

The second test was on a piece of wood from an Egyptian mummy's coffin, which was crafted in the style typical of the period 332–30 BC (we can usually ascertain dates of Egyptian artifacts from their writings). The radiocarbon date for the coffin was 350 BC ± 450. The authors apologized for this blunt instrument (450 years is a sizable variance), explaining that the test was performed when they were just learning the appropriate technique and measures.

The third test sample comprised two pieces of wood from the floor of a palace in what is now northwestern Syria. The primarily historical (i.e. written) evidence suggested a date range of 725–625 BC; the radiocarbon date was 650 BC ± 150. The fourth test examined a section from a giant redwood tree with a segment of rings dating from 1031

Case Study Controversies

Cautionary Note: 'Don't mix radiocarbon and calendar years'

In the correspondence section of the 7 April 2005 edition of the prestigious scientific journal *Nature*, there appeared the following cautionary note, entitled 'Don't mix radiocarbon and calendar years'. In Chapter 11 we will be discussing the name mentioned here: 'Clovis' (once thought to be the material culture of the first people to come to the Americas; and in Chapter 10 we will be discussing 'Kennewick Man' (an ancient skeletal remains found off the American west coast).

Sir—Your News Feature 'Skeleton keys' . . . contains two statements that, together, are misleading. First, it is stated that 'Clovis people had made it to the southwestern United States by 11,500 years ago', and second, that Kennewick man is 'a 9,000-year old human skeleton'.

So Kennewick man lived 2,500 years after Clovis? No, in fact he lived some 4,000 years after. . . . The confusion arises because the Clovis date is in radiocarbon years, whereas the Kennewick date appears to be somewhere between radiocarbon and regular calendar years: 8,400 radiocarbon or 9,300–9,500 calendar. . . .

Radiocarbon years differ from calendar years because the amount of C14 in the atmosphere has varied considerably in the past. So the quoted Clovis date of 11,500 years corresponds to approximately 13,500 calendars years.

Please be consistent, and stick to calendar dates so that non-specialists can understand. (Tyler-Smith, Hurles, & Jobling, 2005)

to 928 BC; the radiocarbon date was 1055 BC ± 165. The fifth sample came from a piece of the deck of the funerary boat of the Egyptian pharaoh Sesostris III, dated 1843 BC ± 50; the radiocarbon date was 1750 BC ± 400. The sixth set of samples came from the royal tombs of Egyptian pharaohs Sneferu and Zozer (not the god from *Ghostbusters*—that was Gozer). Their respective dates from previous research were 2625 BC ± 75 and 2650 BC ± 75; the radiocarbon dating produced a date of 2800 BC ± 250 for both.

In sum, all of the radiocarbon tests fit in the range established by earlier tests. In terms of consilience, the results are considered good, if not absolutely precise.

Potassium–Argon Dating

Another important absolute dating technique using the principles of radioactive decay and the half-life is potassium–argon (K–Ar) dating, also known as radiopotassium dating. It is typically used with volcanic materials. As with radiocarbon dating, it was first devised in the late 1940s—by Tom Aldrich and Alfred Nier of the University of Minnesota—and developed for practical use in the 1950s, by John Reynolds at the University of California, Berkeley, which would continue to be a home for this form of *geochronology* (rock dating).

Potassium (K) occurs in three isotopes: potassium-39 (^{39}K) represents roughly 93 per cent of the world's total; potassium-41 (^{41}K) makes up 6.7 per cent. That leaves potassium-40 (^{40}K), which is the isotope used in this dating method, although it accounts for just .01 per cent (or roughly 1 in 10,000 parts) of the total potassium on earth. Swisher, Curtis, and Lewin describe the processes involved beautifully and simply in the following passage:

> During volcanic eruptions the hot molten lava . . . loses whatever argon it contains as . . . the lava erupts onto the surface of the earth. This effectively sets the argon clock to zero. As time passes, after being expelled by an eruption, a potassium-containing rock will accumulate higher and higher levels of radiogenic argon-40, the product of decay of potassium-40 in the rock. . . . [T]he potassium/argon clock operates on the simple principle that the more radiogenic argon-40 a rock contains, the older it is. The amount of potassium in the rock or mineral sample has to be measured, of course, because the clock is based on the quantity of argon-40 that would accumulate over particular periods of time from a potassium source of a certain amount. (Swisher, Curtis, & Lewin, 2000, p. 22)

The half-life of potassium-40 is huge: 1.248 billion years—not the slowest there is, but still very slow. The traditional age range to which potassium–argon dating may be applied stretches from a minimum of 100,000 years, when about .0053 per cent of the ^{40}K will have decayed or changed into ^{40}Ar, to about 4.3 billion years, a figure that represents, roughly, the length of time the earth has existed. The drawback to K–Ar dating is that the plus/minus range is also rather large. Still, in determining the age of the early pre-humans, even with a variance of, say, one to six million years, this dating method is quite valuable.

Potassium–argon dating has proven itself very useful in the anthropologically famous Olduvai Gorge, a 20-kilometre-long, 100-metre-deep eroded canyon in northern Tanzania, Africa. The gorge is made up of fairly clearly demarcated layers created by successive volcanic flows. It is where, after almost 25 years work, Mary Leakey in 1959 made her first big discovery: *Zinjanthropus*, known at the time as just Zinj (now known as *Australopithecus boisei* or *Paranthropus boisei*). Potassium–argon dating then was new, but when it was used to arrive at a date of 1.75 mya ± .25 (i.e. ±250,000 years; 'mya' stands for 'million

Peter Groenendyk/Photo Researchers, Inc.

Olduvai Gorge, Northern Tanzania, East Africa. The 20 km long, 100 m deep gorge has provided many famous fossil specimens of early humans and human ancestors.

years ago'), it made instant media stars out of Mary and her husband, Louis. With improvements in the technique, the date was changed to 1.85 mya.

As with radiocarbon dating, radiopotassium dating has improved greatly since the 1950s. The next stage or advancement level in potassium–argon dating is known as argon-40/argon-39 dating (or argon–argon dating). The traditional method required two separate measurements of two separate samples, one for potassium, the other for argon. In 1965, Craig Merrihue at the University of California, Berkeley, realized that as potassium contains a large and constant proportion of potassium-39 (see above), which when neutron-blasted becomes argon-39 and which has a similar relationship to argon-40, one process could release from one sample both isotopes of argon. You could determine the amount of potassium-40 by measuring the argon-39 released and playing the proportion game. This made the dating process much simpler.

Profiles in Canadian Anthropology

Derek York (1936–2007)

Derek York was born in Yorkshire, England, and educated at Oxford, but from 1960 until his retirement in 2004 he worked in the physics department at the University of Toronto, which became a world leader in geochronology thanks to his pioneering work in the field. While he practised a number of methods for arriving at dates for rock samples, his primary contribution was in potassium–argon dating. During the 1960s and 1970s he was instrumental in refining the argon-40/argon-39 method, and by the 1980s, he had taken the pulsed laser method, initiated by George Megrue of the Smithsonian Institution to date lunar rocks, and applied it to single crystals on earth, advancing Megrue's ideas greatly along the way. During that period he ran the largest argon–argon laboratory in the world. He applied the single crystal method to a number of important finds in the late twentieth and early twenty-first centuries, including (my favourite) the discovery of feathered dinosaurs in China.

The winner of a bucketful of academic distinctions, including the Canadian Association of Physicists Medal for Lifetime Achievement in Physics and election to the Royal Society of Canada, York was also, and perhaps more importantly, a talented writer. His writing appeared not just in the usual academic journals (where, as we know, things sometimes get lost and forgotten) but also in *The Globe and Mail* and other publications that brought his articles about science to a mainstream, non-scientific readership. Among the books he wrote for a popular or lay audience is his imaginative *In Search of Lost Time* (1997).

A fitting and important posthumous tribute to York's dedication to the open, democratic, and wonder-seeking ideals of science is the online forum *The 49 Society* (named after 49 St George Street, headquarters of U of T's geophysics department and home to the society York himself founded decades earlier to promote the free exchange of ideas on science and other topics). The website contains the following tribute to Derek York:

Derek York, a world leader in geochronology

Mr Delroy Curling, Physics Department, University of Toronto. Courtesy of the York Family.

Following Derek's passionate example, this on-line forum is dedicated to the furthering of our understanding of ourselves, the universe and our place within it, through the monthly presentation of ideas from leading thinkers from all branches of the sciences, technologies and the arts, and through the special re-release of Derek's popular Globe & Mail science columns.

Derek was convinced that science was not boring, tedious and for the intellectual 'elite'. On the contrary, if presented the right way, science was incredibly exciting, relevant, and accessible.

And part of what it means to be human. (www.49society.com)

Take note: a scientist can be a hero.

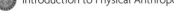

Science Gone Wrong

Tuff Break

All scientists have biases, theoretical and personal, that do inform to varying degrees how they view the data. That much is acceptable. However, when a theory blinds scientists to alternative possibilities, making them unable to see facts staring them right in the face, then that is bad science. Flexibility in the face of data that can't possibly fit into a theory is the quality of a good scientist.

The story of KNM-ER [Kenya National Museum–East Rudolf] 1470 and the push to have a particular date accepted is a story of bad science. KNM-ER 1470 was discovered in 1972 by Bernard Ngeneo, a member of **Richard Leakey**'s archaeological team, at Koobi Fora in Kenya. Leakey's theoretical bias, inherited from his father, was that members of the genus *Australopithecus* (discussed in Chapter 6) were not ancestral to humans, that instead, *Homo habilis*—a species discovered and identified by the Leakeys—was the earliest clear ancestor. The discovery and dating of *Homo habilis* were controversial. For Richard Leakey (b. 1944), assigning a date of nearly three million years to this new specimen, which the Leakeys were calling *Homo habilis*, would provide strong evidence confirming his family's beliefs.

This is a story of vested interests. It occurs when a researcher responds more to the emotional investments, career opportunities, and funding sources involved with a study than to a more objective pursuit of the truth. It's certainly not restricted to anthropology—there is often suspicion whenever a researcher is under lucrative contract to one of the big drug companies or big oil companies, so anything he or she says that downplays the potential harmful effects of certain drugs or drilling practices is highly suspect. Researchers operating under these circumstances are what I call 'embedded scientists', like embedded journalists who report on the campaign of an invading army while living with and being protected by that invading army. Their truth is suspect.

Of course, the story of KNM-ER is a favourite of creationists, since it appears to show evolutionary science in a bad light. (That it also shows the self-correcting, ever-learning, dedication-to-the-truth quality of science is something that creationists have a hard time understanding or accepting.) Much of the present account, though, comes not from a creationist but from the biochemist and eminent science writer Roger Lewin. The relevant discussion is contained in chapters 9 and 10 of his *Bones of Contention: Controversies in the Search for Human Origins* (1987).

In the summer of 1969, a Harvard graduate student by the name of Kay Behrensmeyer (now curator of Vertebrate Palaeontology at the Smithsonian Museum of Natural History) discovered some primitive tools embedded in a layer of rock formed by volcanic ash in the Koobi Fora area of Kenya, near Lake Turkana. These volcanic rock layers are termed **tuffs**, and the one exposed by Kay Behrensmeyer is often referred to as the KBS (*Kay Behrensmeyer site*) tuff.

Richard Leakey, who recalled having seen that same layer elsewhere, at a primitive hippo butchering site, contacted Jack Miller, a Cambridge geophysicist, and Frank Fitch, a geologist, who operated a consulting firm engaged in geochronology. Leakey wanted them to put a date on this tuff to help in understanding the timing of hominid evolution in the area. Miller and Fitch were specialists in potassium–argon dating, and were anxious to try the new method of argon-40/argon-39 dating (see the discussion on p. 40). After one or two misleading false starts, they arrived at a date of 2.61 mya ± 0.26 for the formation of the tool-bearing tuff.

The pair stuck with this date (which they later modified to 2.42 mya) despite evidence of their own and of others to the contrary. At a conference held in 1973, they presented a paper in which they acknowledged that just 7 of 41 determinations of age (ranging widely from 223 mya to 0.9 mya) came with 250,000 years of their date. Yet they did not waver in asserting what they believed was right. And Richard Leakey showed the two scientists tremendous loyalty in supporting their date. A major reason for doing so was the discovery of KNM-ER 1470 in 1972. As noted above, an early age for that fossil specimen was just what Leakey wanted. In his initial article about KNM-ER 1470, which was found underneath the tuff (making it older, in theory), he referred to the tuff as being 'securely dated' at the Miller and Fitch number.

Leakey had good reasons to want to make a great find. He was relatively young and with something to prove. He was the son of two famous anthropologists, who had taught him everything he knew—literally. He had no PhD, no undergraduate degree, had, in fact, never even attended university. He had learned his trade from listening to his parents and the people that worked with them. KNM-ER could make his reputation.

The Pigs Beg to Differ

Enter the spoiler from Dalhousie University, Halifax. **Basil Cooke** was a geologist and paleontologist teaching at the university's Department of Earth Sciences at the time. With a research career of roughly 70 years and a host of honours that includes being appointed a Life Member of the Canadian Association for Physical Anthropology, Cooke remains best known for his *suid* (i.e. pig) research and the role it played in this controversy. It has certainly made him popular on creationist websites.

Cooke was first asked into the dating game by Richard Leakey himself, as he was well known for his skill in the relative dating of what we could call the seriation of pigs. He had noticed that as suids (I love that word—you have to pronounce it like the pig call: *suuuwwWIID*) evolved, their molars got longer: suids in lower strata had shorter teeth than those in higher strata. When, in 1971, he presented his paper on the date of the KBS tuff, he gave an answer that Leakey was not happy to hear. Cooke reckoned that the KBS tuff pigs (as well as the horses, or *equus*, found there) lived less than two million years ago.

Thoroughly displeased with Cooke's findings, Leakey hired another pig team to find a 'better' date for the pigs. The tandem consisted of John Harris, who was head of paleontology at the Louis Leakey Institute and Leakey's brother-in-law, and Tim White. In a 1985 interview with Lewin, Harris would acknowledge the bias in his motives:

> Yes, it sounds pretty stupid when you think properly about it. . . . I can now see that we were seeking ways of justifying the date rather than objectively trying to clarify the evidence. (as cited in Lewin, 1987, p. 199)

At the time, though, the dating controversy sparked a battle royal. Cooke was attacked for using an old-fashioned and highly subjective relative dating technique that was not considered by some to be as modern and objective—in other words, as 'scientific'— as the flash and glam of argon–argon dating (it was, after all, the 1970s—the height of 'glam rock'). By 1973, there were two anthropology 'gangs' that did battle at a number of venues, including conferences and academic journals. On one side was Leakey's crew, whose position we've established. Leakey was opposed primarily by 'Clark Howell Nation', headed by Clark Howell, an early critic of the Koobi Fora dating and the crew chief of another series of sites in the vicinity. Howell found an ally in a geologist from another continent, Garniss Curtis of UC Berkeley, whose dating work was discussed earlier in this chapter. Curtis had worked with Richard's father, Louis Leakey, but the two had split over a young date Curtis had provided when Leakey wanted an older date. (Following the breakup, Curtis was replaced by Jack Miller.) Using the older potassium–argon method, Curtis arrived at a figure of 1.8 mya. Rumours abounded that Curtis had caused the sample he dated to be stolen from a site—that was not true. A student of a Berkeley colleague had been working on a Koobi Fora site and had legitimate access to the material.

Miller and Fitch, meanwhile, were busy coming up with reasons for the contrary evidence against them. One was the faunal barrier, the idea that some barrier to pig interbreeding meant that pig seriation in one area might be different from that in another, that one area's pigs could be 'long in the tooth' and yet younger than the long-toothed pigs at another site. While it is true that geographical barriers between breeding populations of the same species can result in different evolutionary patterns, there was no evidence to support the claim in this case. A second explanation was overprinting, a rare occurrence in which some chemical reaction had effectively re-started the tuff clock for the sample under study, thus producing an artificially younger date. The duo took Curtis to task for using an 'old-fashioned' form of potassium–argon dating.

In 1975 Andrew Gleadow joined the fray, under the close scrutiny of Miller and Fitch, by bringing in a new dating technology: fission-track dating. Not surprisingly, his figure of 2.62 mya ± 0.40 received great support in the Leakey camp. In his own interview with Lewin, Gleadow would later comment on the subtle pressures that came to bear:

> Frank [Fitch] didn't come in and say, 'No, that's not old enough, I want 2.6.' . . . But if you got a number out of a scatter of numbers in your first crude attempts around 2.5, boy, was that praised—great excitement and so on. It all builds up. (as cited in Lewin, 1987, p. 247)

Leakey appeared to be winning, but that would change. In 1977, Harris and White published a paper in which they agreed with Cooke's assessment that the fossil pigs were less than 2 million years old. And Gleadow, worrying about his earlier ability to distinguish genuine tracks when he saw them, redid his study and found that a more likely fission-track date was 1.8 mya. Finally, Australian geologist Ian McDougall came up with a new date of 1.9 mya, derived from his own argon-40/argon-39 dating. His results were published in the same March 1980 issue of *Nature* that featured Gleadow's work. The younger date has been accepted in the field ever since.

Richard Leakey deserves the last word here, as he did come around. In a 1985 interview with Lewin, he stated what he had learned from the whole controversy:

> One realizes that even in the most pure of sciences, which geophysics should be, there is a potential to identify careers, status, and results— and there's a strong political element, too. I should have known this, because I had never really developed the respect that I suppose I should have done for science. But I was upset at times to realize that we may have been given a line that wasn't necessarily secure, even in their own minds. (as cited in Lewin 1987, p. 251)

Since the 1980s, an improvement known as **single-crystal laser fusion** has reduced the size of sample required for testing from several grams of rock to smaller single crystals (see the discussion on **Derek York** in the box on page 41). This refinement has lowered the chances of contamination in the sample, while extending, potentially, the range of dating, so that potassium-rich minerals such as feldspar may now be dated to ages as young as 10,000 years (recall that the minimum age for traditional potassium–argon dating is 100,000 years).

Garniss H. Curtis, an emeritus professor of geology at Berkeley, led the way in the use and improvement of potassium–argon dating. Along with colleague Jack Evernden, Curtis had been recruited by Louis and Mary Leakey to find a date for Zinj in 1960. In 1969, he assigned a date of 1.9 mya to an important Indonesia fossil specimen nicknamed Mojokerto Child. As the conditions for potassium–argon dating were not ideal, and as the figure seemed far too early for a humanlike creature so far east, this date was not generally accepted. Many felt that something closer to 1 million years ago was a much more appropriate date. Determined to get support for his date, Curtis visited Derek York at his Toronto laboratory. As soon as he watched the laser pulse on a single crystal, he knew he had found a method that he should try to prove his figure right. He improved the technology that York was using, and was able to arrive at a date of roughly 1.7 mya, vindication for his belief in Mojokerto Child's great age.

Electron Spin Resonance

Electron spin resonance (ESR) is one of two methods that are dependent on the radioactive decay of uranium-238 (^{238}U), this being the more indirect of the two (the other, fission track dating, is discussed on page 44). Uranium-238, the most abundant uranium isotope, decays into an isotope of thorium; however, it's actually not the thorium but a by-product of this decay—the production of gamma rays—that is key. ESR can be used both for rocks and for the fossils of once-living things. We will discuss the latter here, as it relates best to the goals of physical anthropology.

Once an animal dies, the uranium-238 in the surrounding soil moves in and bombards the animal's remains (notably its teeth, shells, or bones) with gamma rays. Occasionally, these rays knock an electron out of its orbit (that will leave a mark), raising its energy level. Occasionally, some of the higher-energy electrons get trapped in that state. They act like very, very weak magnets. The amount of electrons in that state can be detected by the level of alternating magnetic fields it takes to flip the orientation of those little magnets from north to south (or vice versa). As the rate is considered constant, a date is arrived at by measuring the number of electrons.

An especially important use of electronic spin resonance is to confirm results of radiocarbon dating, with

the two methods coming together to produce consilience. In 2006, an ESR study was performed on a riverside *sambaqui*, or shell mound, at the archaeological site of Capelinha in the state of São Paulo, Brazil. A shell mound (or shell heap) is a site where everyday human refuse was dumped, typically in coastal areas where a community at one time cleaned and processed its catch of shellfish. They make good sites for archaeological study since the calcium carbonate content of shells helps to slow the decay of organic material. The Brazilian *sambaquis* were remnants of an ancient cultural tradition of covering the dead with layers of shells. The mounds could eventually reach about 30 metres in height, and ranged from 500 to 1,900 square metres in area. This *sambaqui* was different from the norm in that it was clear from its stratigraphy that it represented separate occupations of three different groups at different times, rather than being evidence of a single group. A skeleton recovered from the lower of the two levels studied was radiocarbon-dated at between 10,180 and 9,710 years old. To fit with the stratigraphy, the upper level, which had not previously been dated, needed to be significantly younger while still being fairly old. The ESR results indicated that it was roughly 8,140–8,730 years old, which fits well and represents good consilience.

A more recent ESR study was done on another kind of shellfish—barnacles—from the Khyex River in northern British Columbia. As larvae, barnacles cling to rock or other hard substances, and develop their shells right there. In the words of the authors of the study, it is important to assign dates to barnacles for the following reasons:

> They live in intertidal zones and die when exposed. The presence of barnacles thus reflects past sealevels and sealevel changes. Knowing the dates of these changes can constrain the periods of regional hominid occupation and provide a better understanding of local and global paleoenvironments. Barnacles have also been found in archaeological contexts, indicating the use of marine resources by ancient peoples. . . . Further, barnacles adhere to some of the earliest boats. . . . (Skinner et al., 2008)

The British Columbia barnacles turned out to be 15,100 ± 1,000 years old, which correlated well with radiocarbon dates derived from tests on wood samples found at roughly the same level in the area. Had any Aboriginal archaeological material been found there, this would have been a very dramatic find.

Finally, EMS has been involved in an ongoing dispute (actually a lot more than one) concerning the possible coexistence of *Homo erectus* and *Homo sapiens* in Indonesia. Swisher and a team of colleagues published a paper in which they claimed to have found a *Homo erectus* that lived as recently as 27,000 years ago (Swisher et al.,

1996). One of the dating methods used was ESR. When you read the appropriate section of Chapter 7, you will realize how controversial this statement is.

Thermoluminescence (TL) dating is similar to ESR in that it, too, involves the study of trapped electrons. This time, though, they are detected through their stored energy rather than through any magnetic qualities. Time zero in this process is either when crystalline minerals are heated (as with lava flow or the firing of ceramics) or, in the case of sediments, when they are exposed to sunlight. Once the materials have hardened or been shaded from the sunlight, the number of trapped electrons begins to increase at a fairly predictable rate, one that relates to the radiation dose rate of the material in question. The time measurement is taken when the material is heated and the material gives off a rather weak light signal, the thermo- [heat] luminescence [lighting].

The primary use of TL dating for anthropologists and historians is in determining the age of ceramics such as pottery, tile, and brick—artifacts for which there is no other effective means of assigning dates directly. Radiocarbon dating can be used only if there is carbon residue left from corn or some other organic substance.

A more recent advance is the development of optically stimulated luminescence (OSL), which, as the name suggests, examines the luminescence stimulated by light rather than intense heat. The Luminescence Dating Laboratory at the University of Oxford's Research Laboratory for Archaeology and the History of Art has been one of the main labs for TL and OSL research for close to four decades. Members of the lab recently (2007) worked with an international team of scientists to reveal the time period when a section of southern Greenland was forested with evergreens. DNA from plants and insects found more than 2,000 metres below the surface revealed the presence of warmer-weather flora and fauna. The date of more than 450,000 years was obtained from grains of sand (a typical substance for OSL study) that had last seen sunlight around that time.

What age range can OSL dating handle, and how accurate is it? Estimates I have read sharply differ. The Oxford laboratory makes the following claim:

> The age range for pottery and other ceramics covers the entire period in which these materials have been produced. The typical range for burnt flint, stone or sediment (burnt or unburnt) is from about 1000 to 300,000 years. The error limits on the dates obtained are typically in the range ±3 to ±8%, although recent technical developments now allow luminescence measurements to be made with a precision of ±1 to ±2% in favourable circumstances. (School of Archaeology, 2010)

In most instances, though, the circumstances are *not* all that favourable, especially with older dates, and the plus/minus is generally broader than what you find with radiocarbon dating.

Fission track dating is another way of using uranium-238 (^{238}U) and its precisely known rate of decay to assign rough dates to volcanic ash and related substances. When ^{238}U radioactively decays, an atom explodes, tearing its nucleus apart. This is a form of **nuclear fission** that will leave a mark on crystals, such as those of zirconium silicate (or zircon). Researchers then count the

Table 2.6	Absolute dating techniques		
		Date range (years)	
Technique	Material	Minimum	Maximum
writing	paper rock	0	5,000
dendrochronology	trees wooden artifacts	1	8,000+
radiocarbon (carbon-14)	carbon (i.e. once-living objects)	300	70,000*
potassium–argon	volcanic rock	100,000**	4.3 billion
OSL (optically stimulated luminescence)	burnt flint stone sediment	1,000	800,000
ESR (electron spin resonance)	uranium-238	40,000	2 million
fission-track dating	uranium-bearing minerals and glasses	100,000	2 billion

* with AMS
** down to 10,000 with single-crystal laser fusion

Case Study Controversies

The Shroud of Turin and Radiocarbon Dating

The Shroud of Turin is an ideal artifact for those desiring to push an interesting archaeological find to its most sensational limits, for champions of 'the other side' of the story, and for the makers of pseudo-documentaries, who try to balance facts with teasing statements such as, 'But *some* researchers say . . .' or 'Yet scientists so far have been unable to explain. . . .' If you encounter statements like that, or anything that looks like an overreaching attempt to achieve a 'balanced view', be suspicious—be *very* suspicious. This can be (and usually is) an instance of the logical fallacy of the 'Golden Mean', which makes the false assumption that the most valid conclusion is one based on a compromise between two competing positions, even if one of those positions is supported by a large body of evidence while the other is represented by one or two largely unsubstantiated dissenting views. Compromise is useful in marriage counselling, but it follows no valid principles of science. Neither, by the way does the claim that there are two sides to every story. There may be many. Or there may be a scientific one countered by one based in ignorance. Validity is based on proof, on evidence, not on compromise.

So what is the Shroud of Turin? It is a cloth measuring 4.4 by 1.1 metres (14 × 3½ feet) and bearing the faint image, front and back, of a naked man about 1.8 metres (6') tall. He has marks on his wrist and ankles like those of someone who has been crucified. The claim made by some is that it is the death shroud of Jesus, and that the image was somehow created as part of the miracle of the Resurrection after Jesus's death.

The first documented and confirmed appearance of what was then called the 'Lirey Winding Sheet' was in Lirey, France, in the 1350s—some sources attribute the sighting to the French knight Geoffrey de Charney in 1353, others to his widow in 1357. The bishop of the diocese, Bishop Pierre d'Arcis of Troyes, acted vigourously throughout the rest of the century to dismiss the claim that the Lirey Winding Sheet was the death shroud of Jesus, and asserted that an artist had confessed to him that he had been the one who had painted the sheet. The shroud stayed in France until 1578, when it was brought to Turin in northern Italy. Since then, millions have paid to view it on display. The most recent public exhibition was in 2010.

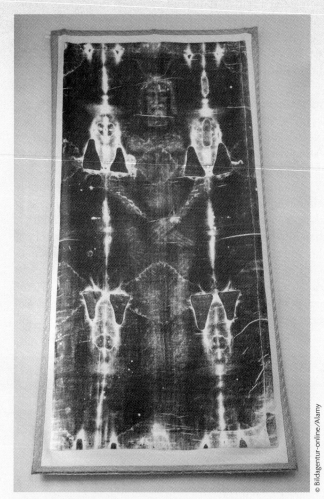

Shroud of Turin

© Bildagentur-online/Alamy

tracks, or marks, on the crystals to determine how old the rock is. This is not a simple process. It took a while for researchers to perfect practices for distinguishing between tracks and 'pseudo-tracks'—marks that were caused by other means. We will see an example of the problems involved in fission track dating in the Science Gone Wrong box later in this chapter.

Summary

We have seen in this chapter a number of the methods used to discover how old things are. We began with relative dating techniques, and then moved on to those that are more absolute but with a little or a lot of wiggle room as to what the precise date is. On the other hand, we have learned that

A historical note is important here. The shroud's first documented appearance, in France, coincides roughly with the most devastating phase of the Black Plague in Europe, when it killed millions (having already wreaked havoc in China). People were looking for miracle cures, and sacred relics—including a forest-load of woodchips said to be bits of the 'True Cross' upon which Jesus had been crucified—were popular as talismans against the epidemic. There was a good living to be had selling relics during the mid-fourteenth century.

Concerns regarding the Shroud's authenticity have long been felt, but the linen sheet was not studied seriously until the second half of the twentieth century. It was examined several times, in 1969 and again in 1973, and most notably in 1978 by the Shroud of Turin Research Project (STURP). However, it wasn't until AMS techniques were perfected that the Shroud's custodians allowed a small strip—1 cm × 7 cm—to be cut off for radiocarbon testing (remember that AMS technology could be applied to much smaller samples). On 21 April 1988, the small strip of cloth was cleaned in an attempt to rid it of any contaminants, then cut into three pieces destined for the three labs—in Arizona, Oxford, and Zurich—that would subject the linen to study. The results of the study were published in a 1989 issue of *Nature*:

> The results of radiocarbon measurements at Arizona, Oxford and Zurich yield a calibrated calendar age range with at least 95% confidence for the linen of the Shroud of Turin of AD 1260–1390 (rounded down/up to the nearest 10 yr). These results therefore provide conclusive evidence that the linen of the Shroud of Turin is medieval. (Damon et al., 1989)

This amounts to a date of 1325 ± 65, which coincides roughly with the time of the shroud's first recorded appearance.

For some, this was just the first shot in a war of papers acting to prove or disprove the authenticity of the sheet. Great sums of money and religious belief were connected with trying to prove its authenticity. Fame would accrue to the person or team who could make a conclusive statement either way. Controversy rose to the occasion. Accusations were levelled (never proven) that someone had replaced the sample with a substitute piece, something that had been used later in the Shroud's life to repair it. Some asserted that a fire that had burnt portions of the sheet in 1532 would make the sheet seem younger than it actually was, though we know that the researchers worked carefully to avoid the burnt parts of the sheet; even if some burnt parts had been included in the sample, their effects would have altered the date only very slightly—certainly not by the 1,000+ years to place it at the time of Christ's death.

Walter McCrone, a member of the STURP team of 1978, who claimed that the rest of the team were 'believers' who conspired to silence his opinions, asserted that he recognized the method used by the painter, and that it included red ochre and vermillion, earth dyes often used during the period. Yet others claim that there was actual blood on the sheet, not earth dyes. Tests performed on some of the supposed blood indicate that it was type AB, which did not exist until medieval times, when Europe and Asia were in significant contact. There was a study that claimed that the sheet contained traces of pollen from a plant known to blossom in Israel in the spring, but the source of the pollen samples was a criminologist known to have faked evidence previously. Finally, there was a claim that carbon monoxide has contaminated the sheet, but a recent study shows that bombarding linen with carbon monoxide does not significantly affect a radiocarbon test.

All told, the evidence is reasonably conclusive that the sheet is a medieval work.

there's a lot to be gained from consilience—that if you have two or more independent dating techniques tell the same chronological story for an artifact, then you have strong evidence for its date.

This chapter comes with a warning about websites that assess the various dating methods. Many of these promote creationist views, and they seize on our 'wiggle room' as though science, to be true, must be perfect, precise.

Typical Student Questions

▶ *How do things get buried under so many strata? Where does all that dirt come from?*

Soil creation happens all the time. Our neighbours have two pine trees. Every year, throughout the year, pine needles fall on our driveway. If we don't sweep the needles off into the garden (where it helps soil building), then within a year, a new layer of soil is created on parts of the driveway. In our backyard we have an apple tree. Every year it sheds blossoms, leaves, apples, the odd twig and the occasional branch. Even with the (very) occasional raking of the lawn, we accumulate a lot of soil there. Our dogs leave small individualized piles of good soil-making material all over the backyard (something we become very aware of in the spring, once the snow melts); they also shred and tear things apart, leaving debris—bones, sticks, rubber dolls with no heads—scattered across the lawn. Then there's human refuse: weeds that have been pulled and cast aside, cigarette butts, Tim Horton's coffee cups, etc., all aiding nature in the soil-production process. Finally, there is 'deep time', which we will discuss in the next chapter.

Review Questions

1. What is consilience? Why is it an important component of dating?
2. How can relative dating and absolute dating methods be used together?
3. Assess the strengths and weaknesses of radiocarbon dating.
4. Describe how dendrochronology works as a method of dating

Discussion Questions

1. Discuss the politics and the science surrounding the dating of the Shroud of Turin.
2. Discuss the politics and the science contributing to the dating problems at the heart of the KBS tuff controversy.
3. How would you (a) explain and (b) defend the methods of dating discussed in this chapter to a skeptic who believes the world is only 6,000 years old? Why might some people find the science so unconvincing?

Answers to Table 2.1

1. The junkyard that opened first is B. Notice how it has the heaviest concentration of early recording devices.
2. The junkyard that opened last is F. Notice how it has the fewest early recording devices.

Key Terms

absolute dating

accelerator mass
   spectrometry (AMS)

bioturbation

calibration

consilience

dendrochronology

electron spin resonance (ESR)

fission track dating

half-life

history

isotopes

law of superposition

meromictic

nuclear fission

paleography

prehistory

principle of cross-cutting
   relationships

principle of lateral
   continuity

principle of original
   horizontality

radioactive decay

relative dating

seriation

single-crystal laser
   fusion

stratigraphy

thermoluminescence

tuffs

varves

Key Individuals

Basil Cooke

Garniss Curtis

Andrew Ellicott Douglass

Richard Leakey

Willard F. Libby

Ernest Rutherford

Nicholas Steno

Derek York

Recommended Print and Online Resources

Arnold, J. R., & Libby, Willard F. (1949, Dec. 23). Age determinations by radiocarbon content: checks with samples of known age. *Science, 110*. Retrieved from http://hbar.phys.msu.ru/gorm/fomenko/libby.htm

Damon, P., Donahue, D., Gore, B., Hatheway, A., Jull, A., Linick, T., . . .Tite, M. (1989). Radiocarbon dating of the Shroud of Turin. *Nature, 337*(6208), 611–15.

49 Society. (Website). Available from from www.49society.com/

Hirst, K. (n.d.(a)) Stratigraphy and seriation. Timing is everything: A short course in archaeological dating. *About.com*. Retrieved from http://archaeology.about.com/cs/datingtechniques/a/timing.htm

Hirst, K. (n.d.(b)) Seriation: A step by step discussion of the dating technique. *About.com*. Retreived from www.archaeology.about.com/od/dating/ss/seriation.htm

Jull, A. J., Donahue, D. J., Broshi, M., & Tov, E. (1995.) Radiocarbon dating of scrolls and linen fragments from the Judean desert. *Radiocarbon, 37*(1), 11–19.

Lewin, Roger. (1987). Chapter 9, The KBS Tuff Controversy: Genesis; Chapter 10, The KBS Tuff Controversy: Denouement. In Roger Lewin, *Bones of contention: Controversies in the search for human origins* (pp. 189–267). New York: Simon and Schuster.

McAndrews, J. H., & Turton, C. E. (2005). The goosing of Crawford Lake with prehistoric corn pollen. *CAP Newsletter, 28*(1), 5–7. Retrieved from www.scirpus.ca/cap/articles/paper035.htm

Nash, S. (2000). Seven decades of archaeological tree-ring dating. In Stephen Nash (Ed.), *It's about time: A history of archaeological dating in North America* (pp. 60–82). Salt Lake City: University of Utah Press.

School of Archaeology, University of Oxford. (2010). Luminescence dating: Age range and precision. Retrieved from www.arch.ox.ac.uk/L.html

Taylor, R. E. (2000). The introduction of radiocarbon dating. In Stephen Nash (Ed.), *It's about time: A history of archaeological dating in North America* (pp. 84–104). Salt Lake City: University of Utah Press.

Tyler-Smith, C., Hurles, M., & Jobling, M. (2005, April 7). Don't mix radiocarbon and calendar years. *Nature, 434*(697). doi:10.1038/434697c

Evolution

Sculptural model of a DNA strand
© Glyn Thomas/Alamy

LEARNING OBJECTIVES

After reading this chapter, you should be able to:

- Present seven forms of evidence for the fact of evolution.

- Distinguish between a genotype and a phenotype.

- Describe Gregor Mendel's influence on evolutionary theory.

- Discuss the different forms of genetic mutation.

- Explain how neutralism, genetic drift, and the founder effect lessen the impact of natural selection in evolution.

Humanity has . . . had to endure from the hand of science two great outrages upon its naïve self-love. The first was when it realized that our earth was not the center of the universe, but only a speck in a world-system of a magnitude hardly conceivable. . . . The second was when biological research robbed man of his particular privilege of having been specially created and relegated him to a descent from the animal world. (Sigmund Freud, as cited in Gould, 1987, p. 1)

What Is Evolution?

A good description of **evolution**, as currently conceived, comes from Jerry Coyne's *Why Evolution Is True*:

Life on earth evolved gradually beginning with one primitive species . . . that lived more than 3.5 billion years ago; it then branched out over time, throwing off many new and diverse species; and the mechanism for most (but not all) of evolutionary change is natural selection. (Coyne, 2009, p. 3)

In Chapter 1, we made the distinction between fact and theory. A **fact**, you'll recall, is an observation that, as far as we know, is true. A **theory** is an attempt to explain facts. We have all read the words 'the theory of evolution'. We may have read or heard that evolution is *only a theory*. These statements are misleading three times over. First, evolution is fact, as we will see shortly. Second, there are *theories* (plural) of evolution, not just one. **Charles Darwin** (1809–1882) had *a theory*, heavily based on natural selection (explained later in the chapter), but he didn't know about **genetic mutation**. In fact, the word 'genetic' wasn't even coined until 1905, well after Darwin's death. His theory of evolution was significantly different from how we theorize about it today. Third, the fact of evolution

is explained by *scientific theories*, not just idle speculation. A scientific theory is 'a well-thought-out group of propositions meant to explain facts about the real world', according to the dictionary definition. In addition, for a theory to be considered scientific, it 'must be *testable* and *make verifiable predictions*. That is, we must be able to make observations about the real world that either support it or disprove it' (Coyne, 2009, p. 15). The current theory of evolution has been tested and proven true. The predictions it makes—for instance, about fossils before they have been discovered, of species like whales with legs and fish with necks—come true.

The young Charles Darwin

What Evolution Is Not

Evolution is not 'survival of the fittest'. Darwin didn't write that. The phrase was coined by a sociologist, Herbert Spencer, to describe the competitive status of societies in the late nineteenth century (with—surprise!—Europeans viewed as the 'most evolved'). Two expressions are better. One is 'survival of the best fit'—in other words, the one that fits best into a specific environment. The other is Coyne's 'survival of the fitter' (2009, p. 13), meaning the one of two (or more) varieties of a species that is better suited to a certain circumstance. It is not about perfection, after all.

Evolution is not synonymous with progress—it isn't about getting better all the time. When advertisers say that computers or cars are 'evolving', they're wrong. New technology doesn't adapt to fit into society; on the contrary, it's society that changes to fit the technology, as manufacturers try to alter people's purchasing habits and create new markets for their products. Evolution is primarily about fitting into a particular circumstance, though these circumstances, like the species that evolve, change over time.

Evolution is not **Lamarckism**, named after French naturalist **Jean-Baptiste Lamarck** (1744–1829). The term refers to the mistaken belief that acquired characteristics— physical traits that individuals have picked up during their lifetimes—could be passed on to the offspring of those individuals. Canadian scientist Karl Giberson explained that if Lamarckism were true, then

> [t]he son of an industrious blacksmith would be destined to develop a robust physique; the daughter of a nimble ballerina would be born with enhanced grace and agility; a politician who learned how to fool people would have similarly talented children. (Giberson, 2008, p. 55)

Inherited characteristics, unlike acquired characteristics (think height versus strength), *are* passed along to offspring: tall parents are likely to produce tall offspring, but there's no guarantee that the children of professional bodybuilders will be born any more muscly than the other kids in the nursery. When tall people seek other tall people to reproduce with and have tall children, this is sexual selection (it's also easier to slow dance with them)—this is an important aspect of theories of evolution that is discussed a bit later in the chapter.

Evolution is not about **orthogenesis**, the belief that evolution tends to move in a particular straight-line direction due to some kind of driving force—think of evolution as travelling without brakes along a long, straight highway. In reality, evolution slows down, speeds up, and even stops to changes direction. Orthogenesis was used to explain the overly large antlers on the now extinct Irish elk, a case that's described in the box on page 54. If orthogenesis were true, we might someday have giraffes that topple over because their necks are too long, or, in humans, pinkie toes that over generations get smaller and smaller until they completely disappear. This little piggie went *wee, wee, wee*, all the way to extinction.

Philosophical Stances and Evolution

There are three philosophical stances that one can take regarding evolution:

- theistic evolution,
- atheistic evolution, and
- scientific creationism.

Theistic evolution entails an acceptance of evolution combined with a belief that a divine creator somehow is involved. It is the official position of the Catholic Church and of most Christian denominations; it also represents the opinion of several prominent biologists, including **Theodosius Dobzhansky** (1900–1975). Born in Ukraine, Dobzhansky moved in 1927 to the United States, where he famously wrote an article called 'Nothing in Biology Makes Sense Except in the Light of Evolution' (1973). The title would lead you to think the author doesn't have a lot of good things to say about religion. And yet, the article contains the following statement:

> It is wrong to hold creation and evolution as mutually exclusive alternatives. I am a creationist *and* an evolutionist. Evolution is God's, or Nature's, method of creation. Creation is not an event that happened in 4000 BC; it is a process that began some 10 billion years ago and is still underway. (Dobzhansky, 1973, as cited in Fairbanks, 2007, p. 164)

Dobzhansky lived and died a member of the Eastern Orthodox Church.

Typically connected with this stance is the position that science and religion are complementary to each other, not opposed. They deal with different, and separate (but not mutually exclusive), spheres of life.

An opposing position involves accepting the fact of evolution and believing that there is no God or divine creator. This ideology can be called **atheistic evolution**. While it's true that scientists likely adopt this position in greater numbers than those who subscribe to theistic evolution, it does not make this position 'more scientific'. Just as there are no scientific proofs for the existence of God, there is no way yet devised to *dis*prove God's existence. Nevertheless, atheistic evolution has some loud and vigorous supporters. Kenyan-born British biologist **Richard Dawkins** (b. 1941) is generally considered the poster boy

Science Gone Wrong

Irish Elk: The Truth of Extinction and the Lie of Orthogenesis

The Irish elk (*Megaloceros giganteus*) was neither an elk nor uniquely Irish. It was the largest deer that ever lived, standing 2.2 metres (7') tall at the shoulder and with antlers that could span 3.65 metres (12'). The species died out perhaps 11,000 years ago in Ireland, slightly more recently on the European continent, and as recently as 3,500 years ago in Siberia.

The Irish elk is famous for two reasons. First, it provided one of the first and most compelling illustrations to persuade early nineteenth-century European scholars, following the French naturalist **Georges Cuvier** (1769–1932), that extinction happened. Before Cuvier held up the Irish elk as proof of species extinction, scholars tended to believe that God would not let any of his created beings die out. Fossil remains of unfamiliar animals were viewed as evidence of species that had long since moved away from the area in which the fossils were found; if you sought long and hard enough, you would eventually find at least a few of any animal that had ever lived. Thomas Molyneux in 1697 wrote the following concerning the Irish elk:

Irish elk: an artist's rendering from the ninth edition (1886) of *Moses and Geology* (Samuel Kinns, London)

Science Photo Library

> That no real species of living creatures is so utterly extinct, as to be lost entirely out of the World, since it was first created, is the opinion of many naturalists; and 'tis grounded on so good a principle of Providence taking care in general of all its animal productions, that it deserves our assent. (as cited in Gould, 1977, p. 82)

Cuvier was right about extinction, but he didn't account for it with anything resembling Darwin's theory of evolution. Instead, he saw it as evidence of an earlier world whose plant and animal species (as well as its geological features) had been wiped out by one or several catastrophic events, a school of thought known as **catastrophism**.

The Irish elk was also used, misleadingly, as an illustration of orthogenesis. The thinking was that once its antlers started growing over the generations, they couldn't stop, driven by some inner force that ignored the practicalities of adapting to local circumstances. It became

extinct, according to this view, when its antlers became so big that they got stuck in trees and bushes, leading to the untimely demise of its proud bearers. Popular science historian **Stephen Jay Gould** (1941–2002) suggested more plausible causes of the Irish elk's large antler size, including sexual selection (the non–antler-bearing does liked them) and ritualized male combat (other bucks feared the dude with the biggest antlers, making him less likely to be attacked by others and more likely to survive to mating age). Gould also raised some possible explanations for the Irish elk's eventual extinction, notably its being one of a number of large animals that didn't cope well with (and couldn't survive) dramatic environmental changes.

Evolution is not all species changing at the same rate, or even all species changing. Humans changed rapidly (relatively speaking—remember, we're still talking millions of years), but the basic blueprint for sharks has changed very little since the time they first came to be as a primitive form of fish. 'Living fossils' such as the fish known as coelacanth (see Chapter 4) would be recognized by their ancestors that lived millions of years ago.

for this position, particularly in *The God Delusion* (2006) and *The Greatest Show on Earth: The Evidence for Evolution* (2009). With this stance there is a tendency to say that not only do the two approaches—science and religion—look at things differently, but that one is right (science) and the other is wrong (religion).

Matching it in opposition is **creation science**, or as it is now known, **intelligent design**. The logic behind intelligent design is summarized nicely by Michael Shermer:

X looks designed

I can't think of how X could have been designed naturally

Therefore X must have been designed supernaturally. (Shermer, 2006, p. 52)

'X' in this case could be any living system whose structure and operations are complex: the human brain, the wings

---

Profiles in Canadian Anthropology

Karl Giberson: Canadian Christian Scholar

Karl Giberson was born in 1957 in Bath, New Brunswick (population of about 1,000), where he received a conservative Baptist upbringing. He left home to study at Boston's Eastern Nazarene College (ENC), a liberal arts college that promotes a Christian perspective; he has been there ever since, joining the faculty as a physics professor in 1984.

When Giberson arrived at ENC, he was a strong believer in creationism, the view that the universe and all living things are products of specific acts of divine creation. His thinking changed, however, once he arrived at the college, which encourages its students to have their faith challenged by controversial ideas in order to make it stronger. Teachers of both science and theology convinced him that Christianity and evolution were compatible, and that the latter was the only way of interpreting biology:

> I have found, . . . after more than two decades as a faculty member at an evangelical college, that the most vigorous opposition to creationism comes from scholars in religion departments rather than in scientific disciplines. As strong as the scientific evidence against creationism has become, the biblical and theological arguments for rejecting it are perhaps even stronger. (Giberson, 2008, p. 119)

Giberson has dedicated his prodigious scholarship to the interface of science and theology, presenting his views on the subject in several influential books and in lectures at such prestigious institutions as the University of Oxford. His *Saving Darwin: How to Be a Christian and Believe in Evolution* (2008) has influenced the writing of this chapter. My favourite Giberson quotation puts to rest the quasi-scientific nature of scientific creationism:

> As of this writing, there is not a single scientist at a major university working within the flood–geology

Karl Giberson, Canadian Christian physicist

> paradigm of scientific creationism. . . . Not a single scientific paper explicitly promoting *any* aspect of this brand of creationism has been published in a scientific journal. . . . Every 'working' creationist is either at a tiny research center like the ICR [Institution of Creationist Research] or at a fundamentalist Christian school, like Liberty or Bob Jones, which do very little research anyway, even outside the sciences. The vast corpus of creationist literature . . . consists almost entirely of popular-level books published by evangelical presses. The 'scientific' output of creationist scholars is a modest bookcase of unnoticed semi-technical works and articles in 'in-house' journals containing little more than miscellaneous sniping at poorly understood details of evolutionary theory. (Giberson, 2008, p. 145)

of a butterfly, the arms of the gibbon (built for swinging through the trees), or the eyes of a hawk. There is arrogance in this intelligent design claim, an arrogance exemplified by its originator, Michael Behe, author of *Darwin's Black Box* (1996). It runs like this: if a feature is too complex for *me* to figure out, then there's no way it could have come about naturally through evolution. In his submission during the Kitzmiller trial (2005), which challenged a Pennsylvania school district's requirement that intelligent design be taught alongside evolution as an intellectually viable alternative explanation for the origin and diversity of life, Behe admitted to not having read the dozens of articles and several books that had critiqued his work (call that not doing his homework); these are *not* the actions of a good scientist. There is also a great lack of imagination in the intelligent design argument. It is like saying, 'I can't imagine a car not run on gasoline, so therefore a car *can't* be run without gasoline.'

Intelligent design is essentially creationism dressed up to look like science, chiefly to make it a more palatable offering in American biology classes. It holds to the traditional creationist view that natural explanations for the origins of life are inadequate, but it appears to take God out of the equation and substitutes an unnamed supernatural agent. The most scientific-looking part of intelligent design is the idea of **irreducible complexity**, championed by biochemist Michael Behe in his *Darwin's Black Box* (1996). The argument runs like this:

X is a complicated system made up of interconnected parts.

If you take any one of those parts out of X, it won't work.

Therefore, X could not have evolved from a simpler state, as anything less than the full set of parts would not work. Nothing intermediate can exist.

A strong argument against irreducible complexity is the principle of **exaptation**. First named by Gould (Gould & Vrba, 1982), exaptation refers to the process through which biological features, developed as adaptations serving one or sometimes several functions, are co-opted for another function or functions later on. That is, a particular feature may adapt again (almost always with structural changes) to serve another purpose. The function a given body part performs now might not be what it has always done. Shermer points out that pre-exaptation stages of a feature should not be seen as poorly developed versions of the feature we know now, but as 'well-developed something else's' (2006, p. 69). Take the feathered wings of birds. They did not just develop fully formed for flight. They performed some other useful function first, before they were used for flying. The theory that most scientists

favour is thermal regulation: feathers help to keep the warm-blooded feathered beasts warm. (I sense that when I scratch my parrot Stanee on that place on the back of the neck that she can't reach without a straw—I feel her body heat and its confinement to the part of the feathers closest to the body.) The feathers first found in the fossil record (see *Archaeopteryx*, on page 65) are quite hair-like, resembling the down feathers of chicks. Like the down-filled jackets that Americans wear when they visit Canada in the spring and summer, a bird's feathered wings, held close to the body, keep the body warm. Spreading feathered wings out with a slight flapping motion cools the body. Flightless cormorants stretch their wings out to gain heat from the sun, demonstrating, perhaps, a pre-exaptation use of feathered wings.

The complicated three-part flagellum of some bacteria, made up of a motor, a shaft, and a propeller, impressed Michael Behe as being irreducibly complex. However, it can be shown as involving the exaptation of simpler forms used for secretion and attaching to other cells. Indeed, there are two-part flagella that perform the same basic function (Shermer, 2006, pp. 67–9).

Natural Selection

Natural selection is the part of Darwinian evolution that has the most stamina. The idea runs like this. Species typically produce more offspring than their environment can comfortably sustain (an idea that comes from the economic theories of Thomas Malthus), so there are winners and losers in terms of evolution. Those individuals that are a better fit within their environment—that have a certain inherited characteristic that gives them a specific advantage in terms of how they get along in that environment—are more likely to survive, mate, reproduce, and have offspring with a good chance of inheriting that favourable characteristic. A key part of this idea is the concept of **sexual selection**: that favourable characteristic has a much better chance of being passed along if it's attractive to the opposite sex and the pool of mating partners. The classic example of this is the male peacock's massive tail. It is not really useful in survival—it attracts predators and makes flying more difficult. But the peahens (the females) really dig a dude with a long, colourful tail.

Jonathan Weiner's *The Beak of the Finch* discusses guppies living in the rivers of Venezuela as excellent examples of both natural and sexual selection. Guppies are pretty much idiot-proof pets: they're small, they breed in high numbers, and if one dies, you can replace it without your children knowing. In the headwaters of these Venezuelan rivers, up in the mountains, the guppies are brightly coloured, but in the lower, larger waters, they are comparatively dull-coloured. The brightly coloured ones reflect sexual selection: individuals prefer to mate with the most colourful individuals of the opposite sex. In this case, the

Science Gone Wrong

Creationism in Canada

Most of the literature describing the evolution vs creationism debate focuses on the situation in the United States, where creation science was born. I want to speak here about the Canadian situation. In part, I think this is important because Canadians sometimes have the bad habit (one that I'm guilty of as much as anyone) of smugly pointing out the flaws of our American cousins (their beer, their military policies, their lack of knowledge about Canadians) without being critical of our own.

Table 3.1 shows the results of an Angus Reid Global Monitor poll, published in June 2007, reflecting the opinions of Canadians concerning the origin of humans. (The figures from a similar poll conducted in 2008 were virtually identical.) While the data suggest that Canadians are more inclined to the prevailing scientific explanation than are Americans, who regularly favour creationism over evolution in similar surveys, we have no reason to feel complacent. When, in the same poll, Canadians were asked whether they believed that dinosaurs and human beings coexisted on earth, only 37 per cent of respondents answered 'No', while 21 per cent said 'Not sure'; the greatest number—42 per cent—were inclined to believe that, to quote American comedian Lewis Black, *The Flintstones* was an animated documentary.

Several factors influence the responses to such questions. The higher the income and the level of education attained, the more likely a person is to accept evolution over creationism. A greater percentage of

Big Valley Creation Science Museum, Alberta

The Big Valley Creation Science Museum in Big Valley, Alberta

younger respondents accept evolution. Political views are also indicative. Supporters of the Conservative Party were the most likely to believe in creationism (29 per cent of those polled) while 83 per cent of Green Party supporters adopted the evolutionary explanation. Geography is another important factor. Alberta is home to not one but two 'creation science museums'—the Big Valley Creation Science Museum and the Bow Island Creation Museum, both established in 2007. This might help to account for the fact that 40 per cent of Albertans polled felt that God created humans in their present form less than 10,000 years ago, an increase of 12 per cent from the previous year.

Table 3.1	Canadians' views on the origins of life on earth*
Which of these statements comes closest to your own point of view regarding the origin and development of human beings on earth?	
Human beings evolved from less advanced life forms over millions of years.	59%
God created human beings in their present form within the last 10,000 years.	22%
Not sure.	19%

* Based on online interviews with 1,088 Canadian adults, conducted 12–13 June 2007.

Source: Angus Reid Strategies

A (male) peacock

bright coloration is a favourable characteristic, and you can imagine how, over time, the best and the brightest will dominate the population. By contrast, the lower waters are home to many more predators. Survival entails not being seen—a bright guppy is an easy target, one that is unlikely to survive to reproduce. Studies of the two strains of the Venezuelan guppy have reproduced (literally) the results of the wild. The key thing is that neither one—bright or dull—is better in any general way. Each variety is simply better for its specific circumstance.

The selection story can get pretty complicated. Northern cardinals show a large colour difference between males (with the bright red plumage and the black mark across the face) and females (with less red, but with some distinctive markings). You would think this has resulted, over time, from sexual selection, that the most successful males are those whose bright colours are most attractive to females. However, this doesn't seem to be the case. I had another theory: that brightly coloured males succeed because they are best able to attract predators and draw them away from the nest, saving the next generation, but this isn't the case either. Male northern cardinals are dedicated parents and have been known to feed the chicks of other species (including those of the parasitic cowbird, which lays eggs in cardinal nests only to have the cardinals raise its chicks). Apparently, the more brightly coloured males are more dedicated feeders than are less brightly coloured ones (Linville, Breitwisch, & Schilling, 1998), giving them a selection advantage. The genetic story must involve some sort of linkage between the two features.

Gregor Mendel and Evolution

There was a lot missing from Darwin's version of evolution. For one thing, he didn't know about mutation and genes. Natural selection is important, but there was more that needed to be discovered.

The first major addition to Darwin's theory of evolution came from **Gregor Mendel** (1822–1884), a monk. In 1856, Mendel (who, in his role as abbot of a small monastery, had a lot of time on his hands) started experimenting with 34 different varieties of the plant genus *Pisum*, otherwise known as the common garden pea (the species *Pisum sativum* is the variety that we eat). He presented the results of his research in a paper called 'Experiments in Plant Hybridization' at two meetings of the Natural History Society of Brunn, Moravia, in 1865. After the paper was published in the Society's journal in 1866, it was soon forgotten or simply not read.

Mendel was 'discovered' in the early 1900s by European botanists who recognized the significance of his studies to the nascent field of **genetics**. His work was championed particularly by the British scientist William Bateson, who coined the term 'genetics' in 1905.

What did Mendel teach the world? There are six basic ideas. The first concerns the independence of each individual trait—in terms of the garden pea, its colour or the smoothness/wrinkliness of its skin—from the others. Mendel also recognized that the male and female partners make separate but equal contributions to their descendants.

Next is the concept of **alleles**, or the differing forms of any gene (often two per trait). An example from Mendel's work is the difference between yellow and green peas, which is caused by different alleles of the same basic gene. The term *polymorphism* (lit. 'many structures') is used to describe a particular genetic location or site that has different alleles. A related idea concerns the principle of dominance, in which one allele is *dominant* over another (which is called *recessive*). For humans, it is generally believed that dark hair is dominant over blond, red, or light hair, that dimples are dominant over no dimples, and that full lips are dominant over thin lips (look at Angelina Jolie's biological children).

An individual's **genotype**—its genetic makeup—may feature just the dominant or the recessive alleles, in which case it would be considered *homozygous*; if it features both a dominant and a recessive allele, it is *heterozygous*. An individual's **phenotype** (like *pheno-* in 'phenomenon')—its physical makeup or appearance—will be determined by its particular genotype. So a heterozygous genotype (one with both a dominant and a recessive allele) will produce a phenotype showing the dominant allele; a homozygous genotype will produce a phenotype displaying whichever alleles it has, dominant or recessive.

The term *Mendelian ratio* is used for the ratio of dominant to passive phenotypes. It works like this: if at a particular genetic locus (or site) you have two different alleles—named, say, *A* and *a*—contributed by the two parents, then you will have a heterozygous genotype (*Aa*) that shows the dominant phenotype. Let's use Mendel's own example of green versus yellow garden peas. We'll say that *A*, the dominant allele, produces green peas, while *a*, the recessive allele, produces yellow peas. If a plant with a homozygous genotype *AA* (green peas) is crossed with another plant with the homozygous genotype *aa* (yellow peas), the plants produced will all have the heterozygous genotype *Aa* and will all have green peas. Simple, right? But what happens to the *next* generation, when two plants with mixed *Aa* genotypes are crossed? That's where things get interesting. As Table 3.2 shows, three-quarters of the plants will have the dominant trait (green peas), while only a quarter of the plants will have the recessive trait (yellow peas). The Mendelian ratio, then, is 3 to 1.

But not every feature is what we call a Mendelian trait. Other features are more complex. A Mendelian trait is genetically simple, in that

- it is generated at one site or locus,
- it is controlled by dominant and recessive alleles but no other factors, and
- it is determined by a simple 'either/or' formula rather than something more complicated.

Table 3.2	The Mendelian ratio: Green *vs* yellow garden peas
Dominant genotype (GD)	*AA* = green peas
Recessive genotype (GR)	*aa* = yellow peas
Dominant phenotype (PD)	*AA, Aa* (green peas)
Recessive phenotype (PR)	*aa* (yellow peas)

First generation: *AA* × *aa*

Parent 1		Parent 2	Offspring
AA	×	*aa*	100% *Aa* (all green)

Second generation: *Aa* × *Aa*

Parent 1		Parent 2	Offspring
Aa	×	*Aa*	25% *AA* (green) 50% *Aa* (green) 25% *aa* (yellow)

This isn't to say that Mendelian traits, being simple, aren't significant: some very serious human diseases and disorders are caused by Mendelian traits, including cystic fibrosis, sickle-cell anemia and Tay–Sachs disease.

By the 1930s, largely through the efforts of biologists Ronald Fisher, J.B.S. Haldane, and Sewall Wright, Mendelian genetics was added to Darwin's simpler formulation of evolution, turning it into what is often called the new synthesis or Neo-Darwinism. It is important to keep in mind that the majority of the academic community

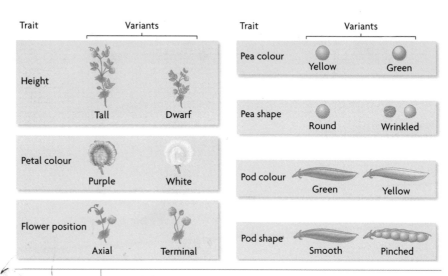

Figure 3.1 | Mendel's traits as expressed in garden peas

does not subscribe to simple Darwinian evolution any-more. Proponents of creation science speak of Darwinism as if Darwin wrote an evolutionary bible that biologists today interpret literally. Scientists do not think like that.

Genetics: The Players

It is useful, when talking about evolution, to go from the biggest genetic units to the smallest ones. The largest is the **cell**, which was named for its likeness to the small, enclosed area or room inhabited by a monk or a prisoner (or a college student); the word derives from the Latin word for 'chamber' or 'storeroom'. There are two kinds of cells. **Prokaryotes** (lit. 'before nut/kernel') were the first kind of life. Most of them are bacteria, though some prokaryotes represent a more recently categorized domain of life, the archaea. The other kind of cell, the **eukaryotes**, came later and are involved in such diverse life forms as lilies, lizards, lemmings, and lawyers. Eukaryotes differ from prokaryotes in having a **nucleus**, which makes genetic and structural complexity much more possible. The prokaryotes are almost all single-celled organisms, with only a few capable of developing into multicellular organisms, and for only part of their lives.

Chromosomes are tightly coiled strands of DNA found in the nuclei of most living cells. The chromosome (lit. 'colour body') was named in 1888 by W. von Waldeyer, who recognized its ability to hold coloured dye well enough to be seen under a microscope. (There is a brief period during reproduction—the stage when duplicates of chromosomes are made—when chromosomes are short and compact enough to be seen when stained.) Humans have 23 pairs of chromosomes, close to the number possessed by chimpanzees, gorillas, and orangutans, which all have 24 pairs. Table 3.3 shows the number of chromosome pairs found in the DNA of a number of species, including our own.

Chromosomes passed on by human parents go through complex procedures of duplication, mixing and matching, and reduction, which results in children having chromosomes carrying elements of both parents' chromosomes but in a unique pattern or combination; the joining of strands of DNA from different organisms is called **recombination**. This ensures that human children are not exact replicas of their parents. Recombination occurs with parental chromosomes that are *homologous*, which means that they have the same structural features and patterns of genes. The sex chromosomes—XX for females, XY for males (see Figure 3.2)—are non-homologous, so they cannot be joined together in recombination.

At the level below chromosome are **RNA** (ribonucleic acid) and **DNA** (deoxyribonucleic acid, having one less oxygen atom than RNA). Their functions were first recognized by molecular biologists **James D. Watson** (b. 1928) and **Francis Crick** (1916–2004) on 28 February 1953.

DNA is important as a carrier of genetic material; RNA essentially acts as a messenger, carrying instructions from DNA for the synthesis of proteins. The process is called *transcription*, and it usually results in the genetic information from the DNA being copied into the new synthesized proteins; however, in the case of certain retroviruses, the RNA makes a DNA copy of itself instead.

The productive relationship between **genes**—sections of the DNA that can act as a unit—and proteins is called **coding**. It is important to know that not all genes are capable of coding; in fact, most aren't. Only about 2 per cent of human genes (called *true genes*) code. Genes are constructed of paired nucleotide bases. Each pair contains a **purine** base—composed of adenine (A) or guanine (G)—and a **pyrimidine** base, composed of cytosine (C) or thymine (T). The base A goes with T, and C goes with G, one big one paired with one small one (see Figure 3.3).

		Number of chromosomes	
Scientific name	Common name	Total	Pairs
Homo sapiens	humans	46	23
Gorilla gorilla	gorillas	48	24
Hoolock hoolock	hoolock gibbons	38	19
Nomascus concolor	black-crested gibbons	52	26
Symphalangus syndactylus	siamang gibbons	50	25
Canis familiaris	dogs	78	39
Mus musculus	mice	46	23
Cambaras clarkia	crayfish	200	100

Table 3.3 | Numbers of chromosomes in the DNA of selected species

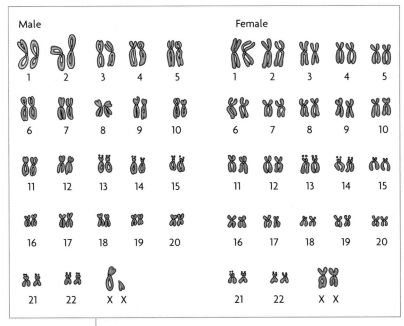

Figure 3.2 | Chromosomes

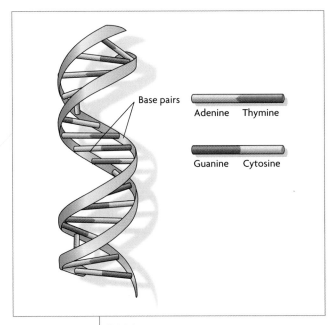

Figure 3.3 | DNA

Mutation: We Are All X-Men

The X-Men were introduced by comic book writer Stan Lee in the early 1960s as a team of young mutants who banded together to use their uncanny powers for the good of humankind. But whether or not you feel like a super-hero, you should know this: we are all mutants. Mutation happens with every human child born. Fortunately, most changes to the human genotype (the genetic makeup) do not result in changes to the phenotype (the physical makeup)—you can jot down words in the margin of a book, but the printed text stays the same. In short, mutation is a

normal and naturally occurring phenomenon (even if it *can* be accelerated by radiation and in certain chemical environments, as we learned from sci-fi films of the 1950s).

Most mutations are **point mutations**. These involve the simple, small change of a pair of nucleotide bases. Usually nothing significant changes, but sometimes, rarely, diseases such as sickle-cell anemia result. Sometimes mutation involves **insertions**, in which strings of DNA or RNA are added to genetic material. This is often achieved through transposable elements. For these discoveries we have the American cytogeneticist **Barbara McClintock** (1902–1992) to thank. In one of the masterworks of pure rather than applied research, McClintock detected these transposable elements in spotted corn kernels; the discovery earned her the Nobel Prize in 1983.

There are two types of transposable elements: transposons and retroelements. **Transposons** exist in about 2.8 per cent of the human genome and, fortunately, are currently completely inactive. They're grounded, confined to (nucleotide) base. They are a kind of genetic cut-and-paste, segments of chromosome that may be taken from one area and moved to another. In contrast, **retroelements**, which make up some 42.8 per cent of the human genome, are more like genetic copy-and-paste. In this more complicated action, a portion of DNA makes an RNA copy of itself, which in turn makes a retrocopy of itself back into the DNA. Remember that retroviruses can develop when RNA makes a DNA copy of itself. Some of these retroelements are still active, and can introduce such destructive elements as cancer. However, they are quite useful to scientists interested in determining the genetic difference between species sharing a common ancestor, as we will see in Chapter 5, on primates.

When we have two or sometimes many more copies of the same gene, with only one needed for transcription, the copies (or the original) are free to mutate with no structural problem. If you have three good goalies and one gets injured, you have fewer concerns than if you have just one superstar goalie and no viable backup. The mutated genes typically alter themselves out of the transcription game and become **pseudogenes**, which are genes that do not code.

This kind of mutation can cause changes (good and bad) by altering what are called *controlling elements*, or *controllers*. These are genetic strings that are ahead of genes in the long strands of DNA. Their best-known function is the turning on and off of genes, a process that happens during the life cycle of the being. Dramatic examples occur in animals that metamorphose (change structure) such as frogs, which have the same set of genes as tadpoles, and butterflies, which have the same set as when they are caterpillars. The difference in genetic structure of tadpoles and frogs is that certain genes are turned on or off during these two major stages of life.

Lactose intolerance—the inability to produce the enzyme lactase, which breaks down the milk sugar lactose—is an example of the role of controllers in the life cycle of humans. It is normal for the controllers in young mammals to have the lactase producing genes turned on when they are of an age to be nursed by their mothers, and to have them turned off at the usual age of weaning. For most of the existence of the species humans were not genetically programmed to drink milk as adults, and lactose intolerance is a holdover from a time when humans didn't drink milk beyond childhood. However, milk, as our ancestors discovered, can be a source of nutrition even among adults. Once human groups began domesticating cows, there was a selective advantage to a mutation that affected the controller in such a way that lactase continued to be produced beyond weaning age.

Insertions can be *microsatellites* (short repetitive sequences of between 2 and 5 nucleotide letters) or *minisatellites* (10–100 nucleotide letters). The latter are particularly interesting to forensic scientists, as there is considerable variation among individuals at certain *loci* (singular **locus**), the specific site on a chromosome where the insertion occurs. When a DNA test is done to determine paternity or to discover the identity of a rapist or murderer, what is studied is not the entire DNA strand (this would be complicated, slow, and expensive) but particularly loci, which are known to be especially variable and, therefore, telling.

Just as there are insertions, there are also **deletions**. These involve the removal of nucleotide bases (or 'letters', as they are commonly called). The most exciting (for humans) example of this is a 32-letter deletion of the gene CCR5, which includes the cellular gateway through which nasty viruses such as HIV/AIDS enter the body. In 1996, researchers discovered that this allele confers resistance to HIV infection. If you receive this allele from both parents, then it completely prevents the entry of HIV. If you receive it from only one parent, then it postpones the onset of AIDS for two to four years longer than is normal.

It's Not How Many Genes You've Got that Counts

A trick I pull on my students begins with giving them the number of genes possessed by certain 'lower' life forms (fruit flies have roughly 13,600; chickens, 23,000) before asking them to guess how many humans have. This is a trick question because it isn't the number of genes that's important but the number of proteins that can be produced. Typically 'higher plants' have more genes than 'higher' animals, but they produce far fewer proteins, and thus are less genetically complex. Why is that?

It has to do with a form of mutation known as **alternative splicing**. Splicing of anything (film, rope, etc.) involves cutting part off. To understand what alternative splicing involves in genetics, you have to know what is spliced.

In more complicated species (including all animals), genes are constructed of exons and introns. **Exons** are DNA segments that can be transcribed into RNA; that means that they have the potential to code. **Introns** do not code—they are spliced out in the transcription from DNA to RNA. Genes with introns are called **interrupted genes**.

Here's where alternative splicing comes in. Some genes have more than one exon, but only one gets transcribed (the other(s) are spliced out). Sometimes, through mutation, an exon that was transcribed in earlier generations is now spliced out, and vice versa. Sometimes an exon from another gene is taken and all the exons from a particular gene are spliced out. Alternative splicing happens a lot, allowing for great diversity of protein production. But it only happens to species whose genes are made up of exons and introns. Fruit flies and chickens may have plenty of genes, but without the capacity for alternative splicing, their protein production is severely limited.

Selective Pressure

You know the expression *Use it or lose it*? That principle works fundamentally (but with some qualification) with evolution. Take the idea of **selective pressure**. When selective pressure is high—i.e. when what a gene codes for is very important to the survival of the species—then any significant mutation is 'punished' by the individual's being selected out of survival. However, when selective pressure is low—as it is for, say, full-colour vision for a nocturnal animal—then mutation resulting in function loss has no significant consequences (i.e. a nocturnal animal doesn't require colour vision for survival).

Here's an example. GULO is a pseudogene in humans and in other primates. When it is a true gene—for instance, with cats and dogs—it enables the body to produce vitamin C. Cats and dogs typically have a low vitamin C diet, so the selective pressure of this gene is high. Among primates, which typically eat a lot of fruits and other foods rich in vitamin C, the selective pressure for GULO is low.

Add to this the idea of **genetic cost**. If something requires a great deal of genes to construct and support it, we say that it has a high genetic cost. A characteristic with a high genetic cost should present a tremendous advantage (i.e. be cost effective) or else it will not be selected. Consider the Mexican tetra (*Astyanax mexicanus*). A form of this species, known as the blind cave fish, spends its life in deep, dark caves, where the need (and, therefore, the selective pressure) for good eyesight is low. Eyes are metabolically expensive to construct, and are easily injured. Given the high genetic cost of eyes, it's perhaps not surprising that some of the cave-dwelling Mexican tetras have lost the power of sight, even though fish of the same species living in different conditions have retained their eyesight. Don't worry about the blind fish: they have developed improvements on a pressure-related sense organ, the **lateral line system**, which helps the tetra feel objects that it cannot see. The genetic cost of the lateral line system may be high, but it's a much better genetic investment for fish that live in the dark, and so its selective pressure is high.

Debates within Evolutionary Theory

Every area of science has room for healthy debate among experts. It's proof that a field is actively developing. Scientific debate concerning evolution fundamentally centres around the question *How strong a role does adaptation have?* The great Japanese biologist **Motoo Kimura** (1924–1994) put forward the idea of **neutralism**, suggesting that many mutations that are perpetuated are adaptively neutral. In other words, they do not confer advantage, and the species receiving them are more or less just fortunate in their time and place. Perhaps the bright colour of cardinal males, found in association with greater parenting skills, is just such an example.

Another debate surrounds the notion of **genetic drift**, promoted by the American biologist **Sewall Wright** (1889–1988), one of the co-founders of the field of population genetics. In this view, small populations moving away from their fellow species members develop unique features that perpetuate because of the relatively small and isolated aspects of the population. It's related to the **founder effect**, which can occur whenever a small population goes to a new location and remains for generations relatively isolated (in a genetic sense). Studies have been done with the Pennsylvania Amish, the Pitcairn, and the Easter Islanders. In Canada, much recent work has been done with respect to the Inuit (specifically concerning their resistance to heart disease), and to the populations of Quebec (in which about 8,500 settlers colonized the St. Lawrence River valley between 1608 and 1759) and of Newfoundland (see Rahman et al., 2003). The principle of genetic drift suggests that chance plays a significant role in what traits are passed on among members of an isolated population. It opposes the natural selection idea that a trait is more likely to be passed on if it's beneficial to the survival of the population. It's worth noting that there is a kind of intellectual genetic drift operating here: American evolutionary biologists, influenced by Wright, tend to emphasize genetic drift more than do British biologists, influenced by J.B.S. Haldane, who took a more adaptationist line.

This debate takes us to the conflict between convergent evolution and contingency. **Contingency** extends the logic of the genetic drift argument to say that small, isolated incidents greatly affected the course(s) of evolution, that other incidents, equally small and equally likely, might have happened instead, changing those courses. Gould was famous for saying that if you wound back the tape of evolutionary history and taped again, the sights and sounds recorded would be different. (That has not made him popular either with traditional evolutionary theorists or with those who look for evidence of intelligent design in everything.) Countering this view is the concept of **convergent evolution**, which holds up species of very different backgrounds that developed very similar physical structures as proof of natural selection. My favourite examples come from comparing the **marsupials** of Australia with **placentals** (that includes us, squirrels, bears, dogs, etc.) in the Americas. A good illustration involves flying squirrels and sugar gliders. The former are placental mammals of the order Rodentia; the latter, marsupials of the order Diprotodontia. They are only very distantly related but, independently, adapted very similar traits to survive in their respective environments, as the photos below show.

Convergent evolution: the flying squirrel (left) is a placental; the sugar glider (right), a marsupial

Proofs for Evolution

The debates *within* evolutionary theory may be healthy, but the ones surrounding it and its viability overall are not. We've touched on this issue earlier, and the campaign mounted by creation scientists to have evolutionary theory debunked. Rest assured, there are multiple proofs for the fact of evolution. That these multiple proofs exist is itself a form of proof, called **consilience** (see Chapter 2). In the final section of this chapter, we'll look at seven proofs for evolution.

1. Direct Observation of Evolution

The signs of evolution are all around us, every week in newspapers, in radio and television news reports, and in the blogosphere. You won't read the 'e' word, but it's there. As I'm writing this, the swine flu scare is big news. An American 'factory farm' operating in Mexico, well out of range of the more stringent environmental controls of its home country, has created the ideal environment for a flu virus to flourish by putting together a huge population of pigs and their feces in unsanitary conditions. The flu has mutated into a form that also affects humans (it has *evolved*), and it has become pan-zooic. Just about every human epidemic, real or imagined, from smallpox (which began with cows) to flu (pigs and chickens), is evidence of evolution. When you read about one year's flu vaccine (made from last year's flu virus) being ineffective against certain 'strains' of flu, you're reading about evolution. We see it, but like Galileo and the rings of Saturn, we don't see it. The implication of this for medicine is well-described by Stephen R. Palumbi, in his powerful (and depressing) book *The Evolution Explosion: How Humans Cause Rapid Evolutionary Change*:

> The rapid change of common diseases . . . has been cataloged in great detail. But the day-to-day impact of these changes has been little noted, and the cause—*evolution*—generally ignored. . . . Many medical journals assert that the ability of bacteria to kill even in the presence of massive antibiotic doses flows from the *development* of antibiotic resistance. But evolution is the root cause, and the failure to make this technical distinction could be fatal to some poor patient. . . . Why? Because if something develops, then it *just happens*, and we may never know why. . . . And if antibiotic resistance just happens, then we have no notion of *how* it comes to be, and no real chance to block the rise of some of the world's deadliest forms of life. But if something evolves, then the science of evolution can chart the answer to why, and perhaps prevent or change it. (Palumbi, 2001, p. 35)

Palumbi focuses on three developments that he highlights as evidence of evolution:

the resistance of bacterial disease to antibiotic cures

the resistance to single drugs of the AIDS retrovirus

the resistance of insects to insecticides.

These are all areas in which we directly observe evolution. In the first case, Palumbi discusses how penicillin and the other fungal-based 'miracle drugs' developed between the 1940s and the 1960s have lost their effectiveness because nasty bacteria like *Escherichia coli* (*E. coli*), *Staphylococcus aureus* (a major cause of postoperative infections), and even the once-conquered tuberculosis have developed resistance to antibiotics. In his eyes, we are new combatants in an old war between bacteria and fungi; this suggests that any fungal trick we use has the possibility of drawing some form of bacterial evolutionary defence. And he gives us the good advice of always taking the full course of our prescribed antibiotics, even if we're feeling better, as we risk leaving some bacteria standing (or whatever it is that they do); it's quite possible that some bacteria have survived longer than others owing to some slight advantage of resistance that we do not want to reward.

If that weren't depressing enough, Palumbi serves up the following numbers:

1908 – first case of insect resistance to insecticide

c. 1965 – 224 insect species are resistant to insecticide

c. 1985 – 447 known insect 'resistors'

c. 2000 – 500+ insecticide-resistant insects. (Palumbi, 2001, p. 139)

Palumbi's news on the HIV front is both good and bad. While it is known that single-drug therapies for AIDS patients can become resisted with as few as four mutations on the part of the virus (p. 120), using an assortment (or 'cocktail') of drugs has been proven effective. Researchers have also discovered a gene in some humans that makes them resistant to HIV—in this case, resistance is on our side (see the earlier section in this chapter on mutation).

2. Transitions and Consistency in the Fossil Record

Fossils are created by a slow and unlikely process in which the parts of a once-living being are replaced by minerals that become stone. Harder objects are more likely to fossilize than softer ones because they last longer. To date, over 250,000 fossil species have been discovered. Taken together, they form a record that can be studied for transitions and consistency. In essence, when we move from lower (and therefore older) geological strata to higher (and younger) strata we can trace species as they move from more ancient to more recent forms. The fossil record is set in stone, so to speak: it's impossible for fossils to turn up out of order in the geological strata. As **J.B.S. Haldane** (1892–1964) once noted, if we found a fossil rabbit in the Jurassic Era (206–144 mya), then we would have to consider abandoning the theory of evolution.

Of course, there is a no-lose intellectual game that supporters of creation science and intelligent design can play. It works like this. A creation scientist will look for gaps in the fossil record, say between a recent species 'A' and its supposed ancestor, species 'B'. This, claims the creation scientist, is evidence that 'B' did not become 'A'. If a paleoanthropologist discovers a transitional fossil 'X' that is something like 'B' but has changed towards 'A', the creation scientist will then say that we have two suspicious gaps: between 'A' and 'X' and between 'X' and 'B'. It is a no-lose but ultimately unscientific argument. There are always gaps, but the path is clear in so many cases.

Archaeopteryx seems not to have been 'first bird'. Transitional species are not necessarily ancestral species to the newer form. This important point was made by Coyne in *Why Evolution is True* (2009):

> A 'transitional species' is not equivalent to 'an ancestral species'; it is simply a species showing a mixture of traits from organisms that lived both before and after it. . . . [T]he dinosaurs that gave rise to birds sported feathers, but some feathered dinosaurs continued to persist well after more birdlike creatures had evolved. Those later feathered dinosaurs still provide evidence for evolution, because they tell us something about where birds came from. (Coyne, 2009, p. 35)

I was hoping that archaeopteryx was ancestral to the Amazon parrot; it could have helped explain the velociraptor sounds our pet parrot Lime makes when I try to put him back in his cage at night.

ARCHAEOPTERYX

A clear example of transitions in the fossil record is the **archaeopteryx**. The first evidence, a fossilized feather, was discovered in southern Germany in 1860, a year after Darwin's *Origin of Species*; the following year, an entire skeleton was found. Since then, nine other partial or complete fossilized skeletons have been found, one of them recently identified in 2005. The original discoverer, Hermann von Meyer, named the specimen *Archaeopteryx lithographia* (lit. 'ancient-wing from printing stone').

Here's what the fossil evidence tells us. The archaeopteryx lived from about 150 to 145 million years ago. It weighed roughly 250 grams and had a wingspan of 58 cm (22.6 inches) and a length of 50 cm. This places it between a blue jay and a crow in size but with a weight about three times that of a jay, with twice the wingspan and slightly less than two times the length (see Figure 3.4).

But don't get caught up in the bird comparisons. Archaeopteryx also had certain reptile features, including manual claws (which no adult bird has) in addition to a separate set of fingers on the wings, a flat sternum or breast bone, a long bony tail, and teeth. But then, it was not completely a reptile: it had too many bird features—feathers, wings, a partially reversed first toe, a furcula (wishbone), and hollow bones (to make it lighter for flight). After reading Pat Shipman's compelling account in *Taking Wing: Archaeopteryx and the Evolution of Bird Flight* (1998), I believe it could fly.

Archaeopteryx lighographia

James L. Amos/Photo Researchers, Inc.

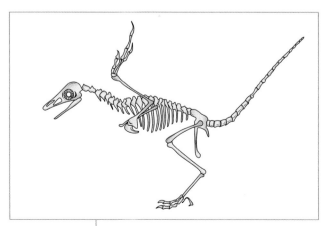

Figure 3.4 | An artist's rendering of the *Archaeopteryx* skeleton

TIKTAALIK

My favourite example of transitions in the fossil record comes from *Tiktaalik roseae* and its fellow fish travellers. For the full story, read Neil Shubin's *Your Inner Fish: A Journey into the 3.5 Billion-Year History of the Human Body* (2009).

The **tiktaalik** (which means 'burbot' or 'large fresh-water fish' in Inuktitut) lived 375 million years ago and is transitional between fish and **tetrapods** (lit. 'four legs') such as amphibians, reptiles, birds, and mammals. Shubin and his crew discovered the first fossil evidence of the tik-taalik in 2004 on Ellesmere Island, in Nunavut. It suggest that the species had several fish features, including scales covering its body and gills with which to breathe. Unlike a fish, it had a mobile head that could move from side to side independently of the body. Losing the bony plates in the gill area, it had, in essence, a neck.

Computer-generated rendition of *Tiktaalik roseae*

Victor Habbick Visions/Science Photo Library

Shubin reckoned that *Tiktaalik* lived in an area with streams and shallow ponds surrounded by broad mudflats. Although large itself (over 2 metres in length), it shared the water with some very heavily armed (or, more accurately, 'toothed') predators. It would have had a distinct advantage in being able to escape to and move around on the mudflats, out of range of its natural predators. To that end, it had a robust ribcage like that of a tetrapod, enabling it to support its body outside of the water. It also had primitive lungs. Hearing needs in water and in air require somewhat different forms of ears: the tiktaalik had a 'hybrid' water-and-air hearing system.

One area of enormous difference between fish and tetrapods is that where fish possess fins with, typically, four thin bones at the base, tetrapods have, on each side, one robust bone close to the body (analogous to the femur of the leg or the humerus of the arm) and two bones farther away from the body (e.g. the radius and ulna of the arm or the tibia and fibula of the leg), followed by the wrist or ankle and the many bones of the hands or feet. *Tiktaalik* shows a transition between the two:

> When we took the fin of *Tiktaalik* apart, we found something truly remarkable. . . . *Tiktaalik* has a shoulder, elbow, and wrist composed of the same bones as an upper arm, forearm, and wrist in a human. When we study the structure of these joints we assess how one bone moves against another, we see that *Tiktaalik* was specialized for a rather extraordinary function: it was capable of doing push-ups. . . . (Shubin, 2009, p. 39)

Table 3.4	Transitional chart: *Archaeopteryx* and *Tiktaalik*		
Species	Old form	New form	Time period
Archaeopteryx	Reptile	Bird	150–145 mya
	manual claws	wings	
	flat sternum	feathers	
	long bony tail	wishbone	
	teeth	hollow bones	
	separate fingers on wing	partially reversed front toe	
Tiktaalik	Fish	Amphibian	375 mya
	gills	lungs	
	scales	robust ribs	
	water ears	air ears	
		neck	
		shoulder	
		one-bone–two-bone progression	
		wrist	

In answer to the question of why a fish would want to do push-ups (other than to impress other fish at the beach), Shubin stated the following:

> With a flat head, eyes on top, and ribs, *Tiktaalik* was likely built to navigate the bottom and shallows of streams or ponds, and even to flop around on the mudflats along the banks. Fins capable of supporting the body would have been very helpful indeed for a fish that needed to manoeuver in all these environments. (Shubin, 2009, p. 40)

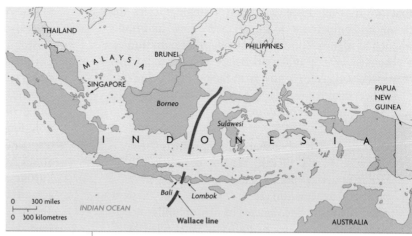

Figure 3.5 | Wallace line

3. Biogeography

Biogeography looks at the way species are grouped across the earth, particularly on islands, and over time in an attempt to determine why certain species are and are not in particularly places. (A subfield, palaeobiogeography, concentrates on extinct species.) The area of study began with **Alfred Russel Wallace** (1823–1913), who noticed a marked difference between the fauna (animal species) of Southeast Asia, where Asian mammals predominate, and those of Australia, home to marsupials not found in Asia. This biological division corresponds to a geographical one, the continental shelf of Southeast Asia. The Wallace line, as the boundary is known, is considered a significant proof of evolution in that species grouping reflects the geophysical conditions (both ancient and present) that permitted contact among certain species and prevented it among others.

Darwin himself discussed the grouping of animal species and geography in *Origin of Species*, using the plants and animals of his beloved Galapagos archipelago as his example. He spoke of how both flora and fauna on the island were like those of South America, and then asked:

> Why should this be so? Why should the species which are supposed to have been created in the Galapagos Archipelago, and nowhere else, bear so plain a stamp of affinity to those created in America? There is nothing in the conditions of life, in the geological nature of the islands, in their height or climate, or in the proportions in which the several classes are associated together, which resembles closely the conditions of the South American coast: in fact there is a considerable dissimilarity in all these respects. On the other hand, there is a considerable degree of resemblance in the volcanic nature of the soil, in climate, height, and size of the islands, between the Galapagos and Cape Verde Archipelagoes: but what an entire and absolute difference in their inhabitants! The inhabitants of the Cape Verde Islands are related to those of Africa, like those of the Galapagos to America. (Darwin, 1959, pp. 39–40)

The answer to his questions is biogeography. As the Galapagos archipelago and South America are close to one another, the similar but different species reflect their shared evolutionary history (the island species are descendants of the mainland species), not special creation. The same situation occurs with the Cape Verde archipelagos and Africa.

Dobzhansky also wrote on the subject of biogeography, noting that of the roughly 2,000 species of drosophilid flies in the world, about a quarter of them lived in the Hawaiian islands, where all but 17 of those 500 or so species were **endemics** (i.e. found only there). He attributed this to the islands' volcanic origin and the fact that they were never part of a continental land mass. He rightfully hypothesized

1. Geospiza magnirostris.
2. Geospiza fortis.
3. Geospiza parvula.
4. Certhidea olivasca.

Adaptive radiation: Contrasting pictures of Darwin's finches that the flies all descended from a single species that, facing no competitors on the relatively 'new' and under-populated islands, could through **adaptive radiation** (see Figure 3.6) develop different species adapted to a variety of unoccupied ecologic niches, to the point of being more diversified than different species of the same group elsewhere. He concluded by noting:

> Oceanic islands other than Hawaii, scattered over the wide Pacific Ocean, are not conspicuously rich in endemic species of drosophilids. The most probable explanation of this fact is that these other islands were colonized by drosophilids after most ecologic niches had already been filled by earlier arrivals. This surely is a hypothesis, but it is a reasonable one. Antievolutionists might perhaps suggest an alternative hypothesis: in a fit of absentmindedness, the Creator went on manufacturing more and more drosophilid species for Hawaii, until there was an extravagant surfeit of them in this archipelago. I leave it up to you to decide which hypothesis makes sense.

The study of palaeobiogeography, which draws upon the fossil record, can be a fruitful source of evolutionary proof. Sahotra Sarkar points to camels as a good example in this regard, since they shed light on life conditions both when the land masses of North America and Europe separated and when the continents of South America and Africa separated, the latter before the former:

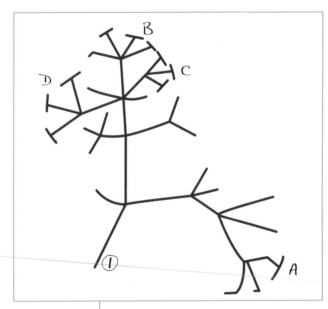

Figure 3.6 | A recreation of Darwin's illustration showing how he believed species ('A', 'B', 'C', and 'D') were generated from starting point (1). A friend of mine has a T-shirt with this on it.

[T]rue camels (*Camelus dromedariius* and *C. bactrianus*) are found only in Asia and Africa. Their closest relatives are the llamas (*Lama glama, L. glama pacos* or alpaca, and *L. guanicoe* or guanaco) and vicuña (*Vicugna vicugna*) of South America. . . . [T]he fossil record includes a large fauna of Tertiary-era North American camels, now long extinct. Camels are believed to have originated in North America and migrated to South America, Eurasia, and Africa. The fauna of Europe and North America are similar because a land bridge connected the two in the early Tertiary era, 40 mya [million years ago]. In contrast, continental drift separated Africa and South America 80 mya, and their fauna have diverged ever since. (Sarkar, 2007, p. 9)

4. Comparative Anatomy

Comparative anatomy involves investigating the similarities and differences in physical structures of different animals. What has long been noted is that certain similarities are striking, particularly with respect to the design of bones. Darwin, for example, made the following comment:

> What can be more curious than that the hand of a man, formed for grasping, that of a mole for digging, the leg of the horse, the paddle of a porpoise, and the wing of the bat, should all be constructed on the same pattern and should include the same bones, in the same relative positions? (Darwin, 1859, p. 434)

Giberson adds his comments about the similarities of our hands and feet to the pedal extremities of bats and mice:

> When we compare these species, we might expect to discover entirely unrelated mechanical configurations of bones. After all, what we do with our appendages bears little resemblance to what bats do with theirs. What we find, however, is the same configuration modified for different purposes. In all these cases we find creatures with five 'fingers' . . . , each of which is segmented into digits. In bats the digits are dramatically extended to make a frame for a large unfolding wing; in mice they are smaller and closer together, adapted to walking; in humans, they are optimized for complex motor skills. There is nothing magic about the number five and yet all these creatures have the same number of 'fingers'. One has only to ponder one's little toe to realize that, for many applications, it would have been fine for 'one little piggy' to have gone to market and not come back. Evolution offers the compelling explanation that a five-fingered ancestral form passed down this

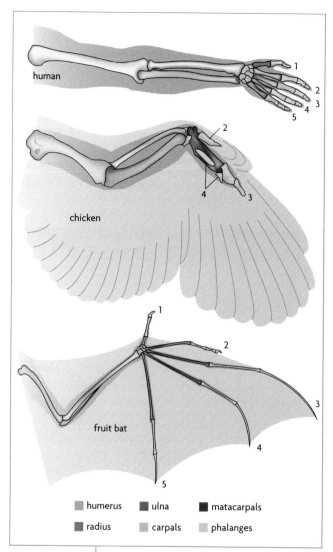

Figure 3.7 | Homologies of the forelimb

has to work with. Mutations for a perfect design may not arise because they are simply too rare. (Coyne, 2009, p. 12)

The great nineteenth-century anatomist **Sir Richard Owen** (1804–1892) is considered the founder of this field. He went into great detail in noting the **homology**, or similarity of structure, of divergent species, and his findings contributed to evolutionary theory. Strangely, Owen's exact position on evolution is unclear. He seems to have accepted evolution on principle, but famously fought with Thomas Huxley (aka Darwin's Bulldog) over the details.

5. Unintelligent Design: Vestigial Structures, Imperfections, and Atavisms

One of the best ways to prove evolution and expose the nonsense of intelligent design theories is to look closely at the poor or *un*intelligent design that exists in many species. Sometimes this exists in vestigial parts—leftovers from a time when the species was something else. Pythons have leftover pelvic bones, and ball pythons have tiny leftover legs from when they were lizards. Humans have tailbones but no tail, a trait we share with gibbons, orangutans, gorillas, chimpanzees, and bonobos, signalling our common descent from a tail-less ancestor. And then there is our appendix, which probably was useful as a kind of second or additional stomach when we were almost exclusively herbivores, before our ability to craft stone tools gave us ready access to meat over two million years ago.

It is the structural flaws of humans that serve as the best examples of poor design. Our big toes are, well, big and get in the way when we walk barefoot, because they used to be grasping thumbs. Karl Giberson, below, points to how badly designed our knees are, something that any long-distance runner (like I once was) or goalie readily understands:

> Our knees are strangely designed and destined for injury. Below them are ankles that can move in several directions; above them are hips attached via 'ball joints' that also permit a wide range of motion, as anyone who has ever played with a hula-hoop knows. But the knees themselves bend in only one direction. No engineer would put three joints in a row and constrain the middle one in the way our knee is constrained. The design is so bad that countless athletes have to wear a special brace to help prevent a knee from bending about the wrong axis. (Giberson, 2008, pp. 162–3)

property to a large number of species. Natural selection worked within this constraint to give up the many variations around this common theme that we see today. (Giberson, 2008, pp. 199–200)

The discussion could be extended to include horses, who lose the outer two fingers and reduce the inner toe, with one long middle finger (imagine their capacity to 'up-yours' to other animals). Seal flippers have a single-bone upper limb (the humerus), and a two-bone lower limb (the ulna and radius), as well as five fingers. An important idea here is that comparative anatomy shows that each animal was not 'designed from scratch' to have a perfect design for how it lives. Instead, old parts are transformed to do new things. Evolution recycles and reworks. Coyne makes this point well when he states this:

> Natural selection . . . doesn't produce the absolute perfection achievable by a designer starting from scratch, but merely the best it can do with what it

Either our ankles should have been designed so that they do not turn, or our knees should have had more pivot built into them.

Our backs are poorly designed for walking upright, and certainly for doing 'unnatural things' such as bending over to pick things up. These actions eventually cause back problems. Human backs are better structured for walking on all fours, sitting up, and moving in trees. The long list of human features that are essentially useless also includes our toenails, male nipples, wisdom teeth, tonsils, body hair, goose bumps, the function of breathing and swallowing through the same tube . . . You get the point.

Atavisms (from the Latin *atavus*, 'ancestor') are ancestral traits that appear occasionally in living species. Roughly one out of every 500 whales is born with rear legs that stick out beyond the body. Some horses are born with three fully developed toes on each leg, rather than the one developed (the hoof) and two vestigial toes that are the usual pattern. And humans sometimes are born with actual tails.

A genetic explanation for atavisms is that there is a team of genes that produced the trait in ancestors. Mutation disabled one or more members of the team, turning it into a pseudogene that no longer codes; because the team was incomplete, the trait was no longer produced. However, in the case of atavisms, a gene mutates

Science Gone Wrong

Of Moths and Men

The topic of this discussion used to be one of my favourite examples of evolution, before one of my students alerted me to the fact that it was so longer considered solid evidence—proof that you can contradict your professor and still receive a very good mark.

This is the story of the British peppered moth, so named because its usual colour is off-white with a sprinkling of black spots. Its scientific name is *Biston betularia*. The usual form or subspecies of this moth was, unsurprisingly, called *typica*, so that it's full name is *B. betularia typica*. First spotted in 1848 and increasing in number over the next half-century, another variety of this moth began to appear and even dominate in heavily industrialized areas: *B. betularia carbonari*. As the name indicates (think 'carbon'), this was a darker version of the peppered moth. The dominance of this form in the industrialized areas such as Manchester and Birmingham was a phenomenon that was given the term **industrial melanism**. While it had been noted and commented on at the beginning of the twentieth century,

it wasn't until 1924, when the great population geneticist J.B.S. Haldane drew attention to what it said about the power of natural selection, that the industrial melanism of the peppered moth was considered important. When, in the early 1950s, **Dr Bernard Kettlewell** (1907–1979), a medical doctor and lepidopterist (one who studies moths and butterflies), published the results of his experiments with peppered moths, the species and its distinct forms became a highly prized example of how natural selection works. Judith Hooper, author of *Of Moths and Men: An Evolutionary Tale: The Untold Story of Science and the Peppered Moth* (2002), presented the following example of how Kettlewell's research was cited in a biology textbook:

> In carefully controlled experiments, equal numbers of peppered and melanic moths were released into the countryside where industry was absent. The trunks of the region are covered by lightly colored lichens. After release, observers

The British peppered moth: dark (left) and light (right) varieties

into a form that can join the other members, thereby making the team whole again.

6. Molecular Biology

Our DNA is remarkably similar to that of a chimp. A brilliant example of that closeness lies in our chromosomes. Chromosome number two was formed by the fusion, or joining, of two separate chromosomes that once occurred in our ancestors and that exist currently in chimpanzees. This indicates that our common ancestor had that pattern,

and part of our evolution from this common ancestor involved this fusion of chromosomes. Two features show this molecular fusion: **telomeres** and **centromeres**. In the illustration of the human chromosomes presented in Figure 3.8b, the joining point, about one million base pairs in length, is the centromere. Daniel Fairbanks explains the role of the centromere:

> When a cell is about to divide in two, the chromosomes condense into compact structures that are easy to observe with the help of a microscope.

with binoculars tried to determine the fate of as many moths as possible. . . . Of the 190 moths that were actually observed to be eaten by the birds, 26 were peppered and 164 were melanic. . . .

The same experiment was conducted in the midland of England, where extensive industrialization had caused most lichens to die and trees became darkened by soot. The results of this experiment were the reverse of the previous experiment. The birds ate many more peppered moths than melanic moths. (as cited in Hooper, 2002, p. xvii)

In Beacon Wood, near the industrial city of Birmingham, years of pollution had, indeed, stripped the lichen off of the trees. Kettlewell set a large number of moths of both varieties onto the trunks of the trees in the morning. The moths were night-fliers—they rested during the day. At day's end, he captured the survivors. After 11 days, 27.5 per cent of the carbonaria had survived, and only 13 per cent of the typicals. Kettlewell then went to Dean End Wood, in unpolluted Dorset, near the southern shores of Britain, and found the opposite result: about three times as many typicals survived, blending in, he believed, with the lighter-coloured lichen-covered trunks of the trees there.

Kettlewell later commissioned an army of 150 amateur collectors to trap moths in various areas to see whether the hypothesized relationship of *industrial area = dark moth/clean area = light moth* would ring true throughout Britain. For the most part it did, although there were some annoying counter-examples of darker moths dominating in rural, relatively unpolluted eastern Britain. He blamed the wind blowing in from the industrialized areas. All of this research seemed to show that camouflage had an adaptive advantage that quite clearly demonstrated how natural selection worked.

By the 1990s, holes in the research had begun to appear, discrediting a once useful textbook example. Tree trunks, it turned out, were not the normal resting places of peppered moths, which preferred the better hiding places underneath branches. The high density of moths placed on the trunks wasn't natural either, creating what Hooper called a 'feeding tray' (2002, p. 266) for the feathered feeders. Kettlewell's studies also suffered from a 'vision problem': birds don't see things the way that humans do, particularly with regards to ultraviolet light. From a bird's perspective, the black moths were *not* more visible in places where the lichens still grew. Then there was the question of whether birds were even the main predators of the moths; perhaps in the moths' natural life experiences, bats were really the moths' principal predators. Birds are often opportunistic (our parrot Finn sometimes steals dog kibble and is a regular raider of my dinner plate) and would have taken advantage of the buffet table that Kettlewell provided.

All of this led researchers to consider other possibilities. Perhaps another difference between the two sub-species was involved. During the pre-adult stage (which includes life as egg, caterpillar, and pupa) the carbonaria moths have proven to be hardier. That's the stage when roughly 90 per cent of moth mortality takes place. Perhaps colour is just one of several changes that differentiate typicals from carbonarias. Maybe *carbonaria* would be better named *robustus*.

All in all, we're forced to conclude that the example is no longer solid. Unfortunately, the disproving of this example—an act of good science—has found its way into the hands of supporters of creation science and intelligent design, who have tried to hold it up not as an illustration of science acting to improve itself (which it is) but as 'proof' that evolution doesn't work.

Ancestral and chimpanzee chromosome 2a	Ta – OS – <u>Ca</u> – OS – <u>Ta</u>
Ancestral and chimpanzee chromosome 2b	<u>Tb</u> – OS – <u>Cb</u> – OS – <u>Tb</u>
Human Chromosome 2	Ta – OS – <u>Ca</u> – OS – ta – tb – OS – cb – OS – <u>Tb</u>

KEY
<u>Ca</u> working centromere from chromosome 2a
<u>Cb</u> working centromere from chromosome 2b
ca mutated disabled centromere sequence from chromosome 2a
cb mutated disabled centromere sequence from chromosome 2b
<u>Ta</u> working telomere from chromosome 2a
<u>Tb</u> working telomere from chromosome 2b
ta mutated disabled telomere from chromosome 2a
tb mutated disabled telomere from chromosome 2b
OS other stuff

Figure 3.8a | Two ancestral chromosomes become one human chromosome

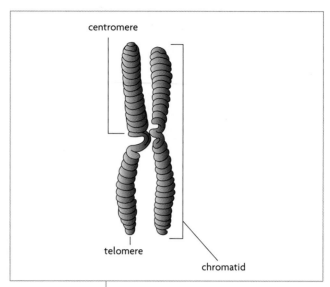

Figure 3.8b | Detail of a human chromosome

Each compact chromosome has a constricted region called a *centromere*, with very specific DNA sequences in it. The centromere serves a critical role in helping direct the chromosome to its proper position when cells divide. The ends of each chromosome are called *telomeres* and they, too, contain highly specific DNA sequences. (Fairbanks, 2007, p. 19)

At some point after chimpanzees/bonobos and humans parted company on the evolutionary path from our common ancestors, two of the telomeres of the two separate chromosomes fused (see Figure 3.8a). The site can be detected. Likewise the centromere of one of the two ancestral chromosomes (now exhibited by chimpanzees) aligns with the centromere of chromosome two. The other centromere aligns with a site in chromosome two where there is no centromere. Its history is seen in how its letters match those of the centromeres of one of the ancestral chromosomes.

We can add to this that our chromosomes share, in the same locations, the dead genes (or pseudogenes) of thousands of ancient infections, or **endogamous retroviruses**, that must have infected our common ancestor, but which became harmless when they mutated (see Coyne, 2009, p. 69).

7. The Embryo: 'Ontogeny recapitulates phylogeny' (in part, anyway)

Darwin believed that embryos offered strong evidence in support of evolution. In 1866, German biologist Ernst Haeckel developed a biogenetic 'law' that is typically summed up as 'ontogeny recapitulates phylogeny' (it would make a nice T-shirt). What it means is that in the development of an embryo (**ontogeny**), there is a recapitulation (or repeat) of the stages of evolution (**phylogeny**) that the being has gone through. The partial truth of this statement can be seen in the fact that a human embryo has, for a while, a tail, one that is almost always reabsorbed into the rest of the developing body. Sharing that future-less fate is **lanugo**, a downy coat of hair resembling that of our hairier ancestors, and the first two kidneys that we grow embryonically, which resemble those of our jawless fish and reptile ancestors. Only the third type of kidney is the 'keeper'.

That much provides evidence for evolution, but the point should not be taken too far. Coyne articulates well the limitations of this form of evidence:

> Embryonic stages don't look like the adult forms of their ancestors, as Haeckel claimed, but like the *embryonic* forms of ancestors. Human fetuses . . . never resemble adult fish or reptiles, but in certain ways they do resemble embryonic fish and reptiles. Also, the recapitulation is neither strict nor inevitable: not every feature of an ancestor's embryo appears in its descendants, nor do all stages of development unfold in a strict evolutionary order. Further, some species, like plants, have dispensed with nearly all traces of their ancestry during development. (Coyne, 2009, p. 78)

While the skeptical reader may question whether one form of evidence or another is really solid proof, the mere fact that all of them support the theory of evolution is overwhelming evidence. There is no scientific way of discrediting all or even most of them.

Summary

Evolutionary theory evolves. We have seen evidence of that throughout this chapter. In fact, the theory was evolving even as this chapter was being written. This year's first issue of *The Scientist* (volume 24, published in 2010) contains a discussion of three new ideas not tackled here—ideas that may prove instrumental in a new synthesis of evolutionary ideas. One is **evolvability**, the notion that some species are more likely to evolve than others. Second is **facilitated variation**, which points to chemical processes that can increase the likelihood of variation. Finally, there is **multilevel inheritance**, an indication that certain heritable features may be passed down in ways other than the usual genetic route. I encourage the reader to have a look at www.the-scientist.com/2010/1/1/24/1.

Typical Student Questions

▶ *Can I believe in evolution and be a good Christian?*

Yes: see the discussion on theistic evolution. Also, I strongly urge you to read Karl Giberson's book, and look at his biography, both of which will give inspiration to anyone attempting to reconcile their spiritual views with science.

▶ *If we evolved from apes, why are there still apes? Why aren't they evolving into us?*

A simple answer is that apes, too, have changed since the time of our common ancestor. They experienced different circumstances and took a different evolutionary path. Remember, too, the lie of orthogenesis—there is no such thing as evolutionary momentum.

▶ *Are human beings still evolving?*

Once humans became cultural beings and we began to adapt by changing our culture, human evolution became less of a factor. We are slow evolvers anyway, as we take so long to reach the age of reproduction. But there remains the potential for evolutionary changes in humans. If, say, there were a worldwide epidemic of parrot flu caused by people not cleaning out their cages sufficiently, causing the deaths of millions, those with some genetically programmed resistance might become prominent in our population. Remember, too, the discussion of lactose tolerance (the ability to break down lactase or milk sugar) in human adults, which is a relatively new feature in our species, coming when we first domesticated cows.

Review Questions

1. Name and explain seven forms of evidence for evolution.
2. Distinguish between orthogenesis and evolution.
3. Distinguish between Lamarckism and evolution.
4. Describe the different forms of genetic mutation.

Discussion Questions

1. Discuss the weaknesses of the Kettlewell experiment involving peppered moths.
2. Why do you think it's difficult for some people to understand and accept evolution?
3. Which of the seven forms of evidence for evolution makes the most sense to you? Which is the most compelling? Which is the least?

Key Terms

adaptive radiation	chromosome	endemics	fossils
alleles	coding	endogamous retroviruses	founder effect
alternative splicing	comparative anatomy	eukaryote	gene
archaeopteryx	consilience	evolution	genetic cost
atavisms	contingency	evolvability	genetic drift
atheistic evolution	convergent evolution	exaptation	genetics
catastrophism	creation science	exon	genetic mutation
cell	deletions	facilitated variation	genotype
centromeres	DNA	fact	homology

industrial melanism
insertions
intelligent design
interrupted genes
introns
irreducible complexity
lactose intolerance
Lamarckism
lanugo
lateral line system

locus
marsupials
multilevel inheritance
natural selection
neutralism
nucleus
ontogeny
orthogenesis
phenotype
phylogeny

placentals
point mutation
prokaryote
pseudogenes
purine
pyrimidine
recombination
retroelements
RNA
selective pressure

sexual selection
telomeres
tetrapods
theistic evolution
theory
tiktaalik
transposons

Key Individuals

Francis Crick
Georges Cuvier
Charles Darwin
Richard Dawkins
Theodosius Dobzhansky

Karl Giberson
Stephen Jay Gould
J.B.S. Haldane
Dr. Bernard Kettlewell
Motoo Kimura

Jean-Baptiste Lamarck
Barbara McClintock
Gregor Mendel
Richard Owen
Alfred Russel Wallace

James D. Watson
Sewall Wright

Recommended Print and Online Resources

Buss, D. M., Haselton, M. G., Shakelford, T. K., Bleske, A. L., & Wakefield, J. C. (1998). Adaptations, exaptations, and spandrels. *American Psychologist, 53*(5), 533–48.

Coyne, J. (2009). *Why evolution is true*. New York: Penguin.

Dawkins, R. (2009). *The greatest show on earth: The evidence for evolution*. New York: Free Press, Simon & Schuster.

———. (2006). *The God delusion*. Boston: Houghton Mifflin.

Dobzhansky, T. (1973, March). Nothing in biology makes sense except in the light of evolution. *The American Biology Teacher, 35*, 125–9.

Fairbanks, D. J. (2007). *Relics of Eden: The powerful evidence of evolution in human DNA*. Amherst, NY: Prometheus.

Giberson, K. W. (2008). *Saving Darwin: How to be a Christian and believe in evolution*. New York: HarperOne.

Gould, S. J. The misnamed, mistreated, and misunderstood Irish elk. In S.J. Gould, *Ever since Darwin* (pp. 79–90). New York: Norton.

———. (1981, May). Evolution as fact and theory. *Discover, 2*, 34–7. Reprinted in S.J. Gould, (1994), *Hen's teeth and horses' toes* (pp. 253–62). New York: Norton.

Hooper, J. (2002). *Of moths and men: An evolutionary tale. The untold story of science and the peppered moth*. New York: Norton.

Linville, S., Breitwisch, R., & Schilling, A. (1998, Jan.). Plumage brightness as an indicator of parent care in northern cardinals. *Animal Behaviour, 55*(1), 119–27.

Sarkar, S. (2007). *Doubting Darwin: Creationist designs on evolution*. Malden, MA: Blackwell.

Shermer, M. (2006). *Why Darwin matters: The case against intelligent design*. New York: Henry Holt.

Shipman, P. (1998). *Taking wing: Archaeopteryx and the evolution of bird flight*. New York: Simon & Schuster.

Shubin, N. (2009). *Your inner fish: A journey into the 3.5 billion year history of the human body*. New York: Vintage.

TheScientist.com: F1000's magazine of the life sciences. (Website). Available from www.the-scientist.com

Evolutionary Pathways

Exhibit at the Grande Galerie
de l'Évolution, Paris
© Oliver Knight/Alamy

Introduction

You have just read about the process of evolution and of the proofs for its acceptance. In this chapter we will look at how the evolutionary process has produced an amazing variety—literally millions—of species. The way the biological taxonomy is written in this chapter reflects both evolutionary history and the sometimes frustrating existence of multiple ways of conceiving of living entities—as discrete, discontinuous entities/species, or as linked, continuous genetic patterns that tie all of us together. All living things on earth share a sometimes disturbingly high amount of genetic similarity. However, it is important to understand the characteristics that make each individual species unique.

In this chapter we take a walk down the evolutionary path with our relatives, some distant and others quite near. The relatively new and rapidly expanding field of molecular biology is teaching us how close even our most distant relatives are to us. I find this very exciting. Even though I have long known that we (meaning plants, animals—all living things) are related, when I read molecular biology (sometimes a daunting task), it powerfully reinforces that fact. Just think of the way certain genes, like the *Hox* genes, can be taken out of one animal (a fruit fly, for instance) and inserted into, say, a frog, where they will continue to function, perhaps adding an extra set of legs. It sounds like science fiction—a cross between Jeff Goldblum's character in *The Fly* and *Rana: The Legend of Shadow Lake*—but it's not fiction, just science.

This chapter is strongly influenced by **Richard Dawkins**'s master work *The Ancestor's Tale: A Pilgrimage to the Dawn of Life* (2004). We will be adopting his term **concestor** to refer to an ancestor shared by 'our group' (whatever that is at any particular point in time). We will also be using his dating method. It's a rough method, but useful for comparative purposes.

Family Trees and Bushes

One of the changes in viewing evolution over the last 30 or 40 years is the way that evolutionary trees are drawn. They are now more like evolutionary bushes. Meaning what? The *tree* (or *ladder*) *model* of evolution is a model with few extinctions, and with long lines of one species being replaced in whole by another, from great-grandparent to grandparent to parent to child, and so on. If you look at this model and you see a trait in the past that is shared with humans, then you quickly label the bearer of that trait an ancestor. A *bush model* of evolution involves a good number of extinctions, and a trait shared by living and extinct species may not be an indication of direct-line ancestry; it may just mean that their common ancestors were cousins. Figure 4.1 shows examples of tree

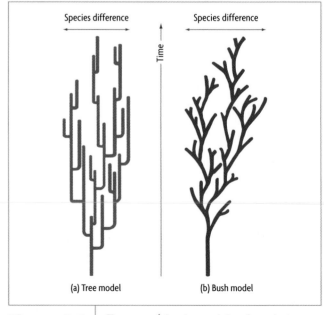

Figure 4.1 | Tree and bush models of evolution

and bush models side by side. Most physical anthropologists think in terms of bushes these days.

Two Concepts that Are Hard to Grasp

1. Deep Time

Deep time, which spans the history of the earth since it formed over 4.5 billion years ago, is perhaps the hardest aspect of evolution to grasp, particularly as we humans haven't really been around all that long. Metaphors help. Here are a few supplied by and borrowed from **Stephen Jay Gould**:

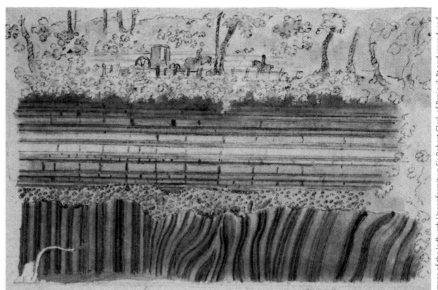

Sir John Clerk of Penicuik's illustration of the Jedburgh unconformity, cited by Hutton as evidence that the earth is shaped by the cyclical process of erosion, deposition, and uplift over time

> An abstract, intellectual understanding of deep time comes easily enough—I know how many zeros to place after the 10 when I mean billions. Getting it into the gut is quite another matter. Deep time is so alien that we can really only comprehend it as metaphor. . . . We tout the geological mile (with human history occupying the last few inches), or the cosmic calendar (with *Homo sapiens* appearing but a few moments before *Auld Lang Syne*). . . . John McGhee has provided the most striking metaphor of all (in *Basin and Range*): Consider the earth's history as the old measure of the English yard, the distance from the king's nose to the tip of his outstretched hand. One stroke of a nail file on his middle finger erases human history. (Gould, 1987, p. 3)

One of the misperceptions that comes from being unable to think in terms of deep time is the sense that everything before the arrival of *Homo sapiens* was just a short prelude for the lead actor on the stage. A good analogy for that particular misperception comes from the American writer Mark Twain, writing over a hundred years ago:

> Man has been here 32,000 years. That it took a hundred million years to prepare the world for him is proof that this is what it [the formation of the earth] was done for. I suppose it is, I dunno. If the Eiffel Tower were now representing the world's age, the skin of paint on the pinnacle-knob at its summit would represent man's share of that age; and anybody would perceive that the skin was what the tower was built for. I reckon they would, I dunno. (Mark Twain, as cited in Gould, 1987, p. 2)

While the term 'deep time' was coined by narrative nonfiction writer John McPhee in *Basin and Range* (1981) and popularized by astronomer/writer Carl Sagan, the best candidate for 'father of deep time' (which he himself called 'immense time') is **James Hutton** (1726–1797), a Scottish doctor, chemist, farmer, inventor, and geologist. In his tough-to-read but beautifully thought-out *Theory of the Earth*, first published in 1788, he argued that there were two forces at play in the shaping of the earth. One was erosion, which wore down the earth and laid down the layers for sedimentary rocks. The other, and opposite, was the dynamic, uplifting force of subterranean heat, which made parts of the earth, as igneous rock, rise above older levels. These two forces he would have had ample experience with, first as a farmer working his rainy country land from 1754 to 1767, then as a citizen of Edinburgh, where he daily witnessed the products of volcanic forces as he looked at the majestic Castle Rock, which looks down over the city, and the mysterious, myth-shrouded Arthur's Seat (named after King Arthur).

Prior to Hutton's writing, it was generally believed by writers on the subject that the earth had been under water (the biblical flood) and had simply eroded since then. Evidence for Hutton's ideas was deceptively simple: (1) vertical (rather than the usual horizontal) strata, which he first saw, famously, off the east coast of Scotland in 1788, just after the first publication of *Theory of the Earth*; and (2) mineral veins that looked to rise like lava through older rock. Remarking on these observations, he concluded:

Time, which measures every thing in our ideas, and is often deficient to our schemes, is to nature endless and as nothing. (Hutton, 1788, p. 215; as cited in Gould, 1987, p. 64)

From James Hutton, the torch illuminating the tunnel of deep time was passed on to geology superstar **Charles Lyell** (1797–1875), who literally cast the idea in stone in his three-volume *Principles of Geology*, published between 1831 and 1833. Evolution hadn't yet been proposed, but Lyell's rejection of the young earth envisioned by many a creationist (then as now) and his assertion that features of the earth's surface were caused by continuous and uniform processes rather than sudden catastrophic events laid the groundwork for a parallel view of the development of life on the planet.

2. Continental Drift

WHERE WERE THE CONTINENTS AT THIS TIME?

When looking at the earlier entries in the timeline of life on earth, it helps to know where the continents were at the time. Continents, remember, aren't fixed in place, but rather move by means of **plate tectonics**. Dawkins describes the process well:

> [T]he entire planet is armoured, covered by plates, sliding over the surface, sometimes diving under another plate. . . . As a plate moves, it does not leave a gap behind it. . . . Instead, the 'gap' is continuously filled by new material welling up from the deep layers of Earth's mantle and contributing to the substance of the plate. . . . In some ways, plate seems too rigid an image: a better metaphor is a conveyor belt, or a roll-top desk. (Dawkins, 2004, p. 242)

The example he gives is both useful and elegant. It involves the **Mid-Atlantic Ridge**. Imagine a huge S-curve, 16,000 kilometres long, running down the middle of the Atlantic Ocean to the east of the American continents—that's the Mid-Atlantic Ridge. Molten rock pushes up at the centre and then flows both east and west 'like two desk roll-tops'. The flow is pushing Africa and South America apart; once upon a time, the two were once connected. Look at a map, and you will see how the western outline of Africa almost perfectly matches the eastern outline of South America. Because of this process, the rock is the youngest at the centre, oldest at the continental edges. The ridge dates roughly from our time, while the rock on the edge of South America (and its African partner) is about 140 million years old; 500 kilometres in from the continental margins, the rock is younger than 100 million years old, 730 kilometres in, it's roughly 65 million years old, and so on, until we reach the ridge (Dawkins, 2004, pp. 202–5).

WHERE IS OUR CONTINENT NOW, AND WHAT, IF ANYTHING, IS CONNECTED WITH IT?

Our continent of North America has changed configuration a number of times over millions of years. Geologically speaking, Canada is part of the continent of Laurentia, which has morphed into different shapes and sizes over time. It has been a separate continent on a number of occasions: 1.5 billion years ago, during the Cambrian period (543–490 mya), and during the Cretaceous period (144–65 mya). From roughly 1.8 to 1.5 billion years ago, Laurentia was part of a supercontinent with India and Australia attached to its west coast and Brazil attached to the east. From 1.1 billion to about 750 million years ago (mya), we were connected with eastern Europe, Siberia, India, Australia, and northern and southern China. Eventually we lost the Indian and Australian connection, but then, from about 600 to 540 mya, we were part of a new supercontinent that joined us with Africa for the first time. After we separated from that tectonic grouping, we drifted along independently again until we collided with eastern Europe about 440 mya, and then with western and southern Europe. Our next move was to link up with the super-supercontinent Pangaea (Greek *pan-*, 'entire', and '-*gaia*', earth'), some 200 mya. We separated again during the Cretaceous, only to collide with South America about 2.8 million years ago.

The continents are still moving, but it is not something that you'd be aware of any more than you can feel the earth turning. The continents are averaging between 5 and 10 centimetres of movement a year—roughly the same rate as the growth of your fingernails. The message here is that some species are linked through a common ancestor that came into being when continents now separate were linked. However, when our genus, *Homo*, came into existence, the continents were in the same basic configuration that they are now.

Taxonomy

Traits

Taxonomy (Greek *taxis*, 'arrangement,' + *nomia*, 'distribution') is the scientific classification of living things according to shared ancestry, characteristics, or traits. An important part of the broader discussion of this chapter is to introduce the distinction between, on the one hand, **primitive** or **generalized traits** (which I sometimes call 'default traits'), and, on the other, **specialized** or **derived traits**. Basically, the significance of this distinction is as follows. The sharing of generalized traits is no indication of close relationships. If humans and tiktaaliks share the generalized pattern of having two bones in the lower limb and one in the upper (i.e. radius + ulna and humerus in the arm; tibia + fibula and femur in the leg), that is not evidence of any close relationship. However, if apes and humans share the more specialized, or derived traits, of a 2:1:2:3 dental pattern, then it argues for close ancestry—closer, certainly, than the relationship between humans and primates with a 2:1:3:3 dental pattern. Many mistakes of interpretation, particularly by journalists reporting anthropological stories, occur when people make a big deal out of sharing generalized traits.

Genus and Beyond

The two-part labels we use for genetic entities consist of a capitalized **genus** (plural 'genera') and a non-capitalized **species**, both conventionally written in *italics* (or underlined, if you're working on a very old Smith Corona typewriter). Our name is *Homo* (genus) *sapiens* (species). A good way to remember the two names is to think of the species name as the one that is most specific, and the genus name as the one that is the most general. The blurry nature of the genus distinction is beautifully captured by Richard Dawkins, when he writes that a genus is 'a group of species that are pretty similar to each other' (Dawkins, 2004, p. 332). How is that for an exact scientific definition?

Genus and species are two of the most specific taxonomic categories (although there are also subspecies). Broader categories, beginning immediately above genus and expanding in size, are **family**, **order, class**, and **phylum**, along with the *sub-* (with *infra-* below that) and *super-* subsets that represent slightly larger or slightly smaller versions of each category. While it is useful to have categories, these ones are often not as concrete as they sound, and biologists are often shifting member species from one to another. To a non-biologist, it must often seem like biologists are making it up as they go along. They are not: they are perfecting their science. Think of it as a sophisticated filing system, with the files sometimes moving from one directory, subdirectory, or folder to another.

Species

The **species** definition is the most concrete one we have, traditionally, of all the forms of taxonomy for labelling living entities. Yet in the light of recent discoveries in molecular biology, it can look rather abstract. In 1942 Ernst Mayr defined *species* as 'a group of interbreeding natural populations that are reproductively isolated for other such groups' (as cited in Coyne, 2009, p. 172). 'Reproductively isolated' means not being attracted to a member of another species or, if somehow attracted to a member of another species, not fully able to produce viable

young (i.e. young that themselves can reproduce). What do you get if you cross a penguin with a pitbull? The point is, you can't, because they're not of the same species (though *frostbite* is the answer to the joke). This is known as the **biological species concept** (bsc), and it is, in essence, the guiding principle used today. In his chapter 'The Origin of Species', Coyne makes strong arguments in favour of keeping it (see Coyne, 2009, pp. 168–89). Yet while I accept the utility of using this concept, I still find that it has serious weaknesses as a solid scientific definition. One reason for questioning it is that there have been many examples of interspecies breeding. Arguably, though, the most famous example of interspecies breeding shows bsc separation: a male donkey bred with a female horse will produce a mule, which will be infertile, incapable of reproducing.

The trouble with the bsc principle is that it is not an either/or situation. There are examples of interspecies breeding that has brought about at least partly fertile young. My wife and I believe that our parrots Louie (a male *Pyrrhura rhodocephala*) and Stanee (a female *P. molinae*) will be able to have viable young—but that's a hypothesis we've yet to test. In Canada, the most familiar examples come from canid interbreeding. *Canis familiaris* (dogs) can mate with *Canis lupus* (wolves) or *Canis latrans* (coyotes) and produce fertile young, although with coyote–dog crosses (called 'coydogs') fertility diminishes somewhat over three generations. There seems to be less of a fertility problem with 'coywolves', crosses between coyotes and wolves, leading some to question whether or not they are really separate species. On the extreme end, there is the beefalo, the product of the mating between a buffalo (*Bison bison*) and a cow (*Bos taurus*). This represents a crossing not just of species but of genera. There are bigger fertility problems here. The beefalo produced by a bull and a female bison is more fertile than the one produced

Beefalo

by a male bison and a cow. Also, the female beefalo are fertile, while the males are not, a common situation with hybrids. The Irish-Canadian lumber baron Mossom Boyd, of Bobcaygeon, Ontario, was the first to successfully breed beefalo, achieving the feat in the late nineteenth century.

The line separating closely related species is often a difficult one to distinguish. Dawkins refers to the concept of **ring species**, where there is a continuum between two species. He uses two species of salamanders and of gulls as examples (Dawkins, 2004, pp. 254–5). Each local subspecies can breed with its closest neighbours, but those that are far enough apart cannot, and the species line never is crossed. It is similar to the linguistic concept of the dialect continuum, in which speakers of each dialect can understand their neighbours, but folks at one end of the continuum cannot understand those at the other end. Take the Dutch–German continuum: native speakers in the Netherlands and Belgium speak a dialect very similar to that of their German neighbours, but they have a difficult time understanding the more distant Austrian and Swiss members of the continuum.

All of this is good to keep in mind when we are trying to distinguish between archaic *Homo sapiens* and relatively recent *Homo erectus*. Of course, applying the bsc concept to fossil specimens is especially difficult, since they can't have sex.

Concerning the species concept generally, we need to stress that it becomes quite messy in philosophical discussions surrounding evolution. Dawkins makes some important points here concerning what he calls the **tyranny of the discontinuous mind** (Dawkins, 2004, p. 252), the tendency to overemphasize the distinctions between species. It operates in opposition to the view of a continuum or continuous model of evolution, which stresses the persistence of some genes throughout all or most species, and with changes to relatively few genes signalling species change. In denouncing the tyranny of the discontinuous mind, Dawkins blames Plato for the contributing philosophy of essentialism, an idea that is violated by evolution. In Dawkins's words:

> Plato . . . believed that actual things are imperfect versions of an ideal archetype of their kind. Hanging somewhere in ideal space is an essential, perfect rabbit, which bears the same relation to a real rabbit as a mathematician's perfect circle bears to a circle drawn in the dust. To this day many people are deeply imbued with the idea that sheep are sheep and goats are goats, and no species can ever give rise to another because to do so they'd have to change their 'essence'.
>
> There is no such thing as essence. (Dawkins, 2004, p. 259)

So Is a Bird a Reptile or What?

Some biologists say that a bird is, in fact, a reptile. Others say that there is no such thing as a separate 'reptile' category. So is a bird a reptile or what? And do reptiles exist as a category? To answer the first question, frustratingly, it is, and it isn't. Just as frustrating is the answer to the second question: they are, and they aren't. In each case, it depends on which form of taxonomy you use.

There are actually two forms of taxonomy used to name and classify biological entities. Both are hierarchical in that they have higher (with more species) and lower (with fewer species) categories. The traditional method is known as Linnaean. It is based on the system developed by the eighteenth-century Swedish physician and botanist **Carolus Linnaeus** (1707–1778) and introduced in Linnaeus's classic work, *Systema Naturae*, first published in 1735. He loved Latin so much that not only did he give species names in Latin, he also changed his Swedish name, Carl von Linné, to a Latin one. The Linnaean system features the **binomial nomenclature** (i.e. two-part names) of genus and species, but it also includes a principle of grouping biological entities together: by morphology. In this method, you look for structural characteristics as clues to who is related to whom. If it has feathers, wings, hollow bones, and a beak, it's a bird. You do not require knowledge or acceptance of evolution to work with this system. Linnaeus questioned traditional religious interpretations of biology, but he was no evolutionist.

A more recently devised system is known as **cladistics**. The principle used in this system (sometimes also termed **phylogenetics**) is an evolutionary one. Living beings are classified according to their evolutionary ancestry. The primary unit of cladistics, the **clade**, includes an ancestor and all of its descendants. The originator of this taxonomic system is generally thought to be the German entomologist (a biologist who studies insects) **Willi Hennig** (1913–1976), who specialized in flies and mosquitoes.

There are a great number of terms that have been generated by cladistics, many of which are contested. You could easily get buried in the terminology. For our purposes, the number of terms will be limited to only those that we absolutely need. (I can almost hear the student reader's sighs of relief.)

The example of the relationship between birds and reptiles neatly demonstrates how the two taxonomic systems differ. In the cladistic system, the **sauropsids** clade includes iguanas, turtles, snakes, crocodiles, and birds, along with one ancestor that lived 310 million years ago. In some of the biological literature, sauropsids are referred to as a 'class'. Within this class are three groups:

the first (called Chelonia) includes turtles and tortoises;

the second is itself composed of three groups—Rhynchocephalia (tuataras), Iguania (iguanas and chameleons), and Schleroglossi (snakes, geckos, and skinks—a lizard found in Canada, as well as in many other places);

the third includes Crocodilia (crocodiles, alligators, and caimans) and Aves (birds).

This seems to violate the way we were taught to think based on the Linnaean morphology approach. We tend to think of birds and reptiles, because of their different features, as belonging to separate categories. Likewise, we usually place snakes and lizards in separate categories—in fact, in some parts of the biological literature, Reptilia and Aves are treated as Linnaean classes. The distinction that cladists (rhymes with 'sadists') would make here is between their own **monophlyletic** grouping, which sees sauropsids treated as a single clade whose members share a common ancestor, and the traditional **paraphyletic** grouping, which sees birds and reptiles treated separately based on their different characteristics. To a 'pure' cladist, paraphyletic is a dirty word.

To sum up, a bird is a bird, and it shares a common ancestor with crocodiles; this common ancestor lived a longer time ago, along with turtles, lizards, and snakes. Does that make a bird a kind of reptile? It depends entirely on how you look at it.

And What Is Louie?

Louie is one of our parrots, a rose-crowned conure. Specifically, using the cladistic system of taxonomy, he is as follows:

Domain:	Eukarya
Kingdom:	Animalia
Phylum:	Chordata
Class:	Sauropsidia
Subclass:	Neornithes
Superorder:	Neognathae
Order:	Psittaciformes
Family:	Psittacidae
Subfamily:	Psittacinae
Tribe:	Arini
Genus:	*Pyrrhura*
Species:	*rhodocephala*

Angelika Steckley

Louie, the rose-crowned conure

The Linnaean system would have Aves as the class, and would not include the subclass or superorder forms. A final note is necessary: Louie does not look like a lizard to me.

Concestors: They've Come a Long Way Since Then

Dawkins's term *concestor*, for a common ancestor, is a cladistic term, and one that we will be using in our discussion that follows. It is very important to keep in mind that our ancestral line is not the only one that has evolved. We may share a concestor with other species, but our evolutionary lines have had a long time to take us in different directions; we are none of us the same. We did not descend from chimpanzees; we share an ancestor. We did not descend from gorillas; we share an ancestor. A lot can happen in seven million years. We did not descend from monkeys; we share ancestors. And if our concestors could see us now, they would say to both of us, chimps and humans, gorillas and humans, even monkeys and humans: 'You've changed!' The same would be true of the common ancestor of crocodiles and birds.

Profiles in Canadian Anthropology

Sir John William Dawson (1820–1899)

John William Dawson was Canada's first homegrown, world-class scientist. He was born in Pictou, Nova Scotia, and it was in that province that he did his most significant paleontological research. As a lad, Dawson collected fossils; as an adult in his early 30s, he would discover the fossil that would eventually receive the rare and somewhat strange honour of being deemed his province's 'official fossil'. (We wonder if it means that the other fossils can't advertise in Nova Scotia.)

Dawson left Canada in 1840 to attend the University of Edinburgh in the homeland of his Scottish ancestors. He went there to study geology but ran out of funds to complete his studies. There were no student loans back then. Undismayed, he returned to Nova Scotia and worked as a geologist for mining companies, in his spare time engaging in his own research.

Education was dear to him, and in 1850, Dawson became Nova Scotia's first superintendent of education, a position he held for three years. But while he invested considerable energy into this work—as an innovator in teacher training and curriculum, school architecture, and even desk design—he did not allow the appointment to keep him from his first love, geology.

Sir John William Dawson

McGill University Archives, PL007387

The Joggins Finds: *Hylonomus lyelli* and *Dendrerpeton acadianum*

Dawson did his most famous research at one major site: the 15-metre-high cliffs by the sea near Joggins, Nova Scotia. The Joggins cliffs are home to the fossilized remains of giant (up to 30 metres in height) club mosses, known as 'scale trees' because of the thick scales visible on the fossils. It was there that, along with his mentor and sometime sponsor, the geologist Charles Lyell, Dawson would find the remains of early tetrapods, both amphibians and reptiles, from the Carboniferous (lit. 'coal-bearing') period, informally known as the Coal Age (indeed, the presence of coal was a major reason for the interest in such sites in the nineteenth century). In 1852, Dawson and Lyell turned up an early amphibian, dubbed *Dendrerpeton acadianum* ('tree-reptile, Acadian'), that would be announced by Lyell in a public lecture in Boston a year later.

Even more significant was Dawson's later find of a reptile he named *Hylonomus lyelli*. The first part of the name—the genus—means 'forest

wanderer'; the second, or species, part represents Dawson's tribute to his friend and mentor Lyell. The creature, which was about 20 cm long if you include the tail, is generally considered the earliest known reptile. It lived about 315 million years ago. When trees of the Coal Age fell and

Hylonomus lyelli

De Agostini Picture Library/De Agostini/Getty Images

began to rot, *H. lyelli* would hunt through the stumps for millipedes and land snails; if the stumps were flooded, *H. lyelli* would be trapped inside, potential fossil fodder for scientists such as Dawson and Lyell.

Dawson's fossil find has been treated with great respect. It was referred to in Darwin's *Origin of Species* in 1859. It was made Nova Scotia's 'Provincial Fossil' in 2002. The site of the discovery was declared a World Heritage site by the United Nations in 2008. *H. lyelli* currently holds the honour of being thought to be the first fully land-dwelling tetrapod (four-legged one).

The Joggins site Fossil Cliffs, Nova Scotia

Dawson *vs* Darwin

Although both agreed that these discoveries were significant, Dawson and Darwin drew different conclusions from them. For Darwin, writing in *Origin of Species*, the appearance of abrupt change given by the diversity of fossil specimens was misleading. The fossil record is, he recognized, necessarily incomplete, owing to the low odds of something fossilizing. Even if paleontologists could view all levels of the Joggins cliffs and all of the fossils they contained, they would never gain a complete picture of the history of life at the site. The apparent abruptness of change, then, was simply a reflection of the incompleteness of the fossil record. Dawson, though, taking the anti-evolutionary line then held by his good friend and mentor Lyell, and being a devout Christian throughout his life, begged to differ. He put forth his own views in 1863, in *Air-Breathers of the Coal Period: A Descriptive Account of the Remains of Land Animals Found in the Coal Formation of Nova Scotia With Remarks on their Bearing on Theories of the Formation of Coal and of the Origin of the Species*. Dawson noted that the Joggins site showed that certain creatures, most notably the early land snail *Dendropupa*, were virtually unchanged over the long period that separated these fossils from current living creatures. For Dawson, this meant that the gradual change that Darwin argued for with his theory of evolution was not happening. Ironically, Dawson's research, which would eventually reveal about 100 early tetrapods, has provided solid evidence for evolution. Discoveries and their later interpretations often have different histories.

Dawson at McGill

In 1855, at the young age of 35, Dawson became professor of geology and principal of McGill University, the latter a position he would hold with great distinction until 1893. During that time, he witnessed and aided the growth of the university from roughly eighty students to well over a thousand. An unattributed obituary featured at the end of Dawson's posthumously published memoirs, *Fifty Years of Work in Canada* (1901), quotes a colleague, one Professor Cox, as declaring that 'McGill was Sir William Dawson, and Sir William Dawson was McGill' (Dawson, 1901, p. 294). Today, many of the specimens collected by Dawson are housed in the Peter Redpath Museum of Natural History, on the McGill campus in Montreal.

A Selection of Dawson's Publications

1855   *Acadian geology—The geological structure, organic remains and mineral resources of Nova Scotia, New Brunswick, and Prince Edward Island*

1863   'Notice of a new species of *Dendrerpeton* and of the dermal covering of certain Carboniferous reptiles'. *Proceedings of the Geological Society, 19*(1–2), 469–73.

1880   *Fossil men and their modern representatives*

1888   *Geological history of plants*

1901   *Fifty years of work in Canada, scientific and educational. Being autobiographical notes by Sir William Dawson* (Ed. Rankine Dawson)

Long Live the Kingdoms (or Maybe Not)

Changing Kingdoms

Traditionally, the **kingdom** was the highest unit in the taxonomic hierarchy, though it is the **domain** that rules today. The number of kingdoms has been a matter of some dispute ever since Linnaeus, in 1735, introduced what he considered the only two biological kingdoms: Vegetabilia (later called Plantae), and Animalia (see Table 4.1). Haekel (1866) added Protista to account for typically unicellular creatures, and the term came to be associated with such familiar figures from grade school biology classes as plankton, slime moulds, and protozoa, but definitely not bacteria. By the mid-twentieth century, Monera had been added to refer to bacteria and blue-green algae, but the term has recently been abandoned. The kingdom Fungi was added in the 1960s, and Eubacteria and Archaebacteria in the 1970s, though the latter two have been reassigned as the domains Bacteria and Archaea, to sit atop the hierarchy alongside (and in opposition to) Eukarya.

To sum up, according to the current convention—which definitely is in flux—there are six kingdoms (Eubacteria, Archaebacteria, Protista, Fungi, Plantae, and Animalia) and three domain system (Bacteria, Archaea, and Eukarya) (see Figure 4.2).

Animalia

Animalia is the kingdom to which humans belong. The discussion here takes a cladistic approach and will be restricted to animals viewed in terms of our evolutionary path. This does not mean that we are not related to bacteria, mushrooms, plankton, or maple trees, but telling their story adequately would complicate this chapter too much.

Running through the Family: From Farthest to our Closest Relations

We'll now take a trip through the Animal Kingdom, looking at the many groups and species that are related to humans. We'll begin the tour with our most distant relatives—the DRIPS—and then make our way gradually to our home order, the primates.

'DRIPS', SPONGES, JELLYFISH, AND OTHERS: OUR FARTHEST ANIMAL RELATIVES

The term DRIPS is an acronym formed from the initial letters of the genera that make up this group. They are all single-celled parasites (most of which feed off of living fish), and in terms of animals, they are the farthest away from us.

The next farthest, choanoflagellates, may be 'free-living' or parasitic. Each one has an egg-shaped body with a collar

Table 4.1	Kingdoms over time						
Time period	**Kingdoms**						
18th century	Vegetabilia	Animalia					
19th century	Plantae	Animalia	Protista				
Mid-20th century	Plantae	Animalia	Protista	Monera			
1960s	Plantae	Animalia	Protista	Monera	Fungi		
1970s	Plantae	Animalia	Protista	Monera	Fungi	Eubacteria	Bacteria
Early 21st century	Plantae	Animalia	Protista	Fungi	Eubacteria	Archaebacteria	

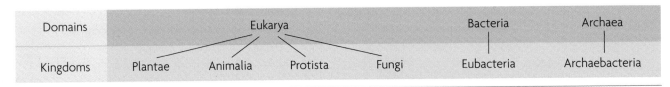

Domains	Eukarya				Bacteria	Archaea
Kingdoms	Plantae	Animalia	Protista	Fungi	Eubacteria	Archaebacteria

Figure 4.2 | The picture today: Three domains, six kingdoms

Table 4.2	Appearance of new features along the human evolutionary line	
New feature	**Seen in (early representatives)**	**Example**
vertebrae	jawless fish	lampreys
jaws	most fish	perch
digits	amphibians	frogs
amniotic egg	reptiles	turtles
hair and milk	mammals	platypuses
placenta	carnivores	dogs
opposable thumbs	primates	monkeys

Although it has no muscles and no nervous system, and does not move forwards or backwards under its own power, a sponge still is an animal. Sponges belong to the phylum Porifera. Our concestor lived perhaps 800 million years ago.

In the phylum Placozoa, there is only one species: *Trichoplax adhaerens*. These are flat, multicellular animals with no fixed shape, generally no more than a millimetre in size.

The term Ctenophora comes from a Greek word meaning 'comb', in reference to the groups of cilia (hairlike parts) whose beating propels the creatures of this phylum (ctenophores) forward in the water. And while the name Cnidaria (from a Greek word for knife) may be foreign to you, the members of this phylum are not: they include jellyfish, corals, and sea anemones. There is no good date for our cnidarian concestor.

at the base (used for trapping and consuming bacteria) and a single flagellum, or tail, used for swimming. They resemble choanocytes, the cells that line the interior chambers of a sponge and that are responsible for ingestion of drifting food particles. So did choanoflagellates come about before or after sponges? It is hard to know for sure, but biologists are studying the relationship, particularly for clues to their earliest common ancestor. Dawkins proposes that our choanoflagellate concestor first appeared 900 million years ago.

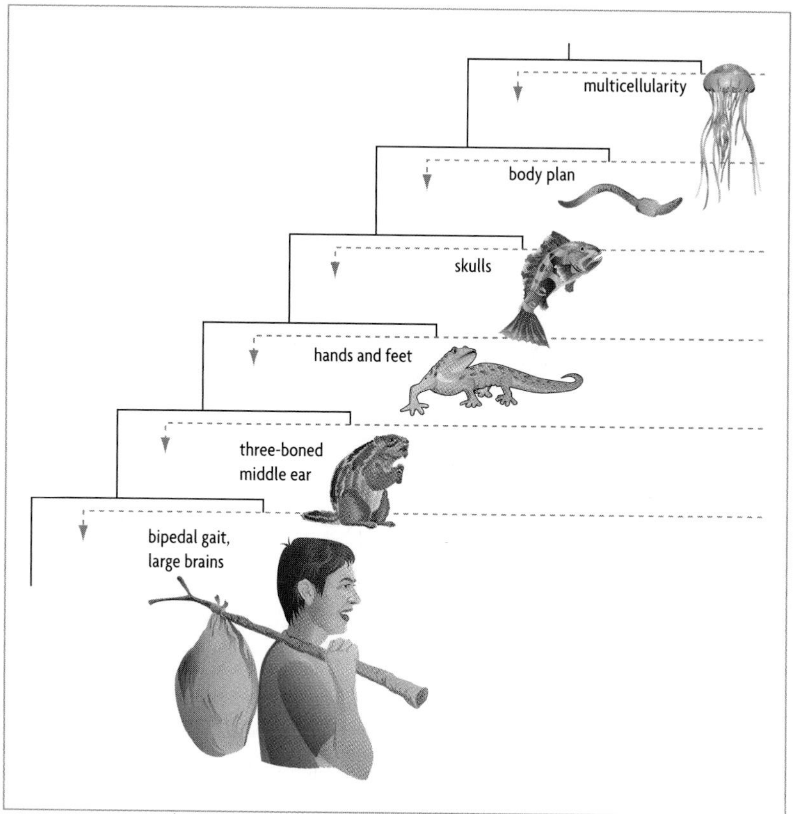

ACOELOMORPHS, FLATWORMS, PROTOSTOMES, AND AMBULACRARIANS

The phylum Platyhelmenithes comprises flatworms, including parasitic flukes and tapeworms, that are characterized by simple, flattened bodies without blood vessels. The acoelomorphs used to be placed fairly securely within this phylum, but recent research strongly suggests that they are on their own (the phylum name is Acoelomorpha). What's interesting about these groups is that many of their species feature bilateral symmetry, meaning that they have a clearly identifiable top and bottom, front and back—they're not just radially symmetrical like jellyfish, which have top and bottom but no sides, or altogether shapeless like the *Trichoplax*. Dawkins takes a fairly wild guess in placing our concestor at 630 million years ago.

Protostomia is a huge group with relatively few fossils. Protosomes (lit. 'mouth first') make up a subkingdom of Animalia in opposition to the deuterostomes ('mouth second'), which includes all the creatures described in the sections that follow. Protosomes include an amazing number of 'worms': arrow worms, flatworms, goblet worms, hairworms, peanut worms, ribbon worms, roundworms, segmented worms, and velvet worms. It also includes, as its largest phylum, the **arthropods**: centipedes, crustaceans (lobsters, crabs, etc.), insects, and spiders. Molluscs, including clams, oysters, mussels, scallops, snails, and other, less edible creatures, are another phylum belonging to Protostomia. Add to these the brachipods and phoronids, which have superficial resemblance to molluscs, and— less known to most people—the rotifers,

multicellularity

body plan

skulls

hands and feet

three-boned middle ear

bipedal gait, large brains

Figure 4.3 | Recent life in perspective

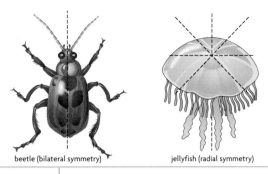

beetle (bilateral symmetry)   jellyfish (radial symmetry)

Figure 4.4 | Bilateral vs radial symmetry

gastrotrichs, brozoans, cephalorhynchs, and tardigrades. The very rough arrival date that Dawkins assigns these creatures—and it's purely through the **molecular clock**—is about 590 million years ago, give or take a lot.

This brings us to the superphylum Ambulacraria. It includes the Echinodermata ('spiny-skinned') creatures, which make up a phylum that has been together for some time. Its members include the closely related starfish and brittlestars, sea urchins (traditionally called 'whore's eggs' in Newfoundland and described as 'warty sausages' by Dawkins [2004, p. 312]), sea cucumbers, and the sea lilies. Newly added by molecular evidence are the worm-like (but still different) acorn worms, the pterobranches, and the xenoturbellids. While it is better to describe entities in terms of what they have than to focus on what they lack, it's worth noting that the ambulacrarians had neither blood nor the bilateral symmetry (clearly defined back/front and right/left sections) of those that we will be describing in the rest of this chapter. Dawkins estimates that they share a concestor with us that lived some 570 million years ago.

SEA SQUIRTS AND LANCELETS

While in its juvenile state, the sea squirt (or tunicate) looks like a tadpole, and it has a very lively life; as an adult, it's pretty much a 'bag filled with seawater' (Dawkins, 2004, p. 306), undergoing a change of state that Dawkins, like others before him, compares to that of a university professor receiving tenure. We share a concestor that lived about 565 million years ago. The estimate for lancets is difficult, maybe going back 560 million years ago. Like the fish, amphibians, sauropsids, and mammals discussed in the sections that follow, the lancet belongs to the phylum Chordata, but it is not quite a vertebrate—it has a **notochord** rather than a backbone running down its back. A notochord affords an organism greater flexibility; vertebrates have them when they are in the embryonic state.

FISH

In evolutionary terms, fish come in a great taxonomic variety. One taxonomically separate group comprises the jawless lampreys and hagfish, with a concestor that lived some 530 million years ago.

Then there are the chondrichthyes, the class of jawed, cartilaginous fishes that includes sharks and such relatives as rays and skates. This class goes way back to an estimated 460 million years ago.

A much more prolific group of fish types is the class Actinopterygii, comprising the ray-finned fish. The actinopterygians includes such good Canadian fish as perch, cod, salmon, trout, pike, eels, and sturgeons, and share with humans a concestor that lived some 440 million years ago. Dawkins makes a very interesting point about this group: many have a **swim bladder**, a modified simple lung that is filled with gas. This is not the kind of bladder we beer-drinking humans empty at the first-period intermission but a specialized bladder used by fish to control their depth. When gas molecules enter the blood from the swim bladder, it increases the bladder's volume, enabling the fish to rise; the reverse process causes the fish to sink. In other fish, where the swim bladder is found by the inner ear, it serves as a kind of hearing aid. Sharks and their kin do not have this. The point Dawkins makes is that the swim bladder was originally a very primitive or simple lung. In some fish (bowfins and gars, for example) it was used for breathing; in others it underwent **exaptation** (to use a term introduced in the previous chapter) to function as a bladder. When you can already breathe perfectly well and have no use for an additional way of breathing, that feature can adapt to perform alternative functions.

COELACANTHS AND LUNGFISH

One of the few species that can reasonably be called a **'living fossil'** (i.e. an organism that has changed very little from the time when its fossilized ancestors lived) is the coelacanth. In 1938, a coelacanth (living, not fossilized) was caught off the coast of South Africa; upon seeing it, the prominent ichthyologist J.L.B. Smith is reported to have said, 'I would not have been more surprised if I had seen a dinosaur walking down the street' (as cited in Dawkins,

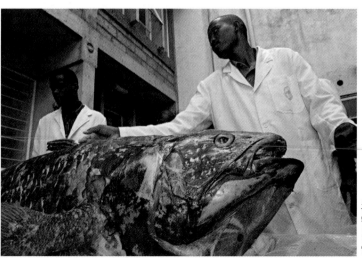

A coelacanth

Case Study Controversies

Eusthenopteron: The Prince of Miguasha, or the Fifty Dollar Fish Fossil

Canada is home to some of the most famous, fascinating, and important fossil discoveries in the world. One of these finds, a fish that lived roughly 380 million years ago, is a transitional organism that provides evidence of the development of the evolving arm and leg structures of tetrapods (amphibians, reptiles, birds, and mammals). Indeed, *Eusthenopteron*, as it's known, is so famous today that it has been featured on both Canadian and Polish postage stamps.

In terms of taxonomy, this fossil fish belongs to the class Sarcopterygii (-*pter*- means 'wing'), the subclass Tetrapodmorpha (referring to a four-footed form), and the order *Osteolepidida* ('bony fin', or 'lobe-finned'). There are four species of the genus *Eusthenopteron*, but the type specimen—the one considered the most representative—is *E. foordi*.

There were three sorts of lobe-finned fish: coelacanths, lungfish, and rhipidstians. *Eusthenopteron* was a rhipidstian. It is the only one of the three groups that didn't survive intact in more or less the traditional form. However, significantly, it did provide ancestors to current tetrapods (including us).

The history of *Eusthenopteron*'s eventual discovery also extends over time. The fossil site of Miguasha, in Gaspé, Quebec, was first explored by Abraham Gesner, New Brunswick's first Provincial Geologist, in 1842. He wrote about the site and the fossil fish he had found there in an issue of a little-read scientific journal that appeared in 1843. This discovery and the site were forgotten (Gregor Mendel could sympathize) until Wheelock Ells 'discovered' Miguasha (much like Christopher Columbus and the Americas) in 1879. Most of the fossil fish recovered from the site were found by geologist Arthur H. Foord, after whom the type specimen was named. British paleontologist J.F. Whiteaves named *Eusthenopteron* in 1881, working on site with William Dawson (see above), whose principal focus was Miguasha's fossil plants.

Eusthenopteron

In 1925, a local farmer, Joseph Landry, went collecting at the site and shipped a crate of his findings to the Swedish Museum of Natural History. Included in the crate was a well-preserved *Eusthenopteron foordi*, about 30 cm long, for which Landry received a prize of $50. It was fortunate that the fossil went there, as it was received by the diligent Swedish paleoichthyologist (lit. 'old fish studier') Erik Jarvik. Over decades he painstakingly and very, very patiently (I would have hit it with a hammer) dissected the fossil. His remarkable work was rewarded when he discovered that, just like the tetrapods, *E. foordi* had a femur, tibia, and fibula in the pelvic or back fin, and in the fore or front fin, a humerus, radius, and ulna. This forms the upper-level (or proximal) one-bone/lower-level (or distal) two-bone pattern that all tetrapods have. The fins were about 5.6 cm long. Unlike the later Tiktaalik (see Chapter 3), it had no bones comparable to our wrists and ankles.

Today, many of the 2,000+ specimens that were found at the site are housed in the Musée d'Histoire Naturelle de Miguasha (Miguasha Museum of Natural History), located in the Parc national de Miguasha (Miguasha National Park, a UNESCO World Heritage Site) on the Gaspé peninsula.

2004, p. 272). Dawkins believes that our concestor lived some 425 million years ago. As for the lungfish, Dawkins assigns a date for our concestor of about 417 mya.

AMPHIBIANS

According to Dawkins, our concestor with organisms of the class Amphibia lived about 350 million years ago. The designation that links amphibians with all vertebrates higher than fish—reptiles, birds, and mammals—is **tetrapod**, meaning 'four-footed'.

Amphibians are classified into three different orders. The first, Anura (lit. 'tail-less'), is a neat and tidy order comprising frogs and toads. The second, which covers salamanders, is a little messy. It's not a question of who is and who isn't a salamander, but of which of two competing terms—Urodela or Caudata—to use. Some argue that the former term be used only for extant (i.e. living) species and the latter for all species, both extinct (fossilized) and extant, while others prefer Urodela for all species. Both terms derive from ancient (Greek and Latin, respectively) words for 'tail', so there's little difference between them. We'll hope the two camps can find a civilized way of settling the dispute (a good, old-fashioned tug of war perhaps, as opposed to more violent forms of combat). The third order is Gymnophiona, which refers to caecilians. These are limbless, burrowing creatures with bad eyesight, native to Southeast Asia, Africa, and South America. They look like earthworms, but they can reach 1.5 metres in length, so you're not likely to find them on the shelves of your cottage country bait-and-tackle shop.

SAUROPSIDS

The name Sauropsida ('lizard face') is probably not familiar to you. To be honest, I did not know the name myself before I started writing this book. The concestor lived some 310 million years ago, and, as we have seen above, includes those creatures that you and I knew just as reptiles and birds. The designation that links these groups to mammals is **amniotes** (adj. *amniotic*), developed from the word *amnion*, which refers to the sac that surrounds, protects, and feeds the embryos of mammals, turtles, and birds, but not amphibians. The result is the amniotic egg. The evolution of the amniotic egg freed the mother laying the egg from having to do so in water—the egg could instead be laid on land or in a nest in a tree. That saved the species from having to undergo separate water and land stages (as the tadpole/frog does), representing a great saving of genetic resources.

Mammals

We're now approaching our closest human ancestors. Mammalia is the class of warm-blooded vertebrate animals characterized primarily by the birth of live young, who are nourished by milk produced by their mothers. Table 4.3 provides an overview of the subclasses, orders, and species that make up this class, while the sections that follow describe some of our closer relatives.

MONOTREMES

Animals of the order Monotremata (Greek 'one hole') are named after a feature they share with reptiles and birds: the materials that pass through the anus, urinary tract, and reproductive tract all exit through the same tube, rather like streams flowing into one river. (It makes you want to wash your eggs before you eat them.) There are three genera of monotremes (and the subclass to which they belong, Prototheria):

1. *Ornithorhynchus.* Its sole extant species, *O. anatinus*, is the duckbilled platypus of eastern Australia and Tasmania.
2. *Tachyglossus.* This genus, too, has just one species: *T. aculeatus*, otherwise known as the short-beaked echidna of Australia and New Guinea.
3. *Zaglossus.* This genus comprises three species of long-beaked echidna, found in New Guinea (*Z. attenboroughi*, *Z. bartoni*, and *Z. bruijni*).

Monotremes have a mixture of mammal and reptile/bird features. Like mammals, they provide milk to their young, are partially covered with fur, and have only one bone in the lower jaw. Like reptiles and birds, however, they lay eggs that provide nutrition to the young before the hatchlings break out, using their beaks/bills. As Dawkins duly notes (2004, pp. 196–7), it would be a mistake to refer to the monotremes as 'primitive', as if they had not evolved over the millions of years since humans and monotremes parted company from our common ancestor. While some of their oldest features have remained fundamentally unchanged over millions of years (note that, genetically speaking, the same could be said of some of our features), monotremes have also evolved along their own path, demonstrating an ability to adapt to their own circumstances. Take, for example, the bill of the platypus, which is very different from the beak of the two genera of echidnas. It has evolved as an extremely effective tool for detecting food (e.g. small crustaceans and insect larvae) in the mud when the platypus's eyes, ears, and nostrils are closed. The surface of the bill contains many electronic and mechanical sensors, which are tuned in to the electrical impulses of the muscles of their prey and to the ripples in the water generated by their prey's sudden movements. The platypus hunts by methodically swinging its bill back and forth while swimming, moving its own little satellite dish for picking up the food channel.

A platypus

Table 4.3 Mammal varieties

Classification					First appearance
Class	Subclass	Infraclass	Superorder	Order	
MAMMALIA	Prototheria			**Monotremata** (monotremes: platypuses, echidnas)	180 mya
	Theria (beasts)	Marsupialia (marsupials)		**Diprotodontia** (diprotodonts: possums, koalas, wombats)	140 mya
				Macropodidae (macropods: kangaroos, wallabies)	
		Eutheria (true beasts)	Afrotheria	**Afrosoricida** (golden moles, tenrecs)	105 mya
				Hyracoidea (hyraxes)	
				Macroscelidea (elephant shrews)	
				Proboscidea (elephants)	
				Sirenia (sea cows: dugongs, manatees)	
				Tubulidentata (aardvarks)	
			Xenarthra	**Cingulata** (armadillos)	95 mya
				Pilosa (anteaters, sloths)	
			Laurasiatheria	**Carnivora** (dogs, cats, hyenas, weasels, bears, seals)	85 mya
				Cetartiodactyla (antelopes, deer, cows, camels, pigs, hippos, whales)	
				Insectivora (moles, hedgehogs, shrews)	
				Megachiroptera (big bats)	
				Microchiroptera (little bats)	
				Perissodactyla (horses, donkeys, tapirs, rhinos)	
				Pholidota (pangolins)	
			Euarchontoglires	**Rodentia** (rodents)	75 mya
				Lagomorpha (hares, rabbits, pikas)	
				Scandentia (culugos, tree shrews)	
				Primates (see below)	70 mya

MARSUPIALS

Marsupialia is an infraclass, which is a category below subclass. Our common ancestor with the marsupials lived some 140 million years ago.

There are three birth strategies used by mammals. Among monotremes, as we have just seen, the embryo is protected in and nourished through an egg. Human fetuses, and those of many other mammals, develop in the placenta, and we use the term **placental** to designate non-marsupial, non-monotreme mammals. The word **marsupial** comes from a Latin word meaning 'pouch'. The baby marsupial (called a 'joey', if it's a kangaroo) makes two major trips that other mammals do not. The first trip seems to me to be fraught with danger. As little more than an embryo, not yet fully formed, it emerges from its mother and then has to crawl through her fur to the pouch, where mom supplies the milk through a large nipple. After pretty much welding itself to that nipple for a period of time, then attaching itself only periodically, it comes out of the pouch to enter the outside world a second time, sometimes returning to the pouch for a safe haven.

Marsupials developed in two main areas. One is Australia, where they had little competition from placental animals (if you discount the ever successful rodents) until dingoes (*Canis lupus dingo*) came to the southern continent from southeastern Asia (and, before that, India) some 3,500 to 4,000 years ago. The other is South America. There marsupials had a prominent place until about 2.8 million years ago, when North and South America came together and the **Great American Interchange** took place: we (North Americans) will give you jaguars, camel-types (ultimately llamas, alpacas, and vicuñas), and pig-types (e.g. tapirs and peccaries), and you (South Americans) give us your armadillos and possums. The marsupials lost out in the exchange. Generally speaking, they do poorly in competition with placentals, particularly placental predators.

AFROTHERES: WHAT'S LIKE A SHREW BUT ISN'T A SHREW?

We share a common ancestor with afrotheres (lit. 'African beasts', although not all of them live in Africa) that lived some 105 million years ago. If I were to name this group, I would probably look in my Latin–English dictionary for a word that means 'weird-looking mammals'. This super-order includes elephants (the brains of the group) and little pointy-nosed creatures misleadingly called 'elephant shrews' (they are not 'true' shrews, and although they are closely related to elephants, their long nose is more or less incidental to that relation), aardvarks (the first mammals alphabetically), the water-dwelling sirenians (named after the Sirens, the dangerous water women of Greek myth, but resembling them only to very nearsighted sailors), manatees and dugongs, hyraxes (from the Greek word for 'shrew'), tenrecs (some of which look like shrews; their

Madagascar home has no 'true shrews'), and golden moles. It is good to remember here, as Dawkins and others have pointed out, that before 65 million years ago, pretty much every mammal ancestor looked like a shrew.

XENARTHRANS

The name used for the superorder Xenarthra comes from Greek words meaning 'alien' (the *Xena-* prefix) and 'joints' (the *-arthr-* part). Our concestor lived some 95 million years ago. The xenarthrans are relatively old and South American in origin, like the marsupials and certain—to quote Dawkins—'strange ungulates' (beasts with hooves) that have become extinct. They include one surviving member that moved north, the armadillo, and two that remained: the sloth and the anteater.

LAURASIATHERES

Laurasia was the northern of two supercontinents formed by the breakup of Pangaea. Laurasiatheria is the super-order comprising mammals believed to have evolved on this land mass. Our concestor among the laurasiatheres ('beasts of Laurasia') lived 85 million years ago.

This grouping covers a great variety of creatures, including seven different orders:

- the newly scientifically established Cetartiodactyla (1) and Perissodactyla (2), both defined by their *-dactyl-*, or hooves;

- Carnivora (3);

- Insectivora (4);

- Microchiroptera (5) and Megachiroptera (6)—little bats and big bats, respectively; and

- Pholidota (7), represented by the pangolin (or scaly anteater).

Several of these orders will receive further comment here.

Cetartiodactyla

Until recently the name for this order was Artiodactyla, referring to ungulates with even-numbered hooves (typically the third and fourth digits); it included such obvious hoof-bearers as antelopes, camels, cattle, the strange little chevrotains (or 'mouse deer'), deer, giraffes, goats, hippopotamuses, pigs, and sheep. The new name reflects recent DNA findings that strongly suggest that this group is closely related to the cetaceans—whales, dolphins, and porpoises. From this research, it appears that the whale's closest land relative is the hippo.

Perissodactyla

The members of the order Perissodactyla are the so-called odd-numbered–hoofed animals (typically having just one digit/hoof on each foot). They include the suborders

Hippomorpha (lit. 'horse structure'), which comprises horses, zebras, donkeys, and the various wild asses of the world (no mention of any politicians, please), and Ceratomorpha (from Greek *kerat-*, 'horn'), including rhinoceroses and tapirs.

Carnivora

This is an odd grouping in that it includes a great variety of creatures, yet, it should be noted, it does not include all of the mammal carnivores. Those that fit into this order are dogs and cats (and their wilder relatives), weasels, skunks, hyenas, meerkats, bears, raccoons, civet cats (which are not really cats), and the various *pinnipeds* (lit. 'fin-feet'), including seals, sea lions, and walruses.

Insectivora

This insect-eating order includes shrews, moles, and hedgehogs. I remember that as an aspiring young biologist of the age of 10, I imagined this group as rodents with relatively small teeth. Add 'predominantly land-dwelling' to the description, and it's not too far off, though please note that rodents have a group all to themselves, and insectivores aren't part of it.

EUARCHANTOGLIRES

Dawkins uses this term to refer to a smaller group, comprising two orders: Dermoptera (lit. 'skin wings') for colugos and the Scandentia ('climbers') or tree shrews, both of which are native to southeastern Asia. He had our concestor with them living at about 70 million years ago. Recent DNA analysis (the bane of textbook writers) now suggests that this is a superorder that includes two clades representing groupings of orders: Euarchonta, which includes Demoptera, Scandentia, and Primatea; and Glires, which includes the two orders Rodentia and Lagomorpha.

Glires: Rodentia and Lagomorpha

Our concestor here lived about 75 million years ago. Rodentia is the most successful of the mammal groups. We are not sure when their diagnostic (i.e. characteristic) big, ever-growing teeth first came into play—the oldest fossils discovered so far are from about 65 million years ago, when the dinosaurs became extinct.

If you could count up individual mammals in the world, most of them would be rodents. With well over 2,000 known species, and new species still being discovered, it's a good area of biology to get involved with if you want to name something. The 2,000+ species also show a tremendous amount of diversity among rodents. Their success lies in their adaptability, their short reproduction cycle, and the large number of young that they have.

Lagomorpha, comprising hares and rabbits, was once considered together with Rodentia but has been separated owing to some subtle differences. Namely, the hares and rabbits that make up Lagomorpha are almost exclusively herbivores, unlike rodents, and their ever-growing teeth include four incisors, not two.

Primates and Their Evolution

We are primates; this is part of our evolutionary story. Roughly 60+ million years ago, some small shrew-like mammals took to the trees. Over time, these creatures adapted to arboreal life in a variety of ways. This order, known as Primates, is our order. So, what exactly is a primate, and what features did primates develop in adapting to a life of dwelling in trees? Read on!

Colin Tudge, in *The Link: Uncovering Our Earliest Ancestor* (2009, p. 194), remarks that it would be better to note a 'suite' of features that primates *tend* to have, rather than trying to identify those features that all primates and only primates have. The suite would include fingernails rather than claws—if you see a mammal with fingernails, you know it's a primate. But not all primates have fingernails. Fingernails themselves do not present an evolutionary advantage. What is important is that they have been diminished from being claws or talons. The elimination of claws affords a primate a better ability to manipulate small objects (say, to obtain food), and exposes the pads on the other side of the fingers, making them very sensitive, for an enhanced sense of touch.

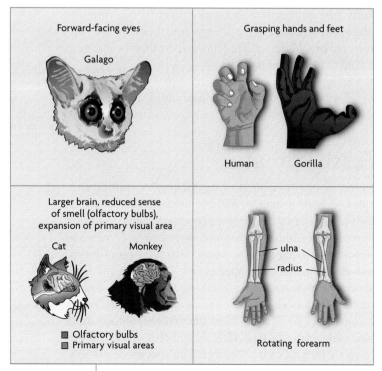

Figure 4.5 | Common primate traits

An **opposable thumb** is a digit situated far from the other digits, flexible enough that it can be used in opposition to a finger or toe to establish a precision grip. This is useful for such activities as picking fruit and climbing. The latter activity is particularly well achieved when along with grasping hands there are also grasping feet, a feature humans lost when they became bipedal (i.e. began walking on two limbs). But here again, not every primate has an opposable thumb. Marmosets and spider monkeys have reduced thumbs that are not very useful for grasping or manipulation. And I have seen a young gibbon grasp a piece of banana with her thumb in opposition to her palm, not her fingers.

Stereoscopic vision (also known as 3-D or depth perception) comes from having forward-facing eyes that are capable of seeing collections of many things from slightly different angles. If you are moving through the branches of trees, particularly very high trees, it's quite useful to be able to tell with a certain degree of precision how far away the next branch is. The reduction of the snout, brought about in part by the diminished dependence on the sense of smell, made it possible to see a greater range of things with stereoscopic vision. Try putting your fist on your nose to see how your central vision is diminished. (Cats find this shared ability quite helpful, and yet they are not primates.) Connected with this ability, and a general increased dependence on the sense of vision over that of smell, is the development of the **postorbital bar**, a strut of bone behind the eye that offers greater protection to primates. This protection is highly developed among monkeys and apes, but such protection for the eye is not unique to primates—cows have it as well.

Arm mobility, made possible to a large extent by the fairly free moving ball-and-socket joint in our shoulder and by the structure of the primate collarbone, enables primates to **brachiate**, or swing through the trees by means of a kind of pendulum motion.

Table 4.4 shows the major divisions of the primate order, with Dawkins's concestor date added. You will notice that the distinctions get finer as we come closer to humans.

Fossil Primates

In the remaining section of this chapter we will be considering fossil primates. We have discussed earlier in the chapter the blurry nature of the species concept. That issue becomes even blurrier when you're dealing with fossils as opposed to living species. Bones, after all, cannot reproduce the way that living species can. This leaves any fossil species open to disputes concerning classification, usually pitting lumpers on one side versus splitters on the other. **Lumpers**, as you'll recall from the first chapter, are those who view relatively small differences between fossil specimens as insignificant with respect to species determination; **splitters** take those differences more seriously.

Proconsul: The Speciating Genus

Proconsul is an extinct primate genus that has been found in Kenya and Uganda. It is believed to have lived sometime between 14 and 20 million years ago, with clear dating for one site setting its age as roughly 18 mya. According to Alan Walker and Pat Shipman, in their interesting and highly readable *The Ape in the Tree*, *Proconsul*'s significance lies in its being 'the best candidate we know of for the last common ancestor of both modern apes and humans' (Walker & Shipman, 2005, p. 10). The genus has features in both general body structure and dental patterns that are associated with features of monkeys (e.g. the strong, long, and flexible back) and apes (e.g. no tail).

Looking at the history of the species assigned to this genus provides insight into the difficulties of identifying species through fossil specimens, a problem we will regularly encounter with species of our own genus (and its immediate predecessors). Initially, just one species, *P. africanus*, was assigned to the genus, but problems arose when researchers detected a significant difference in tooth size across specimens supposedly representing the same species. Some anthropologists, following the lumping

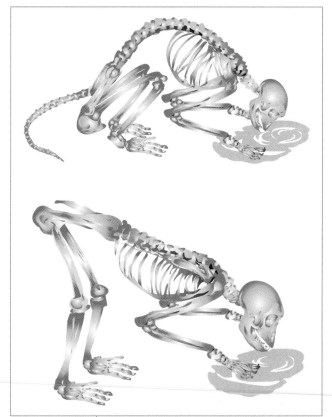

Figure 4.6 Comparing the spine of a monkey (top) with that of an ape (bottom). More lumbar vertebrae means greater flexibility.

philosophy, chalked the differences up to **sexual dimorphism**, differences in size and structure between the sexes. Walker and Shipman raised a reasonable criticism of that explanation, claiming that there were, in fact, two species being represented in the study: *P. africanus* and *P. nyanzae*:

> The fundamental problem in assuming that *Proconsul africanus* and *P. nyanzae* were simply males and females of a single species was that the sexual dimorphism of the teeth of the combined *Proconsul* sample was greater than the dental sexual dimorphism in *any* living species of Old World monkey or ape. To my mind, the idea that *Proconsul* exceeded every other known primate in sexual dimorphism seemed highly unlikely. (Walker & Shipman, 2005, p. 156)

In total, three body size determinations have been made, producing results that suggest that there are at least three different species of *Proconsul* (see Table 4.5 on page 97). Walker and his research team have since this division further divided the *P. africanus* fossil specimens into *P. africanus* and *P. heseloni*, the two different species being located at different sites.

Table 4.4	Divisions within the Primate order

Suborder	Infraorder	Parvorder	Superfamily	Family	Subfamily	Genus/ species	Common name	Concestor date (est.)
Strepsirrhini (prosimians)	Lemuriformes			Lemuridae			lemurs	
	Lorisiformes			Lorisidae			lorises	
				Galagidae			galagos (bushbabies)	63 mya
Haplorrhini (simians)	Tarsiiformes			Tarsiidae		*Tarsius*	tarsiers	58 mya
	Simiiformes	Platyrrhini					New World monkeys	40 mya
		Catarrhinii					Old World monkeys	25 mya
			Hominoidea (hominoids: 'apes')	Proconsulidae (extinct)	Proconsulinae	*Proconsul*	fossil ape	18mya
				Hylobatidae			gibbons	
				Hominidae (hominids: 'great apes')	Ponginae	*Pongo*	orangutans	14 mya
					Homininae (hominins)	*Gorilla*	gorillas	7 mya
						Pan	chimpanzees, bonobos	5 mya
						Homo	humans	

Case Study Controversies

Ida the Darwinus: A Transitional Primate between Prosimian and Simian?

'Missing link', 'Mona Lisa', 'Eighth Wonder of the World'. These are just some of the lavish epithets used in May 2009 to announce the discovery of a primate fossil, nicknamed Ida, who seemed to have a team of PR specialists hyping her story. So what's the skinny on Ida?

Ida was found in the Messel Pit, near Frankfurt, Germany, a site that was declared a UNESCO World Heritage Site in 1995. She was dug up in 1982 by an amateur fossil collector, who had the specimen preserved, probably by a professional, and then kept it hidden away in his private collection until 2006. That was when a photograph of the specimen turned up at the Hamburg Mineral and Fossil Fair—the second-largest of its kind in Europe. Jørn Hurum, associate professor of paleontology at the University of Oslo's Natural History Museum in Norway, saw the picture when it was shown to him by a dealer at the show. He immediately recognized the importance of the fossil, and was determined that the museum should have it. However, the asking price (more like a ransom) was $1 million. Hurum was able to raise the money; he acquired the fossil and named it after his five-year-old daughter, Ida.

Why is Ida (the fossil) so important? She is a fossil primate, 95 per cent complete, and remarkably well preserved, owing to the conditions of the lake into which she fell before she died. No fossil primate has been found in anywhere near as good condition: Ida was found with the fossil remains of her last meal (leaves and fruit) and a rock 'hair shadow' of her fur. The date of her death, about 47.2 million years ago, is also significant. What is of greatest interest to paleontologists and anthropologists is that she may be transitional ('missing link' in the popular press) on the evolutionary path from Strepsiirhini suborder (**prosimians**) to the Haplorrhini suborder (Tarsiidae or tarsiers and **simians**).

From the end of her tail to the end of her nose, Ida would have been about 57 centimetres (23 in) long (her tail accounting for more than half of that length), weighing about 1.3 kilograms (3 lbs). Based on the fact she had both baby (or 'milk') teeth and adult teeth, we know that she was very young—perhaps just six months—so she was not full-grown. We know that she died with a broken wrist and weak left arm, and we can surmise that these injuries left her unable to escape to the trees to avoid the poisonous carbon dioxide cloud that hung over the lake in which she fell.

Martin Shields/Photo Researchers, Inc.

Ida, the Darwinius: fossil primate

Ida was identified as female based on her lack of a **baculum**, or penis bone, something that non-human primates have. Humans use hydraulics (and Viagra). Her features suggest that she was mostly not prosimian. Like a simian, she had a frontal bone and mandible made up of a single bone, not two; a heel bone, or tarsus, that was not elongated; nails instead of claws; and neither toothcomb nor grooming claw. However, her thumbs were relatively small, and her hind legs were much longer than her arms, both prosimian traits. By the fall of 2009, a number of scientists had seriously challenged the assertion that Darwinus was a 'missing link' between Strepsirrhini and Haplorrhini. Basically, their position is that the Adaptiformes (a proposed suborder of early primates to which Darwinus belongs) were pro-simians that appeared after the simians had developed from the Tarsiidae or tarsiers, and therefore were not a missing link. This matter is yet to be settled.

Table 4.5	Three body masses of *Proconsul*	
Species	Weight (range)	Weight (average)
P. africanus	9.3–13.6 kgs (20.5–30.0 lbs)	10.9 kgs (24 lbs)
P. nyanzae	28.0–46.3 kgs (62.0–102.0 lbs)	35.6 kgs (78.0 lbs)
P. major	63.4–86.1 kgs (140.0–189.0 lbs)	75.1 kgs (165.0 lbs)

Source: Walker & Shipman, 2005, p. 163.

Gigantopithecus: King Kong Lives

A member of the subfamily Ponginae and probable cousin of the orangutan's ancestor (both having descended from species of the genus *Sivapithecus*) was the genus *Gigantopithecus* (lit. 'giant ape'). How big was this giant ape? From the three fossilized **mandibles** (lower jaw bones), which are very deep and thick, and more than 1,300 fossilized teeth, it is reckoned that the species *G. blacki* (named after Toronto-born anatomist Davidson Black) stood about 3 metres (almost 10 ft) tall and weighed up to 540 kilograms (approx. 1,200 lbs). Keep in mind that these are estimates based on jaw and tooth size. If any of the larger bones such as a **femur** (thigh bone) or a **humerus** (upper arm bone) are found, we might arrive at a different figure, one that could be even bigger!

Gigantopithecines lived in what are now China, India, and Vietnam. The first specimen was identified by German geologist **G.H.R. von Koenigswald** (1902–1982) in 1935 in a Chinese apothecary shop, the kind of place where fossil bones and teeth were sometimes ground up for traditional Chinese medicine.

The gigantopithecines existed for hundreds of thousands of years. According to Jack Rink of the departments of geography and earth sciences at McMaster University, *G. blacki* lived from about one million years ago to about 300,000 years ago. Some scientists, particularly Russell Ciochon (see Ciochon, Olsen, & James, 1990), believe that in Vietnam and China, *Gigantopithecus* and *Homo erectus* coexisted in the same area for hundreds of thousands of years. This has led to some speculation that the Big Foot, Yeti, and Sasquatch stories may have originated from the near presence of *Gigantopithecus*. Of course, the lunatic fringe cryptozoologists (people who believe in the existence of mythical beings such as sasquatch and the Loch Ness monster) have articulated the belief that *Gigantopithecus* never became extinct.

Why would such a large creature become extinct? It probably relates to both size and diet. A comparison with the panda is useful. Like the panda, whose continued existence in the wild is threatened, *Gigantopithecus* ate a lot of bamboo, a plant that is subject to periodically high die-offs. The big guy may also have been out-competed by more efficient users of bamboo and other vital resources—competitors including *Homo erectus* and the giant pandas of the time.

Science Gone Wrong

Ramapithecus: From Ancestor to Ape in 20 Years

The story of the *Ramapithecus*, once held up as a possible missing link—the earliest human ancestor—before being re-identified as being related instead to a fossil species that was ancestral to modern-day orangutans, not in the human line at all:

Don't overinterpret findings based on little data.

Beware of convergent evolution.

Evolution trees look more like evolutionary bushes.

Don't underestimate the usefulness of molecular biology.

Be careful how you reconstruct fossil parts.

In 1932, G. Edward Lewis, a graduate anthropology student participating in a Yale University expedition to India, found two broken parts of a single primate **maxilla** (upper jaw—think of 'the max' as the top to distinguish this from the mandible) and some teeth. He declared the find a new species and named it *Ramapithecus* after a legendary king and seventh incarnation (or avatar) of the Hindu god Vishnu.

The remains were kept at the Yale museum, where they went essentially unstudied for nearly thirty years before Yale University professor Elwyn Simons, in 1961, declared that the specimen was ancestral to humans. This was a powerful statement, as *Ramapithecus* dated back to 15 million years ago, making the separation between ape

and human at least that old. Joined by his Yale colleague David Pilbeam, Simons continued to advance this claim in a series of articles. Both thoroughgoing lumpers, Pilbeam and Simons boldly incorporated Louis Leakey's *Kenyapithecus wicker*, a species of the same period as *Ramapithecus*, in their analysis. From close examination of its teeth and an upper jaw, the pair argued that Ramapithecus had skilled hands and culture. The key traits they focused on were

> *Ramapithecus*'s small canines (and other anterior, or non-molar, teeth);
>
> its curved dental arcade;
>
> its short face; and
>
> the thick enamel caps on its molars.

The last of these features was identified as primary in 1968, as it seemed to link *Ramapithecus* with the australopithecines and humans themselves. All had thick enamel when compared with the molar caps of modern African chimpanzees and gorillas. In Roger Lewin's words,

> It was naturally—but as it later turned out, wrongly—assumed that the African apes represented the primitive condition and that the 'human line' had developed the specialization of thick enamel caps on the cheek teeth. No one knew at the time that orangutans also have thick-enameled cheek teeth. And no one took the trouble to see what pattern obtained in the rest of the Miocene apes: thick enamel is in fact a common feature, a primitive, not specialized condition. But in 1968, the notion of *Ramapithecus*, *Australopithecus*, and modern humans sharing this 'specialization' fitted nicely with the then-orthodox view of human origins and with the idea that apes are rather primitive creatures. (Lewin, 1987, p. 96)

Their mistakes were multifold. One was to assume that reduction in size of the front teeth and a shortening of the face meant that hands were taking over the processes formerly performed by teeth. The study of the biomechanics of the eating process when small, tough seeds were involved, as was the case with *Ramapithecus*, showed that these changes, along with thicker tooth enamel of the molars, enabled more efficient chewing. A second mistake was viewing the thicker enamel as evidence of convergent evolution rather than of the sharing of specific or derived features. Another mistake

stemmed from the way the jaw had been reconstructed: with only pieces of the maxilla to work with, Simons and Pilbeam reconstructed the whole based on their belief that the *Ramapithecus* maxilla, being different from a modern ape's jaw (which it was), must look more like the jaw of a human ancestor (which it wasn't). When suitable mandibles and more maxilla were found, Simons and Pilbeam discovered that the shape of the dental arcade of *Ramapithecus* was more of a shortened V than either a U or an arch.

Simons and Pilbeam should have listened to what the molecular biologists were saying. Working first with blood proteins, Berkeley's Vincent Sarich and Alan Wilson were arguing for a greater closeness of the gorillas and chimpanzees to humans than the fossil people had ever put forward, claiming 4–6 million years ago as the time of separation between the common ancestors of chimpanzees and humans. As early as the 1960s they were theorizing about the existence of a 'molecular clock', a principle suggesting that mutations occurred like the ticking of a clock.

The difference in approaches here is one defined by the fossil folks' use of morphology (or structure) to infer a genetic relationship. This is like thinking that sharks and dolphins are genetically related, or bats and parrots, based on their morphological similarities. It's an approach that can lead to confusing convergent evolution with a specific line of evolution, and to inferring too broad a hypothetical body structure from only a few small parts. Molecular biology has its flaws, too. I will readily admit that I am not completely convinced that the molecular clock is as regular as many biologists say that it is, knowing as I do that environment, genetic changes themselves, and the varying abilities 'editing' or 'error-correcting' in various species can all theoretically lead to a different mutation rate. Neither approach is always right. Again, consilience leads to proof.

The ultimate demise of the *Ramapithecus*-as-ancestor idea came, ironically enough, from *Sivapithecus* (ironic because Siva is a god of destruction). This was a fossil species often found in the same geological deposits as *Ramapithecus*, and clearly a related species. Two specimens, one found in Turkey, the other in Pakistan, were discovered and described in the early 1980s. *Sivapithecus meteai*, in its teeth and in its maxilla, bears evidence that its closest living relation is the modern orangutan. The same, then, could be argued for its close relation, *Ramapithecus*. After all, if you're related to an orangutan, then your cousin is, too.

Summary

In this chapter we have travelled a long way—in time, in evolutionary history, and in the movement of continents. We have stressed that despite the changes noted, and the changes implied by the identification of different species, genera, and so on, there is a genetic continuity that should not be ignored. We have also pointed out that with our increased perception of linkages owing to the work done by molecular biologists, the taxonomy we are using, and even what we put into specific taxons or categories, is frequently changing. Rather than being perceived as proof of weakness in a disorganized science, this should be taken as a sign that anthropology is a very productive field right now. We'll see these patterns time and again as we get to the work that is more specifically within the research areas of physical anthropologists. It isn't that physical anthropologists can't make up their minds about taxonomy—it's that the more we discover, the more fluid the taxonomy.

In this chapter we have presented basic evolutionary information concerning primates. The next chapter will focus on some of the distinctions between primates, while also taking a close look at the behaviour of other primates and how much that behaviour resembles our own.

Typical Student Questions

▶ *Which came first: the chicken or the egg?*

Fundamentally, you can say that the egg came first, because what gave birth to it was not quite a chicken. This does not strictly mean that the two, parent and child, were a different species. It means that the egg/child emerged further along the chicken path than its almost-a-hen mother and its almost-a-rooster dad. Like all children, they were genetically different from their parents; like some children, the difference was phenotypic; like even fewer, this change persisted.

▶ *What has this got to do with physical anthropology?*

While we haven't talked much about human beings in this chapter, we have been discussing information useful to understanding the physical form that humans take. Some of our features (e.g. four limbs) are very old, which we've seen in this chapter. The chapter also illustrates like no other the way that science—specifically, biology—self-corrects over time. When you look at the later chapters, and you see that same principle operating in relation to our ancestors, it will seem more familiar and 'normal'.

Review Questions

1. What are the two taxonomic systems? What is each based on? Which makes more sense to you?
2. Why is it important to know where the continents were when discussing a fossil specimen?
3. What is meant by the term 'concestor'? Do you find it a useful concept?
4. Why is it difficult to arrive at a consensus on species of fossil specimens?

Discussion Questions

1. Why can the species concept be considered blurry or messy?
2. How can it be said that the categories of 'reptile' and 'bird' are problematic in modern biology?
3. Why is it easy for physical anthropologists and biologists to fall into the trap of declaring a species on a small amount of evidence?

Key Terms

animalia
amniotes
arthropods
baculum
binomial nomenclature
biological species
   concept
brachiate
clade
cladistics
class
concestor
deep time
domain
exaption

family
femur
genus
Great American
   Interchange
humerus
kingdom
living fossil
lumpers
mandible
marsupial
maxilla
Mid-Atlantic Ridge
molecular clock
monophyletic

notochord
opposable thumb
order
paraphyletic
phylogenetics
phylum (plural 'phyla')
placental
plate tectonics
postorbital bar
primitive (or
   generalized) traits
prosimian
ring species
sauropsid
sexual dimorphism

simian
specialized (or derived)
   traits
species (remember
   that this is both a
   singular and a plural)
splitters
stereoscopic vision
swim bladder
taxonomy
tetrapod
tyranny of the
   discontinuous mind

Key Individuals

Richard Dawkins
Stephen Jay Gould

Willi Hennig
James Hutton

Carolus Linnaeus
Charles Lyell

G.H.R. von Koenigswald

Recommended Print and Online Resources

Burgess Shale Geoscience Foundation. (Website). The world's most important fossil discovery. Available from www.burgess-shale.bc.ca

Ciochon, R., Olsen, J., & James, J. (1990). *Other origins: The search for the giant ape in human prehistory*. New York: Bantam Books.

Coyne, J. A. (2009). *Why evolution is true*. New York: Penguin.

Dawkins, R. (2004). *The ancestor's tale: A pilgrimage to the dawn of life*. Boston: Houghton Mifflin.

Dawson, W. (1901). *Fifty years of work in Canada: Scientific and educational*. Being autobiographical notes by Sir William Dawson. Ed. Rankine Dawson. London and Edinburgh: Ballantyne, Hanson, & Co.

Gould, S. J. (1987). *Time's arrow, time's cycle*. Cambridge, MA: Harvard.

Gould, S. J. (1989). *A wonderful life: The Burgess shale and the nature of history*. New York: W.W. Norton

Joggins Fossil Institute. (Website). The Joggins Fossil Cliffs: UNESCO World Heritage Site. Available from www.jogginsfossilcliffs.com

Miguasha National Park. (Website). From water to land. Available from www.miguasha.ca/mig-en/index.php

Natural Resources Canada. (Website). Geological survey of Canada. Available from: http://gsc.nrcan.gc.ca

Repcheck, J. (2003). *The man who found time: James Hutton and the discovery of the earth's antiquity*. Cambridge MA: Perseus Publishing.

Scotese, C. R. (Website). The PALEOMAP project. Available from www.scotese.com

Shubin, N. (2009). *Your inner fish: A journey into the 3.5 billion year history of the human body*. New York: Vintage.

Walker, A., & Shipman, P. (2005). *The ape in the tree: An intellectual and natural history of Proconsul*. Cambridge, MA: Harvard University Press.

5

Primates
Taxonomy and Behaviour

Mother rhesus monkey and baby
© FogStock/Alamy

LEARNING OBJECTIVES

After reading this chapter, you should be able to:

- Distinguish taxonomically the different primates.

- Outline the different ways in which anthropologists study primates.

- Discuss the evidence that at least some primates other than humans have culture.

Preface: The Dread of Apes

The eighteenth-century biologist **Carolus Linnaeus** (1707–1778) saw a connection between apes and humans, but since he did not want to anger the church, he played it safe:

> . . . if I had called man an ape, or vice versa, I should have fallen under the ban of all the ecclesiastics. It may be that as a naturalist I ought to have done so. (as cited in Lewin, 1987, p. 59)

In a 1927 lecture, American zoologist **William King Gregory** (1876–1970) speculated on why Piltdown Man was so uncritically accepted by anthropologists:

> It may be called **pithecophobia**, or the dread of apes—especially the dread of apes as relatives or ancestors. (as cited in Lewin, 1987, p. 59)

It's time to face up to the fact: we are all related to apes (see Figure 5.1).

Introduction

We ended the last chapter by looking at fossil primates and the time periods separating different branches of the primate order. In this chapter we will focus on the taxonomic differences between the branches, and we will look at primate behaviour, which teaches us not just about other primates, but about our species as well.

Primatology

Primatology is the study of primates (don't ever expect that to be on a multiple-choice test). The work is done in several ways. Some primatologists study their research subjects in captivity, often setting tests to see how intelligent the primates are. The studied animals have become well known in the literature: Washo the chimpanzee, Koko the gorilla, and, not quite as well known but equally fascinating, Kanzi the bonobo. Other researchers have studied primates in their natural environment. Some of these figures—Jane Goodall and Dian Fossey in particular—have become famous (the latter even had a movie made about her life, *Gorillas in the Mist*).

Ethology (lit. 'character study') involves the study of animal behaviour, sometimes within one species, sometimes looking at a particular form of behaviour across species. While it is directly linked to biology, particularly evolutionary biology, ethology in many ways can be considered a social science like psychology, sociology, and cultural anthropology. A good number of ethologists have degrees in psychology. Francine (Penny) Patterson, who has worked with and studied Koko since 1972, is a psychologist, as is Anne Russon (profiled in the box on page 123), who works with orangutans. Ethology's research methods are often drawn from other social sciences as well.

In the area of primate research that focuses on animal behaviour and its analogies to or similarities with human behaviour, there are handy labels for the two extremes of the continuum separating acceptance and rejection of apes as our relatives. The two terms were introduced in the opening chapter: anthropomorphism and anthropodenial. The term **anthropomorphism** describes the tendency to inaccurately project human characteristics onto non-humans; it is sometimes called 'projection'—not as in film projection, though it happens to be a common feature of animal movies. You might be guilty of anthropomorphism if you've ever seen a big smile on a chimp's face and thought that meant it was happy. It is more likely that the chimp was showing submission to another chimp (some may parallel this with our own nervous laughter).

Anthropomorphism isn't always a bad thing, as long as the person practising it doesn't assume, uncritically, that animals think and feel exactly the way we do. Ethologists who investigate the thinking and emotional patterns of primates (and other non-human animals) practise what is termed **cognitive ethology**, and when they write about shared thinking or emotional patterns of humans and others, it is called **critical anthropomorphism**. I can present an example from a recent experience I had while researching for a book on gibbons. Penelope is a *Hylobates lar*, or white-handed gibbon, who resides in the Bowmanville Zoo in southern Ontario. She is three years

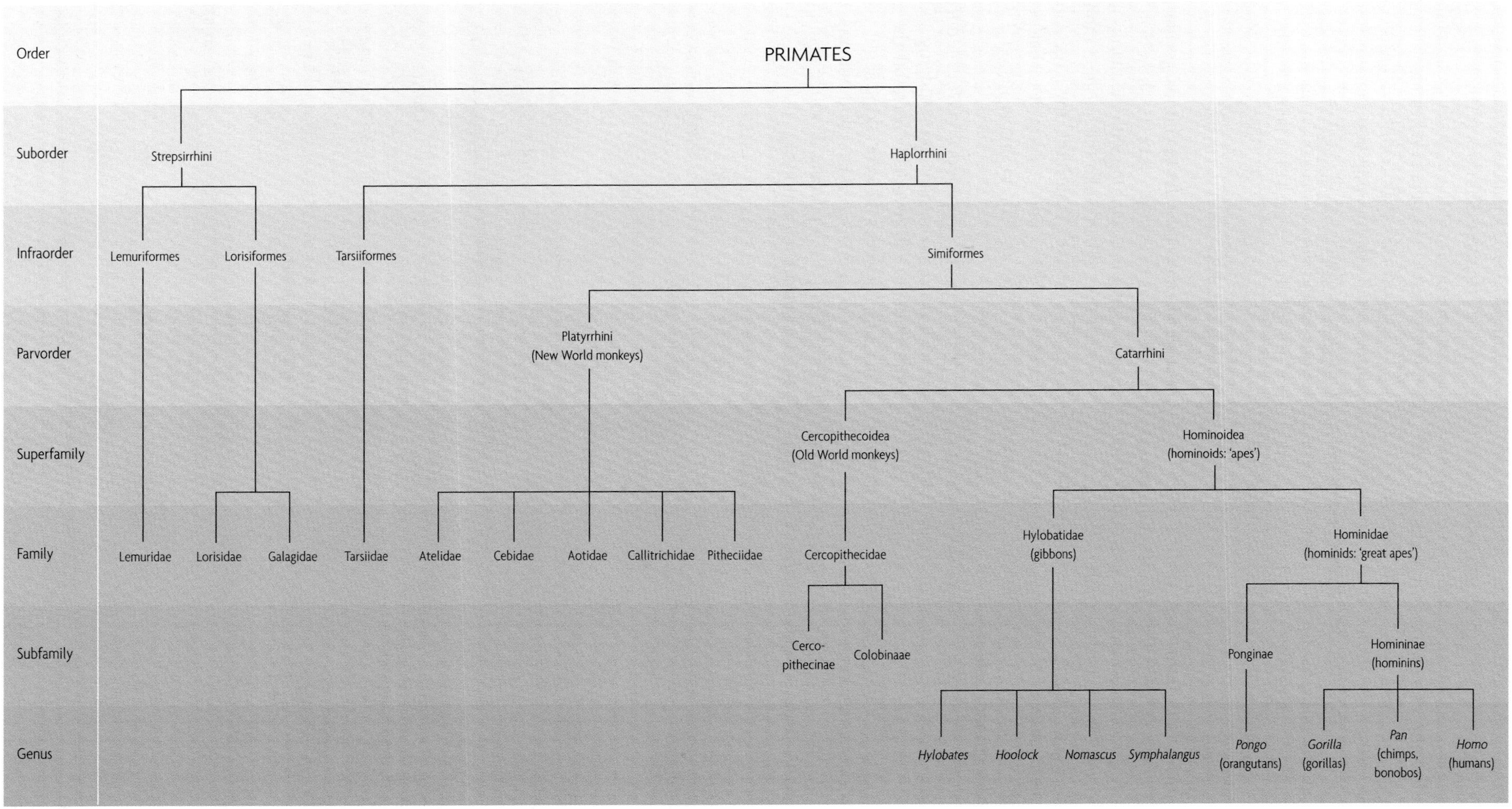

Figure 5.1 Primate taxonomy

old, a juvenile in gibbon developmental terms. When she feels nervous, she sucks her thumb. My wife, Angie, was drawing a picture of Penelope sucking her thumb, but as Penelope became habituated to us, she stopped her thumb-sucking. My wife, wanting to complete the picture, asked me to put my thumb in the same pose. A short time after I began doing this, Penelope reached over and patted my shoulder for the first time. I believe that she thought that I was nervous, so tried to reassure me, a show of compassion, something both humans and apes are capable of. This is what it means to take a critical anthropomorphism perspective.

The more important term for this chapter is **anthropodenial**, which is the failure to recognize those traits that are shared between humans and non-humans. In the first chapter I suggested that Western thinkers seem to suffer from this more than others—owing, perhaps, at least in part, to the general absence of primates in Europe (see Figure 5.2). In *The Sacred Monkeys of Bali*, an important work in the interdisciplinary field of what the author calls **cultural primatology** (study of the relationships between human groups and other primates in different cultural traditions), Bruce P. Wheatley discusses how greatly various Eastern traditions (in India, Indonesia, China, Japan, etc.) differ from Western ones in not separating the human and 'natural' worlds and, specifically, in putting monkeys on par with humans, being equally worthy of respect.

Wheatley writes of Japanese primatologists, influenced by traditional Japanese Shinto and Buddhist faith, holding special ceremonies of thanks to the souls of the dead monkeys that they have studied (Wheatley, 1999, p. 23), and of Japanese people using the honorific title *san* to respectfully address monkeys, the only non-humans to be spoken to in that way. Monkey gods, such as *Hanuman* of India and Indonesia and *Saruto Biko* of Japan, are deities that have traditionally been regarded as being worthy of reverence.

An Ape Is NOT a Monkey

I hate it when people refer to chimps, gorillas, orangutans, or gibbons as monkeys. I love the show *Seinfeld*—even in reruns—but it lost standing with me thanks to the episode in which they called Koko the gorilla a monkey. Koko is no more a monkey than Lime, my large Amazon parrot, is a sparrow. I've wondered why it is that people do that, but I can come up with no simple answer. I don't think it's just ignorance.

When I ask the class one of my favourite questions, 'What can we learn about humans from studying *other* primates?', some students will invariably leave off the all-important 'other', and answer as if I were asking about two very different kinds of animals: primates and humans. Hi, my name is John, and I am a primate. If you're reading this, so are you. But you are *not* a monkey.

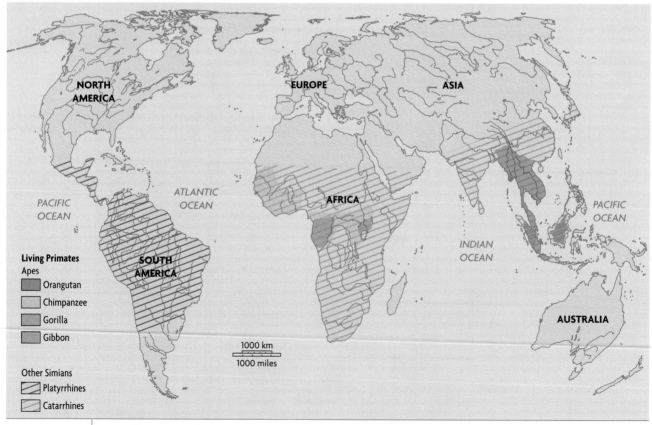

Figure 5.2 | The simian world

Dental Patterns

One way in which different classes of primate differ amongst themselves is in their dental patterns. Fortunately, then, teeth are more likely to fossilize than are most bones. So in this book, from here on in, we will be talking about teeth a lot.

Primates have four different types of teeth. From centre to back they are

- **incisors**, the cutting teeth (think 'scissor')

- **canines**, the grabbing and piercing teeth

- **premolars** and **molars**, the grinding teeth.

Incisors and canines are generally referred to as 'front teeth', premolars and molars as 'back teeth'. Molars are sometimes referred to as 'cheek teeth'. When anthropologists talk about dental patterns, they use the convention of dividing the mouth into four quadrants (top and bottom, left and right), each repeating the same pattern. For a human adult possessing all of his or her teeth, there are four repeats of the following pattern: two incisors, one canine, two premolars, and three molars (wisdom teeth are the third molars). This pattern, which is conventionally written 2:1:2:3, allies us with the group known as Old World monkeys, a group that includes not just monkeys such as marmosets and macaques but the other apes as well.

The term **dental arcade** refers to the shape of the teeth as they are arranged in the upper jaw. Humans are unique in having a dental arcade shaped like an arch; it is more parabolic, or U-shaped, than the dental arcade of apes. That is the official distinction, anyway. I admit that I've always had a hard time visualizing it that way. To me, an ape's dental arcade is long and thin (with the incisors on top), while a human's is shorter and fatter, as though someone took an ape's dental arcade and squashed it down so that it bulged out at the bottom and got shorter (see Figure 5.3).

Taxonomy

These days, the hardest part about discussing primates is probably the **taxonomy**. It certainly is for someone who was raised on an earlier taxonomic system. The terms used to identify and distinguish between various primates are changing. Molecular biology, which focuses on shared genetics, and cladistics, which traces evolutionary paths to common ancestors, are both taking over a tradition that was based on differences and similarities in morphology, or body structure. To avoid your becoming buried in *taxa* (plural of *taxon*), I will be **bolding** the ones I consider the most important of these terms.

It's worth bearing in mind that as foreign and awkward as many of these taxonomic divisions sound, each of the terms means something. Taking a little time to analyze a term to discover the *etymologies* (word origins) of its parts

should help you remember both the term for a given category and the species belonging to it. For example, noses are used as a primary way of distinguishing between primates. Terms built on the Greek-based *morpheme* (meaningful word part) *–rhin–* refer to noses. The English word rhinoceros means 'nose-horn'. (And if you've ever snickered at the thought of rhinoplasty, that would be plastic surgery performed on the nose, not the rhinoceros.) In this chapter there are several words based on the *–rhin–* morpheme; they are introduced in Table 5.1.

Suborders

Primates, taken together, make up an order. Just one step down from the level of order is the suborder. To get some sense of how different suborders can be, consider the following two suborders of the mammal order Carnivora: one is Caniformia ('dog-like'), which includes dogs, bears, pandas, seals, weasels, and raccoons; the other is Feliformia ('cat-like'), which includes cats, mongooses, civet cats, and hyenas.

The order of Primates is divided into two suborders, Strepsirrhini and Haplorrhini (see Figure 5.4), which are described in the sections that follow.

Strepsirrhini

The oldest and largest difference between primates divides the suborders **Strepsirrhini** and **Haplorrhini.** The former can be translated as 'twisted nose', though the suborder is sometimes characterized by the term 'wet nose'. That's because strepsirrhines have moist glandular skin around their noses, similar to what dogs and cats have. This wet patch is called the **rhinarium** (meaning 'of the nose'), and mammals that have it typically have longer snouts and a better **olfactory** (i.e. smell—think of how an old factory smells) sense. Strepsirrhines' upper lips are fixed to their noses and are split, like a rabbit's, rather than being independent but joined as the upper lips are in haplorrhines.

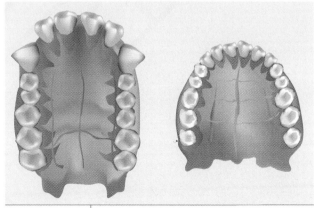

Figure 5.3 | Dental arcades of an ape (left) and a human (right)

Table 5.1	Know the nose		
Term	Translation	Referring to	Examples
Rhinarium	'of the nose'	moist olfactory patch around the nostrils in some mammals	lemurs, lorises, bushbabies; also cats, dogs
Strepsirrhini	'twisted nose'	suborder of the order Primates characterized by having a rhinarium	lemurs, lorises, bushbabies
Haplorhini	'simple nose'	suborder of the order Primates characterized by *not* having a rhinarium	monkeys, apes, us
Platyrrhini	'flat nose'	parvorder of the infraorder Simiformes	spider monkeys, howler monkeys
Catarrhini	'narrow nose'	parvorder of the infraorder Simiformes	macaques, apes, us

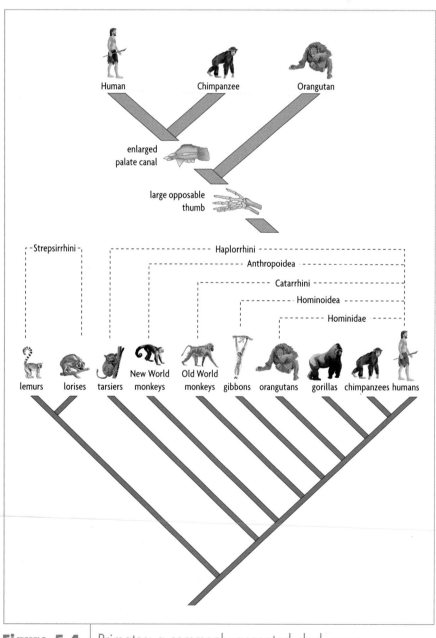

Figure 5.4 | Primates: a commonly accepted cladogram

Strepsirrhines can, again like dogs and cats, and as well as most mammals, produce their own vitamin C. While humans have some genes that used to perform this function, our vitamin C producer has been disabled by mutation, just as it has been in other haplorrhines. That does not hamper haplorrhines, whose diets include vitamin C–bearing fruits, which has reduced the evolutionary pressure to maintain vitamin C production. Apparently guinea pigs have the same problem: unlike most mammals, they cannot synthesize their own vitamin C and must derive it from the food they eat.

Other strepsirrhine features include a ring of bone surrounding the eardrum that is detached from the skull; in haplorrhines, the ring is attached. For grooming purposes—always important to the primate-about-town—strepsirrhines have a **toothcomb**: their front teeth (four incisors and two canines) form a fine, comb-like structure used primarily for cleaning their fur.

There are seven families of strepsirrhines; five are made up of lemurs, a primate unique to the large island of Madagascar, off the southeastern corner of Africa. No monkeys or apes are native to the island, so lemurs diversified in broad ways to fill the niches that monkeys and apes normally would. (Madagascar's unique diversity is not confined to its lemurs: two-thirds of the world's species of chameleons live there, and some researchers think they originated there—see Dawkins, 2004, p. 142.) An extinct family of lemurs, Megaladapidae, included the koala lemur (genus *Megaladapis*), which was larger than a modern gorilla.

In addition to the five extant lemur families, the strepsirrhines include the **Lorisidae** family, comprising the lorises of the Philippines, Indonesia, Thailand, Vietnam, India, and Bangladesh, and the pottos and golden pottos of Africa, and the family **Galagidae**, featuring the galagos, or in plainer English, bushbabies, of Africa.

What I Used to Teach

I used to teach the distinction between **prosimians** (of the suborder Prosimii) and **anthropoids** (of the suborder Anthropoidea). The former group included the strepsirrhines and the tarsiers, while the latter included (as the infraorder Simiformes does now) monkeys and apes. Older textbooks in your libraries will teach this distinction, but even more up-to-date books use the term and category *prosimian* (lit. 'before apes and monkeys') because of its morphologically defined taxonomic merit. For example, most prosimians (except for the strepsirrhine aye-aye, which has a complete set of claws) have **grooming claws**

Lemurs in Madagascar

Chris Mattison/GetStock

(a term I prefer to the synonym 'toilet claw', which conjures up disgusting pictures). Grooming claws are used for that typical primate behaviour, grooming. For the strepsirrhines, the grooming claw is the second or index finger; for the tarsiers, it comprises the second and third fingers. Other traits common to all prosimians include having both their frontal bone (the one that makes up the forehead) and their **mandible** (the lower jaw) divided in half, into two (left and right) bones.

It is important to note that we don't use the term *prosimian* (much) anymore, as tarsiers are more like monkeys and apes (and may share a common ancestor), and less like lemurs and the other strepsirrhines than we once thought.

Haplorrhini

The term Haplorrhini means 'simple-nosed', but the group is sometimes characterized by the term 'dry-nosed' as haplorrhines lack the moist, smell-focused rhinarium of the strepsirrhines. There are two infraorders within Haplorrhini: **Tarsiiformes** (comprising a single family, **Tarsiidae**, and genus, *Tarsius*), and **Simiformes** (a much larger group, formerly known as Anthropoidea).

Tarsiiformes (Tarsiers)

Richard Dawkins, in *The Ancestor's Tale*, describes tarsiers as looking like 'a pair of eyes on legs' (2004, p. 134) and points out that each of its eyes is bigger than its entire brain. This reflects the tarsier's nocturnal lifestyle in its native Southeast Asia. A distinct and defining characteristic of tarsiers is their **tarsus** (otherwise termed the calcaneus, or heel bone), which is extremely elongated, in effect adding another leg joint to their already long hind legs. Their thighs, calves, and heel bones are of approximately

Tarsiers

Holgs/Dreasmstime.com/GetStock

Old World monkeys 'tailed apes'—Dawkins, 2004, p. 121). Among the main morphological differences between New World and Old World monkeys (see Figure 5.5) is the dental pattern: both have a pattern that includes two incisors, one canine, and three molars, but the cercopithecoids differ in having two premolars to the three possessed by the platyrrhines. Another distinguishing characteristic has to do with colour perception, as we'll see next.

COLOUR PERCEPTION

The ability to see colours reflects how many colour-sensitive cells, or **cones**, (typically one, two, three, or four) a species or individual has. The molecular proteins involved are called **opsins**.

Four-cone vision (called 'tetrachromatic') enables the viewer to detect ultraviolet light. This is a useful ability for hawks, as it enables them to see where voles (a prey species) have peed, and for certain other birds that use tetrachromatic vision to detect the UV-reflecting mouths of their young crying out for food. (I don't think that would be necessary for baby parrots: they're quite vocal already.)

Trichromatic vision is what most humans (except for so-called 'colour blind' males) have. While it's common enough among non-mammal vertebrates, it is uncommon in mammals. The ancestors of most mammals lost that capacity when they were evolving during the time of dinosaur dominance. At the time, it paid to be nocturnal, to move about and find food primarily at night, when the mostly diurnal dinosaurs were sleeping. There is low evolutionary pressure among nocturnal animals to have trichromatic vision—it offers no advantage, as colours cannot be seen clearly at night anyway. Add to that the genetic and metabolic cost, and it's not surprising that most mammals (but not Australian marsupials) are dichromatic. That includes bulls, which cannot really 'see red'—waving a flag in front of an agitated bull is just a bad idea, no matter what colour the flag is.

Most platyrrhines follow this pattern in being dichromatic. The exceptions are female squirrel monkeys, howler monkeys, tamarins, and marmosets, which are trichromatic, and the nocturnal night or owl monkeys, which are monochromatic, seeing the world only in black and white (like some Canadian politicians) and shades of intermediate grey.

In contrast, almost all catarrhines have trichromatic vision, with opsins that enable perception of red, blue, and green. People who have investigated this feature generally argue that it enables the possessors of this 'full-colour vision' (as it is sometimes described) to see the reds and pale greens of new shoots and leaves, which are tastier, softer, and contain more nutrients. It should be pointed out that the fact that certain members of the two different parvorders possess trichromatic vision is a product of **convergent evolution**, not the presence of an inherited genetic pattern.

the same length. It should come as no surprise, then, that tarsiers are great jumpers. That ability, and having hind legs longer than their fore limbs, are traits that tarsiers share with their fellow prosimians.

Simiformes: The Primates Formerly Known as Anthropoids

The term **simian** refers to monkeys and apes. The infraorder (Simiformes) is divided into two parvorders (a category that is smaller than an infraorder and larger than a superfamily). The first of these parvorders is **Platyrrhini** (lit. 'flat nose'), otherwise known as New World monkeys. The other is **Catarrhini** (lit. 'narrow nose', with reference to the narrowness of the downward pointing nostrils of the group's members); the catarrhines are otherwise known as Old World monkeys and apes (superfamily Hominoidea). (Note that the term 'New World' is a eurocentric reference to the Americas, a part of the earth that, though well populated, was new to Europeans when Columbus 'discovered' it. Needless to say, 'Old World' and 'New World' should not be used for official dating purposes.)

You should be aware that 'monkey', despite forming part of the nicknames New World monkeys and Old World monkeys, is not itself a clade (i.e. a group with a common ancestor); in fact, it is not truly a group that is unified scientifically—it's like 'reptile' or 'bird' in that way. Old World monkeys are actually more like apes than they are like New World monkeys, despite the fact that they have tails (it might be better, as Dawkins suggests, to call

Platyrrhini (New World Monkeys)

Platyrrhines came to the Americas sometime between 40 and 25 million years ago. There are five families within the Platyrrhini parvorder:

- Pitheciidae, which includes sakis, uakaris, and titis;
- Atelidae, which includes spider, howler, and woolly monkeys;
- Cebidae, with capuchin and squirrel monkeys;
- Aotidae, with night or owl monkeys; and
- Callitrichidae, which includes tamarins and marmosets.

Among these families, the Atelidae are distinctive in that they have **prehensile tails**—a kind of fifth limb that can be wrapped around tree branches, enabling the monkeys to hang.

Catarrhini (Old World Monkeys and Apes)

The Catarrhines comprise two superfamilies, and three families:

- the Cercopithecoidea superfamily, which features the Cercopithecidae family; and

- the Hominoidea superfamily (or hominoids), including the families Hylobatidae (gibbons) and Hominidae (or hominids—the great apes and humans).

Notice that the names of the superfamilies end in *–oidea*, while the names of the families end *–idae*.

Old World Monkeys: Cercopithecoidea (Superfamily) and Cercopithecidae (Family)

Though this superfamily contains just one family and two subfamilies, it features a great variety (and increasing number) of species distributed across 22 genera. The Cercopithecinae family includes monkeys, guenons, vervets, macaques, mangabeys, baboons, geladas, and mandrills. Family Colobinae—the colobine monkeys—includes colobus monkeys, langurs, lutungs, surilis and doucs.

VERVET MONKEY CALLS

All primates are highly vocal creatures, yet traditionally, a distinction has been seen between the purely learned vocalizations of humans and the purely instinctive vocalizations of non-human primates. Recently, though, this view has begun to change. Primatologist **Tom Struhsaker**,

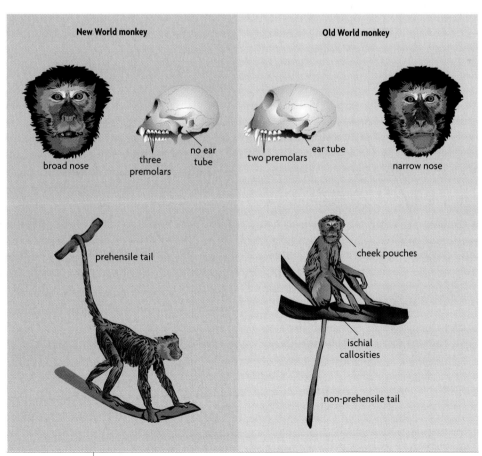

Figure 5.5 | Morphological differences between New World monkeys and Old World monkeys

in his pioneering work with the vervet monkeys (*Chlorocebus pygerythrus*) of Amboseli National Park in Kenya, observed that these monkeys used three distinctly different (and loud) **alarm calls**:

1. *leopard calls*. A loud barking sound made whenever leopards or related big cats were spotted. The response of other vervets was to take to the trees in order to escape the pouncing leopards.

2. *eagle calls*. A two-syllable coughing sound made whenever martial or crowned eagles were seen. Vervets within hearing range of the call would respond by looking up into the air and then running into the bushes.

3. *snake calls*. A 'chutter' sound made whenever deadly snakes such as pythons, mambas, and cobras were seen. The vervet response was to stand on two feet ('bipedally') and peer into the surrounding grass, sometimes with a group of monkeys surrounding the snake so that its location was known to others.

University of Pennsylvania researchers **Dorothy Cheney** and **Robert Seyfarth**, a psychologist, summarized Struhsaker's work and their own findings on the same species in a groundbreaking and insightful volume, *How Monkeys See the World* (1990: see Chapter 4, Vocal Communications, pp. 98–138). One of the main techniques featured in their own research is the use of **playbacks**: they would record the calls made by the animals under investigation and then, with hidden speakers, play the calls back to observe the animals' responses. They did this with the vervet alarm calls that Struhsaker had identified and elicited the appropriate responses. In a few cases in which certain individuals gave the leopard call when there was a potentially conflictive intergroup meeting, Cheney and Seyfarth concluded that the callers were attempting to deceive or disturb the other group.

Struhsaker had also observed some quieter calls produced in less threatening circumstances: these calls announced the presence of minor mammal predators (e.g. jackals, hyenas), unfamiliar humans (used less and less the

Science Gone Wrong

The Hundredth Monkey

This is a story of Japanese macaques, otherwise known as 'snow monkeys', because they are native to the north of Japan, which sees snow fall in winter. In the early 1950s, primatologists studying a troop of 20 macaques on Koshima Island discovered that a young female they named Imo had developed the clever, innovative habit of removing sand off of sweet potatoes by washing them in the shallow waters off of the coast of the island.

You could say that the initial reporting of the practice was embellished until it became a full-blown news story, and not just of a few macaques on an isolated island. The narrative of the 'hundredth monkey' fuelled a widely held belief in a 'tipping point' of monkey brains. According to the theory, once the hundredth monkey learns something, the behaviour quickly takes hold in the entire troop.

The story was first published in a foreword to *Rhythms of Vision*, published by Lawrence Blair in 1975. In 1979 the account was expanded in an emotion-charged two-page passage of the book *Lifetide: A Biology of the Unconscious*, written by biologist and New Age theorist Lyall Watson, who became the figure most closely associated with the story—Watson had been to Japan, where he first heard of the macaques, in 1958. The full explosion of the story happened with the publication of Ken Keyes, Jr's book *The Hundredth Monkey* (1984), which wasn't about monkeys at all but told a hopeful cautionary tale about humans and

the possibility of a nuclear explosion. Keyes made the hundredth monkey phenomenon an illustration of the tipping point, defined by Malcolm Gladwell in his very popular book *The Tipping Point: How Little Things Can Make a Big Difference* as 'the moment of critical mass, the threshold, the boiling point' (Gladwell, 2000, p. 12). It is the point beyond which change is unstoppable.

Here is the story. The practice of washing the sweet potatoes spreads from the clever young female to other youngsters in the troupe. The older members begrudgingly learn to do it as well, like senior citizens on Facebook. The practice becomes compelling when a certain critical mass of macaques are engaged in the practice. The adoption of the practice by the hundredth monkey was seen as the 'tipping point', the time when washing sweet potatoes became a practice that everyone followed.

This is where the story goes straight to the pyramidic heights of New Age thought. The claim was that not only did this become the general practice of Imo's troop, but that it instantly became widespread in troops throughout the area, as though some sort of psych-tsunami had hit the shores of the neighbouring islands, where monkeys adopted a practice without any need for observing others engaging in it. And like that—perhaps like the phenomenon itself—the hundredth monkey became a widely adopted symbol of how change, hopefully for the better, could

more accustomed the vervets became to the humans), and baboons (which occasionally prey on vervets). In the first and last cases, the hearers responded by becoming visibly more vigilant; in the case of humans, they would cautiously and silently move away into the background.

To demonstrate that these calls were learned, Cheney and Seyfarth studied both the calls and responses made by infant vervet monkeys. At first, the very young vervets would over-generalize their use of the calls to mammals, birds, and snakes that were not dangerous, eventually learning to focus in on what the calls true referents were. Likewise, it took some time before they responded appropriately to the calls they heard. Interestingly, the infant vervets also learned the appropriate responses to the calls of a local songbird, the superb starling (*Spreo superbus*). The bird has two different alarm calls: one for terrestrial predators, the other for hawks and eagles. That it was a learned response was proven by the fact that the vervets living in the swamplands, where they would hear the starlings more often, learned to respond appropriately sooner.

A Close Look at Some Catarrhini Species

Baboons

There are five species of **baboons**, which belong to the *Papio* genus of the Cercopithecinae subfamily. The genus used to include the mandrills and the drills, which are larger species, but now they are separated, with their own genus, *Mandrillus* (I wonder who gets the house).

Baboons are highly social, living in groups of up to a hundred individuals. The two sexes live very different lives and exhibit considerable **sexual dimorphism** (having separate body types, especially concerning size). Males, when they become fully adult (at 9 years old), weigh about twice as much as the females when they are adults (by the age of 6) and typically grow huge canines in a dog-like muzzle (one species is called *P. cyncocephalus*, meaning 'dog head'). Male hamadryas baboons grow a white mane.

take place. It became regular fare in magazines, movies, and television shows. It still lives on as a kind of urban legend on the Internet.

Telling the Real Story

Two writers helped to bring this story back to earth: Ron Amundson, a professor of philosophy at the University of Hawaii, who penned 'The Hundredth Monkey Phenomenon' for the *Skeptical Inquirer* in 1985; and Elaine Myers, whose 'The Hundredth Monkey Revisited' came out the same year in *In Context*. Their rightful criticisms were as follows.

First off, there were never a hundred monkeys in the troop when it was being studied in the 1950s. The Koshima Island macaques numbered 20 in 1952, their population growing to 59 over the next 10 years.

The practice did not spread very quickly. By 1958, just 17 of the 30 troop members were engaged in the practice. Of those 17 macaques, 15 were of the younger set, aged 2 to 7; of the 11 adults, just 2 washed sweet potatoes in saltwater, and none of the adult males, who had little contact with the young ones, did. Once the older troop members died and the young ones became more influential, the practice became general.

The spread of the practice outside the Koshima Island troop was not the product of psychic energies. Rather, at least one of the Koshima macaques swam over to another island, where it served as an example to the others. It should be kept in mind here that sweet potatoes were new to the macaques, something offered to them by the human scientists. Sweet potatoes and the washing of these new items arrived at about the same time.

The study of these macaques still is worthwhile as a way of understanding change in all primate society. We can consider the differences between growing up with something new to a culture, and having it introduced to you when you are an adult. As a person who will NEVER use social media such as Facebook and Twitter (as someone, in fact, who still does not own a cellphone and does not plan to in the near future), I can relate to the older monkeys who had never washed their potatoes and didn't see the need to start. We can also see, as Myers pointed out, that 'sweet potato washing led to wheat washing, and then to bathing behavior and swimming, and the utilization of sea plants and animals for food' (1997). The innovation of using seawater to wash food opened up related activities. Once the sea became a part of the macaques' lives, they were intelligent enough to see the potential of the place. It is rather like college and university students initially engaging in post-secondary education in one program, and then learning that their school has other things to offer.

Andrew McConnell/GetStock

Baboons

Baboon 'troops', or communities, are made up primarily of several **matrilines**, lineages determined along the maternal line. Females stay in the communities of their birth, while males leave not long after they become fully adult.

Although both males and females have clear lines of hierarchy, they are established and maintained in different ways. The male hierarchy, made up ultimately of males who have immigrated into the community, is a very combative one that frequently changes. The title of alpha male is rarely held for more than a year. Female hierarchy, which includes adult females and their children, is by lineage, and is therefore much more stable.

Male competition within many species (including humans) is often staged in ways that include no actual combat. Too much male combat is destructive to the group and to the species. In the case of baboons, the males have **wahoo contests**. These are expressively described by Cheney and Seyfarth in their bestselling *Baboon Metaphysics: The Evolution of a Social Mind* (2007):

> Wahoos . . . are extremely loud, low-pitched calls that can be produced only by large, fully adult males. They are costly to produce not just because of their loudness and low pitch but also because males give them in long bouts, often as they race through the group or bounce through

trees, leaping from branch to branch. A wahoo display is therefore an exhausting demonstration of a male's stamina and coordination. . . . [M]ale baboons often engage in wahoo 'contests': one male's wahoo elicits challenges from two, three, four, or many other males, all of whom respond by leaping through the trees and wahooing themselves. The contest continues until, one by one, the males drop out, exhausted. Usually, only one of the most dominant males is left calling at the end. (Cheney & Seyfarth, 2007, p. 52)

Another aspect of male life is more deadly, literally: **infanticide**. When an immigrant male moves into a community, it is very disruptive. As he moves his way up the male hierarchy, the infants of the troop may become collateral damage. In their 14-year study of chacma baboons (*P. hamadryas*), Cheney and Seyfarth recorded that at least 53 per cent of all infants died as a result of confirmed or suspected infanticide (2007, p. 57). The evolutionary explanation they give is that pregnant females have a gestation period of six months, and then they nurse their infants for about a year. This eliminates the majority of a troop's adult females from sexual activity. Killing the infants makes more females sexually available. It is not consciously done as part of a male baboon conspiracy; that is simply the effect.

Talking about evolution in this way tends to fuel creationist fears that evolution is evil, without morality. It also arouses feelings of anthropodenial, the rejection of our primate relations—after all, *we* don't go around killing our infants, right? But before you condemn baboon males, you need to learn about the 'friendships' they develop with lactating females. These are the most enduring bonds between male and female for baboons, lasting for up to a year (as opposed to the few days of the sexual relationship). The friendship involves a male and female foraging for food together, and a lot of mutual grooming. In addition, the male protects the female's infant from potential killing. In fact, the male–infant relationship outlives, in its closeness, the male–female relationship. Cheney and Seyfarth explain what happens once the baboon infant becomes a more independent juvenile (usually by the age of 12–15 months):

> Males . . . continue to defend the now independent juvenile, and if the juvenile is orphaned the male will often take over the role as the juvenile's primary adult companion. Indeed, over time a male's successive friendships transform him into a Pied Piper, and he is accompanied wherever he goes by a troupe of tumbling and twirling youngsters. For an old, slow-moving, low-ranking male, it is as good a way as any to finish out his years. (Cheney & Seyfarth, 2007, pp. 60–1)

Cheney and Seyfarth's central hypothesis is that baboon intelligence is directed mainly towards social knowledge (although environmental knowledge is important as well). This is best seen in the lives of female baboons. They are well aware of the ranking of all other females, and of the nature of the relationships, particularly kinship relationships, between other females. One example used to illustrate this comes from an article by Walter Hoesch (1961) describing an adult female baboon, Ahla, who had worked herding goats for a local farm family. This is not unusual in southern Africa, where Ahla lived. She demonstrated an amazing ability to identify mother-and-child relationships among goats, as Hoesch explained:

> Once she hears from inside the voice of a lamb that is calling for its mother, she will retrieve the correct lamb and jump through the opening between the two enclosures and put it underneath the mother so it can drink. She does this flawlessly even when several other mothers are calling and several lambs are responding at the same time. . . . She also retrieves lambs and brings them back even before mother and infant have begun calling. Apparently, she knows every animal in the herd but it seems unclear how she effectively recognizes them. (Hoesch, 1961, p. 299; as cited in Cheney & Seyfarth, 2007, p. 33)

In one case, Ahla kept reuniting a twin kid with its mother, after the farmer had separated it to be fed by a foster goat mother.

Cheney and Seyfarth found evidence of female baboon social knowledge in several areas. One of these was a practice called **supplanting**. A more senior female approaches a less senior one who is merely sitting, grooming another, or eating from a food supply. Automatically, the less senior female cedes her position to the more senior one. Another show of social knowledge occurred in **kin-mediated reconciliation**. After two female baboons had fought, a relative of one of the combatants would grunt to the other as a way of signalling that the fighting was over. In one of their playback experiments, Cheney and Seyfarth found that when they played a recorded grunt from a relative of a female baboon (call her 'A') who had recently fought with another ('B'), then B would exhibit tolerance of A's presence nearby in a shorter time than she would without such reconciliation (faked, in this case).

Gibbons

Within the Catarrhini parvorder, you have, on the one hand, Old World monkeys (family Cercopithecidae) and, on the other, the great apes (family Hominidae); between them you have the gibbons. As Table 5.2 shows, **gibbons** are intermediate in certain key traits, including the number of their lumbar vertebrae, the presence of maxillary sinuses (located just above the upper jaw), and their minimal sitting pads.

Gibbons used to be called 'lesser apes' because of their small size—the smallest species average just 5 kilograms (11 lbs), while medium-sized species weigh about 7–8 kilograms (15.4–17.6 lbs) and the 'big' siamangs weigh in at about 11 kilograms (24 lbs). Now the preferred term is 'small apes'. Gibbons belong to the family *Hylobatidae*, which contains four genera:

- *Hylobates*, which includes seven species;

- *Hoolock* (formerly included in *Hylobates*);

- *Nomascus*, which includes six species; and

- *Symphalangus*, which features just one species, the siamang.

The distinctions separating the genera, two of which are relatively new, are based on chromosome number (remember that humans have 46). *Hylobates* has 44; *Hoolock*, 38; *Nomascus*, 52; and *Symphalangus*, 50. Gibbons are found in Southeast Asia, and every one of their 16 species (that number is disputed, partly because their hybrid species number more than all other apes, including us, put together) is endangered or threatened.

When I ask students to name different kinds of ape, no one ever mentions gibbons. I have been teaching anthropology classes for almost 30 years. Why don't they know

Table 5.2	Apes, gibbons, and Old World monkeys	
Old World monkeys	**Gibbons**	**Great apes**
tails	no tails	no tails
no sinuses in frontal bone or maxilla	maxillary sinuses only	both frontal and maxillary sinuses
more lumbar vertebrae	intermediate number of lumbar vertebrae	fewer lumbar vertebrae
more flexible backs	less flexible backs	less flexible backs
deep, narrow chests		broad, shallow chests
sitting pads (ischial callosities)	minimal sitting pads	no sitting pads

Source: Adapted from Walker & Shipman, 2005, p. 186.

about gibbons? Their lack of knowledge of this important ape family reflects a broader societal ignorance. A quick Internet search for books about gibbons reveals how very difficult it is to find such a resource. There are none in my local library, which is fairly well stocked, with plenty of books about cute chimps and scary gorillas. When I went to Chapters–Indigo online, and typed 'gibbons' into the search field, I got a whopping 840 results—but almost none of them were actually on this neglected animal.

Gibbons, it seems, have had relatively little research done on them until recently. This I discovered when I found my way onto the website of the Gibbon Conservation Alliance (www.gibbonconservation.org), the source of the following statement:

> Although the gibbons are the most species-rich group of the apes, so far they have been hardly studied. Gibbons are the most under-researched

of the apes. We know nearly nothing about their social life and their cognitive capacities. Several species have never been studied in the wild, have never been filmed, and have never been bred in captivity. Gibbons are highly important in understanding the evolution of humans and apes.

Their characteristics are fascinating. They have the largest arms per body size of any primate. That trait combined with their unique wrist construction—a ball-and-socket joint similar to our shoulder—makes them ideally suited for **brachiation**, swinging through trees, where they can sometimes fly a distance of up to 10 metres from branch to branch. They are **monogamous**, that is, they pair-bond—a rare trait among primates, or among mammals for that matter. They sing songs—another uncommon feature—including duets that mates sing to each other, which are thought to strengthen the pair bond. (To hear samples of this singing, visit the website www.zooschool.ecsd.net/white%20handed%20gibbon.htm.) Their intelligence is shown by the fact that they can develop songs (built on a species-specific base) that can last up to 10–20 minutes long (this is sometimes referred to as the Great Call of female gibbons).

And yet, for all of their fascinating attributes, gibbons are still little known, and they are disappearing. There need to be many more books written about them, much more research done. Knowledge and interest go together well: the more we know about an animal, the more interested we become in its survival. To that end, the feature box on page 117 contains a small piece on two gibbons at the Toronto Zoo, the result of a short visit in late September 2009.

Gibbon in its natural environment

blickwinkel/Schmidbauer/GetStock

Lenny and Holly: Representatives of an Underrepresented Species

This is not a feature on a famous anthropologist, or even a lesser-known anthropologist. Rather, it's an homage to the white-handed gibbon, a remarkable species that, in my opinion, has been the subject of far too little anthropological study.

My wife, Angie, and I went to visit the two white-handed gibbons at the Toronto Zoo, and to learn more about them from the keeper, Bev Carter. Following the signs, we saw directions to the 'African Pavillion', with 'Gorilla' listed as one of the attractions, and to the 'Indo-Malayan Pavillion', whose headliner was the orangutan. No signage for gibbons. No sign is a bad sign for gibbons.

Lenny and Holly had been at the zoo for 26 years, both of them having been bred in captivity. Holly came from the Bowmanville Zoo, east of Toronto, and Lenny came from a zoo in the States. She was 37, he was 35. It is expected that they will live into their early forties (the record in captivity is 60 years). Both of them looked and acted strong and healthy.

Bev informed us that the pair had no children, that they 'didn't like each other very much'—something we could observe by the fact that they never came close to each other at any time during our short observation period. They did sing, at daybreak and day's end, but not in duets. The songs they sang were a mixture of different recorded pieces they had heard. Apparently, they did not play any gibbon songs to Lenny anymore as he would get excited, patrol the gibbon compound, and get quite aggressive with Holly. When we first saw Holly, she sat with her legs crossed, like a short, furry fashion model. She was a brunette; he, an ash blond. If we hadn't been told this, we would not have been able to distinguish their sexes (until he lay on his back sunbathing). Holly is occasionally given a stuffed toy, which she carries with her and grooms. They take it away from her only after she abandons it.

According to Bev, the two gibbons don't show any of the signs of intelligence typically observed among the great apes. They even had trouble with screw-top lids (but then, so do I sometimes). Gibbons, it should be noted, have poorer manipulation skills when compared with other apes because their thumbs and big toes are farther away from their fingers. This helps them in 'brachiating', when they hook their four fingers around a branch, and in the gibbon world, swinging is more important than manipulating for their survival (certainly they aren't required to open jars very often). Bev believed that their intelligence was more covert, hidden behind their apparently passive behaviour. They observe things very carefully. Obtaining food from their cage requires some thought and two hands. Mental stimulus is important with smart animals.

I was not fully prepared for how well and how easily they both walked on two limbs. They seemed to have too much arm for walking, as if they did not quite know what to do with their upper limbs. They moved as though they were trying to walk a tightrope or balance on a skateboard. They occasionally looked like the well-known video clip recording a supposed sasquatch sighting—I found that slightly creepy.

Lenny and Holly become creatures of brachiating beauty when they swing effortlessly from limb to limb to metal arch. They grasp with their long fingers—their shortened thumbs are not part of the process. According to Bev, they occasionally 'brachiate right out of the compound, and have to be shouted back'. The branches of the trees have to be regularly trimmed back to discourage such adventuring, and Vaseline is spread on the Plexiglas around the compound.

During our visit and our own observation, we noted the families that commented on the behaviour of 'the monkeys', and Angie and I would correct them. After we left our furry friends, we went into the gift shop, where there were primate stickers with chimps, orangutans, and gorillas; you could get an orangutan tote bag, and any one of many stuffed monkeys. There was only one stuffed gibbon, which we bought. We call him Lolly.

Great Apes and Humans: A Taxonomic Shift

One of the most significant taxonomic changes concerns the classification of great apes and humans. Humans belong to the family **Hominidae**: we are **hominids**. Apes used to be grouped within a family called Pongidae, making them 'pongids'. However, genetics has demonstrated that we share a tremendous connection with gorillas, chimpanzees, and bonobos. With the last of these, we are approximately 98 per cent the same, genetically speaking, and we're in the high nineties when it comes to similarity with gorillas. The new taxonomy reflects this. The family name Hominidae now includes the subfamily **Ponginae** (**pongins**) for orangutans and the subfamily **Homininae** (**hominins**) for gorillas, chimps, bonobos, and humans. In sum, humans are part of the following taxonomic groups:

Taxonomic level	Name		Members
	Group	Individuals	
Superfamily	Hominoidea[1]	hominoids[1]	gibbons orangutans gorillas chimps bonobos **humans**
Family	Hominidae[2]	hominids[2]	oragnutans gorillas chimps bonobos **humans**
Subfamily	Homininae[3]	hominins[3]	gorillas chimps bonobos **humans**

[1] Note the −oid− spelling for the superfamily names Hominoidea, hominoids.
[2] Note the −id− spelling for the family names Hominidae, hominids.
[3] Note the −in− spelling in the subfamily names Hominae, hominins.

To add to the confusion, the term *hominid* is still generally used to refer to fossil ancestors of humans.

Orangutans

Although for a long time it was believed that there was only one species of **orangutan** (Malay: *orang−*, 'person', + *hutan−*, 'forest'), it has recently been established that there are, in fact, two: one native to Borneo, *Pongo pygamaeus* (although it's no pygmy), and one native to Sumatra, *Pongo abelii*.

Orangutans live most of their lives high in the trees—not an easy lifestyle for a creature so large, but a wise choice for an animal that usually lives in swampland and loves fruits. Their body structure reflects their arboreal life: their arms are 40 per cent longer than their legs, and their hands and feet are quite long, with curved fingers and toes capable of making a hook-like grapple for grabbing tree branches.

Orangutans are very slow to breed. Most females are around 15 when they first give birth, and babies are nursed until the age of 7, when they are finally weaned. There is considerable sexual dimorphism among orangutans. Fully adult females typically weigh between 35 and 40 kg (77–88 lbs), while a large adult male can reach twice that size—up to 80 kg (176 lbs)! A full-sized adult is an imposing figure, with face flanges that look like two sides of a large dinner plate, and a loud, long call. Because of the fierce

Orangutan using a tool

competition among adult males, not all males reach full size. A good number of the male orangutan population are sub-adult—sexually mature but not full-size.

In terms of social structure, fully adult males avoid each other, while sub-adult males tend to be more companionable (there is less to lose by such companionship). The most serious bonding takes place between females (i.e. a female and her daughters and possibly sisters) and between mother and child.

Dutch biologist and primatologist **Carel Van Schaik** (b. 1953), in a recent study of orangutans on the swampland Suaq Balilmbing on the large island of Sumatra, addressed several important issues relating to orangutan behaviour. One of these issues he calls the **orangutan paradox**, though it can also be applied to the rest of the great apes. Research on orangutans in captivity documents the big guy's amazing ability to imitate human actions: washing clothes, sweeping paths, chopping wood, and sharpening axe blades, to name a few. Placed in problem-solving situations by experimenters, orangutans show a striking ability to come up with solutions (not always those that the experimenters had in mind). Yet—and here is where the paradox comes in—it seemed that orangutans demonstrated no such intelligence in their own 'natural' settings in the wild. The solution to the puzzle is that signs of that intelligence do exist—humans just need to know how to look for them.

Van Schaik approached the issue of orangutan intelligence by attempting to address the question: Do orangutans have **culture**? The answer, for orangutans as well as the other non-human hominids, is yes. Following the Japanese primatologist Kinji Imanishi, he stated that 'culture is simply innovation followed by diffusion' (Van Schaik, 2004, p. 139). His proofs for orangutan culture so defined are as follows:

- Orangutan behaviour varies according to geography; in other words, where an orangutan lives and travels has a strong impact on the behaviour that he or she exhibits.

- Orangutan behaviour cannot be explained by environment or habitat: behaviours exhibited in swamp sites are not more similar to each other than they are to behaviours displayed in more upland sites, and vice versa.

- The areas in which orangutans have more association with each other have larger 'cultural repertoires' (i.e. collections of learned behaviours) than those areas where they are more isolated from each other.

Sexual dimorphism among orangutans

Now, the obvious question that needs to be addressed is this: What constitutes orangutan cultural behaviour that varies with geography? Van Schaik's list includes the following behaviours:

1. Using leafy branches:
   - as a swatter to ward off attacks by angry bees;
   - to scoop water from deep holes in trees (when those holes cannot be reached by more direct means).

2. Using a handful or bundle of leaves:
   - to wipe the chin after eating fruit;
   - as a protective covering for the hand when dealing with spiny fruits or branches;
   - as a cushion for sitting on hard branches.

3. Using specially made stick tools:
   - to extract honey, insect larvae, or pupae (cocoons) of ants or termites from tree holes;
   - to extract seeds from the cemengang fruit (a fruit covered with many tiny, stinging hairs);
   - to scratch one's back.

4. Using dead, hollow twigs to suck ants from decaying orangutan nests.

5. Performing such nest-related activities as:
   - hiding under a nest when it rains;
   - building a covering above a nest as a 'roof' to provide shelter against the rain or shade against the sun during midday resting;
   - (among immature or juvenile orangutans) building a nest near one's peers in order to play on it.

Profiles in Canadian Anthropology

Biruté Galdikas

Biruté Galdikas working with primates

Born in 1946 in Germany, but raised in Canada, **Biruté Galdikas** is the third and least-known member of the women primatologists known as 'Leakey's Angels' or the 'Trimates' (a blend of 'primate' and the number three). Jane Goodall studied the chimpanzees, for which she (and the chimps she studied) are hugely famous. Dian Fossey studied the gorillas. Biruté Galdikas, later than the others, studied the orangutans, and, despite winning a number of awards, still hasn't achieved the fame she deserves.

Growing up in Toronto, near the city's large (and at that time wilder) High Park, Galdikas dreamed of living in the wilderness. She would later learn to live in wilderness much larger and harsher than she could have imagined.

At 17, she was accepted into the psychology program at the University of British Columbia, but transferred to the University of California—Los Angeles (UCLA) to graduate in 1966, at the age of 20. Familiar with the work of

Jane Goodall, and finding the subject intriguing, she shifted her major to anthropology as a graduate student. Her life's calling came to her in March 1969, when she heard Louis Leakey give a lecture. She approached him afterwards and spoke of her desire to study orangutans. While she didn't feel that she had made any great impact at first, the next day he called her. He made no promises of immediate financial assistance, but expressed support regarding what she wanted to do.

Galdikas was ready to go, but she would have to wait until 1971—and the receipt of a National Geographic Society grant—before she could set out on the swampy wilderness of Tanjung Puting, in Borneo, an island that makes up part of the country of Indonesia. First impressions were not great:

> My first week at Tanjung Puting felt like a jail sentence. For years I had dreamed of wandering the forest by myself, 'alone with nature', searching for wild orangutans. Instead I was confined to a tiny, airless hut with five men. I had no privacy except after sundown, when oil lamps made from small sardine cans left most of the hut in darkness and I could retreat into the shadows.
>
> My Indonesian escorts would not let me go anywhere by myself, or even with Rod [her husband]. Nor were they eager to venture out with me. They explained that the adjacent forest, which shimmered, mysterious yet inviting, was swamp, or *rawa*. The way they spoke about the swamp, *rawa* sounded like the most feared word in the Indonesian language. (Galdikas, 1995, p. 83)

With her often beautiful writing, she lets the readers of her *Reflections of Eden: My Years with the Orangutans of Borneo* know very well what it was like to follow after

Biruté Galdikas's description of an orangutan's ability as a nest engineer gives evidence of a very impressive activity:

> [They] must select a sturdy support—a strong limb or, less frequently, the tops of two trees brought together or the crotch of a large tree. Next they bend and twist the smaller branches

of the limb at ninety-degree angles, folding them into a springy, circular platform, like a box spring. This rarely takes more than three or four minutes. But they spend up to a half-hour picking loose, leafy branches, and piling them on top of one another to form a mattress almost a foot thick. The end result is a bouncy arrangement, quite comfortable but, for the orangutans, small.

large, orange apes, all capable of travelling skilfully from tree to tree while she trudged through the *rawa*:

> I was in swamp water all day, every day, dragging myself from one pool of slimy mud to the next. I had to constantly watch my feet to keep from tripping, falling, slipping, or sliding. The slick, moss-covered logs were treacherous. More than once I reached for a branch to haul myself out of the muck, only to discover, at the last second, a stout, beautifully green, poisonous viper coiled around the limb. . . . At the same time I had to keep my eyes on the treetops, searching for orangutans, watching for rain so that I could hastily stow my notebook under my clothes. After I began finding and following orangutans more regularly, it 'rained' branches, fruit peels, and bark—and urine as well. My feet in the mud below, my eyes shaded from the deluge from above, I had to take notes as meticulously as if I were sitting in a gleaming high-tech lab, wearing a spotless white coat. (Galdikas, 1995, p. 106)

Still from then until now, a period of more than 35 years, she has weathered the *rawa* while conducting extraordinary work. From her we have learned more about the often solitary orangutans than was ever thought possible. She has informed us about their intricate and culturally learned nest-building (see 1995, p. 111), their making and use of large leaf 'umbrellas', and their acquisition and use of more than 400 'food types' (e.g., plants and animals).

Why do we not know her work better? I spoke of her artistic writing. She is an excellent storyteller as well as a scientist. Near the end of *Reflections of Eden*, she speaks of the battles that the other Trimates waged to gain respect in the largely male scientific community: Jane Goodall's fight to have a paper published in a learned journal with names rather than numbers given to the chimpanzees she was studying, and Dian Fossey's fierce advocacy for the protection of 'her' gorillas. As Galdikas sees it, Jane Goodall was accepted as 'a typically sentimental female', and Dian was admired for 'going native' (1995, pp. 387, 390). Galdikas has done the same—advocated for orangutans (especially with her Orangutan Foundation International), personalized her orangutans, and written in intensely subjective 'feminine' psychological terms of her relationship to her orangutan family, as can be seen in the following:

> Instinctively, as a female primate, I provided the maternal environment for the three orangutans who became members of my forest family. I didn't lick them, I combed them. I didn't suckle them, but I allowed them to suck my thumb and I provided them with milk. I hugged, carried, caressed, and loved them. They became addicted to me and I to them. Their physical warmth, their soft squeals, their expressions, their feel became part of my daily life in the forest. . . . They made me laugh and they made me cry. (1995, p. 212)

I can see many of my male colleagues (and students) cringing when they read this. I like it, but I suspect the reason—the main reason—why I have heard terms such as 'flaky' attached to her is her definitely feminine approach to her work with the orangutans.

In 1999, Canadian writer and editor Linda Spalding published *The Follow*, a critical account of Galdikas's work. It was gutsy of her to take on a conservationist icon (albeit one not well known in Canada) and to cast aspersions on her work. To decide for yourself where you stand on her life, read Galdikas's work together with Spalding's book.

They have to curl up inside or either drape their limbs over the edges or hang them over nearby branches. Young orangutans exhibit a natural tendency to bend branches over and under their bodies, but building nests is not instinctive. . . . Juveniles practice for years, starting with play nests, and only later graduate to the real thing. (Galdikas, 1995, p. 111)

I've marvelled over the resourcefulness shown by our parrot Quigley, who has built nests with chopsticks, but he could clearly learn a thing or two from the orangutans!

Gorillas

Primatologists recognize two distinct and geographically separated species of **gorillas**: the western *Gorilla gorilla* and the eastern *Gorilla beringei*, named after the explorer Robert von Beringe, who 'discovered' and then shot one in 1902. Subspecies include the western lowland gorilla (*G. gorilla gorilla*—that should be easy to remember) and the Cross River gorilla (*G. gorilla diehli*)—both of the western species—and the mountain gorilla (*G. beringei beringei*) and eastern lowland gorilla (*G. beringei graueri*), both of the eastern species.

Sexual dimorphism among gorillas is well documented and easily spotted at any zoo. The fully developed 12+-year-old male (known as a **silverback** because of the area of white hair on his back) can become 1.85 meters (6′) tall and weigh up to 225 kilograms (500 lbs). Females, who reach maturity slightly earlier (a general primate pattern), average about 1.4 metres (4′ 7″) tall and weigh about 100 kilograms (220 lbs).

In terms of social organization, the silverbacks dominate the troop, which can range in number from 5 to 30, and which typically consist of one silverback, a number of mothers and their youngsters, and a few blackbacks (sexually mature males aged 11 and under, which are smaller than the big guys and don't yet have the distinctive silver patch or the larger canine teeth that the boss shows when he's feeling threatened). Blackbacks are also found in small groups of their own, cruising and waiting for sexual and leadership opportunities (think of it as the jungle club scene).

DIAN FOSSEY

> The more you learn about the dignity of the gorilla, the more you want to avoid people. (Dian Fossey, as cited in Smith & Vollers, 1986)

We have learned much about mountain gorillas from **Dian Fossey** (1932–1985), who wrote *Gorillas in the Mist* (1983), a name also given to the movie depicting her life (1988). Her story is one of reaching for a dream. Her first

Dian Fossey

love was animals, but she graduated from college in 1954 with a degree in occupational therapy. She worked for a few years before making a huge life change: she took out a three-year bank loan and used up her life savings to pay for a multinational trip to Africa. Her journey included a stop at Olduvai Gorge, Kenya, for a visit with Louis Leakey, who talked to her about Jane Goodall's work with chimpanzees. Fossey was inspired. Despite having broken her leg at one of Leakey's archaeological digs, she managed to climb mountains to take pictures of mountain gorillas.

Fossey, eventually, had to return to the US, where she took a job at a children's hospital in Kentucky. But in the spring of 1966, she met Louis Leakey again, this time in Louisville, where the latter was speaking as part of an American lecture tour. When Fossey showed him the pictures she had taken, Leakey encouraged her to do fieldwork with gorillas; within eight months he had secured funding for her work.

In 1967, Fossey established the Karisoke Research Center in the Virunga Mountains of Rwanda. During the three years she spent there, before going to London in 1970 to obtain her doctorate in zoology, she bonded with the gorillas, forming a particularly strong attachment with a young male she called Digit. This was quite an amazing feat. She spent more time doing research in Africa from 1974 to 1980, leaving this time to take a teaching position at Cornell University in New York. In 1985 she returned to the Research Center, where she met her tragic end, being murdered with a machete, probably by poachers.

GORILLA BEHAVIOUR AND INTELLIGENCE

Gorillas have recently been observed using tools. In an article published in 2005, Breuer, Ndoundou-Hockemba, and Fishlock recorded the following tool-using event:

> Adult female Leah was seen at the pool edge near where a branch was sticking out of the surface, looking intently at the water in front of her for 1 min before she entered the water. She began to cross the pool walking bipedally, but after her first steps the water quickly became waist deep and she returned to the pool edge. Leah then re-entered again bipedally, but after her first steps the water quickly became waist deep and she returned to the pool edge. Leah then re-entered again bipedally and grabbed the straight branch in front of her with her right hand . . . detached the stick and, stretching forward with it in her right hand, seemed to use it to test the water depth . . .; she grasped the stick firmly and repeatedly prodded the water in front of her with the end of the stick. She then moved further into the pool, holding the detached branch in her right hand and using it as a walking stick for

Profiles in Canadian Anthropology

Anne E. Russon

You cannot discuss the study of orangutans, particularly in the Canadian context, without mentioning the extensive work done by **Anne E. Russon**, who teaches psychology at York University. Her research, especially on orangutans that are being rehabilitated and reintroduced into the wild after having been captured or raised illegally in captivity, has taught us much about the intelligence of these very clever hominoids. One of the elegant aspects of her published material lies in her detailed description of daily orangutan activities, and the complex series of skills and forms of intelligence that they require. The following is an example taken from an article on 'The Nature and Evolution of Intelligence in Orangutans':

> *Enggong* and other orangutans use nearby pole-like trees as vehicles to enter target trees. . . . For palms, they climb a selected vehicle tree, take a position slightly higher than an accessible entry point to the palm, then SWAY the vehicle tree back and forth until they can grab the palm. SWAY represents a behavior complex; *Enggong*'s SWAY was a variable combination of behavior units, including lean out from the vehicle tree to bend it, 'pump' the tree to make it swing back and forth (i.e., as in pumping a swing, dynamically

Figure 5.6 | *Enggong* clambers into a palm's crown while retaining hold of his vehicle tree

change position, posture, and/or degree of lean to induce and modulate the tree's bending), and remove vegetation restricting the swing. *Enggong* iterated [repeated] this behavior complex *until* he attained the palm, qualifying his SWAY as a routine. (Russon, 1998, pp. 495–6)

postural support. (Breuer, Ndoundo-Hockemba, & Fishlock, 2005, Results, para. 1)

Probably the most famous gorilla is Koko. In 1972, a psychologist named **Francine (Penny) Patterson** obtained Koko, then about a year old, for study. Patterson taught Koko to communicate using American Sign Languages (ASL). Proof that she could use the system came from a **double blind test**: Patterson would show Koko an item, and Koko would sign what the item was to someone who could not see it; the recipient of the sign would then report back to Patterson what Koko had signed. Koko passed the test.

Now in her fortieth year, Koko has gone on to learn at least 600 ASL signs. She has demonstrated creativity in her ability to link two signs together to convey a meaning for which she had no sign (e.g. 'white' + 'tiger' for *zebra*; 'eye' + 'hat' for *mask*; 'elephant' + 'baby' for a Pinocchio doll; 'bottle' + 'match' for *lighter*). She has also shown a capacity for deception or lying. When she was five and weighed a mere 90 some pounds, she sat on a sink and broke it.

When asked whether she had broken it, she signed 'Kate [one of the assistants] here bad.' An anecdote I always tell in my classes is about a time when Koko was with some other gorillas that did not know how to sign. Frustrated by their lack of communication skills, she dismissively called them 'stupid animals'.

Chimpanzees and Bonobos

Both **chimpanzees** and **bonobos** belong to the genus *Pan*. The former are known as *Pan troglodytes* and the latter as *Pan paniscus*. Although bonobos used to be referred to as pygmy chimps, they are not noticeably smaller than chimpanzees. Chimpanzees show less sexual dimorphism than do orangutans and gorillas, the female chimps being only slightly smaller than males. The males studied at Gombe Stream National Park in Tanzania are about 1 metre (4') tall when standing upright and weigh between 41 and 52 kilograms (90–115 lbs). Interestingly, a recent study (Foster, et al., 2009) suggests that the biggest chimp is not necessarily the dominant male. Researchers studied

Case Study Controversies

More like a Man than a Monkey: A Lesson in Seventeenth-Century Comparative Anatomy

British scholar and physician **Edward Tyson** (1650–1708), considered the founder of the study of comparative anatomy, gave us the first scientific analysis of a hominoid, in his *Orang-Outang, sive Homo Sylvestris: or, the Anatomy of a Pygmy Compared with that of a Monkey, an Ape, and a Man*, first published in 1695. Despite the name of the book, it was a chimpanzee, and not an orangutan (or an actual pygmy) that Tyson dissected for his analysis. He established that the body and brain of the chimp had more in common with those of humans than with those of monkeys—a groundbreaking, if controversial, finding.

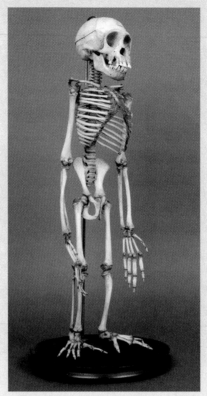

Both images © The Natural History Museum, London.

Two plates from Tyson's *Anatomy of a Pygmy Compared with that of a Monkey, an Ape, and a Man* (2nd edn, 1751): (left) 'The skeleton of the "Pygmy"'; (right) 'External Appearance of "Pygmy"'

the strategies of three alpha males who ruled the troop between 1989 and 2003. The largest one, Frodo (51.2 kg, or 112.6 lbs), employed his size and aggressiveness to maintain his position; Freud, intermediate in size (44.8 kg, or 98.6 lbs), combined aggression with grooming; and undersized Wilkie (37 kg, or 81.4 lbs), 'obsessively groomed both male and female chimpanzees to maintain his top position' (Foster, et al., 2009).

JANE GOODALL

British primatologist **Jane Goodall** (b. 1934) is the superstar in the field of chimpanzee study. She began her work in 1960, a secretarial school graduate handpicked by Louis Leakey to assist him in his work (she was then his secretary; she has since earned a doctorate) at Gombe Stream National Park. Her many discoveries have included the aggressiveness of chimps (although this has not been found among all chimpanzee troops), including murderous battles between troops; the fact that chimps kill monkeys for meat; that rank among

male chimpanzees depends to a significant extent on the rank of their mother; and—the breakthrough point—that chimpanzees use tools (e.g. stripping down sticks to insert in termite hills, withdrawing them covered with yummy termites). Her writings are rich with details of the complex family and other social relationships among chimpanzees. I'm a fan.

An important study to which Goodall contributed focuses on the existence of cultures among chimpanzees. The idea here is that different populations of chimpanzees learn different behaviours or practices that become traditions in the local group or society. The groundbreaking article, entitled 'Cultures in Chimpanzees' (Whiten, et al., 1999), documented 39 cultural variations among different populations. Significant in these variations are ways of cracking very hard nuts against stone anvils using wooden or stone hammers. Juvenile chimps have to practise to get it right. Some troops do this, but others that live where the nut trees grow do not. This proves that the technique for opening nuts is part of a local culture, as opposed to a widespread practice in chimp communities. The list of

Jane Goodall

Upright female and male bonobos show long-legged, humanlike body proportions

culture-specific behaviours includes dragging branches as a display, using a leaf mass as a sponge, walking on sticks over thorns, using a leafy stick to drive away insects or brush off bees, using leaves as napkins or as seats on the ground, slapping branches for attention, and dancing in the rain.

Bonobos

Bonobos aren't known or talked about the way that chimpanzees are. Maybe it's the sex.

I was checking out a book on bonobos, Frans de Waal's *Bonobo: The Forgotten Ape* (1977), at my local library.

There was a lineup at the checkout desk, so I started leafing through it, and soon encountered some bonobo sex pictures. I thought: 'Great, this is what I need—some kid seeing the pictures and saying, "That old bearded man is looking at dirty monkey pictures."' I live in a small town; reputation is important.

Table 5.3	How are chimpanzees and bonobos different?	
	Chimpanzees	Bonobos
Politics	rank is more important	rank is less important
Sexual politics	males rule	females have significant authority
Diet	omnivorous (includes meat)	frugivorous (little meat; more fruit)
Interaction	more aggressive	more sexual

Frans de Waal characterizes this key aspect of bonobo behaviour well:

> The species is best characterized as . . . one that substitutes sex for aggression. Whereas in most other species sexual behavior is a fairly distinct category, in the bonobo it is part and parcel of social relations—and not just between males and females. Bonobos engage in sex in virtually every partner combination (although such contact among close family members may be suppressed). And sexual interactions occur more often among bonobos than among other primates. (de Waal, 1995, p. 82)

Just as interesting, at least intellectually, is the female-centred aspect of bonobo society. Females leave their birth group upon reaching sexual maturity. When they join a new group, they form alliances with other females (others who were originally outsiders), and these alliances give them effective leadership of the group. This is despite the fact that males are bigger and stronger than females.

Summary

This chapter began with a figure outlining the major taxonomic categories that make up the Primates order. In the ensuing discussion we managed to cover the whole thing—except for the tiny box in the bottom righthand corner, the one that applies to us: *Homo*. Beginning in the next chapter, we'll look at the hominoid species that preceded and then eventually became modern humans.

Table 5.4 | Taxonomically speaking, what are humans?

Level	Name
Order	Primates ('first one')
Suborder	Haplorrhini ('simple nose')
Infraorder	Simiformes ('monkey-like')
Parvorder	Catarrhini ('narrow nose')
Superfamily	Hominoidea (hominoids)
Family	Hominidae (hominids)
SubFamily	Homininae (hominins)
Genus	Homo ('man')
Species	Sapiens ('thinking')

If you need a little help remembering all of this, try the mnemonic 'Please help someone categorize 4H seeds', or else, for the first four, 'pretty happy simple cats'.

Typical Student Questions

▶ *If we evolved from apes, why are there still apes?*

We didn't evolve from the apes living today. We have common ancestors, which were different both from them and from us. We evolved in different ways in different situations. Consider this: humans and apes share a common ancestor with fish, yet there are still fish today.

▶ *If we evolved from apes, why aren't we hairy?*

You can tell that we used to be hairy by our 'goosebumps'. Anything that causes goosebumps to rise on our skin would, in a hairy animal, cause its hair to stand up on end. This would make the animal that was feeling threatened look bigger and more intimidating, and it might make a cold animal feel warmer. Our hair loss probably relates to our evolving ways for dealing with cold and heat.

Review Questions

1. What physical traits show that *Homo sapiens* is like other Old World monkeys?
2. What is a dental arcade? (Hint: it's not where dentists hang out and play video games.) How is the dental arcade of a human different from that of the other great apes?
3. How are Old World monkeys and New World monkeys different?
4. How can noses be used to differentiate between primates?
5. Name four features that are shared by prosimians.
6. How do we know that vervet monkeys learn their calls?

Discussion Questions

1. Do you think that primates other than humans have culture? What evidence suggests this to you if you believe it? If you don't believe it, explain why you do not.
2. What can be learned about humans by studying *other* primates? (You had to know I was going to ask this question.)
3. What would you want to learn about primates other than humans if you were to study them?

Key Species

baboons	chimpanzees	gorillas
bonobos	gibbons	orangutans

Key Terms

alarm calls	double blind test	molars	sexual dimorphism
anthropodenial	ethology	monogamous	silverback
anthropoids	Galagidae	olfactory	simian
anthropomorphism	grooming claws	opsins	Simiformes
brachiation, brachiate	Haplorrhini	orangutan paradox	supplanting
canines	Hominidae (or hominids)	pithecophobia	Strepsirrhini
Catarrhini	homininae or (hominins)	playbacks	Tarsiidae
cones	incisors	Platyrrhini	Tarsiiformes
cognitive ethology	infanticide	Ponginae (or pongins)	tarsus
convergent evolution	kin-mediated	prehensile tail	taxonomy
critical anthropomorphism	reconciliation	premolars	toothcomb
cultural primatology	Lorisidae	primatology	wahoo contests
culture	mandible	prosimians	
dental arcade	matrilines	rhinarium	

Key Individuals

Dorothy Cheney	Jane Goodall	Francine (Penny) Patterson	Tom Struhsaker
Dian Fossey	William King Gregory	Anne E. Russon	Edward Tyson
Birute Galdikas	Carolus Linnaeus	Robert Seyfarth	Carel Van Schaik

Recommended Print and Online Resources

Amundson, R. (1987, Spring). Watson and the Hundredth Monkey phenomenon. *Skeptical Inquirer*, pp. 303–4.

Breuer, T., Ndoundou-Hockemba, M., & Fishlock, V. (2005). First observation of tool use in wild gorillas. *PLoS Biology*, 3(11): e380. doi:10.1371/journal.pbio.0030380

Cheney, D. L., & Seyfarth, R. M. (1990). Chapter 4: Vocal communications. In *How monkeys see the world: Inside the mind of another species* (pp. 98–138). Chicago: University of Chicago Press.

——, & ——. (2007). *Baboon metaphysics: The evolution of a social mind*. Chicago: University of Chicago Press.

De Waal, F. D. (1995, March). Bonobo sex and society: The behaviour of a close relative challenges assumption about male supremacy in human evolution. *Scientific American*, pp. 82–8.

——. (1997). *Bonobo: The forgotten ape*. Berkeley: University of California Press.

Foster, M. W., Gilby, I. C., Murray, C. M., Johnson, A., Wroblewski, E. E., & Pusey, A. E. (2009). Alpha male chimpanzee grooming patterns: Implications for dominance 'style'. *American Journal of Primatology*, 71(2): 136. doi:10.1002/ajp.20632+

Galdikas, B. M. (1995). *Reflections of Eden: My years with the orangutans of Borneo*. New York: Little, Brown and Company.

Gibbon Conservation Alliance. (Website). Gibbon Conservation Alliance: Goals. Available from www.gibbonconservation.org/index_engl.html

Goodall, J. (1971). *In the shadow of man*. Boston: Houghton Mifflin.

——. (1990). Through a window: 30 years observing the Gombe chimpanzees. Boston: Houghton Mifflin.

Myers, E. (1985, Spring). The hundredth monkey revisited: Going back to the original sources puts a new light on this popular story. *In Context, 9*, 10. Retrieved from www.context.org/ICLIB/IC09/Myers.htm

Russon, A. E. (1998, October). The nature and evolution of intelligence in orangutans (*Pongo pygmaeus*). *Primates, 39*(4): 485–503.

Spalding, L. (1999). The follow: A dark place in the jungle. Following Leakey's last Angel into Borneo. Toronto: Key Porter Books.

Valley Zoo School. (Website). Songs of the white-handed gibbon. Available from www.zooschool.ecsd.net/white%20handed%20gibbon.htm

Van Schaik, C. (2004). *Among orangutans: Red apes and the rise of human culture*. Cambridge, MA: The Belknap Press of Harvard University Press.

Wheatley, B. P. (1999). *The sacred monkeys of Bali*. Prospect Heights, IL: Waveland Press.

Whiten, A., Goodall, J., McGrew, W. C., Nishida, T., Reynolds, V., Sugiyama, Y., . . . Boesch, C. (1999, June 17). Cultures in chimpanzees. *Nature, 399*, 682–5.

6

Hominins before *Homo*

The Laetoli footprints, Tanzania
John Reader/Science Photo Library

LEARNING OBJECTIVES

After reading this chapter, you should be able to:

- Identify and distinguish the major bones of the arms, legs, and skull.

- Assess critically the savannah hypothesis.

- Discuss the various hypotheses concerning the evolutionary advantage of bipedalism.

- Describe the different pre-*Homo* hominins put forward as species.

- Compare and contrast the various models for defining species.

Every meeting of paleoanthropologists that I have attended has at some point erupted into intense and sometimes unfriendly debate about how many hominid species preceded us. (Johanson & Wong, 2009, p. 101)

Introduction

In this chapter you will learn a bit about the superstars of paleoanthropology and their teams of supporters. These people are brilliant, strong-willed, and often combative. They seek truth, but they're also after the 'Great Find', the elusive, ever-morphing 'missing link'—a goal that is ultimately as unattainable as a trade that will make the Toronto Maple Leafs Stanley Cup contenders.

Primatologists would see behaviour here that they recognize, reminiscent of rivalries between the Leakeys and Martin Pickford or Don Johanson that produce battle scars like those incurred by flange-faced orangutans or silverback gorillas in conflict. Their discoveries sometimes seem more like points of religious dogma or political allegiance than products of science, objects of wonder.

Of course, the chapter is about much more than superstar personalities. Their finds are more than just flags to fly in the face of the enemy, but are characters in their own right. The *Australopithecus afarensis* Lucy is far better known than her famous discoverer Don Johanson.

Finally, a warning: you are about to be presented with a confused picture. No clear lineage of early pre-*Homo* hominins has yet emerged. There are many footprints, but the trail is not yet evident.

Key Anatomical Terms

Bones are the key form of evidence discussed in this chapter, so it is useful for you to be familiar with some of the hominin bones most typically found.

The Arm or Upper Limb

In this brief tour of the arm we will proceed from the **distal** end (like *distant*, the one farthest away from the centre of the body) to the **proximal** end (like *approximate*, the end nearest the centre of the body). We begin with the **phalanges** (*sing.* **phalanx**), the finger or toe bones at the distal end of the hand or foot. In each of your hands there are 14 of these—two in the thumb and three each in digits two to five. These are followed by the **metacarpals** (*sing.* **metacarpus**), the five finger bones at the proximal end of the hand, and the **carpals** (*sing.* **carpus**), the eight hand bones (a proximal and a distal row of four each) furthest away from the fingers. Each hand, then, has 27 bones. That means that your two hands account for over a quarter of the 206 bones in your body. A hint regarding which is which: the farther up the alphabet the first letter of the bone is, the more distal it is (<u>c</u>arpal–<u>m</u>etacarpal–<u>p</u>halanx). The same goes with the <u>h</u>umerus–<u>r</u>adius/<u>u</u>lna combination.

The **radius** is one of the two long bones of the forearm. It is broad at the distal end, narrow at the proximal end, joins the hand on the thumb side, and crosses over the ulna to connect with the humerus at the elbow. The **ulna** is the other long bone of the forearm. In contrast to the radius, it is broad at the proximal end and narrow at the distal end, where it forms the bump on your wrist).

The **humerus** is the single long bone of your upper arm. It is longer and more robust than the two forearm bones, although a slight optical illusion makes it seem shorter.

Your **scapula** is your shoulder bone. It is roughly triangular in shape, with the skinny end down. It has many parts with many labels. The term you should know for this chapter is 'glenoid'. The **glenoid fossa** is the slight depression in which the humerus fits to articulate with the scapula. 'Fossa', like the word '<u>foss</u>il', comes from a Latin word for 'dig'. The **glenoid process** is the head of the scapula. Generally, 'process' refers to a part of the anatomy that sticks out (or <u>proce</u>eds forward). The angle of the glenoid area is closely monitored in studies of pre-*Homo* hominins. If it angles straight up, it's characteristic of a great ape and suggests some arboreal transportation. If it is more sideways, then it is more typical of a human being.

Finally, the **clavicle** is the shallow, elongated, S-shaped bone that connects the scapula with the sternum, the bone that runs vertically down the centre of your chest.

The Leg or Lower Limb

The lower limb begins distally with 14 phalanges. These are followed by five **metatarsals** and then the **tarsals** (*sing.* **tarsus**). The tarsals appear in two rows, but these differ from the two rows of the carpals in that the proximal row has just two bones, while the distal row has five.

The two distal long bones of the lower leg are the tibia and the fibula. The **tibia** (based on a Latin word for 'pipe'), otherwise known as the shinbone, is the second-longest bone in the body. It is longer and more robust than the **fibula**, the calf bone. 'Fibula' comes from a Latin term for 'brooch', and it's a useful analogy: think of the fibula as the thin metal pin that attaches a brooch to your shirt; the thicker tibia represents the broader decorative element of the brooch, the part that is always out in front.

The **patella** (based on a Latin word for 'pan') is the knee bone. The **femur** (*pl.* **femora**; *adj.* **femoral**) is the longest bone in the body, and the most frequently found of the **postcranial** bones (i.e. any bones not forming part of the cranium, or skull) because of its length and thickness.

One of the derived features that contributed to human evolution is the growing length of the legs relative to the arms. This change was necessary as hominins became bipedal. The relative length is calculated with the **humerofemoral index** (or **ratio**): divide the length of the humerus by that of the femur and multiply by 100 to get a percentage. Table 6.1 provides the indices for Homo and selected other hominin species.

Also important is the **brachial index**: divide the ulna length by the humerus length and multiply by 100 to get a percentage. A high percentage is generally an ape feature; a lower percentage is generally a more human feature (see Table 6.1).

Articulating with the two femora is the **pelvic girdle**, consisting mainly of two main bones called the left and right **os coxae** (*pl.* **ossa coxae**; lit. 'bone–hip'), which are literally joined at the hip. Each of these coxal bones comprises the large, blade-like **ilium**, the much smaller, similarly shaped (but upside-down) **ischium**, and the **pubis**, which is joined to the ischium; the three parts of the os coxae are separate at birth but have become fused by adulthood. The shape of the pelvic girdle is important in determining whether an early hominin was bipedal.

The Cranium, or Skull

The **cranium** (skull) is made up of six main bones (see Figure 6.3). The **frontal** is, obviously, at the front. The **occiput** (or **occipital bone**) is not-so-obviously at the back (try: 'Put the ox in the back'). The **parietals**, left and right, are on top (just like parental family members, in theory anyway), left and right. Together these comprise the four bones of the skullcap. On the sides are the **temporals** (think: 'temples'). Particularly important in fossil finds are the upper and lower jaws, the **maxilla** and **mandible**, respectively (remember: the max– is the top).

Sutures are wavy lines like medical sutures that join the bones. The suture that you need to know about here is the **sagittal suture**, which joins the parietals. It is given that name (from the Latin word for 'arrow') because of its slight arrow shape (remember: Sagittarius is the astrological sign of the archer).

Bipedalism

Bipedalism involves walking on two feet. The origin of bipedalism, like that of any species, is not sudden. Physical anthropologists sometimes distinguish between

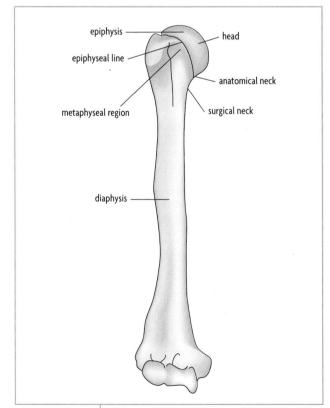

Figure 6.1 | Gross anatomy of a long bone

Table 6.1	Comparative brachial and humerofemoral indices (approximate)	
Species/fossil specimen	Brachial index	Humerofemoral index
Homo sapiens	80%	72%
Australopithecus afarensis ('Lucy')	91%	85%
Pan paniscus (bonobo)	92%	98%
Australopithecus garhi	98%	70%

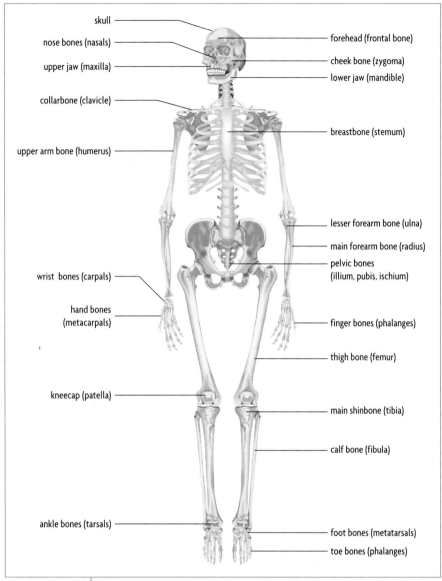

skull

nose bones (nasals)

upper jaw (maxilla)

collarbone (clavicle)

upper arm bone (humerus)

forehead (frontal bone)

cheek bone (zygoma)

lower jaw (mandible)

breastbone (sternum)

lesser forearm bone (ulna)

main forearm bone (radius)

pelvic bones
(illium, pubis, ischium)

wrist bones (carpals)

hand bones
(metacarpals)

finger bones (phalanges)

thigh bone (femur)

kneecap (patella)

main shinbone (tibia)

calf bone (fibula)

ankle bones (tarsals)

foot bones (metatarsals)

toe bones (phalanges)

Figure 6.2 | The human skeleton

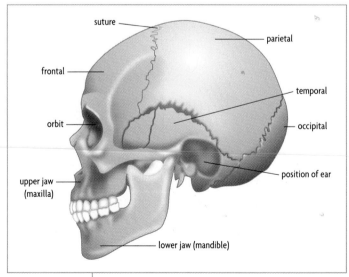

suture

frontal

orbit

upper jaw
(maxilla)

parietal

temporal

occipital

position of ear

lower jaw (mandible)

Figure 6.3 | The human skull

optional bipedalism, where an animal walks on two feet part of the time, choosing when to do so, and **obligate bipedalism**, in which the animal does not have a reasonable alternative (i.e. it is obligated) to walking on two feet. Adult parrots are optional bipeds, while baby parrots, before they can fly, are obligate bipeds. Our ancestors must have experimented with walking bipedally for longer and more frequent periods until, structurally, they had little choice.

Why did we become bipedal? The old theory, the 'savannah hypothesis' (see the box on page 150), is that with the drying and cooling of the planet and the reduction of the number of trees to hang out and swing in, we didn't really have a choice. This opinion has given way to new theories that have more to do with the advantages of bipedalism, which:

- freed the hands to carry infants (particularly useful once we were no longer hairy enough for the infant to grab onto our fur);

- freed the hands to pick and carry food;

- enabled our ancestors to be more likely to see farther and over bushes so that they could better spot and escape faster predators; and

- enabled us to be cooler in the hot sun (with less skin exposed to the rays of the sun).

The Return of Lumpers and Splitters, and the Difficulty of Naming Dead Species

In the scientific messiness of the **species concept**, the **lumpers** *vs* **splitters** debate and the matter of intellectual bias rise to prominence. Rarely do scientists declare *I am a lumper* or *I am a splitter*—this would sound unscientific, although everyone fits into one of the two camps, and some anthropologists go to lengths to promote the values of one side or the other. The paleoanthropologist Donald Johanson is a lumper regarding *Australopithecus afarensis*, though he doesn't declare his position outright. That

---

Case Study Controversies

Bipedalism and the Energy Consumption Theory

On 20 July 2007, a study of chimpanzees trained to use treadmills was published in *ScienceDaily*. The lead researcher, Michael Sockol, had five chimps use the treadmill, first 'knuckle-walking' on all fours, then walking bipedally, on two legs. Oxygen masks and electronic connections were attached to the chimps to measure oxygen and energy consumption during the exercise. The results were compared with those from four humans.

The researchers found that human walking used roughly 75 per cent less energy and burned about 75 per cent fewer calories than both the knuckle-walking and the two-legged walking of the chimps. More important were the results of the two-legged versus all-fours walking among the chimps: three of the chimps expended somewhat more energy walking bipedally; however, for the fourth chimp, the two modes of locomotion were equal in energy consumption, and for the fifth, the two-legged walking was more energy-efficient. These last two had hip and hind-leg skeletal structures more similar to those of early fossil hominins.

To view Jack the chimp doing his best for science, go to the video link http://ngm.nationalgeographic.com/2006/07/bipedal-body/video-interactive.

© National Geographic Society

Jack the chimp on a treadmill

doesn't mean that the evidence he presents is not scientific or carefully collected and thought-out; it's just that the position in a way precedes the evidence. It isn't necessarily wrong; it's just a fact.

I am a lumper. My credo is *One specimen does not make a species.* For me, the burden of proof falls on the researcher to demonstrate clearly that a new find is something significantly different from what has been found before. I will state several times in this chapter that some declared species is just another *A. afarensis* found in a different location, or that a species has experienced postmortem conditions, or mistakes in reconstruction, that make it seem more different from *A. afarensis* than it really was when it was alive. The latter can be revealed through the skilled use of **taphonomy**, 'the study of the processes that influence bone from the time of death to the day of discovery' (Johanson & Wong, 2009, p. 95). It is helpful when an anthropologist declares her or his position from the start; since this happens so seldom, you will need to keep your eyes open for signs of bias in one direction or the other.

Wood and Lonergan, helpfully, have published an article that differentiates the hominin taxa recognized by splitters and lumpers. A summary of their work is presented in Table 6.2; note that it does not include certain taxa discussed in the next chapter (Wood & Lonergan, 2008, p. 355). The splitter taxonomy has 13 species, while the lumper taxonomy has just 5.

The Species Concept and Its Difficulties

Species can be defined broadly, or inclusively, so that there is a lumping together of various diversities. Alternatively, species can be defined narrowly, with more exclusion, so that there is a fine-grained splitting and more variety. Politics are involved. Defining a species narrowly serves many good social and political ends: for example, gaining the financial support of a research institute, promoting the career of a young scientist, recognizing an older scientist, preserving a local variant of a migratory bird by preserving its habitat. On the other hand, defining a species more broadly helps researchers find a common cause, something they can parlay into larger-scale funding requests and bigger publishing opportunities.

Opposing Models of Species Definition

Two primates serve as models of how species can be either lumped or split. The **gorilla model** (not a pretty sight on the runways of Paris or Milan) serves the lumper agenda: within this single species, the animals vary widely, thanks

Table 6.2	A lumper's and a splitter's taxonomy of ancestral hominin species
Lumping taxonomy	Splitting taxonomy
Ardipithecus ramidus	*Sahelanthropus tchadensis* *Orrorin tugenensis* *Ardipithecus ramidus* *A. kadabba*
Australopithecus afarensis	*Australopithecus anamensis* *A. afarensis* *Kenyanthropus platyops* *A. bahrelghazali*
Australopithecus africanus	*Australopithecus africanus* *A. garhi*
Paranthropus boisei	*Paranthropus aethiopicus* *P. boisei*
Paranthropus robustus	*Paranthropus robustus*

Wood & Lonergan, 2008, p. 355.

largely to the tremendous **sexual dimorphism**, as we saw in Chapter 5. Humans present a good model of lumping, too: just look at, say, Shaquille O'Neil and Danny DeVito. On the other hand, the **lemur model** supports splitting. Lemurs with virtually identical skeletal structure can differ significantly in behaviour, fur colour, markings, and vocalizations. They are referred to as different species because they do not interbreed in the wild (although they might in captivity, when breeding choices are limited). According to American Museum of Natural History curator Ian Tattersall:

> Were we to step a million years into the future and study fossils of lemurs and base our assessment on only the skeletal remains, . . . we would greatly underestimate the number of species. (as cited in Johanson & Wong, 2009, p. 103)

In other words, even though we may find skeletal remains that look virtually identical, we could be dealing with different species.

Two opposing definitions of species also serve the two different ends. The **phylogenetic model** of species is a cladistic one. It is broad-based and organizes species around the existence of common ancestors. The **biological species concept** (or **reproductive model**) of species is, as stated earlier, based on Ernst Mayr's notion that a species is a collection of animals that can interbreed and produce fertile young.

Models are analogies; as such, they are imperfect representations of reality. Applying a model does not constitute scientific proof—it merely presents a logical possibility. If species X is like a gorilla, then we have one species; if it is like a lemur, then we may have two or more species. Sometimes the models work well, sometimes they do not. When the species you are trying to define has been dead for over a million years, applying this or that model can be quite difficult. Learning what model a scientist is using makes the reader a better consumer of information in the marketplace of ideas.

Unintelligent Design: Vestigial Structures and Imperfections

Earlier we spoke of how an organism's vestigial structures and imperfections could be viewed as proof of evolution. Of course, there may be no better example of this than the human body, with its many structural flaws. It is important to consider this point as we look at the features of a fossil specimen. They might not all be involved in performing well the functions of the time. They may be remnants of the past, like skeletal versions of our appendix. When scientists take the position that having curved fingers meant that *Australopithecus afarensis* must have spent a lot of time in the trees, it could just be that what they are seeing was a remnant of the past. Of course, it is easy to invoke this for 'inconvenient' features when you want a specimen to be, for example, more hominid than it might otherwise be. This idea is not evidence, but it questions the conclusions of those who automatically assume that primitive feature equals primitive being.

Fossil Hominin Species

Sahelanthropus tchadensis

Sahelanthropus tchadensis was discovered in 2001 by Chadian anthropologist Ahounta Djimdoumalbaye, a member of French paleoanthropologist Michel Brunet's crew in the Toros-Menealla area of northern Chad in central Africa. The story of this potential hominid was first published on 11 July 2002 in *Nature*, with the dramatic headline, 'The Earliest Known Hominid'. It bears the nickname 'Toumai', which means 'hope of life' in the local Goran language. The specimens could not be dated by standard argon–argon dating, so during the initial work, the specimens were dated, at between 7 and 6 mya, by matching the geological layers that the specimens were discovered in with the same layers in East Africa. Shared animal fossils, such as those of elephants, pigs, antelopes, and horses, were used. A recent (2008) development in dating involves the use of beryllium (10Be–9Be testing), which decays with a half-life of 1.4 million years and has a dating range of between 200,000 and 14 million years. When beryllium testing was performed on the specimens

of *S. tchadensis*, it produced a date of 7.04 ± 0.18 mya (ranging roughly from 7.2 to 6.8 mya).

The key fossil specimen is **TM 266-01-060-1** (TM = Toros-Menealla), an adult cranium with a cranial capacity (i.e. brain size) of between 360 and 370 cubic centimetres (cc), around a quarter of the size of the modern human brain. Mandibles and teeth have also been found. Significantly, no postcranial bones belonging to *S. tchadensis* have been found.

The fossil specimens (see Figure 6.4) demonstrate some hominin features, including the following:

- relatively anterior (i.e. forward) placement of the foramen magnum, the big hole through which the spinal cord enters the brain; this is a feature associated with bipedalism, as the head is more efficiently balanced than when the hole is farther at the back

- relatively flat face, or reduced muzzle

- relatively small canines, which display wearing on the apices (*sing.* apex, i.e. the tips) of the teeth; there is no evidence of C/P3 honing (explained below)

- relatively thick tooth enamel (somewhere between that of chimps and the australopithecines).

However, ape features also abound; these include:

- a small skull size

- massive supraorbital torus, or brow ridges

- the lack of forehead

- the bone structure of the back and base of the skull, otherwise known as the nuchal ('back of the neck') area.

Despite the ape features and the early date of the fossil specimens, Brunet claimed that Toumai was an early hominin. His claim was challenged by a team of leading anthropologists—Brigite Senut, Martin Pickford, Milford Wolpoff, John Hawks, and James Aherne—in 'An Ape

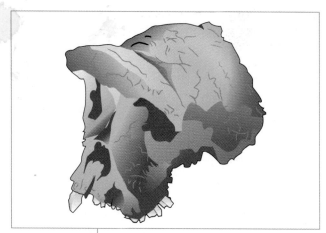

Figure 6.4 | The adult cranium of *Sahelanthropus tchadensis*

or *the* Ape: Is the Toumai Cranium TM 266 a Hominid?' (*PaleoAnthropology*, 2006). They believe it is not a hominin, though it may possibly belong to the gorilla line (as Pickford and Senut argue), the chimpanzee line, or an extinct but closely related ape line.

We need more evidence. As Brunet himself says, we require further findings in which all of the early hominin lines can be directly compared before we can say where *S. tchadensis* belongs.

Orrorin tugenensis

In the fall of 2000 **Martin Pickford** (b. 1943) and **Brigitte Senut** (b. 1954) went fossil hunting in the Tugen Hills of Kenya. What they found there they named ***Orrorin tugenensis***—the genus name means 'original man' in the Tugen language, and the species name is a Latinized version of the name of the hills. The fossil findings were situated between two volcanic layers dated at 5.7 and 6.6 mya, so it was pushing the limit of the date molecular biologists have assigned to the evolutionary fork in the road where chimpanzees/bonobos and humans (or **panins** and **hominins**) went separate ways.

Among the findings were a partial mandible, a finger bone, several isolated teeth, and—most significantly—fragmentary femurs. The most complete of the femoral finds was labelled **BAR 1002'00**. It was found in four pieces, which Pickford and Senut glued together. The pair's reconstruction efforts with the femur would raise a big stink in the anthropological community. Senut had some sense of what would happen, as she recounted to Ann Gibbons: 'I told Martin to throw it in the lake. It would only bring us trouble' (as cited in Gibbons, 2006, p. 195).

Among the remarkable features of BAR 1002'00 was the **femoral neck**, which connects the head of the bone to the shaft. The neck was long, as it is in bipedal hominins, suggesting that *O. tugenenis* walked on two legs. A supporting piece of evidence was a groove where a tendon called the obturator externus presses against the femur in bipeds. Third, there was a thickening of the cortical bone (the dense, hard mass on the outside of all bones, providing support and protection) at the lower end of the femoral neck; this was in contrast to the thinner bone at the upper end (called the superior cortex), that is put under less pressure. Chimpanzees and other quadrupeds have uniform thickness in the cortical bone of their femoral necks.

Based on these apparent signs of bipedalism, Pickford and Senut boldly claimed that *Orrorin* was an obligate biped and—even more daring—that it was more humanlike than either **Lucy** or *Australopithecus afarensis* in general, although the latter was at least two million years younger.

When Pickford and Senut and their research team published their 'External and Internal Morphology of the BAR 1002'00 *Orrorin tugenensis* Femur' in a September 2004 issue of *Science* (Galik, et al., 2004), there was a relatively quick negative reply written by **C. Owen Lovejoy** and **Tim D. White** (b. 1950), with others, published in *Science* on 11 February 2005 (Ohman, et al., 2005). They criticized the CT (computerized tomography, which images bone density) scans used for being unclear (a fair criticism), but then argued that Pickford and Senut should unglue the part of the femur connected to the femoral neck, in order to get a better scan—something that Pickford and Senut refused to do. The critics concluded by saying that:

> We agree that the femur's external morphology suggests some form of bipedality. Yet the more detailed original scans . . . appear to show a distinct superior cortex [i.e. upper femoral neck] being more primitive than that seen in any other hominid, including *Australopithecus*. . . . However, authors B. Senut and M. Pickford have claimed . . . that this femur is so derived as to exclude *Australopithecus* from direct human ancestry. . . . Given the importance of the . . . femur, we urge its discoverers to make available evidence to support

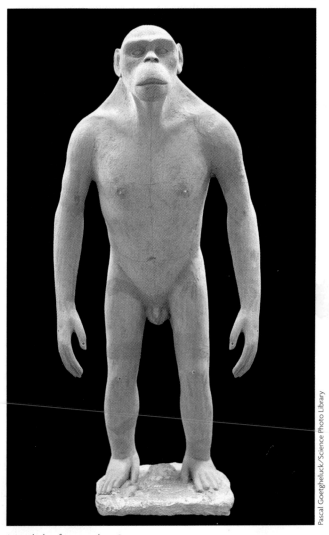

Model of a male *Orrorin tugenensis*

Pascal Goetgheluck/Science Photo Library

their assertions. . . . Exceptional claims demand exceptional evidence; the adjustment of previously published data does not suffice. (Ohman, et al., 2005, p. 845)

That last comment is a shot. But then, in a conference the previous September, Senut had taken White to task over his slow progress in analyzing the *Ardipithecus ramidus* (see below).

In an independent reassessment of *Orrorin* published in 2008, researchers Brian Richmond and William Jungers concluded:

We show that the *O. tugenensis* femur differs from those of apes and *Homo* and most strongly resembles those of *Australopithecus* and *Paranthropus*, indicating that *O. Tugenensis* was bipedal but is not more closely related to *Homo* than to *Australopithecus*. (Richmond & Jungers, 2008)

Most scholars today take this position. *Orrorin tugenensis* was probably obligate bipedal, and possibly preceded the australopithecines.

Ardipithecus kadabba

Ardipithecus kadabba was first named in 2001 as *A. ramidus kadabba*, distinctive only as an early hominin subspecies, but Ethiopian paleoanthropologist **Yohannes Haile-Selassie** and his team renamed it as a separate species in 2004. It is dated at between 5.8 and 5.2 mya—the fossil remains bracketed between two tuffs, found in the Middle Awash area of the Afar depression of Ethiopia.

The findings include an initial 11 specimens discovered in 1997 (five teeth and six postcranial bones, including toe and finger phalanges, part of a clavicle, and a humerus), along with six more teeth discovered in 2002. Indeed, the teeth of *A. kadabba* are key: they show apparent signs of an apelike feature called the **C/P3 honing complex**. Apes literally hone, or sharpen, their canines on the edge of their nearby premolars. Confusingly, the first premolar (or medial), the one closest to the centre, is referred to as P3. It's confusing because, if you'll recall from the last chapter, Old World monkeys and apes have only two premolars; however, the more ancestral mammal pattern features four premolars—over time, we lost the first two.

Along with the apelike curvature of its left foot phalanx, the evidence of C/P3 honing suggests that *A. kadabba* may be more panin that hominin—certainly more so than *A. ramidus*. The toe bone is also problematic in that it was found 25 kilometres east of most of the dental specimens and is some 400,000–600,000 years younger than all of the other fossils (Johanson & Wong, 2009, p. 160). More evidence is definitely required before we can say whether we have (a) a biped, and (b) a hominid.

Ardipithecus ramidus

Ardipithecus ramidus was first discovered between 1992 and 1993 in the Aramis area in the Middle Awash study area in the Afar depression, Ethiopia, by Tim White and his team. The dating method was laser fusion argon-40/argon-39 testing of volcanic ash; it produced a date of 4.5 to 4.3 mya for the *A. ramidus* specimens. The genus name (featured also in *Ardipithecus kadabba*) comes from *ardi–*, 'ground', plus the Greek *–pith–*, for ape; the species name comes from the Afar word for 'root', *ramid*: in other words, 'Ground ape, root'. Eventually 110 fossil specimens were discovered, making up several individuals.

The biggest problem in analyzing *A. ramidus* was the fragile nature of the fossilized skeleton, parts of which

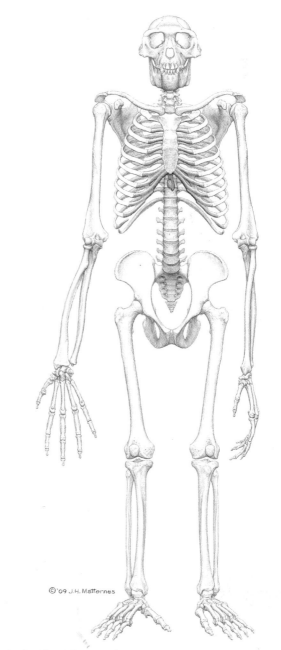

© '09 J.H. Matternes

© Ho/Reuters/Corbis

Ardipithecus ramidus

had long remained encased in stone. The skull, which had been described as being like 'roadkill', was especially delicate. Tim White took some heat for not publishing about *A. ramidus* after announcing the initial discovery in 1994. This was to change. On 2 October 2009 Tim White and his colleague Owen Lovejoy finally announced their findings concerning *Ardipithecus ramidus* in a special edition of *Science*. The long delay reflected the methodical nature of both of the lead researchers (fully aware of what can happen when you announce something 'too soon') and of the time-consuming work involved, especially the piecing together of the badly crushed skull. Eleven articles on *A. ramidus* were published in the special issue (seven of which are referenced among the recommended readings at the end of this chapter).

White and Lovejoy made a number of significant findings, summarized in the list below. Overall, their conclusions reinforce a very important point. When conceptualizing the last common ancestor of humans and chimpanzees/bonobos, it is wise—though admittedly very difficult—to avoid being drawn into thinking of that ancestor as closely resembling a chimp. The 'pull of the present' can be very deceptive that way. Here are the main findings concerning *A. ramidus*:

- small cranium (300 to 350 cc.)

- reduced male canines (and reduced canine/pre-molar complex generally)

- robust thumb musculature

- long, curved fingers

- midcarpal joint in the wrist, allowing for extensive bending backward

- flexible back (more like a monkey's than an ape's), with possibly six lumbar vertebrae rather than the three or four found in the stiffer backs of living apes

- upper pelvis modified for bipedalism

- opposable big toe

- nearly complete female skeleton, indicating a weight of about 50 kilograms (110 lbs) and a height of 120 centimetres (about 4').

These findings suggest that *A. ramidus* was bipedal but perhaps a little awkward in walking. It was still 'good in the trees', with the bending wrist, flexible back, and grabbing toe all particularly useful for arboreal travel. The reduced size of the male canines—in other hominids, a handy weapon and a meaningful threat—suggests that there was little male-to-male conflict, as there is among the apes. The reduced front teeth suggest a relatively generalized diet. Owen Lovejoy, in the authors' summary to 'Reexamining Human Origins in Light of *Ardipithecus ramidus*', speculates as to some of the implications of these changes:

Combining our knowledge of mammalian reproductive physiology and the hominid fossil record suggests a major shift in life-history strategy transformed the social structure of early hominids. That shift probably reduced male-to-male conflict and combined three previously unseen behaviors associated with their ability to exploit both trees and the land surface: (i) regular food-carrying, (ii) pair bonding, and (iii) reproductive crypsis (in which females did not advertise ovulation, unlike the case in chimpanzees). Together, these behaviors would have substantially intensified male parent investment—a breakthrough adaptation with anatomical, behavioral, and physiological consequences for early hominids and for all of their descendants, including ourselves. (Lovejoy, 2009)

This is a bit of a leap, but it gives us a meaningful hypothesis to be tested further by additional data and analysis.

Kenyanthropus platyops

Kenyanthropus platyops was discovered by **Meave Leakey** and her 'Hominid Gang' (including Louise, her daughter) in Lomekwi, West Turkana, Kenya, between tuffs that give dates of 3.5 and 3.3 mya. She presented her discovery to the world in a 2001 publication; the evidence comprised a relatively complete cranium, a maxilla, and 34 other jaw fragments and teeth.

The cranium, **KNM–WT 40000** (WT = West Turkana), is the key feature here, and an object lesson. The object lesson is about **reconstruction**. Typically when fossil specimens of crania or other relatively large parts are discovered, they are found in pieces, and have to be reassembled with great care and time. It's like putting together the ultimate 3-D jigsaw puzzle—without, however, having a picture to guide you in the process, and probably with some of the pieces missing.

K. platyops was reconstructed in such a way that the face was quite flat—**orthognathic**, to use the technical term (think of an orthodontist making your teeth flat). This put it in line with the relatively orthognathic faces of *Australopithecus robustus* and *A. boisei*, as well as with *Homo rudolfensis* or African *Homo erectus* (discussed in the next chapter). In fact, the species name, *platyops*, means 'flat face'. (The opposite term, **prognathic**—think 'protruding'—characterizes the faces of *Australopithecus afarensis* and *A. africanus*.)

Based on this flat face reconstruction, plus, to a much lesser extent, some relatively small but thick-enamelled molars, Meave Leakey and her team declared it a new genus and species, and went on to connect it to another Leakey find, Richard's baby, KNM–ER 1470, that lived some 1.8 million years ago. They were pulling the traditional

Leakey trick of putting australopithecines—*A. afarensis* (a.k.a. 'Lucy') in particular—out of the human line. Of course, any attempt to exclude Lucy will bring her discoverer, Donald Johanson, rushing to her defence:

> By Meave's own admission, the specimen was a mess when she first laid eyes on it. Grass and tree roots had pushed their way into the skull, which itself had become largely encased in a rocky matrix. Those bits of bones that were visible were riddled with tiny cracks. Meave and her collaborators spent more than a year cleaning, reconstructing, and analyzing the cranium, as well as a partial upper jaw found at the same site. And it undoubtedly looks far better now than it did when it was discovered. But I couldn't help questioning whether we can ever know the true shape of a fossil that suffered so much postmortem damage. (Johanson & Wong, 2009, p. 175)

In 2003, Tim White had taken issue with the reconstruction, citing 'expanding matrix distortion'. In essence, he argued that with pieces of the rock **matrix** occupying the cracks between pieces, the reconstruction is very likely to be distorted. Addressing White's argument, Johanson speculated:

> It may well be that if the more than 1,100 pieces making up the face alone—all separated from one another by rocky matrix—were meticulously repositioned in their proper alignment, the specimen would be seen to be nothing more than a Kenyan variant of *A. afarensis*. (Johanson & Wong, 2009, p. 175)

It is best at this point to say that more evidence is required—certainly another skull—before *Kenyanthropus platyops*, under any name, is securely handled by taxonomy.

Australopithecus anamensis

Australopithecus anamensis has been dated at between 4.5 and 3.9 mya, based on the ages of the two layers of volcanic ash it was found sandwiched between. Interestingly, the first fossil specimen, an elbow, of *A. anamensis* was discovered back in 1965. It was discovered in an ancient lakebed at Kanapoi, Kenya, by Bryan Patterson, a self-educated man who had come to hold the prestigious position of professor at the Museum of Comparative Zoology at Harvard. Nearly 30 years later, the species was discovered a second time by Meave Leakey's team in a dry riverbed connected to the same lakebed. In 1995 it received the name *Australopithecus anamensis* (**anam**– meaning 'lake' in the local language, Turkana). Another site nearby, covered with literally millions of fossilized fragments of teeth and bones from a broad variety of creatures from Allia

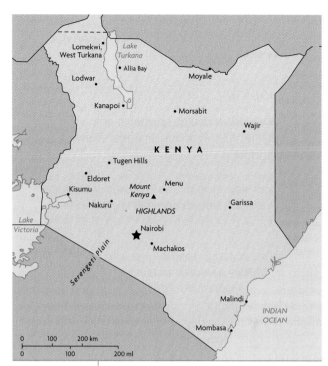

Figure 6.5 | Important paleoanthropological sites in Kenya

Bay, east of Lake Turkana and about 145 kilometres north of Kanapoi, also proved fertile in finds.

Although the two sites produced 78 fossil specimens, the star specimen is the tibia, **KNM-KP 29283**, found at Kanapoi by the amazing fossil finder **Kamoya Kimeu**. There is only a third to a half of the shinbone, the middle of the shaft being missing, but there is good evidence that it belongs to an obligate biped. Parts of the tibia that connected with the ankle and with the knee were thickened for support in ways similar to their human equivalents. In other traits, *A. Anamensis* combines both apelike and humanlike features—the small, narrow, U-shaped mandible being one of the former, the thick tooth enamel being one of the latter (although this may be an example of convergent evolution from eating tough foods such as roots and seeds).

The claim put forward by Meave Leakey and others working with her is that *A. anamensis* is a direct ancestor of *A. afarensis*, and that this could be a case of **anagenesis**: 'the emergence of a species along an ancestor–descendant lineage through time without branching' (Johanson & Wong, 2009, p. 172). Johanson, a proponent of this claim, admits that this 'would be a first in early hominid evolution'.

The evolutionary connection between the two australopithecines was strengthened when Tim White and his team uncovered 31 fossil specimens of *A. anamensis* from two sites (Aramis and Asa Issie) in the Middle Awash study area, nearly a thousand kilometres northeast from the nearest Kenyan site, and with a dating of 4.2 to 4.1 mya. The specimens include the largest hominin canine yet discovered, as well as the earliest *Australopithecus* femur. The proximity of *Ardipithecus ramidus* in the Aramis has led to suggestions of a hominin line descending from *A. ramidus* to *Australopithecus afarensis*.

Australopithecus afarensis

> When I was a teenager, I dreamed of travelling to Africa and finding a 'missing link'. (Johanson & Wong, 2009, p. 3)

> My goal was to make *Australopithecus afarensis* the single best known hominid species from anywhere in Africa. (Johanson & Wong, 2009, p. 92)

Although many people have done research on this species, ***Australopithecus afarensis*** (named after the Afar region of Ethiopia) is most associated with **Donald Johanson** (b. 1943), who found the specimen known as 'Lucy'. Her discovery was the making of Johanson's career.

A. afarensis lived between 4 and 3 mya, according to argon–argon dating of volcanic ash samples. The species is remarkable for its distribution, with specimens having been found in five major locations in Ethiopia, in two locations (Allia Bay and West Turkana) in Kenya, and in Laetoli, Tanzania. Yet it is Lucy who is the best known and most closely studied specimen. We know, from her hip and leg structure, as well as from footprints at an *A. afarensis* site, that the species was bipedal. However, the nature of the hands and curved toes of some of the early specimens and the structure of their shoulders suggest that they were at home moving in the trees as well. Their short stature (see photo) would have made walking long distances rather difficult.

LUCY

The date was 24 November 1974. At the end of a long, hot day in the desert of Ethiopia's Afar region, Donald Johanson found the specimen that would define his career. He would return to camp honking his horn in excitement. The consumption of copious amounts of beer and the playing and replaying of 'Lucy in the Sky with Diamonds', a Beatles song, would go on long into the night.

It took three weeks to take her bones from the rock, but the effort was well worth it. Johanson and his team found hundreds of pieces of fossilized bone that, taken together, represented 20 per cent of the 206 bones of an adult human (unfortunately not including her skull) and 40 per cent (the figure usually given) of the bone mass (many tiny bones of the hands and feet were not found). Lucy was the most complete hominid fossilized specimen ever found, a title she would retain for over 10 years. Judged to have lived 3.18 million years ago, she was proof that bipedalism went deep into our history. The little lady (just barely over a metre tall) quickly became famous, both locally (Ethiopians feel that she is one of them) and internationally.

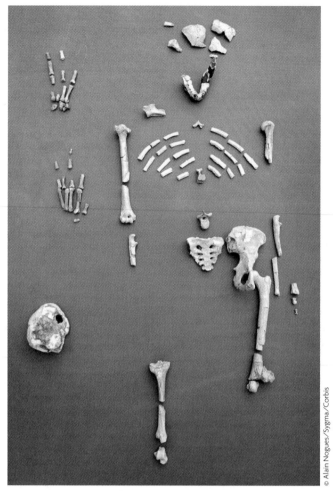

Australopithecus afarensis ('Lucy')

Johanson was now an anthropological superstar, but he soon proved to be more than a one-hit wonder. In 1975, in site A.L. 333 of the Hadar region of Ethiopia, he and his team found a goldmine: between 200 and 300 fossils, including part of a baby's skull, most of an adult male's brain case, parts of faces, portions of a foot, some articulated hand bones, and many mandible fragments and teeth. Johanson and his team reckoned that they had the remains of at least 13 different individuals—9 adults and 4 children. The title of a 1976 article in *National Geographic* would give them their name: the First Family. But Johanson's interpretation of this family would get him into trouble, as we shall see.

LAETOLI FOOTPRINTS

One of the most spectacular finds in the paleoanthropology of Africa was discovered in a most unspectacular way. In 1976, in Laetoli, northern Tanzania, Andrew Hill, a member of Mary Leakey's research team, ducked to avoid being hit by a flying piece of elephant dung thrown at him by another member of the crew (nothing builds team chemistry like a dung bomb fight). While Hill's face was focused on the ground, he noticed some footprints in the stone.

The thousands of footprints eventually uncovered, dated at between 3.75 and 3.5 mya, were found in layers of rock made up of volcanic ash. We can imagine a scene in which soft ash, wetted down by rain, formed a kind of mud in which footprints were easily left. The prints were made by many different animals, almost half of them bovines (cow types), although a significant number of them belonged to ancestors of rabbits, giraffes, and rhinos. Several of the prints belonged to hominins.

There has been a great deal of romanticizing of these footprints, which have been viewed as evidence of hominins walking together (imagine, say, a family of three strolling, hand in hand, in a misting rain). However, there is no clear evidence that the individuals who left the prints walked together. In their initial article on the footprints, Mary Leakey and R.L. Hay wrote about two of the print-making individuals who

> followed the same line or pathway but it is possible that they did not pass by at the same time as there is a noticeable difference in the conditions of the two sets of prints. Those of the smaller individual are sharp and well defined, indicating a firm, compact surface, whilst those of the larger individual, with one notable exception, are blurred at the edges and enlarged, as would be the case if the surface had been dry and dusty. At one point

Closeup of a single adult footprint from the trail of fossilized hominid footprints found by Mary Leakey's team at Laetoli, Tanzania, in 1978

John Reader/Science Photo Library

the smaller individual appears to have paused and made a half-turn to the left before continuing in a northerly direction. (Leakey & Hay, 1979, p. 320)

On what the researchers called trail 1, the length of the footprint was 18.5 centimetres (about 7 inches) and the width 8.8 centimetres (roughly 3.5 inches), while the larger footprints were 21.5 × 10 centimetres (roughly 8.5 inches × 4 inches). Keep in mind that these could be distorted to look larger than they are (like footprints in the snow after a little melting). The distance between steps was 38.7 cm for the first trail and 47.2 cm for the second.

The absence of knuckle prints shows that these hominins were not knuckle walkers. The prints of big toes were in line with those of the other toes, not sticking out as with *Ardipithecus ramidus*, a chimp or gorilla. The walkers had been bipedal for long enough that their feet had evolved significantly. While the discovery of fossilized specimens of *A. afarensis* in the same basic area has led most anthropologists to conclude that these footprints were made by *A. afarensis*, there are some who feel they were made by members of the genus *Homo*.

THE NAMING: UNITING *AUSTRALOPITHECUS* BUT DIVIDING THE ANTHROPOLOGICAL COMMUNITY

One of the most dramatic events in the history of physical anthropology was Donald Johanson's public naming of *Australopithecus afarensis*. The occasion was the 1978 Nobel symposium put on by the Royal Swedish Academy of Sciences in Stockholm. Donald Johanson, as the son of working class immigrants from Sweden, was meeting Swedish royalty and returning to the 'homeland' in triumph.

Johanson was slated to give a talk on the 'First Family'. He had discussed his findings with **Mary** and **Richard Leakey** earlier in Nairobi, and expressed his agreement with their belief that more than one species was evident in Hadar. He even was open to the idea that there might be three different species: *A. africanus*, *A. robustus*, and a species of Homo. The Leakeys, sticking to the Leakey family position, felt that there was a genus *Homo* there, and that any australopithecine found there was less important in the story of human ancestry. But Johanson and Tim White had been talking. White had worked with Mary Leakey at Laetoli, and saw a great resemblance between the fossil specimens there and those recovered at Hadar. White and Johanson approached the collected findings with a gorilla model of species diversity, using the size of the mandibles as their comparative feature. According to Johanson:

> The mandibles fell into three clusters: small, medium, and large. For the gorillas, all the large mandibles belong to males, all the small ones to females, and the intermediate group is where large females and small males overlap. Because

Donald Johanson (left) with two colleagues and a set of hominid fossils found in the Afar region of Ethiopia

the pattern of distribution is identical in both gorillas and *A. afarensis*, this helped corroborate our hypothesis that the hominids at Hadar made up a single species, with the large individuals being males and the small ones being females. (Johanson & Wong, 2009, p. 104)

At the time that Johanson made his presentation at the Nobel symposium, the standard line of anthropology's old guard, including **Phillip Tobias** (b. 1925), Raymond Dart's professorial successor at Johannesburg, South Africa, was that *Australopithecus africanus* was our ancestor. (That is what I learned as a student.) Donald Johanson got up to speak, excited about what he and White had found, and proud to feel able to announce that *A. afarensis* was our human ancestor. The response to his presentation is well described by Ann Gibbons:

As Johanson described Lucy and the Hadar and Laetoli, Mary Leakey was sitting in the audience, visibly perturbed, with a red face. . . . [S]he did not accept placing the fossils in *Australopithecus*, which she still thought was not ancestral to humans. She was angry that Johanson had publicly described fossils her team had discovered, even though she had published an initial description in *Nature*. What bothered her most was that he had labeled them all with the name

Australopithecus afarensis, with its nod to the Afar but not Laetoli. By this time, Mary Leakey's team had found more fossils of the new species at Laetoli, and she thought they belonged in the genus *Homo*. (Gibbons, 2006, pp. 94–5)

Johanson had earlier admired the Leakeys. Louis Leakey's classic piece on 'Zinj' (see page 149) had been influential in Johanson's decision to become an anthropologist. Now he was separated from the Leakeys by a divide as gaping as the Olduvai Gorge. Mary Leakey had previously agreed to have her name added to the list of authors of the scientific paper that Johanson and White would publish on their findings; she had her name removed. When she published her own article on the Laetoli footprints in 1979, she did not name the species of the fossils described, even though she admitted that they resembled the finds in Hadar. Johanson would later engage in a televised debate with Richard Leakey that turned ugly. Tim White, his collaborator, would also eventually split with the Leakeys. The lines were drawn.

DIKIKA BABY

The Dikika Baby (probably less than three years old), also known by the nicknames 'Selam' and 'Lucy's baby', was discovered by Ethiopian scientist **Zeresenay Alemseged** on 10 December 2000. The fossilized specimen was remarkably well preserved, being in Johanson's words 'the most complete early hominid baby thus far discovered'

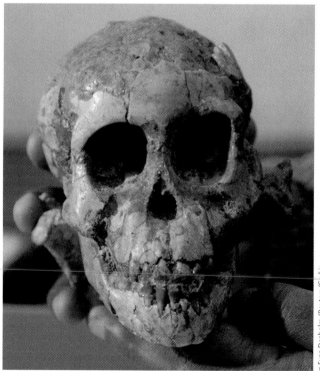

'Dikika Baby' (*A. afarensis*)

(Johanson & Wong, 2009, p. 140), a status once held by her sister *A. afarensis*, Lucy. Included are its cranium (complete with brain endocast), a mandible, and most of the postcranial bones. This even includes its hyoid, the thin bone associated with our voicebox or larynx—a rarely preserved feature.

The specimen was found in Dikika, an area not far from Hadar, in Ethiopia, 35 meters above the Sidi Hakoma Tuff and less than 50 metres below the Triple Tuff, giving it a range of between 3.22 and 3.4 million years old. A CT scan of the specimen's cranial area shows that the permanent teeth, which hadn't yet erupted when the Dikika Baby was alive, were small, suggesting that she was female. Her skull had a cranial capacity of about 330 cc.

Australopithecus bahrelghazali

In 1995 a team of French and Chadian paleontologists headed up by **Michel Brunet** (b. 1940) made an astounding discovery. They found an early hominin (between 3.5 and 3 million years old) in central Africa (in the Koro Toro region of Chad, to be exact). Prior to that, every single find dated at over 2 mya had been found in southern or eastern Africa. Few researchers even suspected that there would be anything anywhere else. The title of the article presenting the find gives a small sense of the surprise and excitement provoked by the discovery: 'The First Australopithecine 2,500 kilometres West of the Rift Valley (Chad)' (Brunet, et al., 1995). The **Rift Valley** spoken of is a long trough, some 6,000 kilometres in length, running from Mozambique in East Africa to northern Syria in the Middle East, and famously including the Olduvai Gorge of Leakey fame. Millions of years of geological and human history may be spotted along its valley walls. Just about every early hominin was found in or near the Rift Valley.

Brunet and his team didn't find much, however, merely the front (or anterior) part of an adult mandible (labeled as KT 12/H1) and an isolated tooth. They named it ***Australopithecus bahrelghazali***, deriving the complicated species name from the area, the Bahr el Ghazal, which in Arabic means 'river of gazelles'. The specimen was nicknamed 'Abel', after a geologist friend of Brunet's, Abel Brillanceau, who had died in 1989 of a drug-resistant strain of malaria.

The evidence supporting Brunet's identification of Abel as a separate species is a little weak: it has thicker enamel on the teeth than *Ardipithecus ramidus* and a very minor structural difference in the mandible; it also has three roots rather than two in the premolars, setting it apart from *A. afarensis*. I would definitely go gorilla model on this one, even human model—humans themselves differ in having either three or two roots in their molars. You can't build a good species case here. Most anthropologists think of *A. bahrelghazali* as a regional variant of *A. afarensis*.

Australopithecus africanus

RAYMOND DART AND THE TAUNG BABY

It was early in the summer of 1924. **Raymond Dart** (1893–1988), an Australian trained in Britain, was in his second year of teaching anatomy at the Medical School in the University of Witwatersrand in Johannesburg, South Africa. It was a new university with very little money, and Dart had few supplies—books, medical equipment, and specimens (bones or fossils)—for his students. So he offered up prizes for the best bones or fossils his students could find. One day, a female student told him about a family friend, a director of the Northern Lime Company (in the business of mining limestone), who had a fossilized skull on the mantel over his fireplace. At Dart's request, the young student borrowed the skull and brought it to him. It turned out to be a fossil baboon—a rare find. A geologist friend of the director's then arranged for the manager of one of the company's mines, located in Taungs, Bechuanaland, to send Dart any interesting fossils the miners might dig up.

One day, two large wooden crates arrived at Dart's house. The problem was, Dart was best man to a good friend being married that same day, at Dart's own home. He was putting on his formal clothes, struggling with his Don Cherry–type collar, when the crates arrived. Dart's wife tried to keep his focus on the soon-to-begin wedding, not the fossils. But he just had to look.

The first crate held nothing of interest. The second one was different. Only Dart's own words can convey his thoughts and feelings concerning what he found there:

> As soon as I removed the lid a thrill of excitement shot through me. On the very top of the rock heap was what was undoubtedly an endocranial cast or mold of the interior of the skull. . . . I knew at a glance that what lay in my hands was no ordinary anthropoidal brain. Here in lime-consolidated sand was the replica of a brain three times as large as that of a baboon and considerably bigger than that of any adult chimpanzee. The startling image of the convolutions and furrows of the brain and the blood vessels of the skull was plainly visible. . . .
>
> Was there, anywhere among this pile of rocks, a face to fit the brain? I ransacked feverishly through the boxes. My search was rewarded, for I found a large stone with a depression into which the cast fitted perfectly. There was faintly visible in the stone the outline of a broken part of the skull and even the back of the lower jaw and a tooth socket which showed that the face might still be somewhere there in the block. (Dart & Craig, 1959, pp. 5–6)

Dart was wondering whether he was holding the missing link when the groom reminded him of his best man's duties:

> Reluctantly, I replaced the rocks in the boxes, but I carried the endocranial cast and the stone from which it had come along with me and locked them away in my wardrobe.
>
> I could scarcely wait for the ceremony to cease and the guests to leave so that I could re-examine my treasures. When the last couple were walking down the drive to their car I was back in the bedroom tearing off my collar and tie and reaching in the wardrobe. (Dart & Craig, 1959, pp. 6–7)

Dart did not have the proper tools for the work. He had no university library with books that would guide his study, and no chisel but his wife's knitting needle to scratch the fossil out of the stone matrix. It took him 73 days before the fossil was freed from the rock. He could see that it had its first set of teeth, with its first permanent molars just beginning to come in. It was a child. Dart estimated that the cranial capacity was 520 cc (a modern assessment puts the figure closer to 400 cc). He astutely noted that the relative forward placement of the foramen magnum (the hole through which the spinal cord enters the skull) indicated that the fossil specimen was more human than ape. Dates have varied since the fossil arrived on Dart's doorstep. The estimate made by Jeffrey McKee, working out of Dart's former department, was that the fossil specimen's location below the faunal remains of animals estimated to have

lived 2.4–2.6 mya makes for a date of something between 2.6 and 2.8 mya.

What came to be called the 'Taung Baby' (or 'Taung Child', since it might have been about three years old at the time of its death) made Dart a local celebrity. According to his account, a popular joke of the time went, 'Who was that girl I saw you with last night? Is she from Taungs?'(Dart & Craig, 1959, p. 38)

Dart named his find **Australopithecus africanus** (lit. 'southern-ape african'; *Austral–* here, as in *Australia*, means 'south'). Few scientists at the time believed that it was a potential human ancestor. With their minds filled with the more modern image of the fraudulent Piltdown Man (see Chapter 1), they could not accept a relationship between humans and something that looked as primitive as the Taung Baby. Many still believed that Asia was where humans first evolved, that Darwin was wrong to tell people to look in Africa. Dart, though, had one strong supporter: **Robert Broom** (1866–1951), a Scottish doctor and palaeontologist, who would look for evidence to back up their common beliefs.

In 1938, Dart was given an adult endocast (size 485 cc) that he classified as *Plesianthropus transvaalensis* (lit. 'near-human, Transvaal region' [of South Africa]). Note that this was during a period of paleoanthropology when splitting was very much in vogue. More significantly, on 18 April 1947 at Sterkfontein, South Africa, Dart and John T. Robinson discovered the most complete skull of an *A. africanus*, although at the time he gave it the same classification as the earlier find. Dart's young co-workers would give it the nickname 'Mrs Ples' (short for *Plesianthropus*), although we can't be certain whether 'she' was female, or even an adult. In celebrity status, Mrs Ples was the Lucy of her day.

That same year, Raymond Dart would have his claim concerning *Australopithecus*'s relationship to humans accepted in the scientific community, after a wait of over 20 years. It came through the refreshing openmindedness of Sir **Wilfrid Le Gros Clark** (1895–1971), one of the anthropologists who would in a few years debunk the Piltdown Man fraud. Le Gros Clark looked at evidence, and based on that evidence changed his mind about Dart and Broom's assertion that the australopithecines were hominids. First he did the background research necessary to assess the fossil specimens:

> I had first made careful examination and taken copious notes of the skulls and teeth of over a hundred anthropoid apes in various museum collections in order to reassure myself of the normal variation that they show in their anatomical features and to compare them with the descriptions of the australopithecines already published. . . . I still remained doubtful of the suggestion that the australopithecines were hominids rather

The 'Tuang Baby' (*A. africanus*)

than pongids [apes], and I still inclined to take the latter view. . . . I felt I was going to South Africa as the 'devil's advocate' in opposition to their claims. (Le Gros Clark, 1967, p. 31, as cited in Lewin, 1987, p. 76)

He then went to South Africa, to look at the actual finds and the locations in which they were discovered:

The results of my studies were very illuminating. . . . They at last convinced me that Dart and Broom were essentially right in their assessment of the significance of the australopithecines as the probably precursors of more advanced types of [humans]. (1967, p. 31; as cited in Lewin, 1987, p. 76)

Australopithecus africanus suddenly received respectability and credibility as our ancestor.

OTHER FINDS AND FINDINGS

The sites for *A. africanus* now include—in addition to Sterkfontein—Swartkrans, Kromdraai, and the more recently discovered Gladysvale. The time period for *A. africanus* is roughly 3 to 2.5 mya (the earlier date is sometimes given as 2 mya), and the cranial capacity ranges from 400 to 500 cc, with a mean size of 460 cc.

There is still much that is problematic about the taxonomic status of *A. africanus*. Based largely on dental evidence, some feel that it evolved into the robust australopithecines (see below). Some, most notably Phillip Tobias, believe that it is ancestral to humans. In some ways—its higher body-to-brain ratio and brain size generally, for example—it seems more humanlike than *A. afarensis*. On the other hand, there is the matter of the high humerofemoral index. As we have seen, we differ from apes in having a lower index score, because our legs are significantly longer than our arms. Chimps have a high index (97 ± 2.1 per cent); ours is typically in the lower seventies. Lucy had an index 84.6 ± 2.8 per cent. Lee R. Berger argues that *A. africanus* (specifically the Sterkfontein specimen Stw 431, discovered in 1987) had a higher and therefore more apelike index. If this is true, it has several implications. One would be that *A. africanus* might not have developed from *A. afarensis*; they would belong to separate lineages, unless it evolved in this specific path independently, which seems unlikely. The other possibility would be that it is not of the human line, unless *A. afarensis* is out, and one of the big changes from *A. africanus* to *Homo* was of the reduction of scores in the humerofemoral index. This, too, seems unlikely. We will return to the index later on.

As I am writing (9 April 2010), there has been announced to the world a new australopithecine species,

Australopithecus sediba, found in a South African cave, believed to have been between 8 and 13 years old and to have lived roughly 2 million years ago. The claim being made by head researcher Lee Berger of the University of Witwatersrand in Johannesburg is that it could be useful in understanding the transition between *Australopithecus* and *Homo*. Time, and a possible second edition of this book, will tell.

The Robust Australopithecines

AUSTRALOPITHECUS (OR PARANTHROPUS) AETHIOPICUS

Australopithecus aethiopicus was discovered twice. The first time was in 1967, by French anthropologists Camille Arambourg and Yves Coppens, in the Omo region of Ethiopia. They uncovered a mandible without teeth and gave it the cumbersome name of *Paraustralopithecus aethiopicus* (lit. 'almost-an-australopithecine Ethiopian'). Not much attention was paid to this fossil specimen until Alan Walker's famous find, in 1985 in West Turkana, Kenya, of the almost complete cranium KNM–WT 17000, also known as 'the Black Skull' because of its colour, caused by its fossilization by manganese salts.

The diagnostic features of *A. aethiopicus* include a robust **zygomatic arch**, or cheekbone, and a substantial **sagittal crest**. This last feature, shared by *A. boisei* and, to a lesser extent, by *A. afarensis* and *A. robustus*, is a bony ridge down the centre of the skull (i.e. on the **sagittal**

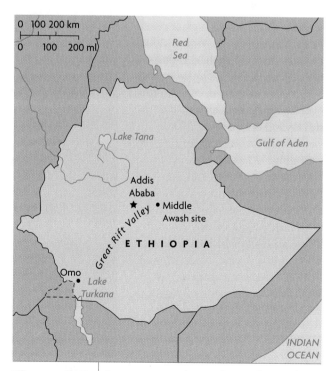

Figure 6.6 | Important paleoanthropological sites in Ethiopia

suture that joins the two **parietal bones** on the top of the head). Picture a bony Mohawk haircut, or a cockatoo with a bony crest. It anchors the muscles on the top of the head that assist the jaw muscles in chewing.

The time period for *A. aethiopicus* is poorly known because of the shortage of finds. Generally anthropologists give a narrow range of something like 2.5–2.3 mya. There also isn't total agreement concerning the connection between *A. aethiopicus* and the other robust australopithecines. Some researchers consider it ancestral to *A. robustus* and *A. boisei*, others see it as related to the latter only. In addition, it is possible (pending verification by further research) that *A. garhi* and *A. aethiopicus* might prove to belong to the same species.

Australopithecus garhi

Australopithecus garhi was discovered in 1996 by the team of Ethiopian **Berhane Asfaw** and American Tim White in Bouri, in the Middle Awash research area. The site is dated at 2.5 mya. The specimen possesses a unique combination of features that makes it hard to place among the other australopithecines. It has a humerofemoral index of about 70.4—low like that of *Homo*—but its brachial index (the ratio of ulna to humerus) is 97.9, higher than that of a bonobo (91.9) or Lucy (90.7) and very much higher than the average human (80). Its teeth are also very distinctive: it possesses a canine larger than that of any other hominin as well as—to use Johanson's words—'giant cheek teeth' (Johanson & Wong, 2009, p. 177) and an apelike U-shaped dental arcade; this is why Wood and Lonergan have put it into the category of 'megadont archaic hominins' (Wood & Lonergan, 2008).

To add to the confusion surrounding *A. garhi*, nearby there have been found several antelope bones showing definite cut marks, the oldest recorded butchering-cut marks made by any hominin. There are claims of local evidence of primitive stone tools as well. This led Asfaw to assert that *A. garhi* was an ancestor of *Homo*, a claim that is contested. Most feel that the species is an offshoot of the human evolutionary line that became extinct, but more evidence is necessary before anything definite can be established.

Australopithecus robustus

Australopithecus robustus is the first discovered member of a group typically called the 'robust australopithecines', otherwise known as Wood and Lonergan's 'megadont archaic hominins' (2008). It has two alternative genus names: Paranthropus and *Australopithecus*. My own preference, as a lumper, is to situate it within the genus *Australopithecus*. I have never read a sentence that began, 'It must be called *Paranthropus* because . . .', although that could be the starting point for a persuasive argument if anyone were to take it up with supporting evidence.

The first *A. robustus* fossil specimens—a maxillary fragment and a first molar—were found in 1938 by a young boy who worked as a guide in a cave in Kromdraai, South Africa. The fossils were purchased by Robert Broom, who had the boy guide him to the place where he had found them. It was there that Broom found a partial cranium and some mandible fragments that went together with the original finds. Broom spent considerable time preparing a report on the new species, identified as TM 1517, as he knew how skeptical the anthropological community would be. After all, they hadn't accepted *A. africanus*.

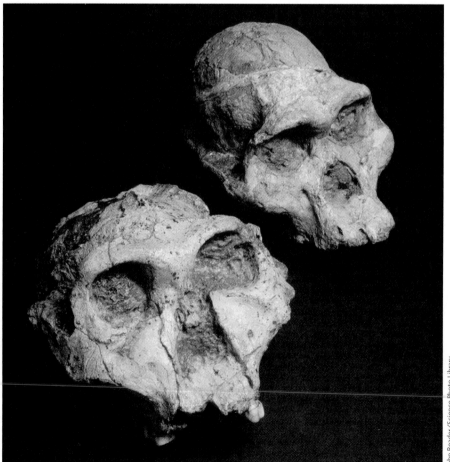

Fossil skulls of gracile (top) and robust (bottom) australopithecines; the gracile skull belongs to the specimen known as 'Mrs Ples' (see p. 146)

John Reader/Science Photo Library

He published the monograph in 1946, and it was well received. With Le Gros Clark's 1947 acceptance of australopithecines, *Australopithecus robustus* readily gained public recognition within the scientific community.

The term *robustus* comes from the fact that the chewing teeth (or back teeth) and face of the species are markedly bigger than those of *A. africanus*. These teeth seem to have developed to be specialized in chewing hard objects such as roots, nuts, seeds, and thick-skinned fruit. One problem in studying *A. robustus* is the general lack of postcranial remains. At first it was thought that the species was robust all over, compared with the gracile (i.e. slender-featured; think 'fragile') *A. africanus*. A whole classroom and popular book mythology developed about 'dumb big dude' (*robustus*) and 'smart little dude' (*africanus*) competing with each other and the smarter, tool-using little guy winning out (see, for example, the works of Robert Ardrey). As it turns out, however, the only things bigger about *robustus* were its chewing apparatus and its brain (530 cc, based on measurements taken from an endocast of specimen SK 1585, found at Swartkrans, a major *A. robustus* site). The dating for *A. robustus*—roughly 2 to 1.5 mya—is based on relative dating carried out with matching mammalian fossils specimens recovered from East African sites.

Australopithecus boisei

The initial discovery of **Australopithecus boisei** in 1959 by Mary Leakey had a huge impact on physical anthropology. The appearance of this well-preserved fossil cranium in *National Geographic* firmly established the Leakey name and reputation, and inspired a generation of future anthropologists (including Don Johanson and Tim White). The cranium was discovered at Olduvai Gorge, Tanzania, and was named *Zinjanthropus boisei* after an ancient name for East Africa (*Zinj–*) and after Charles Boise, a principal sponsor of the Leakeys' work. It provided the first major case for dating using the potassium–argon method, although the initial date of the specimen was later corrected when the techniques improved, to about 1.75 mya. The time period for *A. boisei* generally is thought to be between 2.3 and 1.2 mya.

A. boisei's thick enamel (the thickest of any hominin) and big back teeth (molars covering an area over twice the size of a modern human's) gave Mary Leakey's original find and subsequent specimens of *A. boisei* the nickname 'Nutcracker Man' (Mary Leakey's 'Zinj' was male). At the time it was believed that Zinj's unique features meant that the species lived on a regular diet of 'tough foods' such as nuts, seeds, and roots/tubers. However, more recent research, including studies of the **microwear** (note: *not* tiny underwear) patterns on the teeth suggest that the tough foods were for tough times only. The teeth may have

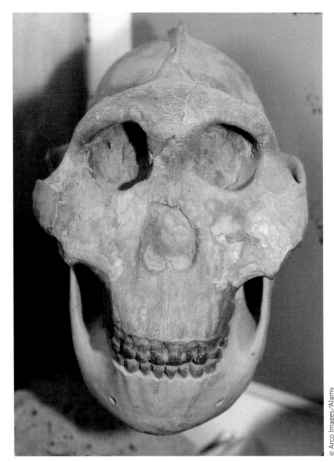

Australopithecus boisei, showing off its sagittal crest

enabled *A. boisei* to rely on fallback food when more preferred foods such as fruits were not available. Being able, however occasionally, to access tougher foods would have given the species an evolutionary advantage, something that has been noted with other primate species.

Other known features of *A. boisei* include a skull size of between 500 and 545 cc, a distinct sagittal crest, and prominent sexual dimorphism: adult males are estimated to have weighed about 68 kilograms (150 lbs) with a height of 1.3 metres (4′), while females likely weighed about 45 kilograms (100 lbs), standing a little over a metre (3′ 6″) tall.

In 1993, Ethiopian anthropologist **Awoke Amzaye** discovered fossil specimens of *A. boisei* in Konso, Ethiopia, the first found outside of Kenya and Tanzania. The finds include a partial skull KGA 10-525, the largest *A. boisei* cranium (roughly 545 cc), and a mandible. The fossils were found between two volcanic tuffs dated 1.41 and 1.43 mya, giving a possible date of 1.42 mya for the specimen. The discovery team has argued that this specimen's unique features offer compelling proof of the species' diversity; they point, for instance, to the specimen's diagnostic sagittal crest, which is emphasized more at the back than at the front, and the lack of the 'visor-like' cheekbones typically found in *A. boisei*. Their claim is contested.

Science Gone Wrong

The Savannah Hypothesis of Bipedalism

Generations of anthropologists were raised with the **savannah hypothesis**. It began with the writing of Raymond Dart, who found the Taung Baby in an area of savannah, or grassland. The story went that a drying climate change beginning about 10 million years ago caused the jungle or forest lands to give way to tall grasses and occasional bushes. The population of hominid species then in abundance diminished greatly; the survivors were those that were very good at what they did in the trees (i.e. gibbons and orangutans), were very big on the ground (i.e. gorillas), or were able to combine adept tree and ground life (i.e. chimpanzees and bonobos). Our ancestors' adaptation was bipedalism, which enabled early humans to stand tall and see above the grasses to spot predators. That was the story, anyway.

People accepted this view for decades without really challenging the specific evidence behind it. We knew, or at least accepted, that climate change generally had taken place. Many of us did not think to question whether climate change affected the specific sites where evidence of the first bipedal walkers was found. Johanson challenged the hypothesis in this way:

> Early hominids were small and bipedalism is a clumsy, very slow mode of locomotion. It seems to be that standing up and announcing that you are on the menu for some marauding carnivore would

have quickly led to our extinction. (Johanson & Wong, 2009, pp. 173–4)

More solid evidence against the savannah hypothesis has come from analyzing the sites on which a number of early hominins were found. Perhaps the best case was made by Ethiopian geologist Giday WoldeGabriel and colleagues in a 1994 publication concerning the Aramis site for *Ardipithecus ramidus*. As reported by Ann Gibbons:

> *A. ramidus* died in the woods of a tropical flood-plain that was teeming with forest-dwelling kudu, antelopes, colobine monkeys, birds, bats, rodents, and even otters. The dominance of colobine monkeys and kudos, in particular, in the same layer of sediment as the hominid fossils was a strong indication that the hominid died in a closed wooded environment, WoldeGabriel wrote in *Nature*. They also found abundant petrified wood, fossilized tree seeds, and even millipedes, which all pointed to the woods, not the open savanna. (Gibbons, 2006, p. 147)

No wonder that, in 1995, senior paleoanthropologist Phillip Tobias declared that the savannah hypothesis was dead. Unfortunately, it still lives on in old textbooks, and in some classrooms.

Summary

These are exciting times for paleoanthropologists: the pace of new discoveries is fast and furious and shows no signs of letting up, particularly now that researchers are not restricting themselves to eastern and southern Africa. And there are some encouraging trends to watch. Paleoanthropology in Africa has been steadily moving away from the old colonial activity it once was. Western non-Africans can no longer just show up, take fossil specimens, and bring them home to museums in Europe or North America. The specimens now reside in museums in their African homelands. If they travel abroad, they are returned after being studied. Just as important, Africans—particularly Ethiopians (e.g. Awoke Amzaye, Yohannes Haile-Selassie, Zeresenay Alemseged, Berhane Asfaw)—are becoming leading researchers in the field, not just lower members of a team, sidekicks who 'find' fossils but do not write about them. Museums and universities based in economically fortunate countries have been active in training Africans.

We have seen in this chapter a great deal of infighting and ego clashing, much of it not for the betterment of the discipline. Hopefully, the field is maturing, and the cowboys are aging.

What we need moving forward is an honest admitting that complete objectivity is impossible. The data do not speak for themselves. A straightforward presentation of what a researcher's biases are should be stated up front. Hi, my name is John, and I am a lumper.

Typical Student Questions

▶ *Do you ever think that we will sort out our early ancestry? Or are we just adding to the long list of finds without actually connecting the dots?*

I predict that within the next ten years we will probably be a lot closer to drawing realistic lines of ancestry. We have a better understanding of how fragile the species concept is, and paleoanthropologists are working towards a consensus on what it means to say that this or that is a distinct species or genus. Bipedalism, though—its advantages and its role in human evolution—is a thorny issue that has been a major focus of journal articles and disciplinary debate over the past twenty to thirty years, with very little clarity or consensus gained. In the concluding chapter, we'll examine a very new hypothesis on bipedalism that has made the picture even more confused. At this point I will merely say, imagine if bipedalism were a feature as old as the very existence of hominoids, or apes.

Review Questions

1. What theories have been proposed for the introduction of bipedalism among hominins? Which ones seems the most compelling?
2. What is the savannah hypothesis? How has it been proven false?
3. Describe the various pre-*Homo* hominins that have been put forward as separate species.
4. Compare and contrast the various models for defining a species. Which seems most useful to you? Do you see yourself as a lumper or a splitter?

Discussion Questions

1. Why is it important for Africans to be trained to be leading researchers in this field?
2. Where do you think that *Ardipithecus ramidus* fits into the picture of human ancestry?
3. Consider one of the splitter species and (a) present the arguments in favour of its being a separate species; (b) present the arguments against its being a separate species.

Key Species

Ardipithecus kadabba

Ardipithecus ramidus

Australopithecus
(or Paranthropus)
aethiopicus

Australopithecus afarensis

Australopithecus africanus

Australopithecus anamensis

Australopithecus
bahrelghazali

Australopithecus (formerly
Zinjanthropus) boisei

Australopithecus garhi

Australopithecus (or
Paranthropus) robustus

Australopithecus sediba

Kenyanthropus platyops

Orrorin tugenensis

Sahelanthropus tchadensis

Key Terms

anagenesis (adj.
anagenetic)

anam

bipedalism

biological species concept
or reproductive model

brachial index

C/P3 honing complex

carpals (sg carpus)

clavicle

cranium

distal

femoral neck

femur

fibula

frontal

glenoid fossa

glenoid process

gorilla model (of species
variation)

hominins

humerofemoral index
(or ratio)

humerus

ileum

ischium

lemur model (of species
variation)

lumpers

mandibles

matrix

maxilla

metacarpals
(sg. metacarpus)

metatarsals
(sg. metatarsus)

microwear

obligate bipedalism

occiput (or occipital bone)

optional bipedalism

orthognathic

os coxae

panins

parietal bones

patella

pelvic girdle

phalanx (pl. phalanges)

phylogenetic model
(of species)

postcranial

prognathic

proximal

pubis

radius

reconstruction

Rift Valley

sagittal crest

sagittal suture

savannah hypothesis

scapula

sexual dimorphism

species concept

splitters (speciose)

sutures

taphonomy

tarsals (sg. tarsus)

temporals

tibia

ulna

zygomatic arch

Key Individuals

Zeresenay Alemseged

Awoke Amzaye

Berhane Asfaw

Robert Broom

Michel Brunet

Raymond Dart

Yohannes Haile-Selassie

Don Johanson

Kamoya Kimeu

Mary Leakey

Meave Leakey

Richard Leakey

Wilfrid Le Gros Clark

C. Owen Lovejoy

Martin Pickford

Brigitte Senut

Phillip Tobias

Tim D. White

Key Specimens

Specimen	Body Parts	Taxonomy
BAR 1002'00	femur	Orrorin tugenensis
KNM–KP 29283	tibia (shin bone)	Australopithecus anamensis
KNM–WT 40000	cranium	Kenyanthropus platyops
Lucy	postcranial	Australopithecus afarensis
TM 266-01-060-1	cranium	Sahelanthropus tchadensis
Zinjanthropus		Australopithecus boisei

Recommended Print and Online Resources

On *Ardipithecus ramidus*

From *Science, 326* (2009, October 2): Special Issue.

Lovejoy, C.O. (2009, October 2). Reexamining human origins in light of *Ardipithecus ramidus. Science, 326*(5949), 74. doi:10.1126/science.1175834

Lovejoy, C.O., Latimer, Bruce, Suwa, Gen, Asfaw, Berhane, & White, T. (2009, October 2). Combining prehension and propulsion: The foot of *Ardipithecus ramidus. Science, 326*(5949), 72. doi:10.1126/science.1175832

Lovejoy, C.O., Simpson, Scott, Matternes, Jay, & White, T. (2009, October 2). The great divides: *Ardipithecus ramidus* reveals the postcrania of our last common ancestors with African apes. *Science, 326*(5949), 73, 100–6. doi:10.1126/science.1175833

Lovejoy, C.O., Simpson, Scott, White, T., Asfaw, Berhane, & Suwa, Gen. (2009, October 2). Careful climbing in the Miocene: The forelimbs of *Ardipithecus ramidus* and humans are primitive 70. doi:10.1126/science.1175827

Lovejoy, C.O., Suwa, Spurlock, Linda, Asfaw, Berhane, & White, T. (2009, October 2). The pelvis and femur of *Ardipithecus ramidus*: The emergence of upright walking. *Science, 326*(5949), 71. doi:10.1126/science.1175831

Suwa, Gen, Asfaw, Berhane, Kono, Reiko, Kubo, Daisuke, Lovejoy, C.O., & White, T. (2009, October 2). The *Ardipithecus ramidus* skull and its implications for hominid origins. *Science, 326*(5949), 68. doi:10.1126/science.1175825

White, Tim, Asfaw, Berhane, Beyene, Yonas, Haile-Selassie, Yohannes, Lovejoy, C., Suwa, Gen, & WoldeGabriel, Giday. (2009, October 2). *Ardipithecus ramidus* and the paleobiology of early hominids. *Science, 326*(5949), 64. doi:10.1126/science.1175802

General

Ardrey, R. (1976). *The hunting hypothesis: A personal conclusion concerning the evolutionary nature of man*. London: Collins.

Agnew, N., & Demas, M. (1998, September). Preserving the Laetoli footprints. *Scientific American*, pp. 44–55.

American Association for the Advancement of Science. (Website). Science/AAAS: Scientific news, research, and career information. Available from www.sciencemag.org

Brunet , M., Beauvilain, A., Coppens, Y., Heintz, E., Moutaye, A., & Pilbeam, D. (1995). The first australopithecine 2,500 kilometres west of the Rift Valley (Chad). *Nature, 378*(6554), 273–5.

Galik, K., Senut, B., Pickford, M., Gommery, D., Treil, J., Kuperavage, A.J., & Eckhardt, R.B. (2004, September). External and internal morphology of the BAR 1002'00 *Orrorin tugenensis* femur. *Science, 305*(6589), 1450–3.

Gibbons, A. (2006). *The first human: The race to discover our earliest ancestors*. New York: Doubleday.

Groves, C. (1999, May/June). *Australopithecus garhi*: A new found link? *Reports of the National Center for Science Education, 19*(3), 10–13. Retrieved from http://noanswersingenesis.og.au/cg_australopithcus_garhi.htm

Haile-Selassie, Y. (2001, July 12). Late Miocene hominids from the Middle Awash, Ethiopia. *Nature, 412*, 178–81.

Johanson, D., & Wong, K. (2009). *Lucy's legacy: The quest for human origins*. New York: Random House.

Lewin, R. (1987). *Bones of contention: Controversies in the search for human origins*. New York: Simon and Schuster.

National Geographic. (Website). *National Geographic Magazine*. Available from http://ngm.nationalgeographic.com

———. (Website). Simian similarities: Bipedal body video. Available from http://ngm.nationalgeographic.com/2006/07/bipedal-body/video-interactive

Ohman, J., Lovejoy, C. O., White, T., Eckhardt, R., Galik, K., & Kuperavage, A. (2005, February 11). Questions about the *Orrorin* femur. *Science, 307*(5711), 845.

Richmond, B., & Jungers, W. (2008, March 21). *Orrorin tugenensis* femoral morphology and the evolution of hominin bipedalism. *Science, 319*(5870), 1662–5.

Senut, B., Pickford, M., Wolpoff, M. P., Hawkes, J., & Aherne, J. (2006). An ape or *the* ape: Is the Toumaï cranium TM266 a hominid? *PaleoAnthropology*, pp. 36–50.

WoldeGabriel, G., White, T. D., Suwa, G., Renne, P., de Heinzelin, J., Hart, W. K., & Heiken, G. (1994). Ecological and temporal placement of early Pliocene hominids at Aramis, Ethiopia. *Nature, 371*, 330–3.

Wood, B., & Lonergan, N. (2008). The hominin fossil record: Taxa, grades and clades. *Journal of Anatomy, 212*, 354–76.

Early *Homo*

Collection of hominid skull casts at the Max Planck
Institute for Evolutionary Anthropology, Germany
Volker Steger/Science Photo Library

After reading this chapter, you should be able to:

- Differentiate between a lumper's and a splitter's taxonomy of early *Homo*.

- Distinguish between *Homo habilis* and *Homo erectus*.

- Discuss whether or not *Homo rudolfensis* is a species unto itself.

- Explain what *Homo floresiensis* might be.

- Outline the impact of *Homo georgicus* on the study of early *Homo*.

Introduction

You might think that as we approach the home stretch (or maybe the '*Homo* stretch') the picture would become clearer: here we are, standing at the current finish line, and we can see the hominins rounding the evolutionary corner. If you thought that, you would be wrong. The picture is very confused, and confusing (see Table 7.1). The purpose of this chapter is more to introduce you to the discussion, the debates, and the controversies, than to resolve issues.

The Brain Game

With early *Homo*, particularly *Homo erectus*, we see a huge gain in the size of the brain. Why would that happen? Brains are expensive. They use up a lot of metabolic energy, and their genetic cost is high. Although they make up only about 2 per cent of our body weight, they use up about 20 per cent of the total metabolic energy available to our bodies. They have to pull their energy weight, then, be self-supporting, not just an energy drain.

Here is one explanation. Brains bought us stone tools. These tools bought us meat. Whether the meat was the product of our ancestors' hunting or their scavenging (see the discussion that follows), the tools gave early humans greater access to this new food source—the ability to

butcher carcasses, remove meat from the bone, and cut it up to be eaten.

Meat helps the **metabolic budget** to balance as well. How? It enabled human ancestors to develop a shorter gastrointestinal tract (i.e. 'gut'). Carnivores have shorter guts than do herbivores, since animal matter is actually easier to digest (no cellulose walls to break down). Humans have guts that are unusually small for a primate of our body size. As the availability of meat and the brain size increased, the diet of vegetable matter (especially the harder-to-digest plants) and the length of the gut decreased. Budget balanced.

Meat also benefits the brain directly as it provides a good source of protein, something the brain and nervous system both require in abundance.

Homo habilis

KNM–ER 1470 and the Great Cranial Capacity Debate

KNM–ER 1470 was discovered in 1972 in the Koobi Fora region, east of Lake Turkana in Kenya. Recall that 'KNM' stands for Kenya National Museum, 'ER' for East Rudolph; the number at the end of a specimen designation is the fossil number for that site. So this specimen, a skull, was the fourteen-hundred-and-seventieth specimen found at the site. Bernard Ngeneo discovered it. He was a member of the 'Hominid Gang' that worked with the second generation of Leakeys, the husband-and-wife team of Richard and Meave.

The specimen has always been surrounded by controversy. It was found slightly below the KBS tuff, so the initial date assigned to it of some three million years ago was way off (see Chapter 2). It is now believed to be about 1.9 million years old. But that was not the only controversy. At 752 cubic centimetres, its (probably his) apparently huge brain gave researchers much to think about. Indeed, the fossil specimen helped convince many anthropologists that **Homo habilis** was not just another Leakey fantasy. After all, here was a fairly complete skull that was neither *Australopithecus* (having a smooth, rounded head) nor *Homo erectus* (lacking the big brow ridges). But in an

Table 7.1	A lumper's and a splitter's taxonomy of the genus *Homo*
Lumping taxonomy	Splitting taxonomy
Homo habilis	*Homo habili*
	H. rudolfenesis
Homo erectus	*Homo ergaster*
	H. erectus
	*H. georgicus**
	H. floresiensis

*This is added to the list.

ironic twist, KNM–ER 1470 was later named as the **holotype**—the type or definitive specimen—for another species: ***Homo rudolfensis***, named by Russian scientist V.P. Alexeev after Lake Rudolf, the old name for Lake Turkana.

The problem with the size of this specimen's cranium, as reconstructed from many pieces, is that it was much bigger than the brain size of its contemporaries that were also assigned to *Homo habilis*. For example, KNM–ER 1813, also found at Koobi Fora, has about the same date but has been given a cranial capacity of 510 cc. The original fossil specimen put forward by **Louis Leakey** and company, OH 7 (Olduvai Hominid) from the Olduvai Gorge in Tanzania, which has been designated the holotype of *H. habilis*, lived about 1.75 mya. Based on measurements of the nearly complete left parietal (top of the head) and a more partial right parietal, its cranial capacity has been given as 363 cc. However, KNM–ER 1813 was a child. Estimates are that had it lived to adulthood, it would have had attained a cranial capacity of about 680 cc. That fits fairly comfortably with the estimated cranial capacities of other fossil specimens assigned to the species *Homo habilis*, as seen in Table 7.2.

How do we deal with the difference between KNM–ER 1470 and KNM–ER 1813? Do we chalk it up to sexual dimorphism? That would accepting that our presumed male (KNM–ER 1470) possessed a larger cranium (and, thus, cranial capacity) because he had a larger body overall than that of an 'average' female of the species. OH 7 has been identified as male, so the two specimens with the largest cranial capacity are both male. Unfortunately, no direct evidence even hints that KNM–ER 1813 is female. When Louis Leakey and his team published 'A New Species of the Genus *Homo* from Olduvai Gorge' in 1964, and projected that the brain size of the juvenile male OH 7, had he reached adulthood, would have been 674 cc, they had to override the previously established (and admittedly arbitrary) line of 750 cc for genus *Homo*. In Johanson's lovely line, 'Louis was convinced that his hominid belonged there and . . . he lowered the cerebral Rubicon [referring to an ancient river whose crossing was considered significant] to 600 cubic centimetres so that OH 7 could make the cut' (Johanson & Wong, 2009, p. 188).

A recent twist on this cranial capacity story came about when **Tim Bromage** (who has both an MA and PhD in biological anthropology from the University of Toronto), questioned the reconstruction of the KNM–ER 1470 cranium. He felt that Leakey's reconstruction of the skull assigned it too human a face (i.e. one that was too flat), which would have affected how the cranium was reconstructed. Bromage took a flexible cast of skull 1470 that could be reshaped (imagine taking a balloon and distorting it). Using computer-generated models with an image of the face that was more prognathic (having a protruding lower jaw), he initially proposed a cranial capacity that was only 526 cc ± 49. This news was not well received by many anthropologists, particularly those in the Leakey camp. Bromage and his co-writers may have bent to the pressures put on him, given that when they presented their findings in the *Journal of Clinical Pediatric Dentistry*, they changed the figure to '~700 cc'. Nevertheless, the episode is a good reminder of the importance of being aware of biases when reconstructing fossils that have been dug up in many fragments. If I were reconstructing the skeleton of an early fossil crow and I strongly believed it was a parrot, it would probably end up looking like a parrot.

Two Recent Finds

The year 2007 added another wrinkle to the already apple-faced profile of *H. habilis*. Spoor and associates (including both Meave and Louise Leakey) published a journal article about two new crania they had found in 2000 in Ileret, east of Lake Turkana: KNM–ER 42700 and KNM–ER 42703. The former belonged to a young adult with an estimated cranial capacity of 691 cc, dated at 1.55 mya ± 0.01. The latter, as represented by the right side of a maxilla (upper jaw) with teeth, is slightly younger at 1.44 mya ± 0.01. While 42700 lacks some of the hallmark traits of *Homo erectus*—the thick cranial bones (or 'vault'), the large brow ridge or **supraorbital torus** (lit. 'above-the-eye bony growth')—Spoor et al. claim that it is a *H. erectus* because it has a **keeling**—a characteristic rounding or gentle curving of

Table 7.2	Estimated skull sizes of *Homo habilis*		
Specimen	Discovery date	Age	Estimated cranial capacity
OH 7	1960	1.75 mya	680 cc (projected; 363 actual)
OH 13	1963	1.6 mya	650 cc
OH 16	1963	1.7 mya	640 cc
OH 24	1968	1.85 mya	590 cc
KNM–ER 1470	1972	1.85 mya	752 cc*
KNM–ER 1813	1973	1.85 mya	510 cc

*Possibly *Homo rudolfensis*, if such proves to be a species

the skull leading to the keel edge of the parietal and frontal bones—and a small cranium size. The problem is, however, it is the smallest cranium attributed to *Homo erectus*, and you don't like to be proposing a specimen that stretches the limit.

KNM–ER 42703 is a different story. Spoor and colleagues consider it to belong to *H. habilis*, largely because the molars are bigger than those of *H. erectus* and similar in size to those of other specimens attributed to *H. habilis*. They concluded that bringing *H. habilis* this far forward in time, with a 500,000-year overlap with *H. erectus* in the same basic area, indicates that locally the two species may have exploited different resources, and, more significantly, that 'an anagenetic relationship between the two is implausible' (Spoor et al., 2007, p. 690). In other words, they do not think that, in that part of Kenya, *H. habilis* became *H. erectus*, although they do think it possible that **anagenetic** change might have taken place elsewhere; confusing.

Dating *H. habilis*

What kind of date range exists for *Homo habilis*? The figure usually given is between 1.9 and 1.6 mya. However, a 1994 find has cast doubt on conventional thinking. The specimen, a maxilla, was discovered in the Afar region of Ethiopia. Labelled AL 666-1, it was analyzed by the team of William Kimball, Don Johanson, and Yoel Rak in 1997. The maxilla had a humanlike parabolic dental arcade (as opposed to the U-shaped arcade characteristic of australopithecines and apes) and was found near stone tools. It has been dated at 2.33 million years old. In Johanson's immodest words, 'My team had found nothing less than the earliest well-dated evidence of *Homo* and the oldest ironclad association of hominid remains with stone tools' (Johanson & Wong, 2009, p. 196).

Final Thoughts on the 'Handyman'

How do we conclude this confusing section? C. David Kreger, a graduate anthropology student at Rutgers University in New Jersey and a prolific online writer whose material is well represented at www.archaeologyinfo.com, offers the following view:

> *Homo habilis* is a very complicated species to describe. No two researchers attribute all the same specimens as *habilis*, and few can agree on what traits define *habilis*, if it is a valid species at all, and even whether or not it belongs in the genus *Homo* or *Australopithecus*. Hopefully, future discoveries and future cladistic analyses of the specimens involved may clear up these issues, or at least better define what belongs in the species. (Kreger, n.d.)

When *Homo habilis* was declared a species by Louis Leakey in 1964, the fact that it was associated with tool use was an important factor in the decision. The name *Homo habilis* means 'handy man'—handy, that is, with the tools found near, and believed to have been used by, the fossil specimen. In fact, the determination of whether a given specimen belongs to *Homo* or to *Australopithecus* (or any other non-*Homo* genus) seems to me to be based more on the specimen's presumed behaviour (i.e. whether it made and used stone tools) than on any specific skeletal measurement. Now that we know that other primates (and parrots) use tools, perhaps we should rethink how we make the decision of who is and isn't *Homo*. Or perhaps we shouldn't make such a big deal about 'the great leap' towards being *Homo*.

Homo rudolfensis

For me, the jury is still out on *H. rudolfensis*. First of all, as noted earlier, the holotype is said to be KNM–ER 1470. That's not good, since it's the very specimen the Leakeys identified as *Homo habilis*. And what else do we have that could be *H. rudolfensis*? There are mandibles from East Turkana discovered in the early 1970s. One of these, a mandible found at Uraha, in northern Malawi (UR 501), was the subject of a 1997 article assigning it to *H. rudolfensis* (Ramirez Rozzi, Bromage, & Schrenk, 1997). It had been recovered from an area where simple stone tools (called 'pebble tools' because of their small size) of a type found in the Olduvai Gorge had been discovered during the mid-1980s. The tools, unfortunately, have a wide range of potential dates, but the age of UR 501 was exciting, as it was dated at around 2.4 mya. The researchers—Fernando Ramirez Rozzi, Tim Bromage, and Friedemann Schrenk—had used the enamel of the back teeth as the main point of analysis. The lateral (side) enamel on the premolars was as thin as that found in early *Homo*, but the enamel on the molars was as thick as that found among the robust australopithecines. However, this doesn't match well with the most complete of the mandibles attributed to *H. rudolfensis*: KNM–ER 1802. In their article, Ramirez Rozzi, Bromage, and Schrenk suggest this is due to variation between populations of *H. rudolfensis*; to me, it seems like more evidence that it is really *Homo erectus* being discussed.

Homo erectus

The story of *Homo erectus* is essentially a tale of three areas: Africa, China, and Indonesia. We will examine the three areas separately, with a few generalizing points at the beginning. The first thing to note, as we did at the outset of the chapter, is the increase in the brain size of *Homo erectus* over that of its predecessors. Generally it ranges from between 850 cubic centimetres and 1,200 cc, the latter being not much smaller than the current *Homo sapiens* average.

Do They Have Thick Skulls?

A number of the diagnostic traits that distinguish *Homo erectus* (and its various subspecies) from the pre-*Homo* species and from *Homo sapiens are* what can be called the 'armouring' of the skull, which makes it appear like the shell of a turtle, a hockey helmet, or, perhaps more closely, the helmet worn by German soldiers during the Second World War. These traits include a thick cranial skull and four features described and named (in Latin) by anatomist and physical anthropologist Franz Weidenreich (1873–1948):

1. the torus supraorbitalis (or supraorbital torus),

2. the torus angularis (or angular torus),

3. the torus occipitalis (or occipital torus), and

4. the crista sagittalis (or sagittal crest).

The Latin word *torus* can mean 'knot', 'bulge', or 'mound', and is usually given in anthropology textbooks as 'ridge'. The **torus supraorbitalis** is the brow ridge, a large bulge over the eyes (*supraorbitalis* means literally 'above the eye sockets').

The **torus angularis** is a thickening or bulging of the top line (or superior edge) of the temporal bone as it approaches the occipital bone in the back. The **torus occipitalis** is a continuation of that ridge along the back of the skull on the occipital bone. The **crista sagittalis** (the Latin word *crista* is a cognate of the English word 'crest') refers here to a sagittal keeling of the parietal bones, which is quite distinct from the sagittal ridge of the robust australopithecines. There is also a thickening of the frontal and the parietals that extends to the torus occipitalis.

Boaz and Ciochon (2004) theorize that the armouring of the skull was useful for protection against attacks by other *Homo erectus*—defence against the famous 'blunt-force trauma' of murder mysteries. The **dura mater** (lit. 'hard mother', as it is the toughest layer) is the outermost of the three layers of the meninges, or the membranes lining the inside of the skull and surrounding the brain. It forms an inner level of protection for the brain. A blunt trauma wound can cause blood to accumulate (as a hematoma) either in the space between the inside of the skull and the dura mater or else underneath the dura mater (making the hematoma *subdural*); either one will diminish the functioning of the brain (at the very least). Thus, having thicker cranial bone in the four skull parts described above would protect the brain from damage that could restrict neural functioning or cause death in the event of a blow to the head.

H. erectus had other forms of protection from typical human injuries, according to Boaz and Ciochon (2004). One of these is the **torus mandibularis** (or mandibular torus), a thickening of the bone on the part of the jaw where it commonly breaks in human injuries. Another is the unique structuring of the middle meningeal arteries

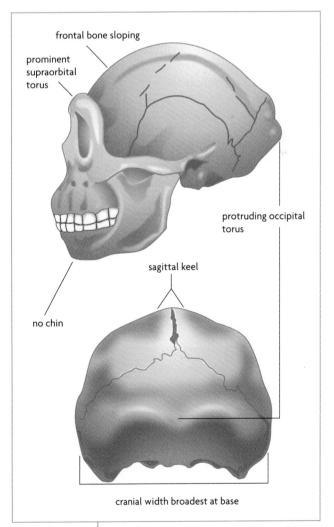

frontal bone sloping

prominent supraorbital torus

protruding occipital torus

sagittal keel

no chin

cranial width broadest at base

Figure 7.2 | Skull of *H. erectus* as reconstructed by Franz Weidenreich

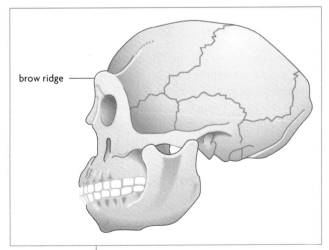

brow ridge

Figure 7.1 | Torus supraorbitalis (brow ridge) of the skull of *H. erectus*

running inside the skull: by having the larger branch of the artery running to the back where there is more protection than in the front (where the larger branch is with humans), bleeding inside the skull could be minimized.

Now, Boaz and Ciochon are just speculating here. They have little real proof that the main reason for the thickening of the skull was to provide protection against attacks by other members of the species. They do note that Weidenreich found 10 depressions or defects in the skull he reconstructed that he thought were due to hominid agency, but they acknowledged that more evidence is needed. Fortunately, as good scientists, they proposed a testable hypothesis:

> We predict that future *Homo erectus* sites will hold evidence, even more direct than the healed depressed fractures of the Longgushan skulls, of past intraspecies conflict and aggression. (Boaz & Ciochon, 2004, p. 171)

The Movius Line

Probably the most distinctive artifacts associated with *Homo erectus* are the **Acheulean** hand axes. The term Acheulean is used to designate the main Lower Paleolithic culture of Europe and Africa, living between 1.65 million and possibly as recently as 100,000 years ago (see also Mode 2 tools in chapter 9). It is characterized primarily by the crafting and use of these stone tools, which are bifacial (i.e. worked on both sides) with a distinctive teardrop shape. The Acheulean tools are typically believed to represent a higher order of thinking than was necessary for making earlier tools. With the hand axes, a definite shape and clear feature are in the maker's mind before the tool is made.

Earlier toolmaking is often thought to have involved simply creating an edge of a rock to produce a crude chopping or pounding tool, nothing quite so advanced as the hand axes of the Acheulean culture. These rudimentary hand tools, the so-called 'pebble tools', were associated initially with the Olduvai Gorge, from which we get the adjective Oldowan (think 'old one'), used to characterize

this earlier (2–1.5 mya) Lower Paleolithic tradition of toolmaking (see chapter 9, Mode 1 tools).

In the 1940s, the American archaeologist **Hallam L. Movius** (1907–1987) noted that the Acheulean hand axes were found in Africa and in Europe but not in China or Indonesia. He proposed a theoretical boundary separating, geographically, the 'haves' from the 'have nots'; the division is commonly known as the **Movius Line** (see Figure 7.3).

Movius's proposal got some people speculating about differences in intelligence on either side of the theoretical east–west divide. Before we let this kind of thinking creep in too far, here are some alternative theories to consider.

THE MINERALOGICAL HYPOTHESIS

Dragon Bone Hill is the nickname given to Zhoukoudian, the cave system in Beijing, China, where Peking Man (*Homo erectus pekinensis*) was discovered. In describing the site, Boaz and Ciochon (2004) explain that the local rock that would have been available to the *Homo erectus* was quartz and sandstone, neither of which could be used to make hand axes of the size and shape of those found in the Acheulean tradition:

> Quartz is an abundant crystalline rock that gives a sharp edge when broken, but it is notoriously poor stone for flaking into large or complex tools. Quartz is shot through with cleavage planes that cause the stone to break into unpredictable shapes, frustrating even the most adept or artistically ambitious of stone-tool knappers. The Longgushan hominids [of the Dragon Bone Hill site] had to settle for small flakes of quartz, which they used for slicing and for scraping muscle off bone. (Boaz & Ciochon, 2004, p. 102)

For work that required larger tools (such as cutting through bone and cartilage, or for chopping generally) there was sandstone, but it, too, was not the best material for making hand axes:

> Sandstone occurs in the Zhoukoudian region, but it is not a crystalline rock—it does not fracture like thick glass or give a sharp edge. But for cracking ribs by using brute force, it is effective. Sandstone just cannot be made into a recognizable bifacial hand ax. (Boaz & Ciochon, 2004, p. 102)

THE ALTERNATIVE TOOL MATERIAL HYPOTHESIS

Another argument is that perhaps we are over-glorifying fancy stone tools. Perhaps the Asian *Homo erectus* populations used sophisticated tools made of other materials.

Generally, the use of bone and antler for making tools is considered a technological step up from stone tool use. The problem is that bone and antler tools are less likely than

Acheulean hand axe

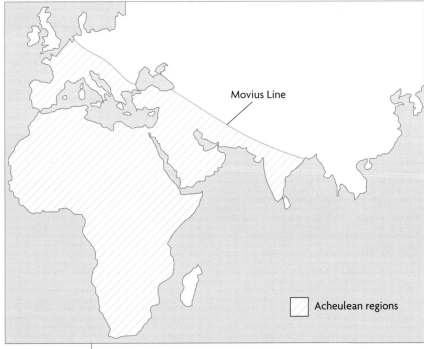

Figure 7.3 | The Movius Line

stone implements to survive the long period since *Homo erectus* died off. Perhaps the Asian populations used bamboo, which was and is readily available for a variety of purposes. The problem with a hypothesis like this is that until the tools made from this material are found, there is no proof.

Dragon Bone Hill: Zhoukoudian, China

© Dean Conger/Corbis

Hunters or Scavengers, and Who Killed the Elephant? A Case of Direct and Indirect Evidence

When archaeologists and physical anthropologists find evidence, they may go through one or several steps to reach their conclusions. They're like detectives at a murder scene. Of course, if you've ever watched or read a good murder mystery, you know that the one-step conclusion—that first accusation levelled by the bumbling representative of Scotland Yard—is often wrong; it's the village busybody who, having conducted her own, far more studious investigation, gets it right in the end.

When cut marks from stone tools appeared on the bones of large animals such as elephants found in association with *Homo erectus pekinensis*, two conclusions followed. The immediate one, based directly on the evidence at the scene, was that *H. e. pekinensis* butchered and ate the animals. The more indirect conclusion, an extension of the first, was that *H. e. pekinensis* had also been involved in killing the animals before butchering them. But could that have been true?

While we can readily picture the hunt scene, the evidence at hand was still indirect—circumstantial, if you will—rather than direct. Bite marks were found on the bones. But while this suggests that *Homo erectus* was not the hunter, this evidence alone is not sufficient—for example, other animals could have scavenged the bones

of *H. erectus*'s prey. It's what happened first that is key. On the other hand, the cut marks appear to overlie the bite marks—in other words, the cuts are on top of the bites. Thus, we have more evidence that the biter was the killer, and the cutter was the scavenger.

Then we need to consider the question of how *Homo erectus* would kill animals as hunters. Walker and Shipman point out that

> . . . everything in the Acheulian tool kit appears to be hand-held. It is difficult to envision how an animal can be killed using a stone tool that you have to hold in your hand. . . . (Walker & Shipman, 1996, p. 195)

If not for killing, then what would these tools be good for? It seems to me that cutting and general butchering would be best served by the Acheulean hand axes that were the diagnostic material cultural trait for much of the *Homo erectus* world. Of course, even this argument has a serious weakness, as Walker and Shipman point out, since as archaeologically 'invisible' tools such as 'wooden spears, snares or traps' could have been employed. We may never know for sure; for now, it's filed under 'unsolved cases'.

THE EARLY DEPARTURE HYPOTHESIS

Swisher, Curtis, and Lewin question the alternative theories we've considered so far (2000, p. 100). They have put forward their own hypothesis, based on the very early dates (e.g. 1.81 mya) that they established for East Asian *Homo erectus*. Perhaps, they suggest, the East Asian population originated in Africa, but their ancestors left the continent long before hand axes were invented there. Having little to no contact with Africa–Europe–West Asia, they simply did not learn about the manufacture and use of hand axes.

THE SOCIAL COMPLEXITY HYPOTHESIS

In 2003, archaeologist David E. Hopwood published a paper in which he stated that about 800,000 years ago, there was a distinct clustering of African *Homo erectus* sites, a pattern he associates with a higher degree of social complexity (i.e. social interaction between groups, and a larger network for the trade of goods and exchange of ideas). By contrast, he found no such clustering for East Asian sites. Thus, the social organization of East Asian *H. erectus* may not have been

conducive to the spread of knowledge about the use and manufacture of stone tools. Associated with this, Hopwood found that the stone used for making tools was transported greater distances throughout the African sites than among the Asian ones. Hence, the African sites were better situated for innovations such as the manufacture of hand axes.

The Big Complication

What if tools that are similar to or 'compatible with' Acheulean hand axes were found east of the Movius Line? In 2000, Yamei and colleagues published a paper announcing the discovery of large stone tools of significant similarity to the Acheulean hand axes in southern China, along the Vietnam border. Granted, the find provides evidence for only one time and place, about 803,000 years ago. The site has been dated by analyzing tektites, glass-like stones thought to be made from meteorites. In fact, a meteorite might have created the opportunity for tool use, as there is evidence of widespread forest fires (such as might be associated with a meteor impact) that, by burning the overgrowth, may have left stony outcrops from which the tools were made.

Homo erectus Finds in Indonesia

Eugene Dubois and the Discovery of *Homo erectus*

The story of **Eugène Dubois** (1858–1940) has been told many ways, but often focusing on the less admirable of his personality traits: paranoia, obsession, bitterness. I will follow the approach taken by Pat Shipman in her biography of the Dutch paleoanthropologist, *The Man Who Found the Missing Link: Eugène Dubois and His Lifelong Quest to Prove Darwin Right* (2001), and portray him in the more heroic light I believe he deserves.

Dubois was an idealistic young man, completely under the spell of the idea of finding the 'missing link'. As a schoolboy he asked himself, 'What is the most important thing yet to be discovered? What is the best proof of evolution? Where can the biggest contribution be made?' (Shipman, 2001, p. 21). In time he would find the answer to these questions, on the Indonesian island of Java.

Initially Dubois travelled a very straight path to success. At 19, he went to study medicine at the University of Amsterdam. By 1881, when he was 23, he was offered an assistantship in teaching anatomy, a prestigious position albeit the lowest rung of the professorial ladder. By 1884, he had advanced to the next step while at the same time qualifying as a doctor; he was successfully pursuing two parallel roads to a bright future. Within another two years, in 1886, he had received a lecturer's position—the second highest rung—and been married. His and his wife's first child was born the next year. Respectability was his.

In her biography, Shipman records a fundamental belief of Dubois': 'To achieve great things, one must cast aside the unimportant and the sentimental, one must follow truth' (Shipman, 2001, p. 20). Scientific truth and his pursuit of the missing link would take Dubois far from the Netherlands, and far from the easy life he was then leading. In December 1887, he, his wife, and their child landed on the island of Sumatra, in what was then the Dutch East Indies (now Indonesia). It was in the tropics, where there were gibbons and orangutans—it felt like the right place to find the missing link.

But a tough route lay ahead of him. He had asked for government sponsorship to subsidize his research, but officials turned him down, citing a depressed economy. He was forced to take a position as an army doctor. In his first half-year, Dubois' medical duties took up all of his time. In May 1888, he was transferred to central Sumatra, where he had readier access to the caves in which he thought he might find fossils. He found bones all right, but none of them were hominid.

Eugène Dubois' reconstruction of the head of 'Java Man'

In March 1889, he was relieved of his medical duties and transferred to the Department of Education, Religion, and Industry, which paid him to pursue the missing link. He had two junior engineers to assist him and had a crew of 50 convicts to do the digging. One engineer died, the other had to be fired, seven convicts ran away or were dismissed as useless, and about half of the rest got too sick to work. And the caves produced nothing remotely resembling a missing link.

But Dubois' luck was about to change. In April 1890 he transferred to the island of Java, away from the unproductive caves of Sumatra but near the Solo River. A little more than a year later, and close to four years after his big career shift, his team of diggers found a skullcap along the river, near the village of Trinil. He had finally found what he'd been looking for. The next year, not far away, Dubois found a left femur that he believed (incorrectly) belonged to the owner of the skullcap. Believing it to be the leg bone of a bipedal walker, he named the specimen *Pithecanthropus erectus*, 'ape-man upright'. His life peaked at that point: he had found his life's dream. In a short time it would become better known as **Java Man**.

Dubois' conviction that he had found the missing link was not generally accepted. The primitive look of the skull didn't help. In 1895, Dubois and his family left Indonesia for good. The University of Amsterdam would reward him first with an honorary doctorate in botany and zoology in 1897, and then in 1899 with a full professorship in geology. Despite having a professorship in the latter area, he continued to pursue his studies in comparative anatomy, doing important (though at the time largely unrecognized) work in **encephalization**, measuring the ratio between brain size and body size. This ratio, even more than brain size by itself, is a good measure of an animal's intelligence. A good illustration comes from comparing the different encephalization ratios of parrots, with their big heads and big intelligence, and mourning doves, with similar body sizes but much smaller heads, and a lot less intelligence.

Stories about Dubois' bitterness at not getting the full recognition that he needed and felt he deserved have some truth but a lot of exaggeration to them. He rejected parallels between what he found and what was found at Dragon Bone Hill. A man of conviction, he is justly called a major and heroic figure in physical anthropology.

Mojokerto Child

'Mojokerto Child' is the nickname given to a specimen—a fossilized skullcap designated 'Perning 1'—that was discovered by **G.H.R (Ralph) von Koenigswald** (1902–1982) in 1936 on the Indonesian island of Java. It has been assigned several different genus (*Australopithecus*, *Pithecanthropus*, *Homo*) and species (*erectus*, *modjokertensis*) identifications. It is my belief that it can be called *Homo erectus modjokertensis*, representing a distinct subspecies of *H. erectus*.

The date for Mojokerto Child was very tentative at first, especially as the exact location of the find was difficult to discover later on, once proper dating methods could be applied. With due diligence and the help of some brilliant insights (particularly those of the geochronology wizard Garniss Curtis) a team of researchers led by Carl C. Swisher III calculated the now accepted date of 1.81 mya ± 0.04 for Mojokerto Child (Swisher, et al., 1996). The cranial capacity and age at death have both proven more elusive. The mostly complete cranium that makes up the fossil specimen is still embedded in the stone matrix that it was found in. Using a computer topography (CT) scanner, a group of researchers (Coqueugniot, et al., 2004) recently reckoned that the brain size was 663 cc.

How big Mojokerto Child's cranial capacity would have become and how old it was at death require two of the most difficult calculations to make with early hominid juveniles. Brain growth rate from time of birth to one year is very different for humans and other primates. Fundamentally, the human brain at birth is about a quarter of its eventual size; that's much smaller than with, say, macaques, whose brains at birth are already 70 per cent of their eventual adult size, or chimpanzees (40 per cent). Human infants maintain the rapid brain growth of their time in the womb until they are about 1 year old. By that time they have achieved roughly 50 per cent of their adult brain size (compared with 80 per cent for 1-year-old chimps); by the age of 10, the figure is about 95 per cent. Thus, in terms of brain growth, it is as though every human is born about a year premature. This makes sense when you consider the opposing needs of being

Figure 7.4 | Important paleoanthropological sites in Indonesia

bipedal (which requires the hips to remain fairly narrow) and giving birth (which is easier with wider hips): the greater the size of the infant's brain, the larger the size of the skull, the more difficult childbirth is for the narrow-hipped, bipedal mother. Put crudely, if the baby is not to get its fat head stuck during birth, it has to have a less-than-full-sized head. Smaller brain size is one reason why human infants are helpless longer than other mammal newborns (another is the sheer volume of things that must be learned to survive as a human).

Back to Mojokerto Child. How can we tell how old it is without teeth—one of the best ways of aging the skeleton of a child (none of the fossil specimen's teeth survived)—and project its adult cranial capacity without knowing the brain growth rate of the species? Coqueugniot's team used three different measures of early skull maturation to arrive at two conclusions. First, they reckoned that Mojokerto Child was about one year old. Second they felt that its brain growth rate was still relatively slow, more apelike than humanlike in speed. Look at it this way: Mojokerto Child was an early *Homo erectus*, so its brain size, had it grown to be an adult, would probably not be more than 900 cc. At a year old, it was already 663 cc, almost 74 per cent of that full size. This, in turn, suggests that *Homo sapiens* may have not yet developed some of the more sophisticated cognitive skills, but this is very speculative.

Sangiran

The Sangiran area of central Java, Indonesia, has been very productive for *Homo erectus* fossil specimens. It began with Ralph von Koenigswald's discovery of a mandible in 1934—the first early hominid discovery in Java since Dubois' finds. In 1937, he found a skullcap, or upper cranium, (designated 'Sangiran 2') at the site. There have been several controversies concerning the Sangiran finds from that period and later, most surrounding the large size of those mandibles classified as *Meganthropus* (discussed on page 173). Another concerns the date of the specimens, which was initially believed to be about 700,000 years ago but was later pushed back to 1.66 mya ± 0.04 by Swisher and

Table 7.3	Sangiran cranial capacity
Specimen	**Cranial capacity (cubic centimetres)**
Sangiran 12	1,059
Sangiran 17	1,004
Sangiran 4	908
Sangiran 10	855
Sangiran 2	813

Rightmire, 1990, p. 195

Curtis. A third controversy surrounds the maxilla known as Sangiran 4. It appears to have apelike features such as projecting canines and the **diastema** (lit. 'space between', a gap separating teeth of different functions) that goes along with that; some researchers have even called it a pongid. But these features are not that unusual for *Homo erectus* in Java. It may just be part of the subspecies development that seems to surround Sangiran.

Ngandong

The romantic-sounding Solo Man was attached to the discoveries in the early 1930s by the Dutch Geological Survey near the village of Ngandong, close to the Solo River in central Java. The find was rich in skulls, including 12 hominid crania or cranial fragments. The brains were larger than usual for *Homo erectus*, and, as von Koenigswald noted early on, there was perhaps too much vaulting in the front of the skull for a *H. erectus*. However, Solo Man seemed rather distinct from *H. sapiens*, having, among other *H. erectus* features, large brow ridges.

Originally dated at between roughly 400,000 and 100,000 years ago, the Solo Man specimen was more recently dated using electron spin resonance of uranium-238 at between 27,000 ya ± 2,000 and 53,300 ya ± 4,000 (Swisher, et al., 1996), which would make it by far the most recent form of *Homo erectus*. It has often been classified as a subspecies of *H. erectus*: *Homo erectus soloensis*.

Homo erectus Finds in China

Homo erectus pekinensis: Peking Man

Homo erectus pekinensis, or **Peking Man** as it was first famously named, is a good source of stories. There are stories of fossilized 'dragon bones' found in Chinese pharmacies, a guaranteed ancient remedy for many ailments. There is the story of civil war threatening the digging and the diggers, the science and the scientists. There is the story of the heroic Torontonian, Davidson Black, who died for science. And then there is the unresolved story of the fossils' disappearance.

Four geographic names are associated with Peking Man. Beijing (Peking) is the city nearest the excavation site, some 50 kilometres away at the time of the research. Zhoukoudian (variously spelled in the literature), meaning roughly 'shop on the Zhoukou [river]', is the closest village. The limestone quarry from which the fossil specimens of Peking Man were retrieved has two specific sites. One was a pillar of limestone and clay called Jigushan, or 'Chicken Bone Hill'. The hill with the cave from which most of the material was retrieved is known as Longgushan, which means 'dragon' (*long–*) 'bone' (*–gu(o)–*) 'hill' (*–shan*).

Enticed by the dragon bone stories to travel to China and having already seen a 'dragon tooth', unearthed in

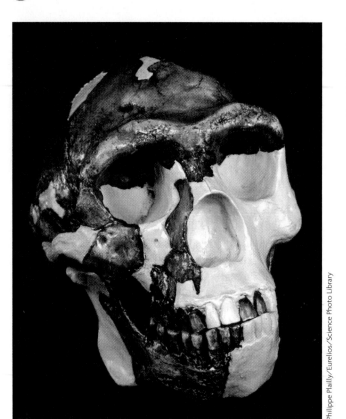

Philippe Plailly/Eurelios/Science Photo Library

'Peking Man' (*Homo erectus pekinensis*)

1899, that proved to be that of a hominid, Swedish geologist **Johan Gunnar Andersson** (1874–1960) responded enthusiastically when he learned of a discovery of fossils near Zhoukoudian in 1918. In 1921 he travelled to China, bringing with him an assistant, the Austrian Otto Zdansky. It was Zdansky who would find the first hominid tooth there, a molar, in 1921; repeating his limited success in 1924 or 1925 with the discovery of a fossilized hominid premolar. For reasons unknown, he did not tell Andersson about his finds. In fact, Andersson would not learn about Zdansky's discoveries until 1926, when he heard about it from a colleague of a colleague.

That year, 1926, was a landmark year for Peking Man. An international conference with attendees that included the crown prince of Sweden was held at the John D. Rockefeller–funded Peking Union Medical College in October. A key player at that conference was Canadian Davidson Black, who taught anatomy at the college. It was there that the name 'Peking Man' was born, and publicized enough to secure funding for research in the years to come.

The following year, Swedish paleontologist Birger Bohlin found the tooth that Davidson Black would name a species after (see box on page 168). The discoveries mounted. In 1928, Bohlin, assisted by **Zhongjian Yang** (1897–1979), newly graduated from the University of Munich as China's first paleontologist, **Wenzhong Pei** (1904–1982), and their team of 60 workers, found

an adult and a juvenile mandible as well as fragments from two crania ('Skull I' and 'Skull II'). In 1929, the two Chinese scientists took over operation of the site. Pei would find the first complete skull ('Skull III') late in the day. Boaz and Ciochon describe its heroic retrieval in flowing prose:

> Pei found himself at the bottom of a long and rough tunnel with a priceless and exquisitely delicate fossil that could break into hundreds of unidentified shards if not handled exactly right. Night was falling and since Pei and the workers had been in the cave since early that morning, they were tired. He would have liked to cover the fossil hominid and come back in the morning when he was fresh, but it was too dangerous. A loose rock could fall on the skull or somebody might even slip in overnight and try to take it. He had to push on. Concentrating and lighting more candles, Pei worked on into the night, removing the skull in two pieces, carefully gluing the fragments and leaving as much of the adhering cave sediment in place as he could for support. He applied plaster bandage supports and waited for them to set. Then the pieces were slowly passed hand to hand up out of the cave. Pei took them to the field building and immediately set them close to the fire so that the glue and plaster would dry and harden. (Boaz and Ciochon, 2004, p. 24)

Finds continued through the early 1930s, but internal political turmoil—the escalation of the Chinese civil war, and the invasion of the Japanese army—threatened the progress of archaeological work in Zhoukoudian. As the Second World War approached, casts were carefully made of all the bones and teeth that had been recovered, and there was a failed attempt to remove the bones to the United States. The bones disappeared in transit and have never been recovered, in spite of the rumours and rewards that have flourished since their disappearance.

However, for one of the finds—'Zhoukoudian V', or 'Skull V'—there has been something of a recovery. Pieces of this skull were first unearthed in 1934 and 1936. In 1966 the frontal and part of the occipital were found during renewed excavations at the site. These more recent finds fit perfectly with casts of the original pieces, enabling researchers to reconstruct a more complete cranium. Because it was discovered in one of upper layers of the site, it is one of the younger specimens. This is interesting, as Skull V has some features that establish it as more modern than the other skulls, notably its sizeable cranial capacity, which, at 1,140 cc, is the second-largest of all *H. e. pekinensis* specimens (see Table 7.4). Tellingly, the expansion of the brain was in areas in which the *Homo sapiens* is larger than *Homo erectus*.

Table 7.4	Cranial capacities of *Homo erectus pekinensis*
Specimen	**Cranial capacity (cubic centimetres)**
Zhoukoudian X	1,225
Zhoukoudian V	1,140
Zhoukoudian XII	1,030
Zhoukoudian II	1,030
Zhoukoudian XI	1,015
Zhoukoudian III	915
Zhoukoudian VI	850

Fire

In 1931, Davidson Black presented four different forms of evidence from Dragon Bone Hill that for him demonstrated that Peking Man had control of fire. Note that *controlling* fire is different from *starting* fire—the evidence that Black cited did not demonstrate that *H. e. pekinensis* could start fire, just that the hominid was able to use it. The fire may have been initiated by lightning, but Peking Man was able to keep it going at a particular location, 'keeping the home fires burning', so to speak.

The four forms of evidence that Black cited were

- carbon deposits,
- ash accumulations in presumed hearths (i.e. small controlled fire sites),
- burned bone, and
- fire-cracked stone.

Boaz and Ciochon (2004, pp. 100–1) dismissed two of the four suggestions as misleading. The carbon deposits, they argue, appear to have been placed there by flowing water. And the presumed ash accumulation lacked *phytoliths*, small bits of calcite from the leaves, stems, and branches of plants—if a lot of wood had been burned at one location, there would be a lot of phytoliths. However, the burned bone and teeth (the fire-cracked upper teeth of a horse) still stand as evidence, as do the fire-cracked stones. The stones were too large to have washed into the cave from the small flowing waters that seemed to have been there. On the strength of this evidence, Boaz and Ciochon accept Black's assertion that Peking Man controlled fire.

Boaz and Ciochon believe that Peking Man did not use fire primarily for warmth, as the Dragon Bone Hill site was in a warm climate when Peking Man was there. They also feel that fire was not then used for cooking. The burned bone is not direct evidence of cooking; it simply points to the simultaneous (i.e. coincidental) presence of fire and bone.

They assert that fire was likely helpful in the scavenging process, as it was 'only in this manner that hominids could ever have displaced larger, fleeter, stronger, clawed and fanged competitors' (Boaz and Ciochon, 2004, p. 107). Picture in your mind a kill site with our ancestor driving away with fire (possibly torches) the animal predator that did all the work. I think this is what Boaz and Ciochon envisioned.

Homo erectus Finds in Africa

Homo erectus/ergaster

The first *Homo erectus* discovered in Africa was found in Olduvai Gorge, Tanzania, by Louis Leakey in 1961. Typical of Leakey, he did not want to connect it to the *Homo erectus* finds elsewhere, as he felt that humans had evolved straight from *Homo habilis*. The fossil remains, OH 9, was dated at roughly 1.4 mya.

The name *Homo ergaster* was given to African *Homo erectus* by Colin Groves and V. Mazak in 1975, when they assigned the species name to 'ER 992', a mandible uncovered in 1971 from Koobi Fora, Kenya, and dated 1.5 mya. There has been debate ever since about whether *H. ergaster* exists as a separate species or as a subspecies of *H. erectus*. As a lumper, I feel that the evidence that it is clearly distinct from other *Homo erectus* specimens is insufficient; a stronger case needs to be made.

Turkana Boy

The most famous and informative fossil specimen of *Homo erectus* in Africa is **KNM-WT 15000**. This exciting find was made in 1984 by **Kamoya Kimeu** (b. 1940), the leading member of Leakey's 'Hominid Gang'. The excitement of the find stems largely from the completeness of the fossil specimen, which includes 67 of the 206 bones of a hominid. For the first time there was a *Homo erectus* skeleton, rather than just isolated pieces.

Because the specimen was found in the West Turkana area, it was nicknamed 'Turkana Boy'. He was deemed to be male based on his rugged, robust skull, his prominent brow ridges, and his pelvic structure, which is convincingly male. Assigning an age at death to the skeleton was a little more difficult, not for the lack of evidence, but because of the simple fact that the rules we work with are based on the typical timing of skeletal and dental events—the life cycle milestones—of a modern human being. We will consider in detail how to date a skeleton in Chapter 12, on forensic anthropology.

Turkana Boy's second molars were newly erupted, something that usually happens at the age of 12 in human males. There were signs of the start of epiphyseal fusion or union (the epiphyses are the articulating ends of the long bones) at the distal end of the humerus (i.e. the elbow), which usually begins at 11 or 12. Next, the innominate (hipbone)—another site of early fusion, where the three

Profiles in Canadian Anthropology

Davidson Black

Davidson Black (1884–1934) came from an elite background. His father, also named Davidson Black, was a Queen's Counsel, a prestigious appointment in the legal profession. The younger Black attended Upper Canada College and then the University of Toronto (following his father's path to that institution), where he began a medical degree in 1903. It doesn't get much more elite than that. And yet Black, by nature, seemed just as comfortable outside the world of upper-class privilege. In his youth he hauled supplies for the Hudson's Bay Company, loading and unloading canoes. It was perhaps this work that facilitated his encounters with Aboriginal people, whom he made a point of meeting with and learning from. He apparently made some progress in the Ojibwe language of northern Ontario. His respect for his Chinese colleagues when he taught and worked in China is often commented on.

Black was fascinated by fossils, and he collected them in his early years from the banks of the Don River in Toronto. The pursuit of knowledge was another driving force in his life. In 1906, after completing his degree in medical science in 1906, which qualified him as a doctor, he elected to stay on to study general arts (as well as comparative anatomy) in order to broaden his mind (a good idea), ultimately receiving a bachelor's degree in biology in 1909. Upon graduating, he parlayed his two degrees into a job as an anatomy instructor at what is now Case Western Reserve University, in Cleveland, within four years reaching the higher position of assistant professor. While there, he pursued he love of fossils in the university's museum.

Fondation Teilhard de Chardin, Paris, France/Archives Charmet/The Bridgeman Art Library

John Reader/Science Photo Library

Canadian anthropologist Davidson Black (with pipe) stands with his colleagues in China, 1929; French paleontologist Father Pierre Teilhard de Chardin is standing to his right (left in the photo); director of excavations Wenzhong Pei stands front and centre. (right) The first two hominid teeth informally labelled by Black as *Sinanthropus pekinensis*

In 1914, Black took a six-month leave of absence to work in Manchester, England, under Grafton Elliot Smith, one of the great believers in Piltdown Man (see Chapter 1). As he furthered his knowledge of anatomy, particularly neuroanatomy (brain structure), his relationship with Smith helped him to become familiar with and passionate about the search for the missing link. As we saw with Dubois, it is not unusual for misinformation (in this case, concerning Piltdown Man) to lead to a passion that trumps the ignorance that generated the passion in the first place.

At the outbreak of World War I, the patriotic Davidson Black attempted to enlist but was turned away because of a heart murmur. In 1917, he was able to join the Royal Canadian Army Medical Corps (RCAMC), stationed in England. Assigned the rank of captain, he provided much-needed treatment for injured soldiers. In 1919, shortly after he left the RCAMC following the end of the war, he was given an appointment at the Peking Union Medical College, funded by the Rockefeller Foundation. A major reason for deciding to go there was his belief in the view, popular at the time, that humans first evolved in modern form in Asia, specifically in China. An influential book that helped popularize that belief was *Climate and Evolution* (1915), written by New Brunswick–born William Matthew, which had a strong impact on Black. The logic behind the belief was fairly simple: Chinese, Indian, and Middle Eastern civilizations were the oldest known to European scholars of the time. It made sense that these civilizations were products of the longest-term human occupations.

Black first taught neurology and embryology, and seemed content to confine most of his energies to those disciplines. However, in 1921 he began to collaborate in fossil research with the Swedish geologist J. Gunnar Andersson. He was hooked. The Rockefeller Foundation was disappointed in the focus of much of his research; they felt it should be medical, not anything as frivolous as fossil collecting. In the words of a letter to Black from one of the Foundation representatives, who had come to China to steer him in the proper direction:

[I have] jollied you, threatened you, and bullied you, even cursed you [to give up anthropology], but you took it well. . . . For the next two years at least give your entire attention to anatomy. Perhaps by that time you will, with your young son, have other interests which will appear more

important than expeditions to mythical caves. (as cited in Walker & Shipman, 1996, p. 47)

Davidson Black showed almost foolhardy courage in staying at the college at that time, in the face of adverse circumstances quite apart from the pressures of the Rockefeller Foundation. It was a time of civil war, and there was near anarchy in the area as Chiang Kai-shek, head of the nationalist Kuomintang Party, launched a military campaign designed to halt the growing Communist forces, led by Mao Zedong, in 1926. Adding to the chaos were the independent warlords operating in the area, as well as the presence of the Japanese army, sniffing around in advance of their occupation of the nearby province of Manchuria in 1931. Many of Black's colleagues at the school left, and Black sent his family away to safety in 1927. Still, though, he persisted in his research and teaching, confident that the good rapport he had established with his Chinese colleagues would see him through any danger.

There is an archaeological adage that you dig up your most spectacular find at the end of the field season. In the fall of 1927, at the end of the archaeological field season, Swedish paleontologist Anders Birger Bohlin discovered a third tooth. He rushed back to Beijing to tell Black. In a letter he later wrote to Andersson, Black breathlessly blurted out the following:

We have got a beautiful *human* tooth at last!

It is truly glorious news, is it not!

Bohlin is a splendid and enthusiastic fellow who refused to allow local difficulties and military crises to affect his work. I was telegraphed on October the 10th urging me to recall Bohlin and Li because of the war but I made careful enquiries here and could see no reason for so doing. So I wrote to Bohlin and explained the situation and told him I didn't want him or Li to run any risks but that as they knew local conditions, they must use their own judgement about coming to Peking.

. . . That night which was October 19th when I got back to my office at half past six in the evening, I found Bohlin in his field clothes, covered with dust, but his face shining with happiness. He had finished the season's work in spite of the war and on October 16th had discovered the tooth, being right on the spot when it was picked out of the matrix! My word, I was excited and elated! Bohlin came

Davidson Black *continued*

here before he had even let his wife know that he was in Peking—he is indeed a man after my own heart. (as cited in van Oosterzee, 1999, pp. 99–100)

Based on this tooth, Black took a huge academic gamble, declaring that a new genus and species had been discovered: *Sinanthropus pekinensis* (lit. 'Chinese-man of Peking').

The year 1927 was not a good time to proclaim a human ancestor species based on a single tooth, as three different such declarations were debunked that year. An article in *Science* showed how *Hesperopithecus haroldcookii*, or 'Nebraska Man', based on a tooth found by rancher/geologist Harold Cook in 1917, was in fact an extinct peccary, or pig. 'Montana Man', identified on the strength of a tooth promoted by naturalist J.C.F. Siegfriedt as being that of a human ancestor, turned out to be an extinct quadruped. And

'Southwest Colorado Man' achieved such public acclaim that perhaps the leading physical anthropologist of the time had to declare publicly, in voice and in print, that crucial evidence, a tooth, was in fact that of a horse. But Black managed to gain credibility for his find, and as a result, his funding grew. It wasn't long before more than teeth would be found.

In 1928, Black found vindication for his bold gamble. Two mandibles, one of a juvenile and one of an adult, and cranial fragments ('Skull I' and 'Skull II') were discovered. Always anxious to quickly publish results of research, Black wrote about all of these fossil specimens. In 1929, when Yang and Pei took over the excavation from the foreign anthropologists, the first full skull was found ('Skull III'; see the account earlier in this chapter). Black saw to it that Pei, the discoverer, got full credit, enabling

previously separate parts of the ilium, ischium, and pubis come together—were still separate. Based on this evidence, Turkana Boy, if he were human, would have been younger than 15. Suspecting that *Homo erectus* would mature slightly faster than humans, in part because of *H. erectus*'s smaller brain size, the investigating team set Turkana Boy's age at 10–12. Based on his height of 1.58 metres (5′ 3″), the researchers estimated that he could have reached an adult height of anywhere between 1.75 metres (5′ 10″) and 1.9 metres (6′ 4″). His brain size was 880 cc, and it is estimated that at full maturity it might have reached 910 cc. He lived about 1.6 million years ago.

KNM–ER 1808

A strange and sad story concerns KNM–ER 1808, the remains of a female who died about 1.7 million years ago in what is now Kenya. The condition of the fossilized bones led Walker, Zimmerman, and Leakey (1982) to declare that she must have died from hypervitaminosis A (an overdose of vitamin A). The most likely cause of this overdose is that she ate the liver of a carnivore. If their diagnosis is true, then this story has two major implications. One is that her group might have been advancing towards being hunters and not scavengers, in order to have access to a carnivore as food. Second, as it takes at least weeks, maybe months, for an individual to die from hypervitaminosis A, during which time she would have been relatively helpless, someone must have been

taking care of her over that time. This is possibly the first archaeological evidence for hominid compassion.

Related Species: *H. georgicus* and *H. floresiensis*

Homo Georgicus

Early *Homo* is found in Africa, and to the far east, in Indonesia and China. So what area is missing? Europe. But it is missing no longer.

Starting in 1991, and with increasing success during the late twentieth and early twenty-first centuries, anthropologists working in the area of the medieval city of Dmanisi, Georgia, found hominid fossil specimens overlying a 1.85 mya volcanic tuff. Dates range from 1.8 to 1.75 mya. In 2002, a Georgian team led by anthropologists David Lordkpanidze and Abesalom Vekua named the species **Homo georgicus** (Vekua, et. al, 2002).

The researchers had essentially three crania and three mandibles to work with. D2280, linked with mandible D211, is an almost complete cranial vault with a capacity of about 775 cc. D2282 comprises a cranium and maxilla fragment, with a size of about 650 cc. D2700, with associated mandible D2735, has the smallest cranial capacity at 600 cc. Mandible D2600 has no apparent surviving cranial connections. All three crania have distinct brow ridges and occipital tori (pl. of *torus*). D2282 and

him to publish his account of the discovery and making sure that when medals were struck to reward the finding and analysis of the skull, Pei would receive the recognition he was due. Black then worked hard in freeing the skull from the limestone matrix and in rearticulating the bones once they were freed, making plaster cast copies of each stage and publishing three preliminary papers on the subject. He worked late into and often through the night.

In 1933, after three productive years of discoveries, Black toured the Middle East and India, returned briefly to Canada, and then made an important presentation in London, all with the purpose of making people aware of Peking Man. When he returned to China at the end of the field season, he was anxious to get to the cave to see how excavation had progressed during his absence. While examining the site, he suffered a mild heart attack,

collapsing onto the cave floor. He had cause to worry: his father had died at the age of 49 of a heart attack. Black was approaching that fatal age himself. Yet he picked himself up and continued his study of the work at the cave.

It wasn't until later that Black went, secretly, to a hospital in Beijing. He did not even inform his wife of the doctor's diagnosis of heart attack. The problem did not go away. In February, Black spent three weeks in hospital. When he was discharged, his immediate desire was to get back to (hard) work. On 15 March 1934 he arrived at his lab at 5:00 p.m., looking forward to pulling another all-nighter. He died in his lab coat, Skull II lying to one side of him, to the other the skull of an early *Homo sapiens* recently excavated from an upper level of the cave he had spent much of his career investigating.

D2700, the ones with the smaller cranial capacity, have less robust, more gracile features, suggesting that they may be females.

The significance of the *Homo georgicus* findings is immense, as Vekua and Lordkipanidze stress in their conclusion to the 2002 article announcing the discovery:

> The Dmanisi hominids are among the most primitive individuals so far attributed to *H. erectus* or to any species that is indisputably *Homo* . . . , and it can be argued that this population is closely related to *Homo habilis*. . . . The presence in Dmanisi of individuals like D2700 calls into question the view that only hominids with brains equivalent in size to those . . . [of] *H. erectus* were able to migrate from Africa northward through the Levantine corridor into Asia. It now seems more likely that the first humans to disperse from the African homeland were similar to *H. habilis*. . . . (Vekua, et al., 2002)

What they are saying is that the old view that *Homo erectus* was the first to leave Africa may be false, that earlier, more primitive hominids may have preceded *H. erectus*.

Homo floresiensis

If you think that *Homo georgicus* messed things up, you need to read this. It was the fall of 2004. It was a Thursday,

and I had a physical anthropology class later that day. I went to *The Globe and Mail* online to check the morning's headlines over a cup of coffee and was immediately seized by a report on an unbelievable hominid find in a cave called Liang Bua (lit. 'cool cave') on one of the smaller Indonesian islands. The discovery of **Homo floresiensis** had been announced to the world.

I knew I had to find out as much as I could, as my students were bound to ask lots of questions. What was so amazing about this find? Just about everything. For one thing, she ('LB 1', from Liang Bua) was just around a metre tall, even though she was an adult. That meant that she was quickly dubbed 'the Hobbit', as *Lord of the Rings* was big that year. Another thing: she was reported to have had a cranial capacity roughly equal to that of a chimp: 380 cc. (Later, though less often mentioned, this figure was adjusted to 417 cc, based on results of a CT scan.) Most surprisingly, she lived only about 17,000 years ago, long after every hominid other than *Homo sapiens* had become extinct.

More facts are necessary. Her height was determined primarily by the length of her sturdy femurs (we will speak more of this sturdiness later): 28 centimeters. That makes her femurs shorter than Lucy's but equal to those of the smallest of *H. habilis*. Although she was shorter than a pygmy (who, being among the shortest peoples, has an average height of 1.43 metres [4′ 9″] for men and 1.35 metres [4′ 6″] for women), her height was calculated using a formula based on pygmy femur-to-total-body-height ratio; it gave *H. floresiensis* a height of precisely 1.06 metres (just shy of 3′ 6″). Australian archaeologist **Mike Morwood**, the lead researcher, felt she could have been even shorter as, unlike the pygmies, she had no forehead to speak of. Her **humerofemoral index** (the ratio of arm length to leg length; see Chapter 6) was 85.4, well outside the range of modern and archaic humans, and about the same as that of Lucy, the *Australopithecus afarensis*.

The small size of the specimen's brain did not mean that she was a dwarf, since dwarves have normal-sized brains. Critics of *Homo floresiensis* were quick to argue that she was probably microcephalic (lit. 'tiny head'), symptomatic of a number of human disorders; in other words, she may not have been typical of her community, and could have been a *Homo erectus* suffering a rare medical condition. But with that brain size, she would not have been able to think well enough to perform basic functions (speak, organize her thoughts, obtain food) if she had been a member of just about any other previously identified *Homo* species. Further, her brain, although small, had the basic structure of a *Homo erectus* brain, in addition to which it exhibited the development of the frontal and temporal lobes, areas of higher brain functions.

And she was not alone. There are claims of at least 8 other specimens of *H. floresiensis*, and perhaps as many as 13 that have been uncovered. Fairly sophisticated **flake tools** (made from hard pieces of stone chipped off a larger block and then further worked to make tools) were found throughout several layers of the Liang Bua cave, as were the fossil remains of other small folks. For example, another adult LB 8, based on its short tibia length, was deemed to be 1.01 metres (3′ 4″) tall, again possibly an over-estimate.

Other features are instructive. Brown and Maeda (2009), having studied the mandibles of three *Homo floresiensis* specimens (LB 1, LB 2, and LB 6), claim that these mandibles 'share a distinctive suite of traits that place them outside both the *H. sapiens* and *H. erectus* ranges of variation', sharing similarities with the australopithecines and early *Homo*. It should be noted that one of the two authors, Peter Brown, was involved in the initial study of LB 1. I've made a point of mentioning which 'side' that Brown is on, because written and verbal battle lines have been drawn in the debate over whether *H. floresiensis* is, to put it bluntly, a 'one-off freak' (or possibly two- or three-off). There has been nastiness involved, which even a brief look at the literature would show. A number of senior Indonesian archaeologists, with the bureaucratic arrogance typical of high-ranking figures in a country conditioned by colonial hierarchy followed by the almost inevitable series of post-colonial dictatorships (you can see which side *I'm* on), were involved with literally stealing the fossil specimens, and damaging one of the mandibles with amateurish cast making. They then lent out specimens to outsiders, who became drawn into the battle on the side of those who had provided them with intellectual opportunities. We will talk in a later chapter on

archaeological ethics of the importance of ensuring that the fate of fossil specimens be determined ultimately by the people of the country in which they were found. However, there is a responsibility to the specimens, to science, and to human heritage, of stewardship, that in some carefully considered cases must override this ethic.

Back to our 'Hobbit'. LB 1's feet are interesting. They are long—more than 60 per cent the length of her femur, which makes them at least 10 per cent larger than the feet of a human of comparable height. However, they lack a well-developed arch and are somewhat curved. This suggests that she was not fast on the ground, but was probably good in the trees. The latter point is supported by a curving of her hands combined with her having sturdy, probably quite powerful (although short) legs and arms. LB 1's face is distinctive, with no chin to speak of, and with brow ridges that don't cross as a single bar but are individualized, like bony eyebrows. And, as mentioned above, she had very little forehead.

A lot of questions need to be addressed, and they certainly haven't been dealt with completely yet. Chief among them is how *H. floresiensis* stands with other hominids. Liang Bua carries fossil specimens that range from 95,000 to 12,000 years ago (when a volcano hit). The island itself has signs of inhabitation that go back possibly to 840,000 years ago. Hominids had to cross sea at least twice (making journeys of 25 km and 9 km; see Morwood and van Oosterzee, 2009, p. 5) before getting to the island, even during the Ice Age, when much of Indonesia was part of a peninsula, not a series of islands. Peter Brown, Mike Morwood, and others directly involved with the discovery and first analysis of *Homo floresiensis* assert that it is not a *Homo erectus*, but is the descendant of a hominid that left Africa earlier than did *H. erectus*, sharing a common ancestor. Although I am a lumper, I *currently* believe that *H. floresiensis* is quite possibly a distinct species.

Just as big a question is that of size. Why was *H. floresiensis* so small? The answer proposed by Morwood and his associates is **island dwarfism**, otherwise part of a more general principle known as the island rule. Which brings us to some Canadian content.

In 1964, **J. Bristol Foster**, a young, Toronto-born, recent PhD graduate from the University of British Columbia, published a paper in *Nature* entitled 'Evolution of Mammals on Islands'. In it he looked at 116 island-dwelling mammals. He noted that of these species, 38 were smaller than others of their kind; 12 were the same size, and 66 were larger. The general principle involved was that members of the orders Carnivora (carnivores; 13 of 15), Artiodactyla (hoofed mammals; 9 of 11), Lagomorpha (rabbits and hares; 6 of 8), and Insectivora (4 out of 5) became smaller, while Rodentia (rodents; 60 of 69) and Marsupialia (marsupials; 3 of 6) became larger. Stated in a somewhat oversimplified way: big got small and small got big. It is the dwarfed ones that interest us here. The idea is that they become smaller because there are fewer resources

Science Gone Wrong

Meganthropus

Franz Weidenreich, whom we met earlier in the chapter, is an important anthropologist whose work in Java, especially his work connecting Peking Man and Java Man as variants of *Homo erectus*, have more than earned him the right to be in the physical anthropology hall of fame. However, he had one big, flawed idea.

When Ralph von Koenigswald found the large Sangiran 6 mandible in 1941, he called it *Meganthropus palaeojavicus* (lit. 'huge-man old-Java'), as it was the largest hominid mandible then known. It has been referred to in the literature as *Meganthropus* A. He had also found huge teeth of an ape he called *Gigantopithecus blacki* ('large ape' plus an honouring reference to his friend, Davidson Black), which we discussed in an earlier chapter.

Weidenreich had a problem to solve that we have addressed earlier in this chapter. Why does *Homo erectus* have thick cranial armouring, while modern apes and modern humans (*Homo sapiens*) do not? His answer to this question was that *Homo erectus* had descended from larger hominids, *Gigantopithecus* first, then *Meganthropus*. The skull armouring represented the retention of an earlier form that disappeared with *Homo sapiens*. He wrote frequently on the subject, culminating in his classic work *Apes, Giants, and Man* (1946). He took the inevitable opposition to this idea rather well, confident that time would find in his favour:

> It took 40 years before Neanderthal Man was recognized as a special human form and not pushed aside as a pathological variant of modern man, and it took 40 years before Dubois's *Pithecanthropus erectus*, originally described as a giant ape, was acknowledged as a normal sized hominid. . . . Therefore there is at least a remote chance that as a consequence of the evolution of anthropological science and scientists, Meganthropus and Gigantopithecus may be admitted to the hominid family at about 2000 A.D. (as cited in Wolpoff and Caspari, 1997, p. 194)

The argument that 'They [i.e. the experts] laughed at Louis Pasteur and he was proven right. They laugh at me so I must be right too,' is a fallacy. When most scientists 'laugh at' a different idea, it is because it's funny.

Weidenreich connected his theory of descent from giants with his idea of multiregionalism (that *Homo erectus* developed into *Homo sapiens* in different areas with specific features being retained; see Chapter 8), with Asians being the ultimate descendants of this line. It is a good thing he never saw Yao Ming of the NBA's Houston Rockets.

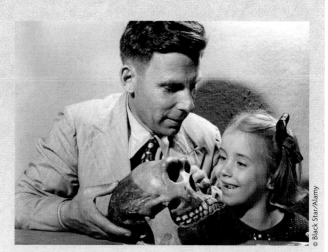

Ralph von Koenigswald and his daughter with the reconstructed skull of *Meganthropus A.*, at the American Museum of Natural History, New York, 1946

The name *Meganthropus* took on a life of its own after Weidenreich's death. When Sangiran 8 was discovered in 1953, it was given the alternative name '*Meganthropus* B'. A recent reconstruction of the mandible by Kaifu, Aziz, and Babo (2005) has demonstrated that it was significantly smaller than first thought. Other claims of *Meganthropus* mandibles have been made: Sangiran 33, a partial mandible, and '*Meganthropus* D', which unfortunately does not have a Sangiran number. None of them have associated remains that could argue for general distinctiveness. Likewise there are three cranial remains for which claims of distinctive status have been made: *Meganthropus* I (Sangiran 27), II (Sangiran 31), and III.

Irresponsible reconstructions of huge size (up to 9 feet tall and weighing 1,000 lbs) to go along with the largeness of the mandibles has encouraged Bigfoot believers, including Grover S. Krantz (1931–2002)—unfortunately famous for being a professor of physical anthropology who believed in the Bigfoot legend—to adopt *Meganthropus* as evidence for their beliefs. Creationists see in *Meganthropus* a Goliath or Nephilim, the latter mentioned as a dangerous group in the Bible, supposedly giants.

There have to be crania, mandibles and—ideally, but not necessarily—postcranial fossil specimens found in association before a species can be claimed. Until that time, the best we can say is that there may be a few mandibles that show differences from others, and a few crania that show differences from others. Until those differences are associated with each other, it is sloppy science to say that we have a distinct species.

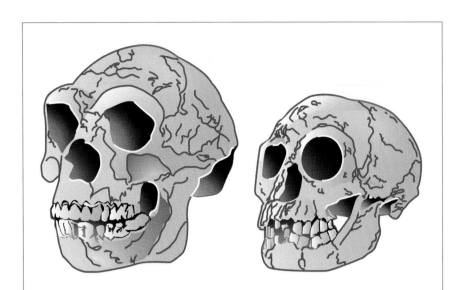

Figure 7.5 | Skulls of *H. erectus* (left) and *Homo floresiensis* (right)

available. Big-bodied individuals are at an evolutionary disadvantage. An absence of predators is important.

Deer are a great example of island dwarfism. I have seen this first-hand when my wife and I went to Gabriola Island, a short distance off the eastern coast of Vancouver Island. When we saw our first deer there, we thought it was a fawn, until we saw that it was actually a mother with a fawn. Elephant types are also well-documented examples of the phenomenon, as has been repeatedly shown in archaeology done on islands in the Mediterranean. The extinct genus *Stegodon* includes species that were among the largest of the Proboscidea (i.e. elephants) order, including adults that were four metres tall, eight metres long, with three-metre-long tusks. Morwood and associates claim that the fossil remains of dwarf stegodons were found on Flores, although this claim has been contested. Morwood's favourite example is that of *Myotragus*, an extinct cave goat that lived on the island of Majorca, which became smaller by about 60 per cent, including an alleged 50 per cent reduction in brain size.

This last point is important. What would be gained by brain reduction? Remember from the start of the chapter that the brain is costly in terms of energy. A smaller brain focused on specific functions might be more efficient. Recently there has appeared in the literature another alleged example of hominid island dwarfism in Palau, Micronesia (Berger, et al., 2008). The finds here are of more

recent hominids than *H. floresiensis* (2,900–1,400 years ago). The individuals studied have small body and face size, little chin, and large brow ridges, similar to *H. floresiensis*. The cranial evidence is not too clear to date, but the researchers reckon that the average will fall below 1,000 cc.

Researchers are continuing to work on the intriguing question of what Homo floresiensis is. In late April 2009, Stony Brook University hosted their seventh annual Human Evolution Symposium. The theme of that year's conference: 'Hobbits in the Haystack: Homo floresiensis and Human Evolution'. Once the papers presented have been published, we will know more, as we will increasingly over the next few years.

Summary

We have seen that scientists are divided on the question of how many early species of *Homo* there are. There is consensus concerning early *Homo habilis* and later *Homo erectus*, but other species names are more controversial. *Homo georgicus* may be one or the other, if it is not altogether distinct. Either way, it points towards an earlier migration from Africa than many had previously thought. And the single most disturbing and potentially revolutionary find in the early *Homo* area is *Homo floresiensis*, the tiny figure that emerged, it would seem, far too late on the historical scene.

Review Questions

1. Why did *Homo erectus* develop such a thick skull?
2. Why do humans maintain rapid fetal brain growth after they are born?
3. How would you explain the Movius Line?
4. Differentiate between a lumper's and a splitter's taxonomy of early *Homo*.

Discussion Questions

1. Discuss whether or not *Homo rudolfensis* should be considered a distinct species.
2. Discuss what you believe *Homo floresiensis* to be—a 'one-off' freak version of *Homo erectus*, a separate and very unique species, something else . . .
3. Discuss the impact of the *Homo georgicus* discoveries on the study of early *Homo*.
4. Is it possible that we have found more hand axes in Africa and western Eurasia because there are more hand axes there, or because we have looked more there?

Key Species and Specimens

Homo erectus pekinensis (Peking Man)	*Homo floresiensis* *Homo georgicus*	*Homo habilis* *Homo rudolfensis*	Java Man KNM–WT 15000

Key Terms

Acheulean	flake tools	Movius Line	torus mandibularis (or mandibular torus)
anagenetic	holotype	torus supraorbitalis (or supraorbital torus)	
crista sagittalis	humerofemoral index		torus occipitalis (or occipital torus)
diastema	island dwarfism	torus angularis (or angular torus)	
dura mater	keeling		
encephalization	metabolic budget		

Key Individuals

Johan Gunnar Andersson	J. Bristol Foster	Louis Leakey	Franz Weidenreich
Davidson Black	Kamoya Kimeu	Mike Morwood	Zhongjian Yang
Tim Bromage	G.H.R (Ralph) von Koenigswald	Hallam L. Movius	
Eugène Dubois		Wenzhong Pei	

Recommended Print and Online Resources

Alexeev, V. P. (1986). *The origin of the human race.* Moscow: Progress Publishers.

Archaeology Info. (Website). Archaeology Info home page. Available from www.archaeologyinfo.com/index.html

Berger, L. R., Churchill, S. E., De Klerk, B., Quinn, R. L., & Hawks, J. (2008). Small-bodied humans from Palau, Micronesia. *PLoS ONE, 3*(3), e1780. doi: 10.1371/journal.pone.0001780

Boaz, N., & Ciochon, R. (2004.) *Dragon Bone Hill: An Ice Age saga of Homo erectus.* Oxford: Oxford University Press.

Bromage, T., McMahon, J., Thackeray, J., Kullmer, O., Hogg, R., Rosenberger, A., . . . Enlow, D. (2008, Fall). Craniofacial architectural constraints and their importance for reconstructing the early *Homo* skull KNM–ER 1470. *Journal of Clinical Pediatric Dentistry, 33*(1), 43–54.

Brown, P. & Maeda, T. (2009). Liang Bua *Homo floresiensis* mandibles and mandibular teeth: A contribution to the comparative morphology of a new hominin species. *Journal of Human Evolution, 57*, 571–96.

Coqueugniot, H., Hublin, J., Veillon, F., Houët, F., & Jacob, T. (2004). Early brain growth in *Homo erectus* and implications for cognitive ability. *Nature, 431*, 299–302.

Foster, J. B. (1964, April 18). Evolution of mammals on islands. *Nature, 202*, 234–5. doi:10.1038/202234a0

Kreger, C. D. (n.d.). Homo habilis. Archaeology Info. Retrieved from www.archaeologyinfo.com/Homohabilis.htm

Morwood, M., & Van Oosterzee, P. (2009). *A new uman: The startling discovery and strange story of the 'Hobbits' of Flores, Indonesia.* Walnut Creek, CA: Left Coast Press.

Ramirez Rozzi, F. V., Bromage, T., & Schrenk, F. (1997). UR 501, the Plio–Pleistocene hominid from Malawi. Analysis of the microanatomy of the enamel. *Comptes Rendus de l'Académie des Sciences, 325*, 231–4.

Rightmire, G. P. (1990). *The evolution of Homo erectus: Comparative anatomical studies of an extinct human species.* Cambridge: Cambridge University Press.

Shipman, P. (2001). *The man who found the missing link: Eugène Dubois and his lifelong quest to prove Darwin right*. Cambridge, MA: Harvard University Press.

Spoor, F., Leakey, M., Gathogo, P., Brown, F., Antón, S., McDougall, . . . Leakey, L. (2007, August 9). Implications of new early *Homo* fossils from Ileret, east of Lake Turkana, Kenya. *Nature, 448*, 688–91.

Swisher III, C., Curtis, G., & Lewin, R. (2000). *Java Man: How two geologists' dramatic discoveries changed our understanding of the evolutionary path to modern humans*. New York: Scribner.

Swisher III, C., Rink, W., Antón, S., Schwarcz, H., Curtis, G., Supryo, A., & Widiasmoro. (1996). Latest *Homo erectus* of Java: Potential contemporaneity with *Homo sapiens* in Southeast Asia. *Science, 274*, 1 870–4.

Van Oosterzee, P. (1999). *Dragon bones: The story of Peking Man*. St Leonard's, Australia: Allen & Unwin.

Vekua, A., Lordkipanidze, D., Rightmire, G. P., Agusti, J., Ferring, R., Maisuradze, G., . . . Zollikofer, C. (2002, July 5). A new skull of early *Homo* from Dmanisi, Georgia. *Science, 297*(5578), 85–9.

Walker, A., & Shipman, P. (1996). *The wisdom of bones: In search of human origins*. London: Weidenfeld and Nicolson.

Walker, A. C., Zimmerman, M. R., & Leakey, R. E. (1982). A possible case of hypervitaminosis A in *Homo erectus*. *Nature, 296*, 248–50.

Wolpoff, M. H., & Caspari, R. (1997). *Race and human evolution*. New York: Simon and Schuster.

Yamei, H., Potts, R., Baoyin, Y., Zhengtang, G., Deino, A., Wei, W., . . . Weiwen, H. (2000, March). Mid-Pleistocene Acheulean-like stone technology of the Bose Basin, South China. *Science, 287* (5458), 1622–6.

8

Transition to Modern *Homo sapiens*

Upper Paleolithic cave painting of
a bison from the Altamira Cave, Spain
© The Art Archive/Alamy

LEARNING OBJECTIVES

After reading this chapter, you should be able to:

- Articulate and distinguish between the three hypotheses concerning the development and spread of modern human beings.

- Distinguish among the four different modes of tool traditions.

- Outline the arguments both in favour of and against the existence of the species *Homo heidelbergensis*.

- Identify the traits that separate *Homo erectus* from *Homo sapiens*.

Introduction

We are now entering the realm of **Homo sapiens**, but in significant ways the mist over the valley of knowledge has not cleared with the rising of the new species. We used to speak in terms of **archaic *Homo sapiens***, partly as a way to avoid acknowledging what the early ones were: *Homo erectus* with a few derived features, or *Homo sapiens* with some primitive features. The species divide gets no easier at this stage than it was in the previous two chapters.

Modern *Homo sapiens*: Lumper's and Splitter's Taxonomies

Again we will use Wood and Lonergan's guide to lumper and splitter taxonomy as a way to view modern Homo sapiens (see Table 8.1). Note that to Wood and Longeran's list we have added a relatively new proposed species. We are also leaving one major figure out of this chapter: Neandertal. It requires a chapter unto itself (see Chapter 9), which will feature a comparison with what is presented in the current chapter.

The Origin of 'Modern' Human Beings

By the end of the chapter we will have reached 'modern' *Homo sapiens* (well, at least some of us will have). But don't expect champagne or fireworks when modernity is reached. There are complicating factors. First of all, we have to consider the distinction between **anatomically modern humans** (AMH) and **behaviourally modern humans** (BMH). The former are hominids that look so much like humans you wouldn't recognize one as something different if it was sitting across from you on the bus. The latter term refers to those engaged in symbolic behaviour (e.g. drawing, carving, or sculpting figures), showing clear signs of belief in an afterlife (i.e. practising deliberate burial and leaving grave goods with the dead), or taking part in other 'typically human' actions or activities. Some question whether these two coincide. Do you have to be an AMH to also be a BMH, or vice versa? And does being either anatomically modern or behaviourally modern, or both, mean that one necessarily *thinks* like a modern human or communicates in a sophisticated language? These are some of the difficulties facing those who engage in this type of research.

Changing from *Homo erectus* to *Homo sapiens*

What are we talking about, exactly, when we're discussing the evolution from *Homo erectus* to *Homo sapiens*? At a very basic level, we are talking about a number of key features, which can best be introduced through the idea of **mosaic evolution**. This is the notion that major evolutionary change in a species take place in stages, with one aspect of the species changing independently of the others. As science historian **Stephen Jay Gould** (1941–2002) explains, this important development in evolutionary theory helped gain acceptance for Dart's Taung Baby:

> When it was discovered in the 1920s, many evolutionists believed that all traits should change in concert within evolving lineages—the doctrine of

Table 8.1	A lumper's and a splitter's taxonomy of modern *Homo sapiens*
Lumping taxonomy	Splitting taxonomy
Homo sapiens	*Homo antecessor*
	H. cepranensis*
	H. neanderthalensis**
	H. sapiens

* I have added this to Wood and Lonergan's list.
** Discussed in the next chapter.

the 'harmonious transformation of the type'. An erect, but small-brained ape could only represent an anomalous side branch destined for early extinction (the true intermediate, I assume, would have been a semierect, half-brained brute). But, as modern evolutionary theory developed during the 1930s, this objection to *Australopithecus* disappeared. Natural selection can work independently upon adaptive traits in evolutionary sequences, changing them at different times and rates. Frequently, a suite of characters undergoes a complete transformation before other characters change at all. Paleontologists refer to this potential independence of traits as 'mosaic evolution'.

Secured by mosaic evolution, *A. africanus* attained the exalted status of direct ancestor. (Gould, 1977, p. 58)

In other words, not all traits evolved at the same rate. Some parts of the fossil specimens described are like their counterparts in *Homo erectus* (e.g. the brow ridges, or the thickness of the cranial vault) while some are like their counterparts in *Homo sapiens* (e.g. the facial skeleton, or the height of the cranial vault or forehead). This reflects mosaic evolution.

Here are some of the major traits that signal a shift from *H. erectus* to *H. sapiens*:

- larger cranial capacity (from a little more than 1,000 cc to a size approaching the adult *H. sapiens* average of 1,350 cc)

- thinner cranial bones (with the increase in the size of the brain, this would, among other things, make the birth of a big-headed baby a bit easier)

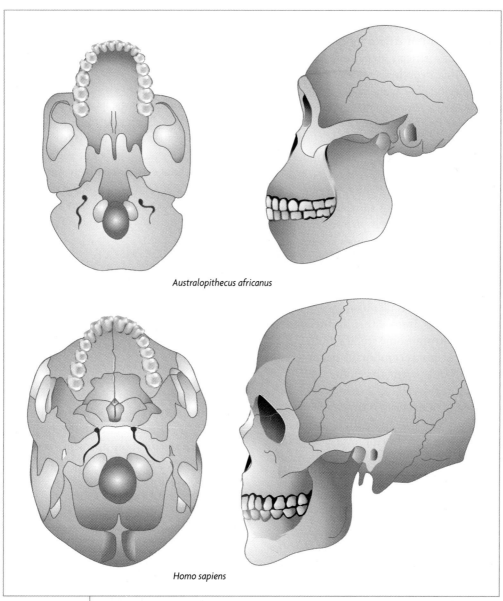

Australopithecus africanus

Homo sapiens

Figure 8.1 | Skulls of *Australopithecus africanus* and *Homo sapiens* compared

- reduction in brow ridges (and replacement with a bipartite structure, as discussed below)

- increase in the height of the cranial vault, or forehead (now we can wear a hat without the loss of vision)

- decrease in the robustness of the face

- decrease in the size of the teeth

- development of a definite chin.

The Origin of *Homo sapiens*: Three Hypotheses

1. Out of Africa

There are three main hypotheses, or models, to explain how anatomically modern *H. sapiens* evolved from earlier *Homo*. Each can be connected with a field of study and with a primary object of scholarly analysis. Currently, the most popular is the **Out of Africa** model, also called 'Recent African Origin', 'African Replacement', or just plain 'Replacement'. This hypothesis states that modern *Homo sapiens* first developed anagenetically from an earlier *Homo* species in Africa. 'Anagenetically', you'll recall from Chapter 6, means that the ancestral *Homo* species evolved entirely into *Homo sapiens*; it isn't as though *H. sapiens* was an offshoot of a species that continued and co-existed with it.

This hypothesis is subject to slightly varying interpretations. Anthropologists differ as to what form of the hypothesis they take depending largely on how recent they put the move out of Africa. Generally there is a tendency to place the origin and spread of modern *Homo sapiens* more recently than with the Multiregionalism hypothesis (discussed next). The field of study most closely identified with this way of thinking is molecular biology or genetics, where the evidence comes from either **mitochondrial DNA (or mtDNA)** or **Y-chromosome** studies. Both involve looking at mutations of **nonrecombinant genes**—genes passed down from only one parent, the mother in the case of mtDNA, the father in the case of Y-chromosome. The 'combination' referred to in the term is the combined genetic input coming from both parents. With nonrecombinant genes, any differences that develop over generations, over time, are purely from accumulated mutations, not from the input of the other parent. Sharing a nonrecombinant genetic pattern means sharing an ancestor.

MITOCHONDRIA

Mitochondria (*sing.* **mitochondrion**) are organelles that exist in large numbers in each cell outside the nucleus. They are involved in converting sugars into energy and exist only in animals, not in plants. Mitochondria are thought to be of evolutionary origin separate to that of the animals that bear them, derived from the circle-shaped genetic structure (or genomes) of bacteria.

Mitochondrial DNA is the genetic content that exists outside the nucleus in a relatively short string of nucleotide bases. In terms of physical anthropology, it is useful for several reasons. One is that it is nonrecombinant: with mtDNA, the mother is the sole provider of genetic content. This is because human eggs contain a significant amount of mtDNA, while human sperm carries its significantly lesser amount in the tail, which is cast away once the sperm fertilizes the egg. Therefore, any changes that occur from one generation to the next must be due to mutation, not to the genetic input from a male.

These mutations enable us to track evolutionary lines, or **haplo-groups**. Everyone who possesses a particular suite of mutations will be connected or related. This will be discussed further, when we look at the migration of modern humans over all of the earth.

Another reason that mtDNA is so useful for our research is that it does not constitute any part of the genetic code that issues the commands as to what to construct. Mutants (those bearing a particular mutation—remember, we are all mutants) will not die off as a line because of a mutation that makes them less 'fit' in evolutionary terms. It doesn't matter what the mutation does; it can't harm the form the physical structure takes. Therefore, mutations persist.

Third, mtDNA mutates faster than nuclear DNA, although the exact rate is not yet known. This enables us to track change over relatively shorter periods of time. Finally, humans have only 16,500 base pairs in the mtDNA, making it easier to sample and study than nuclear DNA, which contains three billion base pairs.

MITOCHONDRIAL EVE

The concept of **mitochondrial Eve** is both controversial and commonly misunderstood. It often pits the 'gene' people versus the 'stones and bones' people, and researcher against researcher. The term refers to the hypothetical woman who first bore the mtDNA pattern that all currently living humans share. This is not the first *Homo sapiens* woman, or the first anatomically modern human. 'Mitochondrial Eve' refers to our **most recent common ancestor (MRCA)** when it comes to a particular ancient mtDNA pattern; she is also known by the term **matrilineal most recent common ancestor**, or **mt-MRCA**. Keep in mind that an MRCA may be different for different genes or genetic patterns.

In the initial study, Cann, Stoneking, and Wilson (1987) took a small portion of the total mtDNA from 147 people originating in five geographical areas:

- 20 sub-Saharan 'Africans' (18 of whom were born in the US)

- 34 'Asians' (from China, Vietnam, Laos, the Philippines, Indonesia, and Tonga)

- 46 'Caucasians' (from Europe, North Africa, and the Middle East)

- 21 Australian Aborigines

- 26 New Guineans.

The researchers soon noticed that there were two primary branches of descent: one that was entirely African, and one that had all five geographical groups. From this they determined that *Homo sapiens* developed first in Africa, as the greatest mtDNA diversity is found there. All other later studies would concur at least on the point that the greatest genetic diversity of any geographical group is within Africa. The 'Out of Africa' origin is another matter.

Cann, Stoneking, and Wilson added a dating component to their study, in order to measure the rate of mtDNA divergence:

> One way of estimating this rate is to consider the extent of differentiation within clusters specific to New Guinea, . . . Australia and the New World. People colonised these regions relatively recently: a minimum of 30,000 years ago for New Guinea, 40,000 years ago for Australia, and 12,000 for the New World. These times enable us to calculate that the mean rate of mtDNA divergence within humans lies between two and four percent per million years. This rate is similar to previous estimates from animals as disparate as apes, monkeys, horses, rhinoceroses, mice, rats, birds and fish. We therefore consider the above estimate of 2% – 4% to be reasonable for humans, although additional comparative work is needed to obtain a more exact calibration. (Cann, Stoneking, & Wilson, 1987)

Based on their estimate that the common mtDNA pattern diverged by an average of nearly 0.57 per cent, the researchers concluded that the common ancestor of all surviving mtDNA types existed between 140,000 and 290,000 years ago. Think of it in terms of rough math: if something changes four times in 1,000,000 years, then it would change once in 250,000 years, assuming it changed at a uniform rate.

In the hoopla surrounding this article, two major points were commonly overlooked—two points that, like the good scientists they were, they had issued as a caution. The first was that the time scale was 'tentative'. Indeed, the date assigned to the earliest common ancestor of all mtDNA types would change. This was owing in part to the fact that once more (and ultimately all) of the mtDNA was analyzed, it became evident that mutations occurred at different rates with different parts of the mtDNA. The figure of 171,500 ± 50,000 ya arrived at by Ingman and colleagues (2000) is as good as we're going to get from this imprecise method of dating.

The second caution the researchers issued was, in their own words:

> [O]ur placement of the common ancestor of all human mtDNA diversity in Africa 140,000–280,000 ago need not imply that the transformation to anatomically modern *Homo sapiens* occurred in Africa at this time. The mtDNA data tell us nothing of the contributions to this transformation by the genetic and cultural traits of males and females whose mtDNA became extinct. (Cann, Stoneking, & Wilson, 1987)

It's important to remember that our mt-MRCA is not fixed in person or in time. Consider this example. Say there was an mtDNA pattern that was shared by all people living today and by 'the last Beothuk', Shanawdithit (*c.* 1800–1829), who died in Newfoundland in 1829. Let's say the mitochondrial 'Eve' of this population is 'B,' and she had two daughters. Shanawdithit descended from the eldest daughter of B, who had a mutation in her mtDNA; we would call this mutation in B's eldest daughter B + 1m. The second daughter doesn't have that mutation, but one of her female descendants has an mtDNA mutation that happens 4 generations later; we would call this pattern M + 5m. At some point all the female descendants of the second daughter carry that pattern. When Shanawdithit was alive, the most recent common ancestor of all human beings would have been B. After she died (without having any children), the most recent common ancestor would have become the descendant of B who was the original B + 5 m mutant. M would still have been an ancestor, just not the MRCA.

Here is an analogy. Ten people all have a memory of a particular event that they all witnessed—say, a hockey game that they all saw. It is the oldest memory that all of them share of something they experienced together. The next year, eight of them see a hockey game that two of them do not. The next day the two that missed the hockey game die in a freak accident. Now the most recent common memory of all the survivors is moved ahead by one year.

How can we have dates for mitochondrial Eve when we can't use the usual dating methods involving rates of radioactive decay? In the initial study, which made a huge popular impact, Cann, Stoneking, and Wilson (1987) dated the mt-MRCA at 215,000 ± 75,000 years ago, a pretty wide range. They did that by assuming that mutation in the midst of the 147 people they studied took place at a uniform rate, an assumption that many non-geneticists find hard to accept with the same confidence they place in methods of radioactive decay. The principle works something like this. Say we knew that the Aborigines had been genetically separated from all other populations for 40,000 years. (As you will see in a later chapter, this point is anything but confirmed. Nevertheless . . .) If you noted that there had been 40 mutations during that period, then you could calculate that the mutation rate was one for every thousand years. Unfortunately, it isn't really that simple.

Y-CHROMOSOMES

You remember that chromosomes exist in 23 pairs. Although we get one of each pair from our mother and father, neither is identical to those of our parents. The chromosomes get reassembled in a way that is usually compared by geneticists to the shuffling of a deck made up of the genes of the two parents. Say the pair that Dad has for chromosome number 10 contains genes A, B, and C in one and D, E, and F in the other. The pair that Mom has contains genes G, H, and I in one and J, K, and L in the other. One child may have A, B, and F in the corresponding chromosome that comes originally from Dad, and G, K, and L (or any similar combination) from Mom; another child may have D, E, and C from Dad and J, H, and I from Mom. That is a major reason why no two children of the same parents are identical.

Now, the Y-chromosome is special. It does not get shuffled. Geneticist **Spencer Wells** (b. 1969) describes it in the following way:

> The Y is found inside cells like any other regular chromosome, but . . . it's all alone—it doesn't have a partner like the other chromosomes. This is because it is actually the presence of the Y, mismatched with its partner the X chromosome, that determines the embryo's sex. If there is a Y present the embryo will be a boy, but if there are two X chromosomes, perfectly paired like all of the others, then it will be a girl. And because of the Y's mismatch with the X [imagine matching Danny de Vito with Queen Elizabeth] . . . there isn't any scope for recombination. The chromosomes can't believe that they've been seated together at dinner and want nothing to do with each other. No recombination takes place, and the Y is passed on intact. That means the Y chromosome gives us an incredible opportunity to trace genetic lineages far back in time. We have a large amount of nonrecombining DNA that we can use to study the distribution and history of genetic lineages. (Wells, 2006, p. 38)

A **Y-chromosome Adam** has been found, one that lived between perhaps 60,000 and 90,000 years ago—much more recently than mitochondrial Eve. This is considered support for the Out of Africa hypothesis. (We will look at this further in Chapter 10, on human variation.)

A worthwhile challenge to those who rely on these two means of studying nonrecombinant genes is the following, from Richard Dawkins's *The Ancestor's Tale*. It does not say that Out of Africa is right or wrong; it just throws down a challenge that must be met. It reflects his gene rather than individual focus, the extreme caricature of which is the idea that an individual is just a gene's method of reproducing itself:

> Different genes tell different stories. It is perfectly possible for some of our genes to have recently come out of Africa, while others have been passed to us from separate *H. erectus* populations. Or to put it another way, we can be both descendants of a recent African exodus and simultaneously descendants of regional *H. erectus*, because at any given time in the past we have a huge number of genealogical ancestors. Some could have recently left Africa. Some could have been resident in, say, Java for thousands of years. And we could have inherited African genes from some and Javan genes from others. A single chunk of DNA, such as a mitochondrion or Y-chromosome, gives as impoverished a view of the past as a single sentence from a history book. (Dawkins, 2004, p. 52)

It needs to be kept in mind that nothing privileges or makes more important the nonrecombinant aspects of our genetics in terms of programming what we are. The nonrecombinant aspects of our genetic makeup are important because of what anthropologists and geneticists can do with them in terms of determining human history. While that is a vital role, we still need to hear from the other genes to tell the story of modern *Homo sapiens'* origins and migrations. We will probably not have to wait long.

2. Multiregionalism

A second hypothesis used to explain how anatomically modern *H. sapiens* evolved from earlier *Homo* is known as **Multiregionalism**. This hypothesis, whose proponents are directly and quite vocally in opposition to Out of Africa, is typically associated with the anthropologist **Milford H. Wolpoff** (b. 1942). The Multiregionalism hypothesis sees modern *Homo sapiens* as having developed at roughly the same time in a number of different parts of the world, most notably Africa, China, and Indonesia.

This model draws support mainly from traditional 'stones and bones' archaeologists and physical anthropologists, who attempt to stress the physical similarities or continuity of features in different regions, from pre–*Homo sapiens* to modern humans. **Franz Weidenreich** (1873–1948), for example, contributed to this hypothesis when he connected the **shovel-shaped incisors** (upper and lower incisors with ridges on the inside like old-fashioned shovels) of *Homo erectus pekinensis* with the same feature found most commonly among East Asians (particularly northern Chinese) and the Aboriginal people of the Americas.

Table 8.2 | Three hypotheses to account for the evolution of *Homo sapiens*

Hypothesis	Primary object of study	Primary method of study
Out of Africa	genes	molecular biology
Multiregionalism	physical features	archaeology/physical anthropology
Clinal Replacement	hominid populations	population genetics

3. Clinal Replacement

A third hypothesis has been called **clinal replacement** by Boaz and Ciochon. Stringer uses the term African Hybridization and Replacement for a very similar (but more confusing) hypothesis. It is an attempt to reconcile aspects of the two previous hypotheses—the relatively recent aspect and ultimate African origin of the Out of Africa model, and the continuity of characteristics and lack of anagenesis (complete replacement of one species by another) that is found in Multiregionalism. Proponents of this model write and speak in terms of **clines**, which are 'geographically defined populations with different gene frequencies (and physical traits) that intergrade at the edges' (Boaz & Ciochon, 2004, p. 149). In the past, clines might have been called 'races'.

The Clinal Replacement model involves a series of slightly different, geographically separate subspecies, rather like the **ring species** spoken of in Chapter 4. Each cline can interbreed with and isn't very different from its neighbouring cline, although there might be significant difference between those at the extreme ends. Clinal replacement involves waves of cline movements over time, with repeated instances in which a cline with a 'newer' evolved trait gradually replaces an older, more 'primitive' cline, although remnant features of the former may persist. The best analogy I can think of is an old drinking game we used to play with a drink called a 'Purple Jesus', a punch made of grape juice and vodka. Whenever the punch bowl needed to be topped up, straight vodka would be added. The drink got stronger as the night proceeded, but there was always some purple left over from the original grape juice. That could be enough for the continuity of some features.

A strength of this hypothesis is that it is not wedded to the species concept the way the other two seem to be. Further, it is not as winner-take-all as the first model, nor does it promote belief in deep-seated 'racial differences'.

Tool Talk, Part 1

When discussing early lithic (i.e. stone toolmaking) traditions, archaeologists often speak in terms of four basic 'modes'. Very simply, the four modes are as follows.

Mode 1

Mode-1 tools are considered typically of the **Oldowan** (remember 'old one') tradition. These tools are pretty much as basic as they can get. You have two stones: one is the hammerstone, and the other is the stone that is struck. The tool stone is shaped through percussion (hitting); little retouching (finer shaping of the stones) is involved. A tool made from the now reduced stone is called a **core tool**; one made from a piece that is struck off is called a **flake tool**. If you want to make the flake tool sharper, you just strike another flake off of it.

Mode 2

Here we have **Acheulean** tradition, characterized by those beautiful bifacial, symmetrical hand axes. A technique that is added here is **pressure flaking**, in which the stone tool-to-be is shaped with gentle pressure or rubbing by wood and bone implements, instead of simply being struck with the hammerstone. This gives the toolmaker greater control over the process. In this tradition, the core is prized as the tool over the flakes.

Mode 3

Mode-3 tools are produced using what is called the **Levallois technique**, in which the core is carefully prepared into a kind of tortoiseshell shape, and then a well-struck blow pops the flakes tool out. The flakes are the most important here. The Neandertal version of this—and it is found from beginning to end of their time as a separate entity—is referred to as **Mousterian**. A good way to remember this standard on multiple-choice tests is to think 'Monsterian'. (Take no offence, Neandertal—it's to help the students.)

Mode 4

The usual (though somewhat *un*usual) name used for this mode of toolmaking is the **Aurignacian** of the **Upper Paleolithic**, and it is typically associated with the **Cro-Magnon** version of anatomically modern humans (AMHs) during the period from roughly 38,000 to 22,000 years ago (figures vary considerably). With these tools, the designated tool part is the relatively long, thin **blade** that is struck off the core. Outside of lithics, or stone toolmaking, the term 'Mode 4' is used to refer to tools made of other materials (bone and ivory, for instance) and to other artistic forms, such as carved stone statuettes, and jewellery (e.g. beaded bracelets and pendants) made out of shell, tooth, and bone. Cave art first appears in association with this tradition.

Homo Heidelbergensis

The term *Homo heidelbergensis* is overused as a kind of dumping ground for what used to be called archaic *Homo sapiens*. It is thought to lie somewhere on the evolutionary path between *Homo erectus* and *Homo sapiens*. Yet nowhere have I read a clear, concise definition of what *H. heidelbergensis* is, although Mounier, Marchal, and Condemi (2009) recently took a noble stab at it. Until such a definition is in place, the transition from *H. erectus* to *H. sapiens* remains foggy.

What we can say, for now, is that those fossil specimens that have generally been assigned to this species possess both modern or derived *H. sapiens* and primitive *H. erectus* features, but not always the same set. *H. heidelbergensis* may stand the test of time, but it could just as easily end up on the taxonomic scrap heap.

Tattersall and Schwartz make an important point in their discussion of *H. heidelbergensis*. Although we throw the term 'anatomically modern human' (or AMH) around a lot, and seem to know one when we see one, it would be scientifically more elegant if there were precisely defined criteria to judge by. They rightfully argue that, for example, 'protruding chin' is not very precise (Tattersall & Schwartz, 2001, p. 203). Thinner skull bones, smoother, rounder skull, and reduced brow ridges—some of the usual characteristics used to identify AMH—could be made a little more measurable, less subjective. One such characteristic that they suggest is having a **bipartite** brow ridge. This is a brow ridge having two distinct parts: a central protruding piece between the eyebrows and extending slightly into the ridge over each eye. There is a slight gap easily perceived if it is there (you can feel it with your finger) followed by a platelike bony structure.

Significant European Finds

Heidelberg Jaw

The great science fiction writer H.G. Wells (1866–1946), author of such classics as *The Time Machine*, *The Invisible Man*, *The Island of Doctor Moreau*, *The War of the Worlds*, and 'The Country of the Blind', in 1922 published *A Short History of the World*. In the section entitled 'Monkeys,

---

Profiles in Canadian Anthropology

Pamela Willoughby

Pamela Willoughby teaches at the University of Alberta and engages in important research in southern Tanzania in the eastern part of Africa. She is a strong and influential proponent of the Out of Africa hypothesis. Her research specialty is the stone tools found in Tanzania during the **Middle Stone Age (MSA)**—a period of stone toolmaking that, according to Willoughby, lasted from about 200,000 to 30,000 years ago—and the transition to the style of toolmaking that characterized the **Later Stone Age (LSA)**, the African version of the Upper Paleolithic. She summarized her research goal in a statement she made in 2007, saying that 'In theory, we're looking for the magical, hypothetical site where MSA tools change into LSA tools,' adding: 'I think one of our sites, Mlambasi, could be that site' (as cited in Stieglitz, 2007).

The MSA tools are 'Mode-3' tools, not all that different from those used by the Neandertals. Willoughby believes modern *Homo sapiens* may have come about during this toolmaking period, gradually improving their

stoneworking skills until they entered the LSA period with the means to leave Africa and have an impact on the rest of the world. She does not subscribe to the view that several large, rapid changes occurring in the LSA made humans physically modern and technologically advanced in a sudden, dramatic shift.

The following University of Alberta publications will introduce you to her work:

Stieglitz, Tara. (2007, October 3). Tools may dig up historical clues. *The Gateway (The official student newspaper at the University of Alberta)*. Retrieved from http://thegatewayonline.ca/tools-may-dig-up-historical-clues-20071003-1005

DiCenzo, David. (2000, April 28). When did history begin? An anthropology professor searches for 'the origin of everyone'—and turns a popular theory on its head. *Folio, 37*(16). Retrieved from www.folio.ualberta.ca/37/17/back.html

The Heidelberg jaw, all that remains of the specimen known as 'Heidelberg Man'

Apes and Sub-men' he had the following to say about the Heidelberg jaw. It captures the way in which the jaw had stirred the popular imagination at the time of its discovery:

> This jaw-bone is, I think, one of the most tormenting objects in the world to our human curiosity. To see it is like looking through a defective glass into the past and catching just one blurred and tantalizing glimpse of this Thing, shambling through the bleak wilderness, clambering to avoid the sabre-toothed tiger, watching the woolly rhinoceros in the woods. Then before we can scrutinize the monster, he vanishes. Yet the soil is littered abundantly with the indestructible implements he chipped out for his uses. (Wells, 1922, p. 33)

The **Heidelberg jaw** was a mandible found in a large sand pit in Mauer, southeast of Heidelberg, Germany. Rough estimates are that it is between 400,000 and 650,000 years old. The jaw was relatively chinless, broad, and massive. The teeth were small, like those of a regular human. The species *Homo heidelbergensis* is named after the jaw simply because it was the first find of the proposed species, and in biology that gives it precedence in naming over what was discovered afterwards. It is not the clearest example of *H. heidelbergensis*, nor does it help in arriving at a definition of what *H. heidelbergensis* is.

Steinheim Skull

In 1933, a skull, now known as the **Steinheim skull**, was discovered in a quarry at Steinheim-an-der-Murr, 20

kilometres north of Stuttgart, Germany. When German paleontologist **Fritz Berckhemer** (1890–1954) analyzed it that year, he gave it its own species name, *Homo steinheimensis*. This would be later changed to *Homo sapiens steinheimensis*, though it is now usually presented as a localized version of *Homo heidelbergensis*.

The skull's features are a mixture of archaic and modern, with significant brow ridges and large eye sockets showing the archaic, but the small teeth and rounding of the skull suggesting its more recent nature. The cranial capacity is usually given as between 1,110 and a little more than 1,200 cubic centimetres, making it slightly smaller than that of a modern adult human. The dating of the skull is problematic, with a range of between 350,000 to 250,000 ya. The somewhat warped shape of the skull has been variously interpreted as a sign either of deliberate mutilation (which, along with a postulated widening of the foramen magnum, suggests cannibalism) or of the earliest known case of meningomes, a benign tumour of the meninges of the brain.

Swanscombe Skull

The interesting story about the **Swanscombe skull** is that it was 'discovered' three times: in 1935, 1936, and 1955, the first two times by Dr Alvin T. Marston, a 65-year-old (in 1935) dentist and amateur archaeologist. The three pieces—the occipital discovered first, followed by the left and then the right parietals—were found to fit together.

The skull, with an estimated cranial capacity of 1,325 cc., is believed to be that of a young female who lived sometime between 200,000 and 400,000 years ago. The tools associated with the site are Acheulean hand axes. The Swanscombe skull is very similar to that of a Neandertal, but without the characteristic occipital bun (a bulge in the back—see Chapter 9). In a recent book, *The Human Lineage* (2009), authors Matt Cartmill, Fred H. Smith, and Kaye B. Brown conclude their discussion of this specimen by saying, 'The boundary between European Heidelbergs and Neandertals is vague and arbitrary' (Cartmill, Smith, & Brown, 2009, p. 306). Tattersall and Schwartz suggest there is reason to believe that the Swanscombe skull belonged to an early Neandertal (Tattersall & Schwartz, 2001, p. 201), although most experts in the field consider it to be a specimen of *Homo heidelbergensis*. That, of course, could change with further analysis.

The Swanscombe Skull was initially treated with some suspicion, in the same way that Raymond Dart's Taung Baby was. Unlike the fraudulent (but then accepted) Piltdown Man (see Chapter 1), it did not look enough like a modern human. The legalistic-sounding Committee of Investigation announced in a 1937 issue of *Nature* that:

> The character of the Swanscombe skull and the conditions of its discovery both in themselves and in relation to the Piltdown skull, suggested that

a certain suspension of judgment was advisable for further consideration of the evidence, before accepting the find at its face value as inferred by Mr A.T. Marston on his announcement of the find. ('Swanscombe skull', 1937, p. 1048)

In other words, it wasn't about to be accepted as a possible human ancestor because it was too dissimilar to the remains of the widely accepted Piltdown Man. If you find the situation highly ironic, you would be bang on.

Petralona Skull

The **Petralona skull** was discovered in 1960 in the cave of Katsika Hill, in Petralona, Greece. It was found by Christos Sarijannidis about 30 centimetres off the cave floor, suspended on a stalagmite. You may remember from high school geography class that a stalagmite is the name given to the stone sticking up from the ground that is formed from dripping water containing minerals. The skull has been hard both to classify—it was first called *Homo neanderthalensis*, then a type of *Homo erectus*, now *Homo heidelbergensis*—and to date—initially set at 70,000 years old, it's now commonly given a date of 160,000–240,000 years ago, or even 300,000–400,000 years old. It possesses a large brow ridge, a massive face, and an estimated cranial capacity of 1,220 cc.

Arago XXI, or 'Tautavel Man'

From Henry de Lumley's archaeological work that began in 1964 in Arago Cave, near the village of Tautavel in southern France, came, in 1971, a cranium known as **Arago XXI**, or '**Tautavel Man**'. The fossil specimen has a sloped, receding forehead, large brow ridges, and a protruding lower jaw, and is robust enough to be thought to have belonged to a male. First identified as *Homo erectus*, it is now generally considered a *H. heidelbergensis*. As the third molar (the wisdom tooth) looks like it was newly erupted, and as the **coronal suture** (the place where the frontal and the parietal bones meet) is still open—traditionally an indication that an individual is less than 25—it was reckoned that Tautavel Man was around 20 years of age when he died, roughly 400,000 years ago. His cranial capacity has been estimated at 1,100 cc.

Sima de los Huesos (the Pit of Bones)

The Gran Dolina cave (Trinchera Dolina, or TD) in Spain, which has been excavated since 1981, has two strata, or levels, of interest to our discussion here. The more recent of the two strata (TD 10) is known usually as **Sima de los Huesos** (lit. 'the pit of bones') because of the thousands of fossils found there, including the largest collection of ancient hominid fossils (more than 2,000) in the world. It

is estimated that there were 32 different individuals whose remains have been found there. In some cases, we have the very smallest of bones that are found in very few ancient hominid sites. This includes the bones of the middle ear, the **malleus** (more popularly known as the **hammer**—think 'mallet'), the **incus** (otherwise called the **anvil**), and the **stapes** (the **stirrup**). They have recently been dated (by Bischoff, et al., 2003 and 2007) as possibly being between 400,000 and 600,000 years old, and are generally assigned the species of *Homo heidelbergensis*. Tattersall and Schwartz, however—you'll remember they like to split—claim that the Sima finds represent a separate species:

> Because the Sima skull [Skull 5] can be readily distinguished from the skulls of other hominids, we think it will eventually be accepted by palaeo-anthropologists as representing a distinct species. (Tattersall & Schwartz, 2001, p. 201)

While they could be right, conclusive prove is still lacking.

From the extensive fossil remains, we know a great deal about the individuals who lived at Trinchera Dolina. They were robust and reasonably tall; males averaged about 1.68 metres (5′ 7″). The birth canals of the females were larger than those of modern-day women. One individual, 'Skull 5', who is amazingly well preserved and complete, we can tell suffered several hard blows to the head during his life, and may have died due to infection when a tooth was broken off (see www.amnh.org/exhibitions/atapuerca/simahumans/index.php).

There are still some intriguing mysteries about the Sima de los Huesos. Why has only one tool been found there, an Acheulean or 'Mode-2' hand axe? Why were so many hominid bones there when it does not appear that they lived in that cave (when, for instance, there are no signs that they built fires there)? Why are all the bones those of young adults? Only 3 of the 32 individuals are thought to have been over 30, yet there are few children and people in their early adolescence. Jean-Pierre Bocquet-Appel and Juan Luis Arsuaga suggest that this may be evidence that there was some kind of 'ecological crisis' that could have been survived only by the strongest of the population (Bocquet-Appel & Arsuaga, 2001, pp. 230–1).

Boxgrove Man

At the Boxgrove site, in West Sussex, England, some early Acheulean (or Mode-2) tools and faunal remains—some of them butchered—were found between 1983 and 1994. In that last year, a massive tibia was found, gnawed at both ends, that is estimated (using tibia-to-body-size formulae) to be that of a male about 1.8 metres (6′) tall and weighing in at around 80 kilograms (176 lbs). The site has been estimated as being roughly 500,000 years old, perhaps somewhat younger.

Gran Dolina

Excavating in 1994 and 1995 at the strata known as TD 6 (or 'Aurora'), José María Bermúdez de Castro and his team of archaeologists discovered the fossil remains (80 fragments) of at least six hominids. These show two fascinating aspects. The first and foremost is the age of the fossils—between 780,000 and 850,000 years old—which makes them the oldest European fossils next to the *Homo georgicus* finds. This has caused some anthropologists to assign them to a new species, **Homo antecessor**, which would be ancestral to *H. heidelbergensis*.

Homo cepranensis

They exhibit the general robustness of *Homo erectus* while showing some signs of the facial structure of *Homo sapiens*.

Second, cut marks on the bones, consistent with a defleshing of the bones, have pointed to the earliest evidence that we have for hominid cannibalism. This includes the most complete remains—those of El Niño de la Gran Dolina (the boy of Gran Dolina), thought to be 10 or 11.

The hominids of Gran Dolina are estimated as having had a cranial capacity of over 1,000 cc, but probably not by much. They did not have any bifacial tools of the Mode-2 type. Their tool types were restricted to Mode 1, and yet in the time in which they lived, their African contemporaries had long been making Mode-2 tools.

Homo cepranensis

Recently adding to the growing jumble of *Homo* species is **Homo cepranensis**, the name given to a calvaria, or skullcap, with an almost complete frontal, fragmented parietals, and well-preserved occipital, found in 1994 near the town of Ceprano, Italy, almost 90 kilometres southeast of Rome. Distinctive features of this partial cranium (which took five years to reconstruct) include a brow ridge (supraorbital torus) of variable thickness and an expanded frontal area. From such features and the dating of between 800,000 and 900,000 years ago, the researchers claim that *H. cepranensis* represents 'a morphological "bridge" between the *H. ergaster/erectus* group and that composed by specimens commonly referred to [as] *H. heidelbergensis*' (Manzi, Mallegni, & Ascenzi, 2001). Time and additional specimens will help to establish its status. My money is against its being a separate species, but then I am a lumper.

The Schöningen Spears

A short piece in *Nature* entitled 'The World's Oldest Spears' began with the following sentences:

> Just occasionally, an archaeological discovery leaves one speechless. Reasons usually concern

Table 8.3	Significant European finds			
Site/specimen name	Country	Date (thousands of years ago [kya])	Cranial capacity (cubic centimetres [cc])	Taxonomy
H. cepranensis	Italy	800–900	—	*H. cepranensis*
Gran Dolina	Spain	750–850	—	*H. antecessor*
Heidelberg jaw	Germany	450–650	—	*H. heidelbergensis*
Boxgrove Man	Britain	500	—	*H. heidelbergensis*
Sima de los Huesos	Spain	400–600	—	*H. heidelbergensis*
Arago XXI	France	400	1,100	*H. heidelbergensis*
Petralona Skull	Greece	300–400	1,220	*H. heidelbergensis*
Swanscombe Skull	Britain	200–400	1,325	*H. heidelbergensis*
Steinheim Skull	Germany	250–350	1,110–1,200	*H. heidelbergensis*

the degree of preservation, the unexpectedness of the find, and its wider implications. The discovery of complete, unambiguous throwing-spear 380,000–400,000 years old at Schöningen in Germany, . . . meets all these criteria. (Dennell, 1997, p. 767)

What are the **Schöningen spears**? Why are they so special? They are four spears, each one about 2 metres long (imagine, in terms of total length, the collective height of an NBA starting lineup, minus one), found in association with stone tools and thousands of horse bones. One of the spears was found embedded in the skeleton of a horse, and cut marks serve as clear evidence that some of the animals were butchered. Each spear is made from the trunk of a 30-year-old Norway spruce. The sharpened edge is made from the base of the trunk where the wood is the hardest. Each of the spears has a centre of gravity about one-third of the way up from the sharp end, like a modern javelin, for maximum throwing distance and force. Each of the spears weighed a little over two kilograms, and so would have required a muscular thrower (a modern Olympic javelin is longer but weighs just 800 grams, less than half the weight).

The conditions in which the spears were preserved are special. They were found in what was a peat bog. You will discover in later chapters that peat bogs provide great conditions for preserving wood and skin, as they generally lack oxygen, making it difficult for bacteria and insects to aid in their decomposition. Unearthing one such find is hugely significant, as it points (no pun intended—maybe) to much more widespread use of wooden spears in places and instances in which they were not preserved. The absence of stone spear heads or projectile points at any site does not necessarily mean that there were no throwing spears. It simply means that instead of using spears with stone points, people may have been using completely wooden spears.

One further implication of the find is that the spears tell us a lot about their makers. It would take patience, planning, and experimentation to develop these tools and then make them on a regular basis. It might even require language in order for spear-makers to work together or pass on their knowledge. The spears tell us that between 380,000 and 400,000 years ago, there were some pretty smart and diligent hominids walking around in what now is Germany.

The Upper Paleolithic

The **Upper Paleolithic** (lit. 'old stone'), a period from roughly 40,000 to 10,000 years ago, was a near-magical time of cultural transitions in Europe. It seemed that everything began happening at once, the sometimes spectacular cave paintings and figurines being only the most visible signs. During a remarkably short span, early humans went almost literally from zero to sixty, from no cave paintings at all to the detailed and lifelike images of animals that feature so prominently in the popular notion of 'cavemen'. *Homo sapiens* came alive during the Paleolithic, without a doubt. What remains disputed are the factors that brought about this sudden transition. Was it a combination of numbers and culture? Did the numbers reach high enough levels that greater interaction among groups and the sharing of ideas became inevitable? And what might have led to the growth in numbers? Was it the combined effects of a series of innovations that came together to a prehistoric 'tipping point' that made for sudden-seeming change? Or was it biology? Did some subtle change in the brain, not yet (or perhaps not ever) detected in the skulls of those who lived in the Upper Paleolithic period, bring about a cultural revolution? Some scholars suggest that language as we know it know—a vocabulary of tens of thousands of words with well-formed grammar and a capacity for more abstract and symbolic thought, so wonderfully illustrated in the cave paintings and figurines—came about only during the earliest part of this period. With our increasing knowledge of the human genome and its connections with the brain, and with our increasingly sophisticated ways of retrieving and replicating ancient DNA, we may someday be able to determine whether or not this is true.

Cro-Magnon

Since they were first discovered and named, both in the second half of the nineteenth century, **Cro-Magnon** (lit. 'great cavity') and Neandertal seem to have been doomed to share a story of close comparisons. In this chapter we

Artist's rendering of Cro-Magnon

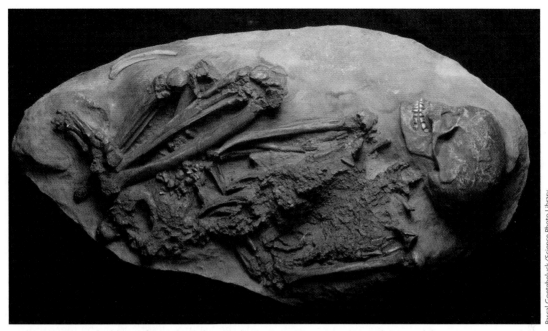

Cro-Magnon skeleton found at the Dolni Vestonice site, Czech Republic

will be writing about Cro-Magnon; in the next, we will take up the other side. Cro-Magnon was first discovered in a cave or rock shelter of the same name at Les Eyzies, in southwestern France. The discoverer, French geologist **Louis Lartet** (1840–1899), found five fossilized remains, with the holotype (or type specimen) being **Cro-Magnon 1**. Unlike Neandertal, here was an ancient hominid that people could readily identify with as an ancestor. It was fundamentally gracile (slender) rather than robust, and had a high, vaulted forehead, not one that sloped back like that of Neandertal. It looked like us—a little more robust perhaps, and with a slightly larger cranial capacity (an estimated 1,600 cc in the case of Cro-Magnon 1), but it was still one of us. We had arrived.

The time period that is generally attached to Cro-Magnon is 40,000–10,000 years ago although the oldest clearly identified and dated Cro-Magnon, **Oase-1**, discovered in Romania in 2002, is dated as 34,950 ± 890 ya.

The culture felt right, too. The Aurignacian, a Mode-4 tool tradition, lasted in Europe roughly from between 40,000 and 35,000 years ago to between 28,000 and 26,000 ya. The tradition was named for the initial find, in 1860, at Aurignac, by Louis Lartet's father, Edouard. The inventory of cultural artifacts includes shells and animal teeth made into necklaces (some of which were buried with the dead), antler-tipped spears, bone needles for sewing, and the famous cave paintings—all nice and modern.

Ancient mtDNA

In 2008, geneticist David Caramelli and his team published the results of a unique study based on the analysis of mitochondrial DNA taken from the tibia and skull of a Cro-Magnon, **Paglicci 23**, found in an Italian cave. The findings were twofold. First, after checking the mtDNA against that of the seven people who handled the bone, Caramelli found that the pattern of Paglicci 23 was clean of contamination from its handlers, something he could not prove in an earlier mtDNA study done from findings in the same cave; that guaranteed that the results of any mtDNA analysis performed on the bone would be genuine. Second, the mtDNA pattern was found to be still a common one—shared, in fact, with the **Cambridge Reference Sequence (CRS)**. CRS was the first mitochondrial DNA pattern ever sequenced, at Cambridge University in 1981, corrected in 1999. It should be noted that the sequence is a European one, of the haplogroup H2a2. The larger group of which it is a part, haplogroup H, comprises about 30 per cent of all Europeans. This means at the very least that there is a genealogical continuity across the 28,000-odd years (28,100 ± 350) from the time that Cro-Magnon lived to now.

Significant African Finds

Kabwe (Broken Hill) Skull

In 1921, a cranium was discovered by a Swiss miner, Tom Zwiglaar, in what was then known as Broken Hill, Northern Rhodesia (now it's Kabwe, Zambia). Its first classification was *Homo rhodesiensis*, or 'Rhodesian Man'. After the cranium was unearthed, it was followed by the discovery on the same site of a maxilla from another individual, two femur fragments, a tibia, and a sacrum. The dating is very rough as the site was destroyed, but the figures usually given are between 125,000 and 300,000 years old. The skull, which has a cranial capacity of an estimated 1,100 cc, is very robust and Neandertal-like, with huge brow ridges, a low, sloping forehead, and a long, oval cranial vault. The Kabwe skull is now typically classified as a member of *Homo heidelbergensis*.

Florisbad 1

Florisbad 1, which includes pieces of the frontal and right parietal bones as well as part of the right side of the face, was discovered in 1932 in South Africa by T.F. Dreyer, a zoology professor at Grey University College in Bloemfontein. Dreyer felt that it was distinctive enough to have its own species name, *Homo helmei*. That name was long ago rejected, and it is now generally felt to be *Homo heidelbergensis*. G. Philip Rightmire (2005) analyzed the specimen and believed it to be significantly similar to the Kabwe skull. It has been dated at 259,000 ± 35,000 years ago.

Salé Cranium

The **Salé cranium** was discovered in 1971 by workers at a rock quarry in Morocco. This adult cranium has been assigned a date of roughly 400,000 years ago, and like others of its time period, it exhibits traits of both *Homo erectus* (thick vault bones and a cranial capacity of 860 cc) and *Homo sapiens*. Even more interesting, researchers speculate that it may have suffered from a severe handicap from birth. An unusual thickening at the base of the cranium, as well as muscle attachments standing out on the left side while being diminished or atrophied on the other side, suggest to French anthropologist Jean-Jacques Hublin that the individual suffered from a fetal abnormality, caused by a twisting of the head in the womb, that would have significantly limited the person's head and limb movement. That the individual survived into adulthood suggests that the people around him must have shown **compassion**.

LH 18 (Ngaloba)

In 1976, **Mary Leakey** led a team that discovered a hominid skull, **LH 18**, in the Ngaloba Beds of Laetoli, Tanzania. It is dated at between 129,000 and 108,000 years old. Initially the skull was estimated to have had a cranial capacity of 1,200 cc, but it has been more recently supersized to 1,367 cc. The bone of the vault is quite thick, especially in the occipital region. Middle Stone Age tools have been found at the same level.

Bodo Cranium

The nearly complete **Bodo cranium** was discovered in Bodo, Ethiopia, in 1976, in a strata that also contained Acheulean tools. It resembles *Homo erectus* in its large facial skeleton, its low braincase, the thickness of the bone,

and its keeling effect. However, it has a larger cranial capacity, which depending on how it is reconstructed, could be anywhere from 1,200 to 1,325 cc. There is a lack of consensus on the typology of the skull. Some experts say that it represents *Homo erectus* with a few *Homo sapiens* characteristics, while others assign it to *Homo heidelbergensis*. It is dated at roughly 600,000 years ago.

UA 31 (Buia)

This fossil remains was discovered at the Uadi Aalad site, in the Buia Basin of the Afar Depression in Eritrea (a country that was formerly part of Ethiopia). The site has been excavated since 1995, when UA 31 was found. It comprises a well-preserved cranium with a cranial capacity of roughly between 750 and 800 cc, the facial skeleton (without the mandible) found close to two teeth, and three pelvic portions thought to have belonged to the same individual, likely a female. The cranium is asserted to be a million years old. The features described in the literature so far include marked brow ridges and a long, narrow, high braincase, a mixture of old (*H. erectus*) and new (*H. sapiens*).

Omo I and II

In 1967, a National Museums of Kenya team directed by **Richard Leakey** found hominid remains in the Kibish Formation, near the Omo River in southwestern Ethiopia. The Kibish Formation is divided into four layers called **members**. In Member I, the lowest layer, were found two specimens, **Omo I** and **Omo II**, from different sites. Omo I came from Kamoya's Hominid Site (KHS), and Omo 2 from Paul's Hominid Site (PHS).

Of Omo I we have much of the skeleton, including most of the cranial vault, parts of the face and mandible, and a good number of postcranial bones. It has a cranial capacity of about 1,400 cc, close to the modern human average. As it is said to look 'more modern' than Omo II, it is now typically referred to as the 'earliest anatomically modern human'.

The Omo II remains include a virtually complete calvaria, or skullcap. Early dates of over 100,000 assigned to these remains were often challenged until recently, when argon-40/argon-39 dating was used to date Member I at an estimated 195,000 ya. Lithic (stone) technology found in the Member I layer give the appearance of what is called Middle Stone Age, which in Awoke's Hominid Site (AHS) includes cores on which the Levallois technique has been applied. We can say, generally, that Mode-3 technology was employed here.

H. sapiens idaltu (Herto)

In 2003, a series of articles were published concerning discoveries made in the Upper Herto Member of the Bouri Formation, Middle Awash, in the Afar Rift of Ethiopia. The focus of the investigation was three crania: BOU–VP

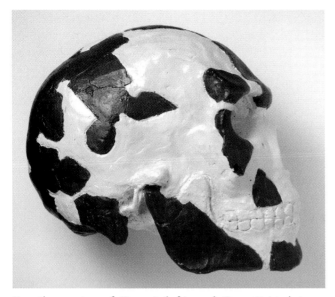

Fossil remains of Omo I (left) and Omo II (right)

16/1, BOU–VP 16/2, and BOU–VP 16/5 (the BOU comes from Bouri). The first of these is considered the most significant as it is almost complete (missing only a part of the left side) and adult, with an estimated cranial capacity of 1,450 cc, and an exceptional (when compared with modern human beings) cranial length. BOU–VP 16/2, which is less complete, would appear to have an even larger cranial capacity. The third cranium is that of a child, perhaps 6 or 7. The estimated date of the findings, based on argon–argon dating, is between 160,000 and 154,000 years ago. Tools found at the level are Mode-2 Acheulean and some Mode-3 Middle Stone Age.

Tim White has argued that the differences in the Herto crania are sufficient to warrant a new classification, ***Homo***

sapiens idaltu (*idaltu* means 'elder' in the local Afar language), and believes that the subspecies is transitional between *H. heidelbergensis* and anatomically modern humans (White, et al., 2003). Not surprisingly, this view was contested in the same issue of *Nature* in which White presented his analysis. The dissenter was Chris Stringer (2003), who expressed his opinion that there was not enough difference in the Herto finds to constitute a subspecies designation.

Klasies River

At the **Klasies River** sites, a series of five closely situated caves in South Africa, the various members, or levels, date from over 110,000 years ago (some say as old as

Table 8.4	Significant African finds of early *Homo sapiens*/*H. heidelbergensis*			
Site/specimen name	Country	Date (thousands of years ago [kya])	Cranial capacity (cubic centimetres [cc])	Taxonomy
UA 31	Eritrea	1,000	750–800	?
Bodo	Ethiopia	600	1,250	*Homo heidelbergensis* or *H. erectus*
Salé cranium	Morocco	400	860	*H. heidelbergensis* or *H. erectus*
Floresbad I	South Africa	259	—	*H. heidelbergensis*
Rhodesian Man	Zambia	125–300	1,100	*H. heidelbergensis*
Omo I, Omo II	Ethiopia	195	1,400	*H. sapiens*
BOU–VP 16/1	Ethiopia	145–160	1,450	*H. sapiens*
Ngaloba	Tanzania	108–129	1367	*H. sapiens*
Qafzeh IX	Israel	90–100	1508	*H. sapiens*
Skhul V	Israel	90	1520	*H. sapiens*

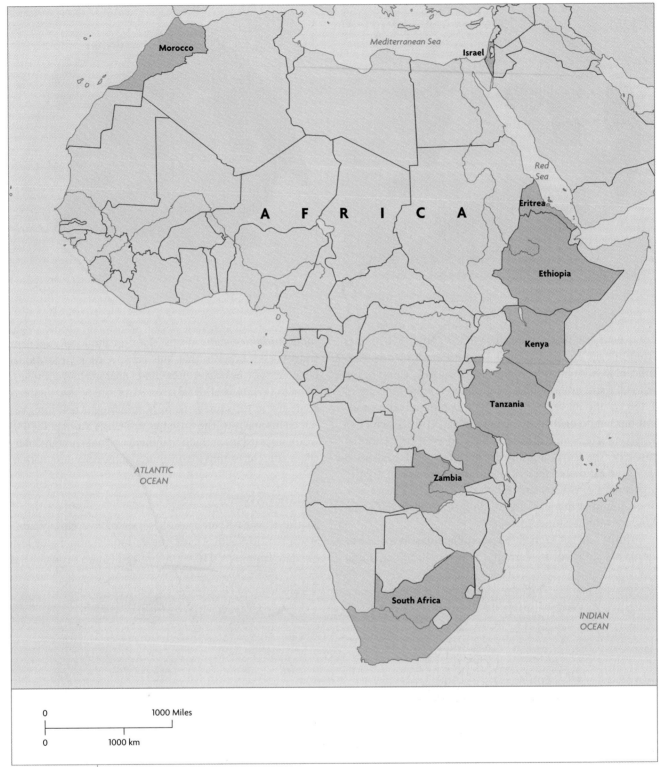

Figure 8.2 | Transitional hominid sites, Africa and the Middle East

125,000 ya) in the lowest level to around 70,000 ya at the Howiesons Poort level. The material discovered here tells us more about the culture of these people than it does of their bodies, there being but a few cranial fragments, mandibles, an ulna, and five metatarsal bones. However, the cut, tear, percussion, and burn marks on the cranial fragments suggest that the individuals were part of a community that practised cannibalism. Their stone tools show a progression from long (10 cm) quartzite blades made from prepared cores to Levallois-prepared flakes (Mode 3) and smaller Mode-4 blades obtained from as far as 20 kilometres away. The people cooked plants and animal foods on hearths, and used a great diversity of food sources, including fish, shellfish, and a variety of plants.

South African archaeologist Hilary John Deacon's useful online *Guide to Klasies River* entertains some interesting ideas about symbolism and the people of Klasies River. While they did not make use of shells or animal teeth to make the perforated beads so celebrated by scholars of the European Upper Paleolithic, they did use **red ochre**, an earth dye traditionally used by peoples in many locations to decorate their bodies with symbols. Red ochre is present at all Late Pleistocene, Middle Stone Age, and Later Stone Age sites in southern Africa, and it is still used extensively throughout Africa for body decoration. Red ochre is more than simply a cosmetic. Together with white clay and black earth it forms a triad of colour symbols used to mark events like rites of passage, from birth to death. The occurrence of ochre, especially together with examples of ground, pointed 'pencils', is convincing evidence of symbolic communication in the Middle Stone Age. Thinking in symbols is a hallmark of the modern mind, and communicating the meaning of symbols presupposes language as we know it (Deacon, 2001).

Interesting as well is Deacon's attempt to connect the Klasies River people not with those more northern Africans who left the continent in the Out of Africa hypothesis, but with the local San or Bushmen, who are distinct both linguistically (incorporating 'clicks' and other sounds in their speech) and physically (gracile, with light brown skin and an epicanthic fold over the eyes). Studies of mtDNA have shown them to have long ago broken off from other African groups, conceivably as far back as 150,000 years ago. More discoveries of human bones at Klasies River may support this claim.

Border Cave

The Border Cave site of South Africa was discovered in 1933 and first excavated in 1940. Traditionally, the cranium known as **Border Cave 1**, with its high, vaulted forehead and generally rounded shape, was considered anatomically modern, though the dating was considered suspect. Recently that has changed. The oldest level of the site has been relatively reliably (but still roughly) dated at 74,500 ± 5,000 years ago. The main reason for disputing the modernity of Border Cave 1, reconstructed from a good number of fragments, concerns the brow ridges, which feature a thickening that resembles the protruding brow ridges of earlier hominids, with no clear bipartite brow ridge. Is possessing that archaic feature enough to challenge Border Cave 1's anatomically modern status? It is difficult to say. The stone tools found at the site are useful for analysis: they reflect a transition from Middle Stone Age (MSA) to Later Stone Age (LSA), ranging from between 50,000 and 30,000 years ago.

John Reader/Science Photo Library

Archaeologists extract fossilized hominid remains from the rock at a site at the Sterkfontein caves, South Africa

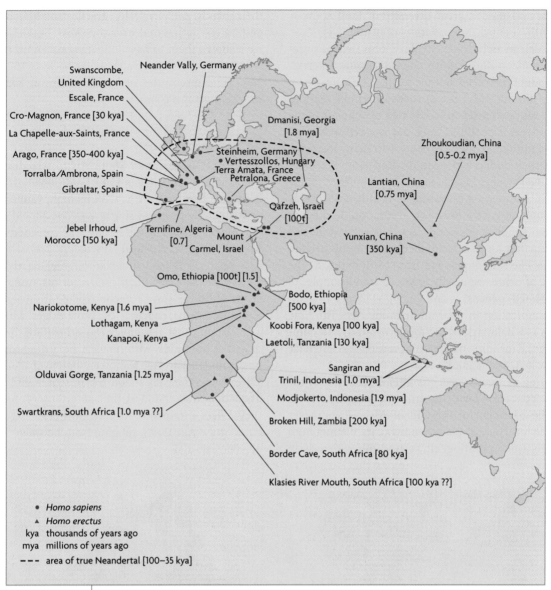

Figure 8.3 | Dates and localities of fossil *H. erectus* and *H. sapiens*

Skhul V

In 1932, T. McCown and H. Movius discovered **Skhul V**, as well as the remains of nine others, in Skhul Cave, near Mount Carmel in Israel. The cranium, like the others in the collection, exhibits a mixture of archaic and modern traits, which made early researchers believe that the population there was transitional between Neandertal and modern humans. On the modern side, Skhul V presents a high, vaulted forehead and a modern occipital (with no occipital bun—see Chapter 9), but there is no bipartite structure to the well-developed brow ridge, and a relatively protruding face, both archaic features. The cranial capacity is large, about 1,520 cc, which is within the range of both modern *Homo sapiens* and Neandertal. The individual is male, judging by its robustness. When the site was thought to be only 40,000 years old—younger than Neandertal remains found in the area—the argument was

put forward that it was transitional. However, it has since been more reliably dated as being about 90,000 years old, diminishing that argument.

Qafzeh Cave

Also discovered in the early 1930s, and very close to Skhul Cave, were the hominid remains of Qafzeh Cave. When research began again in earnest during the 1960s, Bernard Vandermeersch uncovered the most significant find in the cave, **Qafzeh IX**, a skull and nearly complete skeleton, found 'buried' beside the skeleton of a small child. The individual was male (the pubic bones tells us that) and died at about 20 years of age. Like Skhul V, Qafzeh IX has the archaic features of a prognathic lower face (meaning his lower face sticks out) and supraorbital ridges (a prominent brow ridge), and the modern features of a high, vaulted skull, with an estimated cranial capacity of 1,508 cc.

Again the early dating had the finds being relatively young, living about 50,000 years ago, and it was suggested that they were transitional from Neandertal to modern humans. More recent (and more accurate) dating has moved the probable time period back to about 100,000–90,000 years ago. By the 1980s the Skhul and Qafzeh finds were being referred to as proto-Cro-Magnons, meaning that they were transitional between African *Homo sapiens* and the early Europeans who would gradually take over the continent from Neandertals in the period between 40,000 and 30,000 years ago.

In 2006, Conrad Quintyn published a book challenging the proto-Cro-Magnon identification (Quintyn, 2006). Quintyn used a 42-dimensional study of 400 crania (more precisely, 'craniofacial data') from Africa, western Asia, eastern Asia, and Europe to determine which populations the Qatzeh and Skhul crania were most like. It was not Cro-Magnon, so he rejected the proto-Cro-Magnon label. They turned out to be most like both archaic and early modern African and Levantine (i.e., Middle East) crania.

As with the Klasies River finds, researchers have found red ochre on the Qafzeh sites, in this case in the form of 71 unused pieces, in addition to some red-stained bones and shells. As some of the stained bones are thought to have been deliberately buried, it is possible that the ochre was part of a ritual behaviour, which suggests something of the symbolic thought patterns of modern *Homo sapiens*. Interestingly, while the evidence suggests that ochre use was a feature of the early period, the pigment was not used in the area again until about 13,000 years ago.

Significant Chinese Finds

Findings in China are not as extensive or as frequently discussed in scholarly circles as the African and Eurasian finds. However, more and more archaeological work is

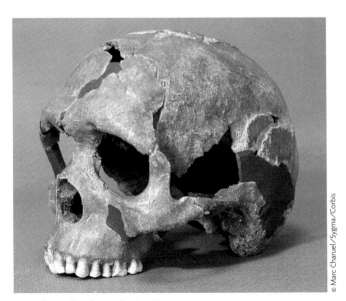

Fossil skull of Qafzeh IX

being done in parts of the country, and Chinese scholars are emerging as more vocal contributors to the anthropological discussion of this period of human history.

Dali Cranium

The **Dali cranium** was discovered in 1978 in the Shaanxi province of China. It was dated at 209,000 ± 23,000 years ago by means of ox teeth also found at the site, although the relationship between teeth and skull is still not absolutely clear. Principal researcher Wu Xinzhi found the vault shape and facial skeleton to be intermediate or transitional between *Homo erectus* and *Homo sapiens*. The vault is more typical of *H. erectus*, while the facial skeleton has a **zygomatic arch** (or cheekbone) that is more characteristic of *H. sapiens* in its relatively delicate nature. The brow ridge is well developed but unique in that it bulges in the centre, between the eyes, and thins as it extends laterally to the side of the head. The cranial capacity is within the 1,100–1,200 cc range. This is slightly larger than the Peking Man finds, as well as the Yunxian *Homo erectus* skulls found by Li Tianyuan, estimated at 350,000 years old with a cranial capacity of roughly 1,065 cc, and the Hexian *Homo erectus* cranium, usually presented as 150,000–190,000 years old though possibly older, with a cranial capacity of 1,025 cc.

Jinniushan Man

Jinniushan Man was found in 1974, named for the mountain where it was discovered in Liaoning province, China. The age of the site is contested: Lu Zun'e (1989) dated it at roughly 280,000 years ago, while Chen, Yang, and Wu (1994) placed it nearer 200,000 ya. The principal fossil specimen is thought to have belonged to an adult male, aged between 30 and 40 years, with a cranial capacity of roughly 1,390 cc. The skull, which was damaged during excavation and required extensive reconstruction, is not as robust as the Dali cranium, especially in the brow ridge area, but the facial skeleton is quite similar in its delicate nature.

Maba Cranium

In 1958, a cranial vault fragment was discovered in a cave near Maba Village, in Guandong province, China. Dated at between 135,000 and 129,000 years ago, it shows the thinning of the cranial vault that typifies the evolution of *Homo erectus* towards *Homo sapiens*. The brow ridges are prominent, but not to the extent of those of the Dali cranium. Unfortunately, not enough of the cranium has been found to calculate a cranial capacity.

Liujiang

Discovered in 1958 by farm workers digging in a cave in Liujiang County in the south of China, this cranium is now the subject of a major controversy. It seems to be that of a modern *Homo sapiens*, with a cranial capacity calculated by CT scan (there is a heavy rock matrix still connected to it) of 1,567 cc. The controversy comes from the dating, with rather disparate estimates of between 111,000 and 139,000 years old. The dating is precise for the material that was analyzed; the problem is that the exact context in which the cranium was found is not clear. The material dated may not be from where the cranium was lodged. If the exact context is discovered, this find could be a major victory for champions of the Multiregionalism hypothesis.

Table 8.5	Significant Chinese finds			
Site/specimen name	Country	Date (thousands of years ago [kya])	Cranial capacity (cubic centimetres [cc])	Taxonomy
Jinniushan	China	200–280	1,390	*Homo heidelbergensis* (?)
Dali	China	209 ffl 23	1,100–1,200	*H. heidelbergensis* (?)
Maba	China	135–129	—	*H. heidelbergensis*
Liujiang	China	139–111 (?)	1,567	*H. sapiens*

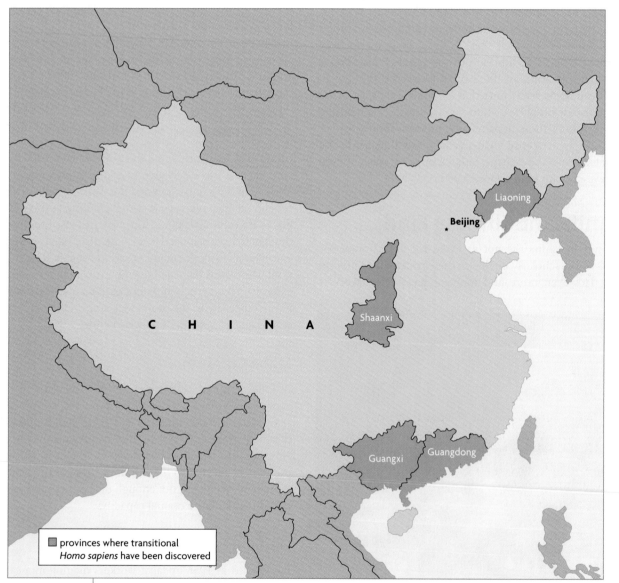

Figure 8.4 | Transitional hominid sites in China

Science Gone Wrong

Beware of Eurocentrism

The cultural explosion of the Upper Paleolithic has been a fruitful subject for paleoanthropological investigation and debate, yet there is one avenue that is usually not explored: the Eurocentric approach that has long dominated the field.

The first region studied archaeologically in any great depth (pun intended) was Europe. The more you study, the more you find, and, naturally, what is found in abundance can take on the appearance of being more important than what is not found in equal measure, or even looked for. A certain locality might have received a great deal of attention not because it was especially significant in terms of world history, but because it was important to Europeans. Archaeology as a scientific discipline first developed in Europe, and Europeans researched first and most extensively sites in their own backyard, along with a few sites in areas of special significance to Western culture (e.g. Egypt). It's also important to remember that archaeology requires a well-developed university system to train and fund archaeologists so that a critical mass of scholars develops in a country; Europe, of course, has long had a rich and supportive academic infrastructure. But compare that with the situation in, say, China, a country with a tremendous amount of rich archaeological ground to cover. After the Communist Revolution succeeded in 1949, China became essentially out of bounds to Western researchers, and it took some time for the country to develop its own archaeological community (Davidson Black broke from the academic norm in his active support of Chinese archaeologists) and to open its doors to Western scholarship. In other words, it is entirely possible that the sudden change that is characteristic of the European Upper Paleolithic is characteristic only of Europe and not of eastern Asia; we can't be certain until more research has been done at sites in China and other parts of the Far East. It would be misleading, in the absence of related findings from sites in Asia, to generalize based on evidence from just one area.

This is but one form of **Eurocentrism**, which is the tendency to see things only from a European perspective, or to generalize from a European experience as though it were a universal phenomenon. A classic example that resonates particularly with North American students is the Eurocentric notion that Christopher Columbus 'discovered' two continents that had long been populated by millions of people already very much aware of the continents' existence.

So is 'culture' rooted in traditions that began with early humans in Europe? Researchers who have looked at the artistic work done in the Klasies River Caves of South Africa see evidence of symbolic thought long before the 40,000 ya start-date of the Upper Paleolithic in Europe. Writing about the Klasies River and related sites, Deacon states:

> From the perspective of the evidence from Klasies River and other African sites, rather than marking universal stage in human progress, the Upper Palaeolithic can be seen as a still impressive but specifically regional behavioural response to unique environmental conditions in the northern high latitudes. The roots of modern humans are African and long predate the Upper Palaeolithic. (Deacon, 2001)

Summary

This chapter introduces several alternative ways of viewing our early human history. These alternatives can be assessed through simple questions. Do you accept the splitter's multiplicity of species, or the single species of the lumper taxonomy? Is trait or species a more scientifically meaningful focus during this time? Are behaviourally modern traits just as important as or even more important than anatomically modern traits during this time? Which of the three hypotheses surrounding the development and spread of modern human beings will ultimately prove the most insightful? There are no definitive answers yet, but at least you know the questions.

Typical Student Questions

▶ *So when, really, did modern Homo sapiens first appear?*

Good question. At least skeletally, 195,000 years ago: that's a sound date for the first AMH, or anatomically modern human. The first BMH (behaviourally modern human) or 'CMH'—culturally modern human, if we could coin the term—is a whole other matter that has yet to be firmly established.

Review Questions

1. What are the Schöningen spears? What is their significance?
2. What are the four different modes of tool technology outlined in this chapter? What are the characteristic, or diagnostic, tools for each?
3. Identify the traits that distinguish *Homo erectus* from *Homo sapiens*.
4. Explain what is meant by mosaic evolution.

Discussion Questions

1. '*Homo heidelbergensis* is a dumping ground for fossil hominids that are neither clearly *Homo erectus* nor *Homo sapiens*.' What does this statement mean? What does it imply? Do you agree with it?
2. Name some of the apparently sudden cultural changes that occurred during the Upper Paleolithic. Describe some of the theories concerning what brought about these rapid changes. How might proof for one of the different theories be established? How might the information now coming from outside Europe (e.g. Africa and Asia) condition your answer?
3. Which of the three models or hypotheses of the origin and spread of modern *Homo sapiens* do you think is the most credible? Explain with an outline of each hypothesis.

Key Specimens and Species

Arago XXI (or Tautavel Man)
Bodo cranium
Border Cave 1
Cro-Magnon 1
Dali cranium
Florisbad I
Heidelberg jaw
Herto (or *Homo*) *sapiens idaltu*

Homo antecessor
Homo cepranensis
Homo heidelbergensis
Homo neanderthalensis (or Neandertal)
Homo rhodesiensis (or Rhodesian Man)
Homo sapiens

Jinniushan Man
Kabwe (or Broken Hill) skull
Klasies River
LH 18
Oase-1
Omo I
Omo II

Paglicci 23
Petralona skull
Qafzeh IX
Salé cranium
Sima de los Huesos
Skhul V
Steinheim skull
Swanscombe skull

Key Terms

Acheulean
anatomically modern humans (AMH)
archaic *Homo sapiens*
Aurignacian
behaviourally modern humans (BMH)
bipartite

blade
Cambridge Reference Sequence (CRS)
Clinal Replacement
Clines
compassion
core tool
coronal suture

Cro-Magnon
Eurocentrism
flake tool
haplo-group
incus (or anvil)
Later Stone Age (LSA)
Levallois technique
malleus (or hammer)

members
Middle Stone Age (MSA)
mitochondria (sing. mitochondrion)
mitochondrial DNA (or mtDNA)
mitochondrial Eve
mosaic evolution

Mousterian
MRCA (Most Recent
  Common Ancestor)
mt-MRCA (matrilineal most
  recent common ancestor)

Multiregionalism
nonrecombinant genes
Oldowan
Out of Africa
pressure flaking

red ochre
ring species
Schöningen spears
shovel-shaped incisors
stapes (or stirrup)

Upper Paleolithic
Y-chromosome
zygomatic arch

Key Individuals

Fritz Berckhemer
Stephen Jay Gould
Louis Lartet

Mary Leakey
Richard Leakey

Franz Weidenreich
Spencer Wells

Pamela Willoughby
Milford H. Wolpoff

Recommended Print and Online Resources

American Museum of Natural History. (Website). Atapuerca: The humans of Sima de los Huesos. Retrieved from www.amnh.org/exhibitions/atapuerca/simahumans/index.php

Bischoff, J. L., Shamp, D. D., Aramburu, A., Arsuaga, J. L., Carbonell, E., & Bermudez de Castro, J. M. (2003). The Sima de los Huesos hominids date to beyond U/Th equilibrium (>350 kyr) and perhaps to 400–500 kyr: New radiometric dates. *Journal of Archaeological Science, 30*, 275–80.

Bischoff, J. L., Williams, R. W., Rosenbauer, R. J., Aramburu, A., Arsuaga, J. L., Garcia, N., & Cuenca-Bescós, G. (2007). High-resolution U-series dates from the Sima de los Huesos hominids yields 600 kyrs: Implications for the evolution of the early Neanderthal lineage. *Journal of Archaeolical Science, 34*, 763–70.

Bocquet-Appel, J-P, & Arsuaga, J-L. (1999). Age distributions of hominid samples at Atapuerca (SH) and Krapina could indicate accumulation by catastrophe. *Journal of Archaeological Science, 26*(3), 327–38.

Boaz, N., & Ciochon, R. (2004). *Dragon Bone Hill: An Ice-Age saga of Homo erectus*. Oxford: Oxford University Press.

Cann, R., Stoneking, M., & Wilson, A. C. (1987). Mitochondrial DNA and human evolution. *Nature, 325*(6099), 31–6.

Caramelli, David, Milani, L., Vai, S., Modi, A., Pecchioli, E., Girardi, M., . . . Barbujani, Guido. (2008). A 28,000 years old Cro-Magnon mtDNA sequence differs from all potentially contaminating modern sequences. *PLoS One, 3*(7). e2700.

Cartmill, M., Smith, F. H., & Brown, K. B. (2009). *The human lineage*. Hoboken, NJ: Wiley.

Chen, T., Yang, Q., & Wu, E. (1994). Antiquity of *Homo sapiens* in China. *Nature, 368*, 55–6.

Dennell, R. (1997, February 27). The world's oldest spears. *Nature, 385*, 767–8.

Deacon, H. J. (2001). Guide to Klasies River. Retrieved from http://academic.sun.ac.za/archaeology/KRguide2001.PDF

Ingman, M., Kaessmann, H., Pääbo, S., & Gyllensten, U. (2000, December 7). Mitochondrial genome variation and the origin of modern humans. *Nature, 408*, 708–13.

Lu, Z. (1989). Date of Jinniushan man and his position in human evolution. *Liaohai Wenwu Xuekan, 1*, 44–55.

Manzi, G., Mallegni, F., & Ascenzi, A. (2001, August 14). A cranium for the earliest Europeans: Phylogenetic position of the hominid from Ceprano, Italy. *Proceedings of the National Academy of Sciences of the U.S.A., 98*(17), 10011–6.

Mounier, A., Marchal, F., & Condemi, S. (2009). Is *Homo heidelbergensis* a distinct species? New insight on the Mauer mandible. *Journal of Human Evolution, 56*(3), 219–46.

Oppenheimer, S. (2004). *The real Eve: Modern man's journey out of Africa*. New York: Carroll & Graf.

Quintyn, C. B. (2006). *The Morphometric Affinities of the Qafzeh and Skhul Hominans: Method and Theory*. Bloomington, IN: AuthorHouse.

Wells, S. (2006). *Deep ancestry: Inside the Genographic Project*. Des Moines, IA: National Geographic.

White, T. D., Asfaw, B., De Gusta, D., Gilbert, H., Richards, G., Suwa, G., & Howell, F. C. (2003, June 12). Pleistocene *Homo sapiens* from Middle Awash, Ethiopia. *Nature, 423*, 742–7.

Zin'e, Lu. (1989). Date of Jinniushan man and his position in human evolution. *Liaohai Wenwu Xuekan, 1*, 44–55.

Neandertal

Fossilized Neandertal skeleton, found at
the Kebara Cave, Mount Carmel, Israel
Volker Steger/Nordstar—4 Million Years
of Man/Science Photo Library

After reading this chapter, you should be able to:

- Outline the basic arguments that support both the position that Neandertal has made a contribution to our modern human gene pool and the position that Neandertal is merely the last extinct hominid species.

- Outline and assess the basic hypotheses concerning why Neandertal became extinct.

Over the century and a half since the labourers offhandedly shovelled out the bones of the man from the Neander Tal, scholars have sought to understand him and people like him. This striving has led them to reshape, even to re-create, their notion of the world. It has been a restless quest for a reality that can accommodate both these ancestors and their descendants, the humans who study them. Through the years, the shifting paradigms have drawn as much on the scientists and scholars themselves, and the times and social climates in which they lived, as they have on the ancient bones.
(Trinkaus & Shipman, 1994, pp. 6–7)

Introduction

One of the greatest mysteries in physical anthropology is the relationship between Neandertal and anatomically modern humans (or AMH). The standard story currently is that AMH arrived in Europe from Africa about 40,000 years ago, and within 10,000 years or so (i.e. about 30,000 ya), Neandertal had disappeared from Europe (and, of course, from the world altogether). The question we ask is simply, 'What happened?'

Before we address this question, I should point out that I am using the term **Neandertal** here for a specific reason. It allows me to take an intellectually neutral position (although you probably know by now that that I am rarely 'intellectually neutral'). If I were to use the taxonomic designation *Homo sapiens neanderthalensis*, I would be taking the position, associated with the **Multiregionalism** hypothesis, that Neandertal was a subspecies of *Homo sapiens*, and thus most likely in some way a significant contributor to our modern human gene pool. The remains of comparable subspecies are just waiting to be found in China and Indonesia, the argument goes. On the other hand, if I were to use the term *Homo neanderthalensis*, I would be signalling a belief that it represents a separate species that is most likely out of our gene pool. Of course, we could subscribe to a third view—that it could be in the gene pool but a separate species. This is consistent with the **Out of Africa** hypothesis, which states that local variants—different species—disappeared.

By now, you should know my position that the species line does not necessarily preclude reproduction with fertile young of individuals from different species. If there is to be a second edition of this textbook, I might be backing this claim up with the example of the parrot offspring of Louis and Stanee, two different species of conure. I should point out, as well, that following current convention, I am spelling the name Neandertal without the –h–, which is not pronounced in modern German.

Unlike the case of species discussed in earlier chapters, we have a lot of evidence here. The remnants of more than 400 Neandertal have been unearthed over the last century-and-a-half, more than what we have for any other group of fossil hominins. Still, as they are scattered over time and space, and as we do not have anything like a cemetery that would present a good cross-section of the population, interpreting all of this data is not an easy matter.

As I say in class during my lectures on Neandertal, this whole unit is dedicated to helping you, the student reader, put together an answer to two important questions. The first is whether or not Neandertal is in our gene pool. In other words, did Neandertal interbreed with AMH to a significant enough extent that *some* of their genes still exist in at least *some* humans living today? You may read in a number of textbooks and other source materials that, swayed by studies of **mitochondrial DNA (or mtDNA)** genetic evidence, the jury hearing the case of Neandertal *v.* Modern Human has returned a verdict of 'not part of our gene pool'. But I think that such a position is premature, that what we have, really, is a hung jury. Certainly, many anthropologists—the majority, in fact—believe that Neandertal is not even a partial ancestor to some of us. But truth is not a popularity contest. There are good reasons to take the opposite position.

The second question, one that is the focus of a very lively intellectual discussion right now in the anthropological literature, is why Neandertal disappeared as a (sub)species. A sub-question here is what role, if any, anatomically modern *Homo sapiens* played in this disappearance—conqueror? superior? competitor? non-player? Remember that we are most likely looking at an evolutionary competition here. The physical conquering of one group by another is typically at best a minor player, and usually a non-player, although this was proposed by German physical anthropologist Hermann Klaatsch in *The Evolution and Progress of Mankind* (Klaatsch, 1923). There is no strong evidence for such a 'genocidal' war. If life among the early humans were a UFC match, the much bigger and stronger Neandertal would whump the AMH opponent nearly every single time. When considering an evolutionary competition, the answer is typically tied to which one of two varieties is more 'fit' for the environment that they are trying to survive in. As we will see, **Cro-Magnon** probably had better weapons for hunting animals, and ones that, unlike the Neandertal weapons, could kill from a distance. Maybe, rather than genocide, there was simply enough back-and-forth violence between the two that the underdog Neandertal was forced to move when and where it was not wise to move, just as other large animals do when their environment is crowded out by humans. We have a bad track record in that regard.

One of the problems in reviewing the literature concerning early humans is that the position different anthropologists take (and have taken) on the matter of Neandertal is often influenced heavily by theories they hold generally about evolution, race, and a whole raft of social issues. It is hard to let the facts speak when theory is shouting so loudly (like a parrot that wants to be fed—Stanee is making jungle noises as I write this). A second warning is warranted here. We will be returning to the dizzying world of genetics in this chapter. Hold onto your occipital bun (see below for explanation) if you have one.

Neandertal in Popular Culture

Unlike most of the species and subspecies discussed to this point, Neandertal is well known. You've heard the name, at least—it's been used for decades as an insult term for someone oafish and unsophisticated. *I* have been called a Neandertal, though I'm informed enough to take it as a compliment (and, yes, I may be big and unsophisticated, but I am not stupid). *A. afarensis* never had its own comic book, but Neandertal has figured prominently in a good number of novels and movies that have portrayed our hulking forebear either negatively, as big, dumb and vicious, or positively, as mystical, telepathic, and tragic.

Robert J. Sawyer is a Canadian science fiction writer who put together a well-researched and well-written trilogy called the Neanderthal Parallax. The three books—*Hominids*, *Humans*, and *Hybrids*—deal with the anthropological, technological, and philosophical consequences of the existence of parallel worlds: one, our world, in which *Homo sapiens* became the sole surviving hominid, with all the mistakes and kinds of foolishness that that implies; and the other, in which Neandertal became a surviving species that has systematically made better choices than we have. The parallel worlds were created during (and as a result of) the emergence of 'consciousness' about 40,000 years ago during the 'cultural revolution' of the Upper Paleolithic. The main character is Ponter Boddit, a physicist in the Neandertal world, who travels to the world of *Homo sapiens*, falls in love with *H. sapiens* geneticist

'Neanderthal Man': Anthropology's pop cultural bad boy

Mary Vaughan (despite her presence of chin and absence of brow ridges and occipital bun). Among the interesting issues they have to deal with are their concerns that the child they intend to have will become religious because it is part *H. sapiens* (who have religion, possibly as part of their genetic baggage, unlike the case with Neandertals) and their worry over how to stop humans from pillaging the relatively unspoiled world of the Neandertals.

Sawyer's series shows how an awareness of the existence of Neandertal in our past inspires human creativity, just as learning about physical anthropology can do. Physical anthropologist and mystery writer Aaron Elkins makes the hybrid possibility the background of a murder mystery in his 2008 book *Uneasy Relations*. Here, in a cave high in a mountain in Gibraltar, scientists find the fossil remains of a *Homo sapiens* woman holding a part-*H. sapiens*/part-Neandertal child nicknamed 'Gibraltar Boy'. Trust me: it's a good book.

John Darnton's 1996 bestseller *Neanderthal*—a title that's hard to forget—explores the popular theme that surviving Neandertal have special powers. On the back cover, enticing the potential reader, are the following words:

> In the remote mountains of central Asia, an eminent Harvard archeologist discovers something extraordinary. He sends a cryptic message to two colleagues. Matt Mattison and Susan Arnot—once lovers, now academic rivals—are going where few humans have ever walked, looking for a relic band of creatures that have existed for over 40,000 years, that possess powers man can only imagine, and that are about to change the face of civilization forever. (Darnton, 1996, back cover)

And if you think that Sawyer's, Elkins's, or Darnton's stories are a bit far-fetched, you should read (well, no you shouldn't) how the theme of Neandertal secret knowledge is picked up by members of the lunatic fringe and New Age crowd, such as 'paranormal researcher' Stan Gooch, who has recently published *Neanderthal Legacy: Reawakening Our Genetic and Cultural Origins* and *Dream Culture of the Neanderthals: Guardians of the Ancient Wisdom*. I don't know about 'ancient wisdom', but I do know modern ignorance when I see it in the title of a book.

Location and Time Period

So far, fossilized Neandertal remains have been discovered in western Europe (Belgium, Britain, Germany, France, Gibraltar, Spain, Italy), eastern Europe (Czech Republic, Croatia, Hungary, Russia, Romania, Ukraine, Uzbekistan), the Middle East (Iraq, Israel), and possibly as far east as southern Siberia (see Table 9.1).

It would be an understatement to say that the extent of the time period during which Neandertal lived is contested. The figures given for the start and end dates vary significantly. I used to tell my students that they lived from around 130,000 years ago to 30,000 ya, but the beginning date is now often given as 250,000–200,000 ya, while the ending date has been set as low as 28,000–24,000 ya. The former is highly difficult to assess, as there is a fine line between being a Neandertal and having Neandertal-like features (which could be said of *Homo erectus* and certainly of *H. heidelbergensis*). Remember that an argument can be made to have some European fossil specimens (e.g. the Swanscombe Skull) now classified as *Homo heidelbergensis* relabelled Neandertal.

It is important to note two significant facts about the span of Neandertal on earth. One is that it intersected with some significant glacial episodes during the Ice Age, so Neandertal had to adapt, physically and culturally, to living in a kind of subarctic/arctic climate. Their adaptations to environments of extreme cold are sometimes referred to in the literature as *hyperarctic*. A second fact is that there was a significant period of overlap—at least 10,000 years—in which Neandertal and our brand of *H. sapiens* both lived in Europe and southwestern Asia. They were neighbours of a sort.

The Physical Description

Give me five Neandertals on my team and we could beat any football team on earth. We would have no great speed to speak of (short legs) and there would be no passing (arms are too short, and there is a lack of height), but our ground game (our running backs would be more like 'aggressive walking backs') would rule. Helmets would be optional, unless we are playing against a *Homo erectus* team—then we might change strategy and start passing.

To begin with, it should be pointed out that Neandertal came in a small range of sizes and builds, although this range would have been significantly less than what we find among modern humans, as their numbers and locations were significantly less. Think of it more like the range in appearance you'd find among a particular group of people, say the Inuit (an example I've chosen for good reason as you will see). In a way, in terms of their numbers, that is what they were: a group. There could have been thousands of Neandertals at any given time, but probably not tens of thousands.

In helping you understand what Neandertal's body build was like, it is useful to say that they were well adapted to cold climate living, and that they reflected well two biological rules or principles developed in the nineteenth century: Allen's rule and Bergmann's rule. The former was put forward by American zoologist **Joel Asaph Allen** (1838–1921) in 1877. The rule relates to animals that are **endotherms**, that is, animals such as mammals, birds (including, of course, parrots), certain fishes, and insects that are capable by internal means (*endo-* means 'inside') of generating enough heat to maintain a relatively high core body temperature. It is similar to, but not the same as, the more familiar term 'warm-blooded', which technically is not accurate. **Allen's rule** states that endotherms living in colder climates have shorter limbs (legs, arms, or wings) and appendages (beaks, ears, fingers, toes, or tails) than do those native to warmer climates. It has to do with surface area and the conservation or dissipation of heat: the smaller the surface, the greater the conservation of heat, and vice versa. My favourite examples of this principle are lagomorphs (rabbits and hares). You see pictures on television of the jackrabbits that live in the southern US and you see huge ears—Bugs Bunny ears. The eastern cottontails that I am familiar

with, and which I often see when I walk my dogs along the streets of small-town Bolton, Ontario, have much smaller ears—you can barely see them. Most comparisons for humans contrast the Inuit (of Siberia, Alaska, Arctic Canada, and Greenland), who traditionally would have shorter legs and be just a few inches over five feet tall, with the Maasai of Kenya and Tanzania (think of people with the arms and legs of erstwhile Toronto Raptor—now traitor—forward Chris Bosh, or google up an image of former NBA centre Manute Bol). Neandertals as a cold climate (sub)species had short lower leg bones (tibia and

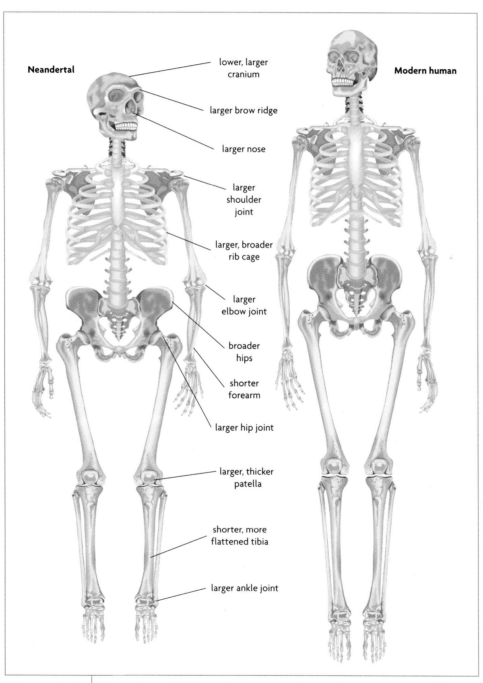

Figure 9.1 Neandertal (left) and anatomically modern human (right) skeletons compared

Artists' impressions of Neandertal: (left) early twentieth-century artist's rendering of Neandertal based on Boule's reconstruction of remains found by the Bouyssonie brothers at La Chapelle-aux-Saints (see page 209); (right) a more contemporary rendition of adult male Neandertal

fibula) and lower arm bones (radius and ulna) relative to their (and our) femora and humeri.

Bergmann's rule, which, like Allen's rule, applies to endotherms with a long north–south geographic range, was developed by German biologist **Christian Bergmann** (1814–1865) in 1847. It deals with body shape (stocky versus slender, robust versus gracile). Some controversy surrounds the application of the rule, such as whether it can apply between species or only within a species. While the former works when comparing polar bears with any other more southerly bears (e.g. *Ursus americanus*, or black bears), the version we will use here is the subspecies

or variant version. It asserts that the subspecies or variants of larger and stockier size are found in the colder parts of the species' range, while those that are smaller and slenderer are found in the warmer parts of the range. This works for koalas, those cuddly-looking (I hear they have a nasty temper) Australian marsupials. In the cooler climate of the southern state of Victoria, males typically weigh 12 kilograms and females 8.5 kilograms. By contrast, in the more tropical state of Queensland, males average 6.5 kilograms and females just over 5 kilos. So if you're asked to coach an all-koala Australian Rules football team, recruit from Victoria, not Queensland.

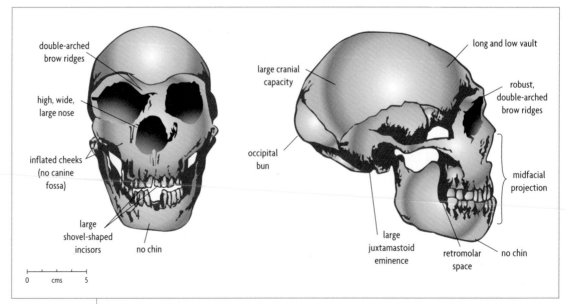

Figure 9.2 | Neandertal skull and teeth

Marcellin Boule's Reconstruction of the Holotype of Neandertal

Between 1911 and 1913, **Marcellin Boule** (1861–1942), a prominent French anthropologist and editor of the journal *Annales de Paléontologie*, reconstructed Neandertal in a way that would influence the thinking of anthropologists and the popular perception of the species for decades to come. Most unfortunately, it was a very distorted image, one that reflected overwhelmingly Boule's preconceived notions of what Neandertal should look like: more ape than human. The main subject of Boule's reconstruction was the fossil remains discovered in 1908 by the Bouyssonie brothers (and fathers—they were then recently ordained priests), Amédée and Jean, in a small cave near the village of La Chapelle-aux-Saints in central France. It was, at the time, the most complete Neandertal skeleton ever found. Where parts were missing or too diseased to descriptively reconstruct the species, Boule chose parts from the La Ferrassie, Spy, or Feldhofer specimens. The composite picture he produced was, to use Trinkaus and Shipman's apt words, 'astonishingly wrong':

> At every turn, Boule emphasized the physical similarities between Neandertals and apes and distanced the fossil humans from living people. He reconstructed the vertebral column of Neandertals as much straighter than that of modern humans, giving rise to a stooping posture and slouching gait. Boule's Neandertal had a forwardly thrust head—which somehow highlighted its elongated shape, protruding face, and large brow-ridges—perpetually bent knees, and a widely divergent big toe. The drawing in his monograph imprinted itself on the minds of anthropologists everywhere. It was the perfectly imperfect troglodyte: the brute, the savage. And, of course, it was not Boule's or anyone else's ancestor. (Trinkaus & Shipman, 1994, p. 190)

Trinkaus and Shipman give a good account of the elements Boule got wrong. Consider, for example, his reconstruction of the big toe of Neandertal. Boule had it sticking out from the other toes, the way it would appear on the grasping foot of a gorilla or chimp. There was no evidence to suggest that this was the way the toe should be positioned. Boule's good friend Henri Martin had recently found, at La Quina, the ankle bones of a Neandertal, and had judged (rightly) that the way the head of those bones was positioned indicated that the big toe was in line with the others. Martin had already published his work on that

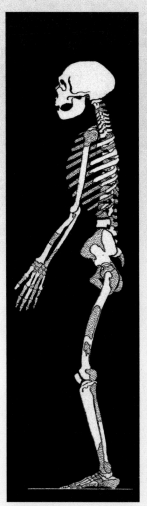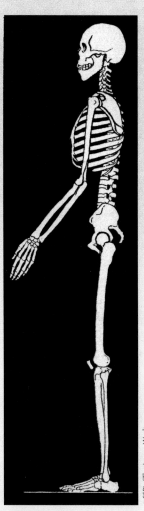

Bent knee, stooped posture: Boule's reconstruction of skeletal Neandertal (left) compared with the skeleton of an Australian Aborigine, from an article that appeared in *Nature*, August 1913

© SSPL/The Image Works

reconstruction when Boule wrote his articles, but the latter either ignored or overlooked it.

As for the bent knees, there had never been any solid evidence behind them. Belief that Neandertal walked with bent knees is just a holdover from a misconception held by previous researchers and based, to a degree, on the large size of the Neandertal patella (knee bone) of one of the finds at Spy d'Orneau in 1886. The current thinking is that those big patella would have made Neandertal a good power lifter. If I were moving and I had Neandertal friends, they would be the first people that I would call.

The cold climate–adapted Neandertals had a compact body with significant weight for their height (like that old joke about being short for your weight). According to the Smithsonian Institution, males averaged about 1.7 metres (5′ 5″) tall, while weighting an estimated 84 kilograms (185 lbs). Females were about 1.5 metres (5′) tall, weighing an estimated 80 kilograms (176 lbs) (Smithsonian National Museum of Natural History).

The head is a clear identifier. It is both longer and broader than that of a typical human. In the front we would see significant brow ridges, a low forehead, a big nose and cheekbones, and no pointed chin. The skull is swept back in a shape reminiscent of the engine of a big Harley-Davidson. At the back is an **occipital bun**, simply a bulge of the occipital bone at the back of the head. There is an old and somewhat fruitless debate about whether a Neandertal, if you dressed it up, could travel unnoticed on a subway or bus (sounds like the sort of question asked by guys at a bar with too much time on their hands, waiting for the hockey game to come on). I think it would be noticed.

Different-looking or not, it should be pointed out that Neandertal's brain size is impressive. Based on those crania from which we can get a good approximation, the average cranial capacity of an adult Neandertal is about 1,450 cubic centimetres, about 100 cc more than that of the average modern human. This does not mean that Neandertal was any smarter than we are (just as we cannot argue that modern men, judging by average skull size, are smarter than women). It reflects more the large build of the body. Not surprisingly, the context-cheating, cherry-picking, creationism-toting websites claim that this brain size difference contradicts evolution (i.e. that we are somehow going backwards from bigger to smaller); needless to say, it does not.

Early Neandertal Finds

The Feldhofer Find: Naming the Other

The fossil specimen that would give Neandertal its name was unearthed in August 1856 from the Kleine Feldhofer Grotte (Little Feldhofer Grotto), a cave by a limestone quarry (where have we heard that before?). The valley

Table 9.1	Timeline of significant Neandertal fossil finds		
Year	Site	Country	Discoverers
1829–30	Awir Cave II, Engis	Belgium	P.C. Schmerling
1848	Forbes Quarry	Gibraltar	Lt Edmund Flint
1856	Little Feldhofer Grotto	Germany	Johann Fuhlrott
1866	Trou de la Naulette Cave	Belgium	Edouard Dupont
1874	Pontnewydd	Wales, UK	W. Boyd Dawkins
1876	Rivaux	France	(?)
1910	La Quina Shelter	France	Henri Martin
1924	Kiik-Koba	Ukraine	Gleb Bonch-Osmolovskii
1925	Fischer and Kämpfe quarries	Germany	F. Weidenreich, O. Kleinschmidt, E. Vlcek
1926	Devil's Tower Cave	Gibraltar	Dorothy Garrod
1929	Saccopastore	Italy	Mario Graziosi
1930	Mount Carmel, Tabun	Israel	Dorothy Garrod
1933	Mount Carmel, Skhul	Israel	team
1935	Saccopastore	Italy	H. Brenne, Alberto Blanc
1939	Guattari Cave, Mount Cicero	Italy	Alberto Blanc
1953	Shanidar Cave	Iraq	Ralph Solecki
1964	Vertesszollos	Hungary	Laslo Vertes
	Verdouble Valley, Tautavel	France	Henry de Lumley
1965	Jebel Qatzeh Cave	Israel	Bernard Vandermersch
1972	Bilzingsleben	Germany	Dietrich Mania
1979	Saint-Cesaire	France	Francois Lévêque
1983	Kebara Cave	Israel	Lynne Schepartz
2000	Mezmaiskaya Cave	Russia	Igor Ovchinnikov, Kirsten Liden, William Goodman
2002	Neander Tal	Germany	Fred Smith

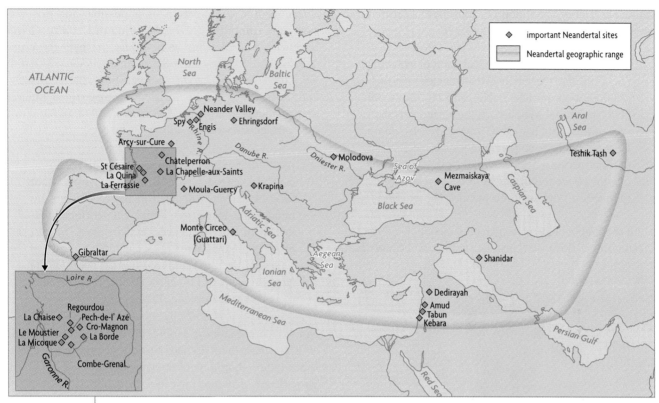

Figure 9.3 | Sites of notable Neandertal fossil finds

(–*t(h)al* in German) was named after a local seventeenth-century composer, Joachim Neumann, who 'Grecofied' ('Greekized'?) his name to Neander (lit. 'New-man'). Workmen presented a local teacher and amateur scientist (as so many educated Europeans were at that time), **Johann Karl Fuhlrott** (1803–1877), with what they thought was a 'cave bear'. The specimen, known as Feldhofer I or Neandertal I, comprised:

- a skull cap;
- two femora;
- the right humerus, ulna, and radius;
- the left humerus and ulna;
- the ilium (part of the hipbone); and
- fragments of scapula and ribs.

Fuhlrott, well aware of the significance of the find, enlisted the aid of Hermann Schaaffhausen, an anatomy professor at the nearby University of Bonn, to announce the find to the world, in 1857. Schaaffhausen believed the remains represented a distinct species, normal for its type. This would become a minority view, as more prominent German scientists, notably the academically powerful **Rudolf Virchow** (1821–1902), said that it was a human afflicted with a pathology that made him appear different. He focused in particular on the slender left humerus and damaged left ulna, which Schaaffhausen had correctly identified as a healed fracture. Many Neandertal fossil specimens have healed fractures, as they lived a life of frequent danger and risk. For Virchow, however, this showed that the fossil specimen was of a pathological human.

Virchow's conclusions concerning the specimen's differences are not surprising. He has been recognized as the founder of the field of pathology, so, just as they say that to a person with a hammer every problem is a nail, to this pathologist every difference was a defect. Virchow's view of Neandertal as pathological human with defects is sometimes found today on creationist websites, which claim that he was ignored or overlooked because of his anti-evolution stance on this and other matters.

Another prominent German scholar, August Franz Mayer, assigned the structural differences in the Feldhofer skeleton to lifestyle and disease. Mayer thought that the curvature of the femora and the structure of the pelvis indicated that Feldhofer had been riding horses from when he was very young, making him possibly a Mongolian Cossack of the Russian army. He also argued that the specimen had the weak bones of someone with rickets, a disorder that weakens the bones' ability to support weight, and causes a curving of the support bones. This, though, came in spite of the fact that Feldhofer's femora were very solid and thick, the very opposite of rickety bones, which are quite fragile. The large brow ridges Mayer assigned to a lifetime of using the facial muscles of the forehead (e.g., as with frowning). In 1864, English biologist Thomas Henry Huxley (sometimes known as 'Darwin's bulldog') beautifully caricatured this position, saying:

> Given a rickety child with a bad habit of frowning (say from the internal disturbances to which such children are especially liable), and the result will be a Neandertal Man! . . . [T]he Neandertal man was nothing but a rickety, bow-legged frowning Cossack, who, having carefully divested himself of his arms, accoutrements, and clothes (no traces of which were found), crept into a cave to die. . . . (as cited in Trinkaus & Shipman, 1994, pp. 58–9)

It should not be surprising that many scientists rejected Neandertal at this point, just as Virchow and Mayer had

done. Its first presentation as a possible ancestor was two years before the 1859 publication of Darwin's *Origin of Species*. But Darwin's work would not help the cause very much. Not long after Schaaffhausen's description of Feldhofer was translated into English, C. Carter Blake—aptly described as an 'amateur geologist/geographer/anthropologist' (Trinkaus & Shipman, 1994, p. 82)—was quick to see difference as idiocy:

> All these characters are compatible with the Neanderthal skeleton having belonged to some poor idiot or hermit, who died in the cave where his remains must have been found. They are incompatible with the evidences which must be left in a Westphalian [part of Germany] bone-cave of a normal healthy uninjured human being of the *Homo sapiens* of Linnaeus. (Blake, 1862, as cited in Trinkaus & Shipman, 1994, p. 83)

Not all commentators would be quite so critical. For example, prominent French scholar and 'brain expert' Paul Broca (after whom one of the language areas of the brain, 'Broca's area', is named) rejected the automatic assessment of idiocy:

> As we have never seen such a skull, and do not wish to admit that it belonged to a race of which no other vestige remains to us, we are forced to seek a morbid origin for the peculiarities which it presents. . . . Idiocy, competent to produce a cranium of this kind, is necessarily microcephalic; now this skull is not microcephalous, therefore it is not that of an idiot. (Broca, 1864, as cited in Trinkaus & Shipman, 1994, p. 83)

In 1864, Irish anatomist Thomas King, a one-time student of the great geologist Charles Lyell, officially named the fossil specimen *Homo neanderthalensis*.

Other Early Finds

It should be noted that Feldhofer was not the first Neandertal dug up. In late 1829 or early 1830, Belgian physician and anatomist Phillipe-Charles Schmerling found the partial skull of a 2- to 3-year-old Neandertal child in Awir Cave II, near Engis in Belgium. This skull was not recognized as that of a Neandertal until 1936. Likewise, a cranium discovered in 1848 in Forbes Quarry, Gibraltar, was recognized as Neandertal in 1864.

In 1880, schoolteacher Karel Maška, who would become an important foundational figure in eastern European anthropology, discovered the mandible of a Neandertal child possibly 8 or 9 years old in Šipka, Czech Republic (then in what was known as Moravia). Rudolf Virchow was so certain of his Neandertal-as-pathology

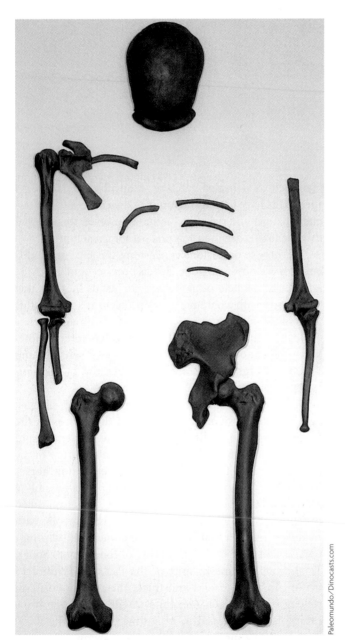

Skeletal remains of Feldhofer I (Neandertal I)

hypothesis that he took the apparently contradictory evidence of a robust jaw and a child's dental development state to say that it was the jaw of an adult with an abnormal, pathological dental eruption pattern. In the words of Mark Twain, 'how empty is theory in the presence of fact.'

The Spiritual Cannibal: Remaking Neandertal into a Human

False images of Neandertal have not been confined to pathologists' reports and caricatures of a slouching, knee-bent, oafish, apelike hominid. Positive images formed from hypotheses of Neandertal 'humanity' can be just as untrue if they run roughshod over 'the facts' that more objective eyes cannot fail to see.

In 1939, on Monte Circeo, Italy, Neandertal emerged from its dark cave in a new (and more favourable) form that was brought to light in a publication that caused quite a stir. **Alberto Blanc** (1906–1960) was presented with a Neandertal skull in the Guattari Cave (I prefer the Italian *Grotta Guattari*) by a worker who had found it, picked it up, and replaced it somewhere on the cave floor (thereby altering the specimen's precise context, akin to messing up a crime scene). Blanc looked upon the skull *in situ*, pictured the context as a stone circle deliberately created with the skull as centrepiece, and began to hatch a story. The next clue related to damage that had been done to the base of the skull, around the **foramen magnum**, that 'big hole' through which the spine passes en route to the skull. Suspicions of cannibalism had surrounded the broken Neandertal bones from the early sites, so connecting that practice to presumed Neandertal ritual, as envisioned by Blanc, was not hard to do, particularly when Blanc's

university museum possessed skulls that had been cannibalized by Melanesian headhunters. Blanc's story was of a two-stage burial. In Trinkaus and Shipman's words:

> Blanc believed that the Guattari cranium had been separated from the body and deliberately placed within a crown of stones inside the cave. Then, the skull had been intentionally broken open and the brain extracted for a ritual cannibal feast. The empty braincase itself had perhaps been used as a cup or chalice in a further ritual, involving the sacrifice or consumption of the animals (mostly red deer, fallow deer, and aurochs, a type of wild ox) whose bones were scattered about the floor. This implied not simply caring for other individuals to the extent of disposing of their remains, but a belief in the afterlife and even, perhaps, the concept of inheriting the qualities of the dead person by consuming the brain. (Trinkaus & Shipman, 1994, pp. 254–5)

This story had, as they say, legs, until they were finally hobbled for good in the April 1991 issue of *Current Anthropology*, which contained a set of articles grouped under the heading 'The Cultural Significance of Grotta Guattari Reconsidered'. The most damning evidence came from taphonic work conducted by professor and anthropologist Mary Stiner (Stiner, 1991). Remember that **taphonomy** refers to the study of what happens to bones or other remains from death up until the time they are found and analyzed by anthropologists. Stiner demonstrated clearly that the supposed Neandertal cannibalism of the skull and other bones was really the work of spotted hyenas. The tooth marks and typical scatter patterns of those scavengers/predators are all over the crime scene,

Profiles in Canadian Anthropology

Eugène Morin

The study of Neandertal has not been a major area of research for Canadian anthropologists—you would think that as a cold-weather people, we might relate well to our Ice Age ancestors. An important exception to this rule is **Eugène Morin**, a young Canadian anthropologist affiliated with both Trent University in Peterborough and Laval University in Quebec. He holds a controversial theory (see Morin, 2008) asserting that Neandertal did make a contribution to the gene pool of anatomically modern *Homo sapiens* (AMH). His theory stems from his work on fossilized animal bones unearthed at Saint-Césaire in western France. He noted that during a severe cold snap lasting from

roughly 40,000 years ago to 35,000 ya, the populations of large herbivores such as bison and horses decreased significantly in the area, whereas animals better adapted to cold climate, such as reindeer (near relatives of our caribou), became more widespread. This decrease in the diversity of sources of food readily available to the Neandertal caused a crisis situation of population decrease that led them to expand their social networks to include anatomically modern humans, a group that were combating the same difficulties. The comparatively small numbers of the Neandertal meant that this genetic contribution was fairly low, but, argues Morin, it was still there.

so to speak. This is not to deny that cannibalism may have taken place at other sites, but before this can be declared with any certainty, hyenas and other predators, as well as other natural processes, must be eliminated.

Still, San Felice Circeo has become a favoured resort town, to a significant extent on the basis of the skull and the story. Upon request, you can, on weekends, visit the cave, which is enclosed on the grounds of the Hotel Neanderthal.

Flowers and Compassion: Shanidar Cave, Iraq

Part of the same shift in the perception of Neandertal came from the work of Ralph Solecki. In 1953, 1956, 1957, and 1960, he oversaw excavations of Shanidar Cave in the Kurdish area of northern Iraq. (Fortunately, in those more peaceful times in southwestern Asia, Western researchers were permitted to do work in countries such as Iran and Iraq.) The cave is huge: it stretches out to a width of 50 metres and a depth of 45 metres, and was then in use by Kurdish herders as temporary dwellings for several families and their horses, cows, goats, and chickens. During the period of Solecki's research, nine different Neandertal individuals were found. There is no hard-and-fast dating on the site, but the likely range is between about 50,000 and 80,000 ya.

Among the fossil specimens of particular significance are Shanidar I and Shanidar IV. Shanidar I was a male between the ages of 30 and 45. His survival to what was a relatively ripe old age for a Neandertal was against the odds of personal fitness. Making his survival all the more remarkable was the fact he was severely disabled: the bones of the right shoulder blade, collarbone, and upper arm were considerably thinner and less developed than their counterparts on the left side, indicating that they had not been used for a long time. No remains of the right elbow, lower arm, or hand were found, and these are suspected to have been amputated or to have atrophied and fallen off sometime in Shanidar I's past. In addition, his right knee, lower leg, and foot were damaged, with the arch having suffered a fracture. The left side of the body showed a fractured eye socket, probably severe enough to have displaced the eye, causing blindness on that side. That Shanidar I survived as long as he did shows that another individual or a group took care of him. This was interpreted by Solecki and others to follow as demonstrating that Neandertal of this time and place felt and expressed compassion. There are other Neandertal whose fossilized remains show that they survived serious (although not as serious as the wounds of Shanidar I) injuries and probably required the assistance of others to help them through the tough times of recovery. On the strength of this evidence, the argument for Neandertal compassion seems fairly solid.

The intriguing part of the story of Shanidar IV, a male possibly in his forties, relates to his treatment after death. In doing his excavation, Solecki took soil samples from the grounds where the bodies were found. These samples were then sent to a **palynologist**, one who studies pollen (a substance that lasts an incredibly long time). A palynological analysis is usually done to discover the nature of the local environment, to find out what was growing at the time. In this case, the palynologist on the case, **Arlette Leroi-Gourhan** (1913–2005), found a significant amount of wildflower pollen, more than what would naturally exist if the pollen had merely blown into the cave. Her interpretation, shared by Solecki, was that Shanidar IV had been buried with flowers, another sign that Neandertal was perhaps more like us than anthropologists had previously thought. Solecki expressed this opinion in his 1971 summary of his research, tellingly entitled *Shanidar: The First Flower People*:

> The death had occurred approximately 60,000 years ago . . . yet the evidence of flowers in the grave brings Neanderthals closer to us in spirit than we have ever before suspected. . . . The association of flowers with Neanderthals adds a whole new dimension to our knowledge of his humanness, indicating that he had 'soul'. (as cited in Trinkaus & Shipman, 1994, p. 340)

There has been some criticism of this position over the years. The most damning critique involves a small, gerbil-like rodent, *Meriones persicus*, otherwise known as a Persion jird. It has been speculated that the flowers came not from grieving humans but from this industrious little animal, which is native to the area and which stores flowers and seed in parts of its burrows. More recently, anthropologists have questioned an assumption made by Solecki and others that Neandertal found in a flexed or fetal position were deliberately placed that way in burial. While it is true that many later human cultures buried people in this way and that Neandertal probably did bury at least some of their dead, it is not enough to prove the 'buried with flowers' hypothesis. It should be noted that Solecki's view has not yet been disproven, just questioned; there ought to be a way of determining who brought the flowers to the grave.

Why Did Neandertal Become Extinct?

A good number of theories have been put forward over the years to explain why Neandertal went extinct. Some have suggested that they made inefficient use of fires, based on evidence that they did not put stones around their hearths to preserve heat and prevent the spread of flame—this would have made surviving a glacial episode especially difficult. Others have argued that, being larger than their svelte Cro-Magnon neighbours, Neandertal required more food and were therefore less able to survive when times got tough. Some have speculated that Cro-Magnon visitors may have spread warm-weather diseases to the unprotected Neandertal, which affected them much the way common European viruses were fatal to Native North Americans who, never having been exposed to the diseases, had never developed any immunity to them. A recent hypothesis suggests that because Neandertal women were actively involved with men in the hunt, they made poorer mothers. Criticizing a Neandertal's mom is a pretty low blow.

A certain segment of the anthropological community—probably the same set that enjoys watching UFC fights on television—has found the idea of a genocidal conflict between Cro-Magon and Neandertal compelling as an explanation for the latter's demise. That's where we'll begin our investigation into why Neandertal became extinct.

Theory 1: Violence Between Cro-Magnon and Neandertal

Was there violence between Cro-Magnon and Neandertal? Would this have contributed to Neandertal's extinction? This was the subject of William Golding's 1955 bestseller,

The Inheritors. In physical anthropology, we try to encourage students and other interested parties to think of evolution as a survival race in which the victor is the best fit, not necessarily the biggest, meanest party or the one with the most effective weapons (natural and/or cultural). Would we be doing our students a disservice by pushing the arm-to-arm combat story too hard here? No, fighting is a possibility, and one that should show significant evidence if it is true.

So what evidence do we have of Cro-Magnon *vs* Neandertal conflict? Here is one case, reported recently in an online periodical (Duke University, 2009). A 40- to 50-year-old male, probably white, with signs of arthritis, has a deep cut on his left ninth rib. And he is dead, possibly owing to complications following his rib injury, which has only partially healed. Time of death: 50,000 years ago, or thereabouts. He is known as Shanidar III. Although the injury is 'consistent with' (to use forensic speak) a self-inflicted accidental injury or accidental stabbing by someone in his hunting party, Steven Churchill, a professor of evolutionary anthropology at Duke University, suspects foul play at the hands of a rival gang. Shanidar III is a Neandertal; his possible assailant belongs to the Cro-Magnon.

The weapons advantage here goes to the Cro-Magnon, as anthropologists suspect they had throwing spears and possibly darts (this was long before lawn darts were

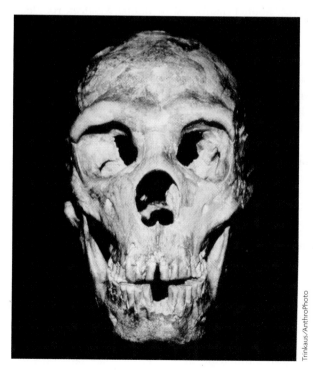

Cage match—Neandertal *vs* Cro-Magon:
(left) fossil skull of the Neandertal 'Shanidar III';
(right) fossil skeleton of a Cro-Magnon adult
male, found in Perigeux

Trinkaus./AnthroPhoto

© Charles & Josette Lenars./Corbis

banned), while the Neandertal had only thrusting spears. Both had sharp knives.

Churchill set up experiments to help determine what weapon might have killed Shanidar III. In doing so, he made a distinction between high and low kinetic energy. Apparently, high-kinetic-energy injuries—ones in which there is considerable momentum, such as those associated with being stabbed with a knife or a spear—have a large impact on the general area (e.g. smashing, rather than cutting, ribs). Lower-kinetic-energy injuries, produced by blows involving less momentum, have a more targeted impact. Churchill concluded that Shanidar III's rib injury, which is consistent with the impact of a throw spear or dart, was most likely produced by a high-kinetic-energy blow. He based his assessment in part on his experience in forensic anthropological experience, supplemented with an experiment involving pig carcasses and a crossbow firing projectiles with points modelled on those used during Shanidar III's time. The damage done to the pig ribs was similar to that done to Shanidar III's rib.

Then there is the trajectory, or angle, of the wound to consider. It was at about a 45-degree angle downwards. If Shanidar III was standing when injured (he stood about 5′ 6″ tall), then this is consistent with a projectile on its downward slope after having been thrown or flung.

Is this reading too much into too little evidence, or does it strongly suggest that there was conflict between the two (sub)species? You are the jury.

Theory 2: Differences in Hunting Practices

Were Neandertal Bad Hunters ...

Early on in the study of Neandertals, anthropologists assumed—with not a lot of evidence—that Neandertal only scavenged, and did not hunt. Now, Neandertal's status as hunters is well secured, but what kind of hunters were they?

According to one prevailing belief, they were poor hunters—so poor, in fact, that it caused their extinction. Proponents of this view believe that Neandertal failed as hunters to the extent either that they were unable to compete with Cro-Magnon or that they were simply unable to hunt effectively enough to survive. Think about it. Hunting was very important to them, as there was little plant food available. They often were up against some pretty nasty prey animals: bears, the woolly rhinoceros, the woolly mammoth. And this without having throwing spears. You wouldn't want to be the first person to stab a rhinoceros—that would only make him mad.

There is dental evidence to show that Neandertal children were not always provided with sufficient nutrition. **Dental enamel hypoplasia** (DEH) is an indicator of nutritional stress or deficiency during the time in which a child grows up and the teeth come in and develop. Striations or grooves in the tooth enamel are indicative of periods of time in which the individual has experienced a scarcity of food, some physical trauma, or a disease. In a study of 669 Neandertal dental crowns, published in 1988, Ogilvie, Curran, and Trinkaus found that 36 per cent of the teeth and 75 per cent of individuals showed signs of DEH. The fact that the numbers came from 41.9 per cent of permanent teeth and 3.9 per cent of deciduous (or 'baby') teeth suggests that the nutritional deficiency began after children were weaned and extended into adolescence. I don't believe that this necessarily means that Neandertal were bad hunters; it is more the case that they might have been inconsistent providers.

Or Were They *Turrific* Hunters?

On the other hand, a number of recent studies have suggested that the Neandertal were quite efficient hunters, faring just as well, at the same locations and with the same species, as Cro-Magnon. In 2003, an American–French team led by Donald Grayson and Françoise Delpech published their study on Grotte XVI, a site near the Dordogne River in southwestern France, which was occupied from roughly 65,000 years ago to about 12,000 ya. Neandertal were first on the scene; Cro-Magnon appeared for the first time only about 30,000 years ago. After analyzing more than 7,200 bones and teeth from a variety of ungulates (i.e. hoofed mammals, including reindeer, roe deer, red deer, chamois, and horses), the researchers concluded that the two hominid populations were equally effective in hunting these animals. Differences in which species were killed depended on the climate, with reindeer making up a higher percentage when the weather got colder.

In 2006, a team of American and Israeli anthropologists released the results of their study of animal bones found at the Ortvale Klde Rockshelter in the Republic of Georgia, a site that was active from about 50,000 to 20,000 years ago. The researchers did a comparison study of how effectively Neandertal (who were there first, before they disappeared) and anatomically modern humans (AMH, the Cro-Magnon who were there afterwards) hunted the Caucasian tur, a species of mountain goat. They found that there were no differences in the hunting effectiveness of the two varieties of hominids. Examination of the tooth wear on the bones of the often elusive and talented steep-slope–climbing tur showed that about two-thirds of the animals killed were adults in their prime, the most difficult to kill. It was also possible to determine that the goats were hunted primarily between late fall and early spring. This is the time of year when they would have been together in herds, rather than roaming individually (which would make hunting less efficient). Neandertal appeared to have been well aware of the seasonal habits of their prey.

A further note is necessary here. I am never comfortable with the debate concerning whether or not Neandertal had language such as we know it today. Direct proof has so far evaded us, and indirect proof often seems weak. Yes, a Neandertal **hyoid** bone has been found. This is the horseshoe-shaped bone that attaches to the muscles of the floor of the mouth and the tongue above, and the larynx or voice box below. It facilitates human speech, but it does not constitute direct proof for language. One source of indirect 'proof' relates to hunting. Animal predators have basic strategies of hunting that they use to good effect over and over again. These are to a large extent learned, but they depend on the predators' natural gifts of speed, power, big teeth, and sharp claws. Humans, having no such gifts, must rely on other advantages, advantages that may include speech. After all, to hunt successfully, plans need to be made and changes in situation have to be communicated. It can be argued, then, that once you have successful hunting (particularly of large and deadly prey) among hominids, you have fundamental language; the one is indirect evidence for the other.

Theory 3: Differences in Tool Use

Tool Talk, Part 2

Weapons are a kind of tool. Two of the main theories put forward for the extinction of Neandertal have to do with their tools. One of them runs like this. Neandertal made tools from more localized stone than did Cro-Magnon, the latter utilizing stone that came from further away. This would have produced two advantages for Cro-Magnon:

1. They would have had better rock to choose from. I will use a baseball analogy here. We were very pleased when our son Justin made the Bolton rep team. However, as Bolton is a small town, we knew that it was a lot easier to make the Bolton team than, say, the team in Newmarket, which, being a lot bigger than Bolton, has a much larger pool of talent to draw from.

2. A second benefit would have been how the more exotic stone was obtained. Stone brought in from a distance of, say, a hundred kilometres might have been obtained through trade. With the social interaction of trade comes an interaction of ideas, always beneficial for the survival of a people. Cro-Magnon were part of such a relatively broad exchange of ideas, while Neandertal were probably much more limited in their range of contacts and exposure to different ideas.

The second reason is perhaps a stronger one. Neandertal had essentially the same kind of stone tools for tens of thousands of years. They do not appear to have been tool innovators, unlike Cro-Magnon, who showed signs of

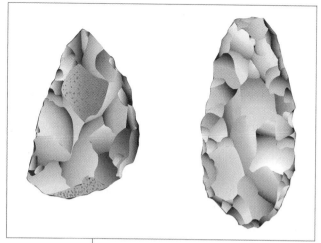

Figure 9.4 | Mode-3 (Mousterian) stone tools of the Upper Paleolithic

progress in toolmaking over the years. In order to understand this argument more fully, we will discuss the kinds of tools involved.

What's in the Toolbox

It's time for a little review. Remember from the previous chapter that archaeologists speak of early stone tools (or 'lithics') in terms of modes. **Mode 1** (or **Oldowan**) includes very basic tools with simple **cores** and **flakes** that have been knocked off the stone. **Mode 2** (or **Acheulean**) includes bifacial, symmetrical tools often made through **pressure flaking** rather than percussion, with the core prized over the flakes. **Mode 3** involves the careful core preparation of the **Levallois technique**. The Neandertal form of this mode is known as **Mousterian**, and almost all Neandertal tools are of this type (see Figure 9.4). **Mode 4** relates to the **Aurignacian** of the Upper Paleolithic, typically associated with Cro-Magnon. **Blades**, bone and ivory work, and artistic works are associated with this mode.

Theory 4: The Acculturated Neandertal

The Châtelperronian Tradition

We've just stated that Neandertal are associated with the Mode-3 tool tradition of the Mousterian. This statement is both right and wrong, a partial truth and a partial lie, for two fundamental reasons. One is that the Mousterian tradition was not strictly limited to Neandertal: it is also found in North Africa and in the Near (relative to Europe) East on non-Neandertal sites. I have never read an explanation for that; no one, to my knowledge, has suggested that AMH were influenced by the Neandertal in the adoption of this tradition. Was it independently invented, the tool equivalent of convergent evolution? Did Neandertal learn to use this tradition from their neighbours? It would be a good doctoral thesis for someone.

Of greater significance in the literature is the Neandertal adoption of a Mode-4 tool tradition known as **Châtelperronian**, found in central and southwestern France and adjacent parts of northern Spain. The first Neandertal fossil find associated with this tradition was the fragmented right side of a skull and mandible, part of a face, and some postcranial bones at Saint-Césaire in western France, at a site dating about 36,000 years old. The Neandertal, a young adult male, was identified by François Lévêque and Bernard Vandermeersch in 1979. He appears to have been buried with some tools and more artistic works. Points made from bone and perforated teeth, probably from a necklace, were discovered at the same stratigraphic level.

Châtelperronian was definitely a Neandertal tradition, as the evidence has been found at sites where Mousterian elements were also found. Flake tools developed using the Levallois technique have been found alongside more complicated blade tools struck from cylindrical cores. How did it develop? Did the Neandertal learn this toolmaking tradition from Cro-Magnon? It seems to be the case that a kind of **acculturation** occurred, in which the Neandertal were influenced by Cro-Magnon they encountered, or at least by tools the Cro-Magnon had made.

A major portion of the evidence supporting Neandertal acculturation comes from Grotte des Fées de Châtelperron in east-central France, a site first discovered by Henri Delporte. Roughly four layers, or strata, of the site bear Châtelperronian tools, while one features Aurignacian tools. Mellars, Gravina, Ramsey use the term *interstratification* in their 2006 article on the subject. The lowest Châtelperronian part of the Châtelperron site, labelled B5 in the article, was radiocarbon dated twice, at 40,650 ± 600 BP (years before present) and 39,150 ± 600 BP. Level B4, which contains some Aurignacian material (e.g. perforated animal teeth), was dated at 39,780 ± 390 BP and 35,540 ± 280 BP, and the upper levels, B1–B3, have been assigned seven dates that cluster between 36,340 ± 320 BP and 34,550 ± 500 BP. To simplify these figures (for myself and for you, the reader), we have a sequence of 39,000–40,000 BP for the lowest level (B5), 36,000–37,000 BP for

B4, and 34,000–36,000 BP for the strata nearest the surface (see Table 9.2).

The changing culture in the later, post-AMH period is not unique to the western part of Neandertal's territory. In Vindija Cave, Croatia, for example, we seem to have evidence of a progressive AMH acculturation. In the lowest hominin-bearing level (labelled G3) of this site (within a range of 38,000–45,000 BP), the tools are Mousterian but the choices of source materials for tools is of a broader variety than what was usual for Neandertal, more like what the subsequent AMH would draw upon. By the most recent Neandertal level, G1, ranging from 32,000 to 34,000 BP, there is a mixture of Mousterian and Mode-4 artifacts.

The stratigraphy of this site together with the fairly conclusive evidence of an Aurignacian influence in the area from 37,000 to 38,000 BP suggests contact and cultural borrowing to me. One reason that this interesting at a larger level is that it could be viewed as indirect evidence of possible interbreeding of Neandertal and Cro-Magnon in this part of Europe.

Anatomy versus Culture: Who was Better in the Cold?

One of the most intriguing set of European sites for Upper Paleolithic hominids and anatomically modern humans is at Kostenki, located by the west bank of the Don River in Russia, about 400 kilometres south of Moscow. While it is difficult to be absolutely certain from the fossil hominid specimens (two teeth) that the people of this location were AMH, the cultural evidence strongly suggests that they were. Recently (see Anikovich, et al., 2007) it has been demonstrated that the oldest levels at Kostenki are between 45,000 and 42,000 years old, as they lie beneath a volcanic ash horizon, known as the CI (Campanian Ignimbrite)/Y5 tephra, that has been dated by argon-40/argon-39 dating as 41,000–38,500 years old. The youngest level dates to about 30,000 years old.

The artifacts discovered at Kostenki reveal a very sophisticated Aurignacian culture. The lowest levels include stones perforated with a hand-operated rotary drill (not a Black & Decker type, but a long, narrow tool that could be spun between the hands), bone points and awls, worked ivory and perforated shells and arctic fox teeth. An important question to ask is how these northern residents survived in this environment without being as well adapted physically adapted to the cold as Neandertal. One answer, put forward by Arsuaga (2001, p. 302), is that some of the bone material found there was used as needles that would have enabled the humans living there to tailor their furs and skins well enough to keep them warm in the cold. This was one of the master strategies of the Inuit, whose tailored furs enabled them to survive well in the Arctic. The argument here is that this technology, appearing in the northern part of eastern Europe, would have enabled

Table 9.2	Interstratification of the Châtelperron site	
Stratum	Time period (BP)	Tool tradition
B1–B3	34,000–36,000	Châtelperronian
B4	36,000–37,000	Aurignacian
B5	39,000–40,000	Châtelperronian

these early humans to outcompete the Neandertal when the weather got really cold in that area about 25,000 years ago, just as the Neandertal remnant groups were disappearing.

Gorham's Cave

Gorham's Cave is an archaeological site in the British-administered territory of Gibraltar, a rocky strip of land connecting with Spain that looks like a long, narrow arrowhead aimed to strike across the 14 kilometre–wide Strait of Gibraltar separating Europe from Africa. When it was first excavated, during the 1950s, archaeologists found Mousterian tools but no fossilized Neandertal remains. In 2006, Clive Finlayson published an article concerning the time in which Neandertal inhabited the region and the variety of resources open to them. The three dates that he and his team produced came from dating charcoal found at the sites of three hearths (fire pits), one on top of the next: 24,010 ± 320 BP, 26,400 ± 440 BP, and 30,560 ± 720 BP (Finlayson, et al., 2006). The authors of the paper characterize the site as a relatively warm refuge with abundant resources, not far (about 100 kilometres) from the nearest yet-discovered Cro-Magnon site. Unlike the Châtelperronian sites, it presents no evidence of the mixing of tools. And interbreeding can't be proven or disproven until some Neandertal fossil remains have been discovered.

Excavating Gorham's Cave, home of Neandertal

The Famous Fossil Femur Flute

In 1997, anthropologist Ivan Turk announced to the world that a 43,000-year-old femur of a bear cub found two years earlier at the Divje Babe I site in Slovenia was, in fact, a flute, crafted by Neandertal and similar to the bone flutes made more recently by AMH. The femur had four holes in it that appeared to be in line with each other, with what was possibly a hole for the thumb on the femur's opposite side. The two holes at either end were partial as the ends had been damaged, consistent with chewing. Saskatchewan musicologist Bob Fink says that the spacing of the holes would have produced a basic *do–re–mi–fa* pattern, in other words that the notes were in a diatonic sequence, as is characteristic of Western music.

This was a bold statement about Neandertal capabilities, and it was soon to be attacked, with critics saying that the holes were randomly spaced, just bite marks on a bone that had obviously been chewed, and that the femur was too short to produce the notes claimed. It is difficult to assess conclusively whether the Divje Babe 'flute' is a

Disputed Neandertal flute, found at Divje Babe, Slovenia

musical instrument or not. Finding other flutes like it would help. I myself believe that the evidence for it is, at this point, stronger than the evidence against. Fink's point about the unlikelihood that such evenly spaced holes could have been made randomly, accidentally, is well taken.

Hybrids

In anthropology as in the automotive industry, the search is on for hybrids. You would think that a hybrid Neandertal–AMH specimen would provide solid evidence for interbreeding. But before we look at the material, we should again consider that there can be a relative independence of traits, that there is the possibility that an AMH can have what are usually considered 'Neandertal' (or at least 'Neandertal-like' traits) without having been the product of any interbreeding. Likewise, a Neandertal can at least hypothetically have AMH or AMH-like traits without being a hybrid (or ever touching ethanol). Declaring conclusively that a fossil specimen is a hybrid is a difficult task. If you asked ten physical anthropologists what a Neandertal–AMH hybrid would look like, you would probably get ten different answers. Proof that you had found such a hybrid would be difficult to establish. It is also important to keep in mind that accepting that we have hybrids is not the same as accepting that Neandertal is part of our gene pool. Hybrids could just as well be dead ends with no descendants as they could be the ancestors of someone living today.

There are strong views on the matter of hybrids, with influential anthropologists on each side of the debate. **Milford Wolpoff** (b. 1942), for a long time the leading proponent of the Multiregionalism hypothesis of human evolution, is probably the one who for the longest time has leaned furthest towards the existence of hybrids and towards Neandertal being part of our genetic heritage. **Erik Trinkaus** and Fred Smith have also looked favourably on the possible existence of hybrids. Dead set against the idea are **Chris Stringer**, **Ian Tattersall**, and **Jeffrey Schwartz**. Tattersall and Schwartz again demonstrate that bias of presentation can exist with even the best researchers. Students need to be able to recognize bias when they read it; here's a taste:

> . . . although many students of human evolution have lately begun to look favorably on the view that these distinctive hominids merit species recognition in their own right as *Homo neanderthalensis* (e.g., refs. 4 and 5 [here the authors cite themselves and Stringer]), at least as many still regard them as no more than a strange variant of our own species, *Homo sapiens* (6, 7 [Wolpoff and Smith]). This difference represents far more than a simple matter of taxonomic hair-splitting. For, as members of a distinct species, of a completely individuated historical entity, the Neanderthals demand that we analyze and understand them on their own terms. In contrast, if we see them as mere subspecific variants of ourselves, we are almost obliged to dismiss the Neanderthals as little more than an evolutionary epiphenomenon, a minor and ephemeral appendage to the history of *Homo sapiens*. (Tattersall & Schwartz, 1999)

In other words, they seem to argue, either you side with us and respect Neandertal, or you take an opposing stance and fail to give Neandertal its due. That is not a scientifically reasoned position.

Lapedo Child

The first clear contender for Neandertal hybrid came in 1998, when Portuguese anthropologist **João Zilhão**

discovered the fossilized remains of a child of about 4 years old in Lagar Velho, Portugal, about 145 kilometres north of Lisbon. The **Lapedo Child**, as he has been nicknamed, is believed to have lived about 24,500 years ago. It appears, based on the presence of pierced shell and red ochre around him, that he was buried in a shallow grave.

The popular media tend to enjoy any claim of Neandertal's connection with modern-day humans. An article in the *New York Times* (25 April 1999) followed the somewhat misleading headline 'Discovery Suggests Man Is a Bit Neanderthal'. Following the journalistic convention, the sensational headline was followed by a bold assertion that was immediately qualified in the article's second paragraph:

> Neanderthals and modern humans not only coexisted for thousands of years long ago, as anthropologists have established, but now their little secret is out: they also cohabited.
>
> At least that is the interpretation being made by paleontologists who have examined the 24,500-year-old skeleton of a young boy discovered recently in a shallow grave in Portugal. (Wilford, 1999)

The article goes on to report the speculation that this find *may* be a Neandertal–Cro-Magnon hybrid, and that there *may* have been others elsewhere, and that *perhaps* these hybrids are distantly related to humans, that Neandertal were not superseded by Cro-Magnon but in fact merged with them. It quotes Fred Smith, Milford Wolpoff, and Erik Trinkaus in support of the hybrid hypothesis, and even quotes a skeptical Chris Stringer as being prepared to view the Lapedo discovery with an open mind, though 'He told The Associated Press that the current evidence was not sufficient to convince him of Dr Trinkaus's hybrid interpretation' (Wilford, 1999).

What features of Lapedo Child lend themselves to the hybrid interpretation? The boy had a prominent chin and generally the facial features and tooth size of an anatomically modern human. His body, however, was stocky, and he had short, robust leg bones, consistent with the hyperarctic (i.e. cold-adapted) body proportions of a Neandertal. To be conclusive, there would need to be more evidence of specimens presenting this blend of features, perhaps a small group of fossil specimens at one site.

The Romanian Evidence

The southwestern Romanian locality Peştera cu Oase (lit. 'the cave with bones') is the site of some intriguing specimens dated around 35,000 years old. These include a mandible ('Oase 1') and a cranium ('Oase 2'). Both combine Cro-Magnon characteristics (e.g. no occipital bun or brow ridges with Oase 2) with features that are more common among Neandertals than among AMH (e.g. large molars,

Mircea O. Gherase

Three anthropologists—Stefan Milota, Hélène Rougier, and Oana Modovan—exploring the interior of Peştera cu Oase

Mircea O. Gherase

Stefan Milota holding the recently discovered cranial remains of Oase 2

and a progression of molars increasing in size from the first to the third). Are they hybrids, or do they merely have archaic features for other reasons? More specimens from this locality are needed to be sure.

Perhaps more compelling evidence comes from another Romanian site, Peştera Muierii (lit. 'cave of the old woman'), with a date of approximately 30,000 years ago. Again we find a mixed suite of characteristics, the most significant of which, in terms of Neandertal heritage, is evidence of a large occipital bun (Soficaru, Dobos, & Trinkaus, 2006).

Spanish Teeth

Investigating Neandertal mandibles from the 40,000–43,000-year-old site of Sima de las Palomas del Cabezo Gordo, a team of researchers (notably including Erik Trinkaus) discovered that these findings bore features that were both 'typical Neandertal' and/or more closely associated with AMH. They came up with three possible scenarios for this mixing of features:

> i) late Neandertal genetic drift in the direction of modern human morphology through isolation-by-distance in the cul-de-sac of southern Iberia,
>
> ii) an adaptive shift to local environmental constraints in some of these features, and/or
>
> iii) the product of gene flow from early modern human populations to the immediate north.
>
> (Walker, et al., 2008)

The first scenario they deemed unlikely, as these Neandertal from southern Iberia (referring to the peninsula that includes Spain and Portugal) showed some common features with other Neandertal that would not exist if the southern Iberian population had been isolated for a long time (i.e. thousands of years). Concerning the second possible explanation, they could find no specific evolutionary advantage for the 'modern' features. To the researchers, their elimination of the first two possibilities made the third possibility, that there was some **gene flow** from AMH, the most likely explanation, one that has been well received in anthropological circles.

Mitochondrial Evidence

A number of mtDNA studies support the belief that Neandertal did *not* have any genetic input into modern humans. It still seems like a scientific miracle to me that mitochondrial DNA can be retrieved from something that died tens of thousands of years ago. What does this evidence look like?

Mezmaiskaya Cave

Ovchinnikov and Goodwin (2001) performed a study done with mtDNA extracted from the rib of a Neandertal infant found at the Mezmaiskaya Cave in southern Russia. The strata from which the findings were recovered were dated at 29,000 years ago. For comparison purposes, DNA is studied relative to one particular mtDNA sequence known as **Cambridge Reference Sequence** (**CRS**), the first DNA sequence ever established, in 1981 at Cambridge University. So the CRS and Mezmaiskaya sequences were aligned so that any mutations—such as base substitutions, insertions, or deletions—could be noted. Ovchinnikov and Goodwin recorded 22 substitutions and one insertion. When compared with 300 modern European sequences, the average number of substitutions was 25.45 ± 3.27. With groups of modern Asians and sub-Saharan Africans, the averages were 23.27 ± 4.06 and 23.09 ± 2.86, respectively. However, when the Mezmaiskaya DNA was compared with DNA sequences for other Neandertal groups, there were far fewer substitutions. For instance, a comparison with a similar sequence taken from the Feldhofer specimen showed that the two differed by just 12 substitutions. Over the same sequence, a Vindija Cave sample differed from the other two by only 6 substitutions. In addition, all three sequences shared a specific insertion (at nucleotide base position 16,263—remember that DNA is ordered in a line). All of this led the authors to conclude that:

- '. . . it appeared that the DNA extracted from the rib was not closely related to any modern mtDNA,' and

- 'the Neanderthals formed a group distinct from all modern humans' (Ovchinnikov & Goodwin, 2001).

Pitfalls of mtDNA Studies

As of 2008, 17 different mtDNA sequences had been obtained for Neandertal, ranging from short sequences (of about 350 base pairs) in the first studies to the first complete sequencing of the mtDNA genome (16,565 base pairs).

There are a couple of potential holes in this kind of research. One is that in a way it is like comparing apples with oranges—in this case, ancient hominids with modern ones. To produce an exact comparison, there should be an equivalent set of ancient AMH or Cro-Magnon DNA in the comparison mix. This is on the way, and it will be interesting to see what it produces.

Second, mtDNA lines have the potential to die off quickly. Say that a particular mtDNA line of a Neandertal mother and an AMH father had been established, and their union produced only sons: that line would not be continued in the AMH population, but there still would have been Neandertal genetic input, just not that marked by mtDNA. The same thing could happen if, in a local population, a large number of people died and that number included all the women carrying that line.

A recent comprehensive study of Neandertal mtDNA concluded the following:

> The studies of Neandertal mtDNA do not show that Neandertals did not or could not interbreed with modern humans. However, the lack of diversity in Neandertal mtDNA sequences, combined with the large differences between Neandertal and modern human mtDNA, strongly suggest that Neandertals and modern humans developed separately, and did not form part of a single large interbreeding population. The Neandertal mtDNA studies will strengthen the arguments of those scientists who claim that Neandertals should be considered a separate species which did not significantly contribute to the modern gene pool. (TalkOrigins Archive)

Concerning the first point, on the lack of diversity in Neandertal mtDNA sequences, a recent study (Orlando, et al., 2006) has shown that Neandertal had more mtDNA diversity 100,000 years ago than they possessed more recently.

The Neandertal Genome

In May 2010, a scientific team led by Svante Pääbo of the Department of Evolutionary Genetics, at the prestigious Max-Planck Institute for Evolutionary Anthropology, published an article entitled 'A Draft Sequence of the Neandertal Genome' (Green, et al., 2010). A genome, if you remember, is the entire genetic blueprint of a species or individual. The researchers compared four billion nucleotides from three Neandertal with the genomes of five present-day individuals from different parts of the world. This very tentative study suggested that:

> . . . Neandertals shared more genetic variants with present-day humans in Eurasia than with present-day humans in sub-Saharan Africa, suggesting the gene flow from Neandertals into the ancestors of non-Africans occurred before the divergence of Eurasian groups [i.e. the ancestors of Europeans, Asians, and the indigenous people of the Americas] from each other. (Green, et al., 2010, p. 710)

Summary

We can see that over the more than 150 years since the discovery of the Feldhofer Neandertal, the views concerning the appearance and behaviour of Neandertal have swung back and forth from the largely negative appraisal of a brutish ape or a pathological human to a more favourable view of a musically inclined, compassionate flower child. Neandertal elicit powerful emotions in anthropologists and in the general population. They are our nearest relative, and yet they look, at least superficially, very different from us. We can say that they were more intelligent and better adapted to their environment than the earliest Neandertal researchers believed. They were hunters, and despite the seemingly primitive nature of their weapons, they were very efficient at hunting, at least during the latter stages of their existence on earth.

There are thornier issues yet to be fully answered. Do Neandertal live on in the genes of some of us? Do humans with heavy brow ridges (I think of major league baseball player Johnny Damon) have Neandertal ancestors? Some people are reported to have something resembling an occipital bun. Did their great, great, great . . . grandfathers make Mousterian tools? Is it more a function of having relatively large brains but a narrow skull, as some have suggested? Are people from the Middle East, with noses that are typically more prominent than those of Asians or Europeans, descendants of the Neandertal who lived in present-day Iraq and Israel (the larger nose being a characteristic feature of Neandertal)?

The answers to these questions lie, I believe, in finds that have yet to be made and research areas that have yet to be fully explored. We are getting closer to the answers, though. I have made a point in this chapter of exposing the flaws on both sides of the debate about whether Neandertal lives on in our gene pool. My own leanings are on the pro side, as I hope I have made clear, though I acknowledge that I could be wrong.

Typical Student Questions

▶ *So what do you think: is Neandertal our ancestor?*

Being something of a romantic as well as a scientist, I tend to favour the possibility that Neandertal is the ancestor of some of us. Certainly there is some promising evidence to support this view (see especially the recent work of Svante Pääbo noted just above). But I could be wrong. I am sure, however, that we will find more evidence to help us settle the matter one way or the other.

▶ *If an animal or species (such as Neandertal) becomes extinct, doesn't that mean that it is inferior?*

Not when you consider that the vast majority of all animals that have ever lived have become extinct. It is more the case that a species that becomes extinct was not able to adapt to a specific set of circumstances at a particular time.

Review Questions

1. Distinguish between the various tool traditions, and identify the significance of Châtelperronian tools.
2. Outline and assess the basic hypotheses concerning why Neandertal became extinct.
3. Outline the ways that Marcellin Boule distorted Neandertal in his work.
4. What were the flaws in Alberto Blanc's work on Neandertal?

Discussion Questions

1. Do you think that Neandertal is in or out of the modern human gene pool? Present your reasons for the position that you take. (You may even take the position—provided that you support it—that the evidence is not yet adequate to make that decision.)
2. Do you believe that Neandertals made and played flutes? What are the intellectual implications of their possible capability to play (and, presumably, appreciate) music?
3. Why do you think that Neandertals disappeared? What might have made Cro-Magnon or anatomically modern *Homo sapiens* a better fit for the evolution competition in Europe? Present several reasons.
4. You receive a flyer in your mailbox for The Neandertal Moving Co. Do you hire them to help you move into your new apartment? Consider everything we know about Neandertal—build, behaviour, intelligence, and so on.

Key Terms

acculturation
Acheulean
Allen's rule
Aurignacian
Bergmann's rule
Blades (blade tools)
Cambridge Reference
  Sequence (CRS)
Châtelperronian
cores (core tools)
Cro-Magnon

dental enamel
  hypoplasia (DEH)
endotherms
flakes (flake tools)
foramen magnum
gene flow
Homo neanderthalensis
  or Homo sapiens
  neanderthalensis or
  Neandertal
Homo sapiens

hyoid
Lapedo Child
Levallois technique
mitochondrial DNA (or
  mtDNA)
Mode 1
Mode 2
Mode 3
Mode 4

Mousterian
Multiregionalism
occipital bun
Oldowan
Out of Africa
palynologist
pressure flaking
taphonomy

Key Individuals

Joel Asaph Allen	Johann Karl Fuhlrott	Chris Stringer	Milford Wolpoff
Christian Bergmann	Arlette Leroi-Gourhan	Ian Tattersall	João Zilhão
Alberto Blanc	Eugene Morin	Erik Trinkaus	
Marcellin Boule	Jeffrey Schwartz	Rudolf Virchow	

Recommended Print and Online Resources

Anikovich, M., Sinitsyn, A., Hoffecker, J., Holliday, V., Popov, V., Lisitsyn, S., . . . Praslov, N. (2007, January 12). Early Upper Paleolithic in eastern Europe and implications for the dispersal of modern humans. *Science, 315*(5809), 223–6.

Arsuaga, Juan Luis. (2001). *The Neanderthal's necklace: In search of the first thinkers*. New York: Four Walls Eight Windows.

Duke University. (2009, July 22). Human spear likely cause of death of Neandertal. *ScienceDaily*. Retrieved from www.sciencedaily.com/releases/2009/07/090720163729.htm

Finlayson, C., Pacheco, P., Rodríguez-Vidal, J., Fa, D., López, J., Pérez, A., . . . Sakamoto, T. (2006, October 19). Late survival of Neanderthals at the southernmost extreme of Europe. *Nature, 443*, 850–3. doi:10.1038/nature05195

Fossil Man 1. (1913, August 28). *Nature, 91*, 662–4. doi:10.1038/091662b0

Green, R., Krause, J., Briggs, A., Maricic, T., Stenzel, U., Kircher, M., . . . Pääbo, S. (2010, May 7). A draft sequence of the Neandertal genome. *Science, 328*(5979), 710–22.

Marean, C. (1998, August). A critique of the evidence for scavenging by Neandertals and early modern humans: New data from Kobeh Cave (Zagros Mountains, Iran) and die Kelders Cave 1 Layer 10 (South Africa). *Journal of Human Evolution*, pp. 111–36.

Morgan, J. (2009, February 12). Neanderthals 'distinct from us'. *BBC News*. Retrieved from http://news.bbc.co.uk/2/hi/health/7886477.stm

Morin, E. (2008, January 8). Evidence for declines in human population densities during the early Upper Paleolithic in western Europe. *PNAS, 105*(1), 48–53.

Ogilive, M., Curran, B., & Trinkaus, E. (1998). Incidence and patterning of dental enamel hypoplasia among the Neandertals. *American Journal of Physical Anthropology, 79*(1), 25–41.

Orlando, L., Darlu, P., Toussaint, M., Bonjean, D., Otte, M., & Hänni, C. (2006, June 6). Revisiting Neandertal diversity with a 100,000 year old mtDNA sequence. *Science Direct*, pp. R400–2.

Ovchinnikov, I., & Goodwin, W. (2001, January). The isolation and identification of Neanderthal mitochondrial DNA. *Profiles in DNA*, pp. 7–9. Retrieved from www.promega.com/profiles/402/ProfilesinDNA_402_09.pdf

Science Daily. (Website). Science Daily: News & articles in health, science, environment & technology. Retrieved from www.sciencedaily.com

Smithsonian National Museum of Natural History. (Website). Human evolution by the Smithsonian Institute's Human Origins program. Retrieved from http://humanorigins.si.edu/

Soficaru, A., Dobos, A., & Trinkaus, E. (2006). Early modern humans from the Peştera Muierii, Baia de Fier, Romania. *Proceedings of the National Academy of Sciences USA, 103*, 19266–71.

Stiner, M. (2001). The faunal remains from Grotta Guattari: A taphonomic perspective. *Current Anthropology, 32*(2), 103–17.

TalkOrigins Archive. (Website). The TalkOrigins Archive: Exploring the creation/evolution controversy. Retrieved from www.talkorigins.org/

Trinkaus, E., & Shipman, P. (1994). *The Neandertals: Of skeletons, scientists and scandal*. New York: Vintage Books.

Walker, M., Gibert, J., Lopez, M., Lombardi, A.V., Perez-Perez, A., Zapata, J., . . . Trinkaus, E. (2008, December 30). Late Neandertals in southeastern Iberia: Sima de las Palomas del Cabezo Gordo, Murcia Spain. *PNAS, 105*(52), 20631–6.

Wilford, J. N. (1999, April 25). Discovery suggests Man is a bit Neanderthal. *The New York Times*. Retrieved from www.nytimes.com

10

Human Variation

Human diversity
© David Colbran/Alamy

LEARNING OBJECTIVES

After reading this chapter, you should be able to:

- Assess the evidence for and against the existence of biologically defined race;

- Discuss the relationship between skin colour and racial categorization;

- Outline the mtDNA and Y-chromosome model of human migration;

- Explain the relationship between earwax type and geographic population;

- Describe the role of dental anthropology in categorizing human variation.

In its 'Statement on Biological Aspects of Race' (1996), the American Association of Physical Anthropology (AAPA) has this to say on the subject of human variation:

> There is great genetic diversity within all human populations. Pure races, in the sense of genetically homogenous populations, do not exist in the human species today, nor is there any evidence that they have ever existed in the past. (AAPA, 1996)

This is the position taken in this chapter.

Introduction

We have seen in previous chapters what a leaky concept taxonomy can be. Items contained in a particular genus or species can seep into a neighbouring one or be taken out and moved into an entirely separate taxonomic category, based on scientific evidence. In Gould's words, 'One man's genus may be another man's family' (1977, p. 232). Even the kingdoms of living things have been re-categorized over the years, as we saw in an earlier chapter. We can only imagine, then, how much messier we'll find the concept of 'race' as it is proposed to exist at the sub-species level.

How real is race? Variation certainly exists within the human species—there's no questioning that fact. We differ quite visibly in hair, eye, and skin colour; in hair texture and straightness; in height and breadth (again, picture Danny DeVito and Shaquille O'Neal standing side by side); in build (we can be robust or gracile); and cranium shape and size. Minor dental patterns and—believe it or not—earwax type (wet/gooey *vs* dry/flaky), as well as skin and eye colour, fairly neatly follow geographic patterning, with some intermixing. Categorizations based on either traits or association with a particular **geographic population** seem to be more meaningful than divisions of race in that they are easier to identify. A trait exists at both the genotypic level (detectable by a molecular biologist) and the phenotypic level, so it can be readily perceived as either being there or not being there, and it can be given a precise definition. Due to the genotypic existence of trait, meaningful terms such as **generalized** (i.e. ancient and existing 'by default') and **derived** (i.e. new) can be assigned. Identifying a particular geographic population is somewhat less precise, with vague terms such as 'Northern Chinese' and 'Mediterranean', but this approach aligns well with single traits and is easy to identify roughly. Still, biological race can in some ways be meaningful—a 'necessary evil', as we shall see.

Variation within (Non-human) Species

Phases, Mutations, and Breeds

Various names are used by biologists to talk about variation within a species. The black-coloured squirrels found aggressively begging for food in Toronto's Queen's Park are actually members of the grey squirrel species (*Sciurus carolinensis*). They are called 'black-phase grey squirrels'. The black-coloured (or 'melanistic') members make up a **phase**, and, unlike adolescence or eating nothing but Kraft Dinner, it's something the black squirrels will never grow out of. The percentage of the southern Ontario grey squirrel **population** made up of black-phase members has been growing over the last few hundred years. In the 1600s, the Huron highly prized robes made from the pelts of these melanistic grey squirrels. When I worked for the Ontario Ministry of Natural Resources thirty years ago, some government biologists speculated that the darker colour was somehow genetically linked to more aggressive behaviour.

Other animals show similar phases. Ontario is also home to three different phases of *Vulpes vulpes*, the red fox. The generalized, or default, pattern is the red phase, and the two derived phases are the cross fox (named for the dark cross in its fur) and silver fox.

Quaker parrots (which are *not* native to Ontario) come in two varieties, known as **mutations** (which has a slightly different meaning from the usual). The blue mutation,

Earwax: Are You Wet and Gooey or Dry and Flaky?

A great genetic tool for mapping human migrations is cerumen, otherwise known as earwax. Earwax types can be classed as either 'generalized' or 'derived', which is helpful because there is a geographic point where the derived pattern begins.

There are two genetically determined patterns of earwax. The wet and gooey type is genetically dominant and can be considered the generalized, or 'default', pattern. The generalized pattern is found among the vast majority of African and European descendants (the figure is usually given as 97 per cent or more). Dry, flaky earwax is the derived type. It has its highest percentage among Koreans and northern Chinese, with the rates falling as you move south through Asia and into Australia. Aboriginal people commonly have it as well, further evidence of their Asian heritage (a subject that will be explored in the next chapter). A classic study of Aboriginal migration as tracked by earwax type is summarized in Bass and Jackson's 1977 article 'Cerumen Types in Eskimos'.

Dry earwax has never been—and probably will never be—demonstrated to have an evolutionary advantage. It is the result of one gene's having the substitution of one nucleotide base (an 'A', or adenine, replacing a 'G', or guanine, base). It is, however, correlated with armpit odour (now *there's* a conversation icebreaker). People with dry earwax sweat less, and thus have less armpit odour. Perhaps in the cold climate of North Asia, sweating less had an adaptive advantage sometime over 15,000 years ago.

---

which is less usual, is the derived pattern; the green mutation is the generalized variety. There may be characteristics other than colour that are genetically linked to each mutation. For instance, blue-mutation Quaker parrots are slightly smaller and are generally thought to be less aggressive (something we have noticed in distinguishing Quigley, our green Quaker, from Finn, our blue Quaker).

All dogs are of the species *Canis familiaris*, but they come in different **breeds** that are grouped into broader distinctions such as terriers and hounds. But even individual breeds come with their own distinctions. Take border collies, for example. We have one 'pure' border collie and a border collie cross. Purebred border collies may be either robust (the ones with the big hind quarters for bodychecking sheep) or gracile (for chasing tennis balls, I suppose). Some are mostly black with a bit of white so they cannot be confused with small, dark wolves; others have a great deal of white; while others still are light brown and white. Most have a very intense look to them, possessing the famous 'collie stare', but others are known as 'loose-eyed collies' because they do not have that stare. However, the border collie was bred mostly not for its appearance but for its behaviour, which is easy to identify: basically, a dog that runs in circles trying to herd every moving creature from sheep to small humans and geese is very likely a border collie.

Subspecies

The great evolutionary biologist **Ernst W. Mayr** (1904–2005), who had a very profound influence on the way in which biological categories are defined, put the following scientific description on subspecies:

> The subspecies, or geographical race, is a geographically localized subdivision of the species, which differs genetically and taxonomically from other subdivisions of the species. (Mayr, 1942, as cited in Gould, 1977, p. 232)

That seems to make sense. However, the solidity of that definition is more apparent than real. It can fall apart when specific examples are presented. By that definition, I could qualify as a subspecies. I have a height, girth, weight, eye colour, etc., all of which can be determined scientifically and which serve to distinguish me from others of my species. I am genotypically unique. I tend to be geographically limited to southern Ontario. My nephew Todd Dias and I belong to the same subspecies (as my sister Ann jokingly asserts). If you saw us standing together, you would readily recognize from our builds that we are closely related. Like me, he lives and tends to stay in southern Ontario.

My point is fairly simple. In cases where making a distinction is important (say, in distinguishing geographic or historical varieties), the subspecies concept can be a useful way to link specific traits that differ, perhaps from region to region, within a species. However, the concept itself is not a hard-and-fast one.

A Brief Look at the History of the Race Concept in Science

So what about **race**: how sound is it as a concept from a scientific standpoint? When physical anthropologists believed in the validity of race as a biological concept, they worked hard at developing a comprehensive list of what the different races were. But they were much better at assigning people and groups to different racial categories than they were at coming up with precise definitions of those racial categories. **Racial time** (the period of time in which humans have been classified according to race) is well represented by the work of Western scientists. **Carolus Linnaeus** (1707–1778), the eighteenth-century biologist who initiated the binominal taxonomy of genus and species, was the first to subdivide members of the human population, identifying five human subspecies, or races: *afer* (African), *americanus* (Aboriginal people), *asiaticus* (Asians) *europaeus* (Europeans), and *monstrosus* (lit. 'monsters', but essentially everybody else: Australian Aborigines, Pacific Islanders, South Asians, and so on). This was not a good start, involving too much lumping (and this from a self-confessed lumper!).

Defining the Races: Blumenbach, Morton, and Coon

Johann Friedrich Blumenbach (1752–1840), a German physician and scientist, whose *De generis humani varietate nativa* (*On the Natural Varieties of Mankind*) was published when the author was just 24, coloured-coded his racializing of human beings in the following way:

Caucasian = white

Mongolian = yellow

Malayan   = brown

Negroid   = black

American   = red

A few comments on Blumenbach's terms (which also feature in other classifications). The term **Caucasian** (or **Caucasoid**) was (and still is) used to refer to white people, because Western scientists, early on in the study of human variation, believed that people living in the area of the Caucasus Mountains of eastern Europe were the 'pure type' (we would use the term **holotype** or **type specimen** today). **Mongolian** (also **Mongoloid**) was used to refer to people of East Asia, possibly because of the historical impact of the Mongols and Genghis Khan. Unfortunately, during the early twentieth century, some medical researchers classified different kinds of 'idiots' using racialized terms. Individuals with Down's syndrome were known as 'Mongoloid idiots' for quite some time, because of a superficial resemblance of head shape and eye configuration (the term was still in use when I was growing up).

The early American skull-measuring doctor/scientist **Samuel George Morton** (1799–1851), in his 1839 opus *Crania Americana; or, A Comparative View of the Skulls of Various Aboriginal Nations of North and South America*, declared that there were four races: Caucasians, Mongolians, Americans (i.e. Aboriginal people), and 'Negroes'. In addition to describing physical types, as was typical of the 'scientific racism' of the day, he attached prejudiced notions of the intellectual capacities of each group, participating in an ill-conceived practice that continued well into the twentieth century. Here is a snapshot of Morton's views:

- on Caucasians:

  This race is distinguished for the facility with which it attains the highest intellectual endowments. . . . The spontaneous fertility of [the Caucasus] has rendered it the hive of many nations, which extending their migrations in every direction, have peopled the finest portions of the earth, and given birth to its fairest inhabitants. . . .

- on 'Mongolians':

  In their intellectual character the Mongolians are ingenious, imitative, and highly susceptible of cultivation [i.e. capable of learning]. . . . So versatile are their feelings and actions, that they have been compared to the monkey race, whose attention is perpetually changing from one object to another. . . .

- on 'Americans' (i.e. Aboriginal people):

  In their mental character the Americans are averse to cultivation [education], and slow in acquiring knowledge; restless, revengeful, and fond of war, and wholly destitute of maritime adventure. They are crafty, sensual, ungrateful, obstinate and unfeeling, and much of their affection for their children may be traced to purely selfish motives. They devour the most disgusting [foods] uncooked and uncleaned, and seem to have no idea beyond providing for the present moment. . . . Their mental faculties, from infancy to old age, present a continued childhood. . . . [Americans] are not only averse to the restraints of education, but for the most part are incapable of a continued process of reasoning on abstract subjects. . . .

- on 'Negroes':

  In disposition the Negro is joyous, flexible, and indolent, while the many nations which compose this race present a singular diversity of intellectual character, of which the far extreme is the lowest grade of humanity. . . . The moral and intellectual character of the Africans is widely different in different nations. . . . The Negroes are proverbially fond of their amusements, in which they engage with great exuberance of spirit; and a day of toil is with them no bar to a night of revelry. Like most other barbarous nations their institutions are not infrequently characterized by superstition and cruelty. They appear to be fond of warlike enterprises, and are not deficient in personal courage;

but, once overcome, they yield to their destiny, and accommodate themselves with amazing facility to every change of circumstance. The Negroes have little invention, but strong powers of imitation, so that they readily acquire mechanic arts. They have a great talent for music, and all their external senses are remarkably acute.

Carleton Coon (1904–1981), a prominent American physical anthropologist and one-time president of the American Association of Physical Anthropologists (1961–2), was perhaps the last reputable physical anthropologist to try to systematically classify the races of the world. Unfortunately, and somewhat unfairly, this has become a lasting part of his legacy, overshadowing a lifetime of anthropological fieldwork, not to mention some wartime espionage for the US Office of Strategic Services, all of which is chronicled in several books he wrote for a popular audience. Long before his name began to appear on websites associated with white supremacists, Coon's writings were embraced by white segregationalists in the United States, who, during the 1960s, used his work to support their argument that institutions such as schools should be segregated, or separated, according to race.

Coon's thinking was essentially a very extreme version of the Multiregionalism hypothesis. He believed that *Homo sapiens* evolved from *Homo erectus* five separate times deep in time, producing five basic races:

- **Caucasoid**, which he subdivided into a number of separate subraces, such as the Upper Palaeolithic (an interbreeding of *Homo sapiens* and Neandertal) and Mediterraneans (whose origins supposedly lay in North Africa);

- **Negroid**;

- **Capoid**, including the click-carrying Khoisan-speaking peoples;

- **Mongoloid**, including most Asians; and

- **Australoid**, the Aborigines of Australia.

The similarity of this view to Multiregionalism (with, however, fairly clear distinctions) has been used to tar the Multiregionalism hypothesis with the brush of racism that has stained many scientific references to Coon today, given his unscientific view of differing intelligence levels of the races. Contemporary multiregionalists, it must be noted, do not subscribe to this view.

The Race Game: Other Problems

Even apart from its racist overtones, Coon's work was flawed. He ended up with a problem similar to that of a person who takes a machine apart and, after putting it back together, finds that there are pieces remaining: the system won't work when parts are left over. Coon's was not an uncommon problem—all systems of racial classification have had parts left over. In Coon's case, he had problems with classifying people in South Asia, whom he linked with North Africans and Europeans as Caucasian. The people from the south of India and Sri Lanka, and from the Andaman Islands, not far away to the east of India, never neatly fit into his or anyone else's classification of races. There were other problematic groups as well, many of them living in the Pacific islands. Some of them, who were grouped under the unfortunate designation 'Negritos', are the indigenous or apparent first peoples of South Asia, the Andaman Islands, Malaysia, the Philippines, and New Guinea. They are very dark-skinned (hence the *Negr–* part, from the Latin word for 'black') and somewhat small (signified by the Spanish diminutive *–ito*), with fuzzy hair. We will see, later in the chapter (page 242), that in terms of their Y-chromosomes and mtDNA, they have an interesting genetic story to tell. They are genetic relics on the path from Africa to Australia. Meanwhile, scientists are still working hard at trying to ascertain the genetic identity of the **Ainu**, the indigenous people of Japan, who are hairier and typically lighter-skinned than the more recent immigrant Japanese, and who speak an unrelated language.

Another problem with classifying the global population by racial type, beyond the fact that not all groups fit

Table 10.1	Different systems of racial classification				
Anthropologist (date of work)	Proposed racial categories				
Linnaeus (1735)	*afer*	*americanus*	*asiaticus*	*europaeus*	*monstrosus*
Blumenbach (1776)	Negroid	American	Mongolian	Caucasian	Malayan
Morton (1839)	Negroid	American	Mongolian	Caucasian	
Coon (1962)	Negroid		Mongoloid	Caucasoid	Australoid (Aborigines)	Capoid (Bushmen/San)

San Bushman couple near Kalahari Gemsbock National Park, South Africa

Case Study Controversies

Does Race Exist?

In 2000, the PBS series *NOVA* aired an episode called 'Mystery of the First Americans'. The documentary, which was produced largely to explore the recent discovery of the 'Kennewick Man' remains (discussed in the next chapter), featured a panel debate on the question 'Does Race Exist?'. Two of the panel members were prominent physical anthropologists George W. Gill (on the 'yes' side of the question) and C. Loring Brace (on the 'no' side). Their arguments will be summarized here. Keep in mind that my own position is that race does not exist as a biological entity; that does affect my summary.

Gill, an emeritus professor at the University of Wyoming with an interest in osteology (the study of bones), noted the divisive nature of the issue, pointing out that scientists involved in research on blood types typically reject the biological entity 'race', while forensic anthropologists are 'overwhelmingly in support of the idea of the basic biological reality of human races'. Blood type, like eye colour, is a clinal characteristic, meaning that a blood type may occur with increasing or decreasing frequency across different populations, but there are no 'breaks' along racial boundaries: not all white Europeans have type-A blood, just as they don't all have blue eyes. This kind of variation supports the views of those who believe that race is a purely social construct. But Gill argues that the reason that clinal characteristics

cut across racial lines is that they are not shaped through natural selection by the climatic factors that influence such 'racial' traits as skin colour, hair form and colour, lip shape and, of course, bones. In other words, there are still plenty of traits that can be used to distinguish one race from another, even if certain traits occur across these supposed races.

The strongest argument that Gill, along with other forensic anthropologists, has in his pro-race arsenal is the high rate of accuracy in predicting racial origins from forensic evidence. He claimed that by using a diverse battery of measures, such as mid-facial measurements and femur traits, he could achieve better than 80 per cent accuracy in predicting the ethnic origins of a skeleton. Racial differences, Gill would contend, run deep.

You may have seen news features or even TV shows (of the *CSI* ilk) in which forensic anthropologists talk about identifying skeletons by race, particularly in the United States. It's true that they can do this with some success in a country like the US, where the vast majority of people are either black (tracing their origins to a relatively small area of Africa), white, or Asian (mostly from a relatively limited part of Asia). Based on a person's skeleton or, especially, their skull, there are certain broad generalizations that can be made about this narrow set of origins. The techniques are not as successful in a more

neatly into the available 'races', is that there was often as much or more variation *within* racial categories than there was *between* racial categories. In Spencer Wells's words concerning Africa:

> The range of variation in Africa is extraordinary, containing both the tallest (the Maasai) and the shortest (Pygmies) people on Earth. Facial characteristics are similarly diverse. The heavy features of the West Africans contrast with the finer-boned features of the San Bushmen. The San also have an epicanthic fold—the extra layer of skin above the eyes that characterizes people from East Asia—while other African groups lack this feature. The skin colorization also varies enormously, ranging from a deep brown in the West Africans to the lighter skin of the San. Overall, the main uniting feature is a generally darker skin color for Africans relative to people from northern latitudes such as Europeans. (Wells, 2006, p. 150)

The **epicanthic fold**, associated with the once common racial category 'Oriental', is messy as a tool of classification. The San have it, as do some (but not all) Aboriginal groups (e.g. the Inuit). The Andaman Islanders, who are sometimes referred to as 'living fossils' because they possess a number of ancient genetic traits, also are said to have this feature. This has led Wells to hypothesize that the epicanthic fold is an ancient trait.

In many systems of racial classification, South Asians—some of whom are darker than many African groups—were labelled Caucasoid, or Caucasian. Less often commented on but true nonetheless, the Aboriginal peoples of the Americas likewise show tremendous differences in height, skin colour, and build, though they were typically lumped together. Races were great dumping grounds of difference, cognitive wastebaskets used the way the disputed species *Homo heidelbergensis* is used today.

Individual variations, not group variations, account for most differences between people. In a groundbreaking article published in 1972 in the journal *Evolutionary*

diverse population. Also, these generalizations do have a margin of error. They are statements of probability, and (just like an 80 per cent chance of rain) they can be wrong. We'll look more at this issue in Chapter 12, on forensic anthropology.

Weaker, though, is Gill's other argument, that 'race denial' is politically motivated, a dangerous part of the political correctness agenda that could, he warns, lead to a denial of racism. Gill's charge would be hard to prove conclusively. Furthermore, you don't need the biological concept of 'race' to accept the existence of and to fight against racism; all you need is awareness of the social process of **racialization**, or racializing. Around the world, people gain social advantage in areas such as employment, housing, and respect for culture and language, all based on the social identification of some visually prominent features, particularly skin colour. Non-visible features such as blood or earwax type don't stir emotions the same way. Just imagine if you had to provide a sample of earwax for a job interview! But skin colour the world over jumps over the borders of the traditional races. What darker-skinned people have in common, whether they're Europeans, Aboriginal peoples, Asians, or Africans, is the proximity of their traditional territory to the equator. Lighter-skinned people tend to live in the northern and southern extremities of the earth.

Human beings have long used small differences as excuses to judge one another. When it isn't differences in physical appearance, it's something else. Minor variations in major religions comes to mind, cleverly satirized by Jonathan Swift in his early eighteenth-century novel *Gulliver's Travels*, where the small people from Lilliput and Blefuscu violently dispute over which end of the soft-boiled egg the faithful should open. The point is, you don't need to establish races for people to engage in, or oppose, racist tendencies.

Brace, a University of Michigan anthropology professor who has done a great deal of forensic work himself, accepts the use of what he calls 'geographic labels' (such as 'South Asian' and 'West African'), but argues that there is 'no such thing as a biological entity that warrants the term "race"' (Brace, 2000). He points out that few people (and this would include many physical anthropologists) have experienced, first-hand, the tremendous diversity of human variation. Who among the readers of this book have stood beside an Ainu, an **Austronesian**, and/or an Aborigine? That experience, he suggests, would quickly dispel anyone's tendency to lump the three individuals together under a single racial label. He also contends that the process of American racism made race awareness stronger than it might otherwise have been, although he is quick to note that racism is not an American invention.

Biology, biologist **Richard Lewontin** (b. 1929) noted (using the genetic differences then known, those of genetically produced proteins) that approximately 85 per cent of human variation was individual, occurring *within* populations; the remaining 15 per cent was distributed roughly equally between variation within and between races (Lewontin, 1972).

Skin Colour as a Racializer

In 1833, an observant German naturalist and ornithologist (one who studies birds) by the name of **Constantin Wilhelm Lambert Gloger** (1803–1863) published some observations concerning widespread populations of warm-blooded animals (i.e. endotherms—mammals and birds). The crux of these observations is known today as **Gloger's rule**. Basically, it states that among these populations, the darker subspecies occur in warmer and/or more humid climates, while light-coloured subspecies are native to cooler and/or drier climates. The rule works, for the most part, with birds, which is what Gloger studied, but also has a certain relevance for black bears (*Ursus americanus*), where the populations exhibiting lighter colours (e.g. the derived colours 'honey' and 'cinnamon') live in the dry areas of the American southwest, while the bears that are truly black (exhibiting the generalized colour) live in damper areas. It also works with the eastern fox squirrel, *Scirius niger*: populations living in the northeastern United States show grey on the back and a yellow or rusty colour below (like an old Ford); those

exhibiting dark brown or black fur with white bands live in the more southerly Appalachian Mountains; the uniformly black version (presumably the one for which the species is named) lives in the warmer climates of the American south.

As skin colour has, for centuries, been used to classify and judge people, it's important here to deal with some basic questions and answers. Remember that the differences in colour all come down to **melanin** (see Figure 10.1). To be more technical, we are talking about the melanocortin receptor (MC1). And, as we will see shortly, it all has to do with how long and how direct the rays of the sun are in the area in which particular human populations live.

Skin Colour: Frequently Asked Student Questions and Rarely Given Answers

QUESTION 1: WHAT COLOUR WAS *HOMO SAPIENS* ORIGINALLY?

Answer: The truth is, we don't really know. However, as it is most likely that the first anatomically modern *Homo sapiens* came from Africa, and as the best 'one colour fits all' adaptation for different environments is brown, I would speculate that *H. sapiens* started off brown.

QUESTION 2: HOW QUICKLY DID COLOUR VARIATION BEGIN?

Answer: Current genetic research suggests that changes in skin colour happen quite rapidly, so we can imagine that AMH populations changed from darker to lighter and lighter to darker soon after they began moving to different parts of the world. According to Nina Jablonski of Penn State University, for many lineages, the change in skin colour has taken place over 100 to 200 generations (i.e. 2,000–3,000 years). Far back, she adds, 'almost all of us were in a different place and we had a different color' (as cited in Krulwich, 2009).

QUESTION 3: WHY ARE SOME PEOPLE 'WHITE'?

Answer: While it's tempting to answer with something facetious (e.g. 'Someone has to eat white bread and buy all those tacky velvet paintings of Elvis'), it actually has to do with the production of vitamin D. Canadians have been hearing a lot about this important vitamin lately, and the fact that, living in the north, we may not be getting enough of it. That's because the trigger for the skin to produce vitamin D, which is necessary for the development of strong, healthy bones (particularly in children), is exposure to ultraviolet radiation, i.e. sunlight. Light skin benefits the production of vitamin D because it filters sunlight less than darker skin does.

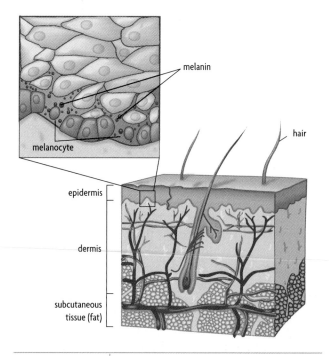

Figure 10.1 | Melanin and the structure of skin

In places with weak or insufficient sunlight, such as Britain (home to some of the whitest people imaginable), there is a significant danger of **rickets**, which weakens the bones and can cause death. Rickets was a severe problem for many British children, up until the early twentieth century. Later on in the twentieth century it made a bit of a comeback, which has been attributed to immigration from India. Oppenheimer explains:

> Rickets, or osteomalacia [*osteo*– means 'bones'], came back in a small way to Britain in the second half of the twentieth century, but affected the children of families from the Indian subcontinent. Part of the reason for this was their darker skin colour, which filtered out some of the already meagre sunshine. (Oppenheimer, 2004, pp. 198–9)

A recent study examined 104 confirmed cases of vitamin D–deficiency rickets in Canadian children from 1 July 2002 to 30 June 2004. Of the 104 cases, 98 children (89 per cent of the sample) had 'intermediate or darker skin' (Ward, Gaboury, Ladhani, & Zlotkin, 2007).

There may also be a connection between light skin colour and protection against frostbite. In experiments conducted on light-and-dark–skinned guinea pigs (experiments in which guinea pigs were actually used as guinea pigs), the darker skin was found to be more affected by frostbite. To my knowledge, no one has explained why this might be so.

QUESTION 4: WHY ARE SOME PEOPLE 'BLACK'?

Answer: Again, it comes down to our relationship with the sun. The darker people of the world—the indigenous peoples of Africa, the Americas, Eurasia, and the Pacific Islands—have all spent significant time living near the equator, where the rays of the sun hit most directly and with the greatest heat. Darker skin provides greater protection against skin cancer and prevents UV rays from breaking down the body's stores of folic acid, thereby reducing the risk of anemia and other nasty things. Folic acid deficiency is of particular concern to pregnant women, who are at greater risk of miscarrying or developing neural tube defects in the fetus if their folic acid levels are low. It is for this reason that many Western countries (including Canada) require certain foods (flour, for instance) to be fortified with folic acid, and pregnant women are encouraged to take folic acid supplements to combat the breakdown of folic acid to which fair-skinned people are especially vulnerable.

Late nineteenth-century artist's rendering of the five human races (clockwise from top left): American, Malayan, Mongolian, Ethiopian, and Caucasian (centre)

Ann Ronan Picture Library/HIP/The Image Works

QUESTION 5: WHY IS THE 'WHITENESS' OF SKIN HIGHLY VALUED IN THE AMERICAS, EUROPE AND SOUTH ASIA?

Answer: Because in each of these regions, the conquerors (and, ultimately, the people with power and influence) were more lightly coloured than the conquered people were. In India, that happened twice: first when lighter-coloured Indo-European speakers (whose languages now bear names such as Hindi and Punjabi) moved in to live amongst darker Dravidian speakers (whose languages now bear names such as Tamil), and then when European powers such as Britain, France, and Portugal conquered India beginning in the late seventeenth century.

QUESTION 6: WHY ARE 'INDIANS' OR NATIVE PEOPLE CALLED 'RED'?

Answer: We have seen in an earlier chapter that an earth dye, **red ochre**, was once used in Africa as personal decoration (with, perhaps, more practical functions as well). The same was true for some Aboriginal peoples of North and South America. One particular example is the Beothuk, the extinct tribe that lived on the island of Newfoundland until the early nineteenth century. They were one of the first peoples that European explorers encountered, and they were big users of red ochre. Aboriginal people have been called 'red' ever since. If Aboriginal people had 'discovered' Britain 2,000 years ago, they would have seen Celtic tribes (Welsh, Irish, and Scots) painted blue with wode dye; it's possible we'd now refer to all Europeans as 'blue men'.

QUESTION 7: WHY ARE ASIAN PEOPLE, PARTICULARLY CHINESE PEOPLE, CALLED 'YELLOW'?

Answer: This is very hard to tell, as Asian people are, of course, not yellow. The idea was common among European thinkers during the eighteenth century, so it has been around for centuries. I suspect—but this is just speculation—that the main reason is that it was simply a convenient way of distancing Asians as a 'race' from western Europeans.

Eye Colour

While brown is the generalized eye colour for humans, lighter eye colours have shown up as derived in a number of different areas: grey among the northern Mongolian people (in various languages, Genghis Khan's people were known as 'grey-eyed devils'), blue among Scandinavian people (e.g. Swedish, Norwegian, Finnish, Icelandic, Danish), and green among the Irish. It has been hypothesized that this lighter colouring might give some sort of advantage in the poor light and lack of light of the north and in the fog-bound areas of Ireland, but to my knowledge this has never been seriously tested.

Three genes have been found to have some bearing on eye colour. Chromosome 15 has two of them:

- 'bey1' (for 'brown eye' with one **allele** or genetic variant), which produces brown eyes; and
- 'bey2' (for 'brown eye' with two alleles or genetic variants, brown and blue).

With this gene, the brown colour is dominant, the blue recessive. Chromosome 19 has the third gene, 'gey' (for 'green eye'), with two alleles, green and blue; the green is dominant. These genes do not completely explain all colour variations, for example grey and hazel.

The Cephalic Index: Race and Plasticity

Craniometry, the measurement of skulls, has a long history of use in physical anthropology, for purposes both good and bad. One of the most favoured craniometric measures in the history of the discipline is the **cephalic index** (sometimes also called the **cranial index**). This is calculated by multiplying the maximum width of the skull by 100, and dividing this result by the maximum length, front to back. The results are classed in three categories, or types:

- **dolichocephalic** ('long-headed'), with a low cephalic index of about 65 for males, 75 for females;
- **mesocephalic** ('middle-headed'), with an intermediate cephalic index of 65–75 for males and 75–80 for females; and
- **brachycephalic**, with a high cephalic index greater than 75 for males and 80 for females.

The cephalic index is used today primarily for detecting problems in the crania of fetuses (where a rapid decrease in cephalic index is considered dangerous). It is also used in the comparison of dogs breeds: dachshunds, greyhounds, and borzois (Russian wolfhounds) are considered dolichocephalic; German shepherds and golden retrievers are considered mesocephalic; and bulldogs, boxers, and pugs are referred to as brachycephalic.

During the nineteenth and early twentieth centuries the cephalic index was often used to justify racist views. Dolichocephalic brains were considered to represent superior intelligence over the brachycephalic brains, particularly by northern European (primarily Germanic) scholars, whose heads tended to be longer than those of the more brachycephalic Europeans farther south. Similar use was found for measurement of the **facial angle** (i.e. whether the lower part of the face protruded or was relatively flat).

One way in which this craniometric measure has had an important history in physical anthropology is with respect to the **plasticity** of the cephalic index of a lineage—in other words, the potential that it has to change, owing to environmental conditions, over time. Probably the most famous study of this phenomenon in physical anthropology—with findings that continue to stir debate today—was published in 1912 by German-born American anthropologist **Franz Boas**. As part of an ongoing attack on the use of racial categories to define and judge people, he set out to disprove the racializing notion that skull shape and size correlated with intelligence. Anthropologists of Boas's era had identified differences in the cephalic index of Americans and immigrants to the country, suggesting that these differences reflected differences in intelligence among different 'races'. In his report on the study, entitled 'Changes in the Bodily Form of the Descendants of Immigrants' (Boas, 1912), Boas wrote about how the

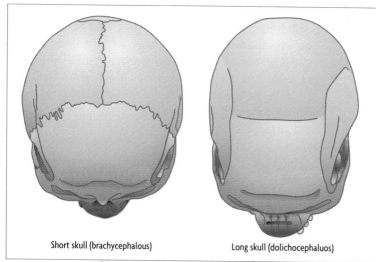

Short skull (brachycephalous)   Long skull (dolichocephaluos)

Figure 10.2 | Short-headed ('brachycephalic') skull (left) compared with a long-headed ('dolichocephalic') skull (right)

cephalic index actually changed from immigrant parents to their born-in-America children, suggesting that nutrition and general social conditions, not innate racial differences, were key factors in skull plasticity.

Dental Anthropology

Dental anthropology is an important area of physical anthropology, especially when it comes to tasks such as differentiating between groups and finding out how old a skeleton is (see Chapter 11).

A leader in this area of study is **Christy Turner**. When he started working among the Inuit in the 1960s, he gave particular attention to their teeth. He eventually broadened the scope of his research to include all Aboriginal people, as well as East and Southeast Asians. The two key terms that we owe to Turner are **sinodont** (lit. 'Chinese teeth') and **sundadont**, which are used to characterize two distinctive dental patterns. The latter term is derived from the Sunda shelf, which includes the islands of Indonesia, where the sundadont features are found (as well as in Malaysia and Southeast Asia). Turner developed an eight-point comparison of the two different patterns. The clearest differences relate to three derived features in sinodont teeth:

- shovel-shaped upper incisors (or maxilla)—these have a scooped-out shape towards the back, like an old coal or snow shovel;

- single-rooted maxilla premolars; and

- three-rooted lower first molars.

The **shovel-shaped incisor**, as a trait shared by East Asians (most notably Chinese people) and the indigenous people of the Americas, was first formally noted by **Aleš**

Hrdlicka (1869–1943) in his classic 1920 work 'Shovel-Shaped Teeth.' The feature is an example of what is called **non-metric variation**, meaning that it is not something that can be directly measured; it's more of a question of 'either/or'—you either have it or you don't. That being said, it is one of the kinds of variation that can be scored on a scale. As we will see in Chapter 12, on forensics, scoring is a great tool in determining the sex and age of a skeleton. An illustrative example comes from Thailand, in an article called 'Talking Teeth: Dental Non-Metric Traits of Wooden-Coffin People in the Pang Ma Pha Cave Sites, Northwestern Thailand' (Nakbunlung and Wathanawareekool, 2004). Sixteen sites sampled were dated at between 2080 ± 60 BP and 1240 ± 90 BP. They were located between the main areas of sundadonty in Southeast Asia and sinodonty in Northeast Asia. The researchers used a 6-point grading system to classify their findings (see Table 10.2).

One of the study's main conclusions was 'that the wooden coffin people rarely had shovel-shaped . . . incisors' (Nakbunlung and Wathanawareekool, 2004, p. 140). This finding could prove useful in tracing migration patterns. First, as the authors note, it shows that sinodont populations recently had migrated into the area from the north. Studies such as this over time could tell the researchers something about how the ethnic mix of the people changed, and when migrations occurred. Now, if they had earwax samples too, they might have consilience or confirmation of their findings from another historical method.

Table 10.2	Grades and percentages of upper central incisor shovelling among the wooden-coffin people
Scoring grade	Percentage found
O (= 'none')	28.1
1 (= 'faint')	17.1
2 (= 'trace')	12.8
3 (= 'semi–shovel-shaped')	33.1
4 (= 'semi–shovel-shaped')	4.2
5 (= 'shovel-shaped')	0.5
6 (= 'markedly shovel-shaped')	4.2

Nakbunlung & Wathanawareekool, 2004, p. 140.

The Genetic Approach

In talking about Y-chromosome and mtDNA distinctions we will be entering the world of **phylogeography**. This involves taking the mutations that define particular **haplogroups** (the *phylo–* part) and placing them in specific times at specific places (the *–geography* part). The key to locating the time and place of a specific mutation or a specific haplogroup lies in the DNA of people living in particular places today. Two characteristics are important: frequency and diversity. Frequency relates to the percentage of the population sampled in a particular area that belongs to a particular haplogroup. Diversity refers to the different versions of that haplogroup—the new mutations that are in addition to the haplogroup-defining ones—that are found in a particular area.

Mitochondrial DNA (mtDNA)

What follows is a greatly simplified version of the research synthesis performed by Stephen Oppenheimer in 2004.

The dates are rough, in part because 'solid proof' for them is still questionable. The story begins with Mitochondrial Eve, who has been given the label 'L1'. The dates for this individual vary; Oppenheimer prefers 190,000 years ago (see Table 10.3). All the subgroups that begin with L1 are African only. The same is true of the subgroups that begin with 'L2', who may have lived some 100,000 years ago. 'L3', head of another daughter line, is referred to as 'Out of Africa Eve' and is said to have been born in Africa some 83,000 years ago.

All non-African lines come from L3. The daughter groups of Out of Africa Eve include 'M' (Oppenheimer's Manju), which has all Asian descent groups. One descendant group from Manju is found in central Asia by about 74,000 years ago, one in New Guinea 75,000 years ago, and one in Australia by at least 60,000 years ago, a date with which at least some archaeologists agree. An interesting implication of this date is the suggestion that there were people in Australia before there were in Europe.

Another daughter group of L3 is 'N' (Oppenheimer's Nasreen), who he believes began farther west and a little

Table 10.3	Mitochondrial Eve (L1) and her descendants according to Oppenheimer			
Scientific name	Oppenheimer's name	Relationship	Area	Time period
L1	mtDNA Eve	mother	Africa	~190,000 years ago
L2		da of L1	Africa	~100,000 years ago
L3	Out of Africa Eve	da of L1	Africa	~83,000 years ago
M	Manju	da of L3	Asia	>75,000 years ago
N	Nasreen	da of L3	Western Eurasia	~70,000 years ago
R	Rohani	da of L3	South Asia	~65,000 years ago
U5		grda of R	Europe	~50,000 years ago

Adapted from Oppenheimer, 2004.

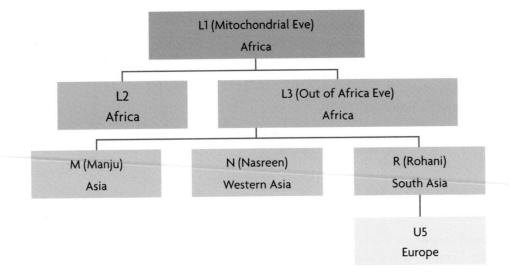

later, possibly about 70,000 years ago. All those groups from western Eurasia—meaning Europe and a bit of Asia—are descendants of Nasreen.

A daughter of Nasreen is 'R' (Oppenheimer's Rohani), who is supposed to have lived some 65,000 years ago, born in South Asia. One of her granddaughters is U5, who at 50,000 years ago heads the earliest line into Europe (some 10,000 years after humans were first in Australia).

The oldest line follows 'the beaches' from south of the Red Sea into the Levant, and along the south of Asia to Australia. One reason for the poor archaeological record of this trip is that, as with the proposed West Coast migration down the Americas, the sites are now quite a bit under water.

It should be noted that there is definitely not consensus on the chronology or actors proposed by Oppenheimer. Adcock and colleagues (2001), for example, put together a plausible counter-scheme based on mtDNA taken from 10 'ancient Australians' ranging from Lake Mungo 3 (who may have lived in Australia as far back as 60,000 years ago) to specimens living just 8,000 ya. Adcock's team argue that when fossil specimens are included in the research, there is a significant divergence of lineage between earliest anatomically modern human and the proposed Mitochondrial Eve. They argue that the Lake Mungo 3 lineage:

> probably diverged before the most recent common ancestor of contemporary human mitochondrial

genomes. This timing of divergence implies that the deepest known mtDNA lineage from an anatomically modern human occurred in Australia. . . . Our results indicate that anatomically modern humans were present in Australia before the complete fixation of the mtDNA lineage now found in all living people. (Adcock, et al., 2001, p. 537)

Y-Chromosome: The General Picture

The Y-chromosome picture is not quite as well developed as that of mtDNA. A much greater number of nucleotide bases is involved. It was only in 2002 that genetic biologists reached consensus on the letters A to R for the major haplogroups. Another problem is that the dates for the Y-chromosome story are much more recent that those of the mtDNA tale. This is usually explained by saying that it is easier for a male line to die off than for a female line to do so, as a higher percentage of females breed—males in some early societies resemble gorillas in that the leaders procreate more than others. The example typically cited to illustrate this is the incredible fertility of the Mongol leader Genghis Khan (1162–1227), and of course his sons. It has been claimed that nearly 8 per cent of the men living in what was the tremendous expanse of the Mongol Empire carry Y-chromosomes that come from Genghis and his family; that accounts for roughly 16 million descendants. He must have enjoyed Father's Day.

Table 10.4	Y-Chromosome Adam and his descendants according to Oppenheimer			
Name	Mutation number	Relationship	Area	Time period
Y-Chromosome Adam	—	father	Africa	60,000 – 70,000 ya
Out of Africa Adam (OAA)	M168	son	Africa	>60,000 ya
C (Cain)	M216	son of OAA	?	~60.000 ya
D/DE (Abel)	M145	son of OAA	?	~55,000 ya
F (Seth)	M89	son of OAA	?	~50,000 ya

Oppenheimer, 2004.

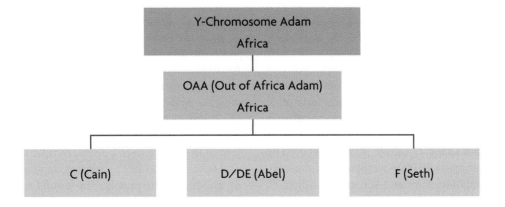

Three children from the Andaman Islands

AP Photo/Anthropological Survey of India, HO

50,000 years ago. It includes in its descendants all groups from 'G' to 'R'. This, of course, includes 'Q', which is the major Y-chromosome haplogroup from which Aboriginal men are said to descend. However, although it is a major non-African haplogroup, it does not include males from central Asian countries such as Tibet and Kazakhstan, from 'Far East' countries Mongolia and Japan, or from the many islands in which Polynesian languages (e.g. Maori, Hawaiian, Samoan, Fijian) are spoken.

Race and Measurement: A Modern Perspective

As we have seen, some anthropologists view racial categories as a necessary evil in some circumstances. One of those circumstances is the contemporary use of anthropometry. **Anthropometry** is used today to determine the nutritional and general health status of specific populations (or 'races') of people. A recent example comes from longitudinal study (i.e. a study of the same people over a significant length of time) carried out by the World Health Organization (WHO) on the Onge. The **Onge** are one of the tribal groups of the Andaman Islands, a very isolated group of islands in the Bay of Bengal in the Indian Ocean. The islands are administered by the government of India, very far away. The Onge, who are typically classified as 'negrito', became minor celebrities during the tsunami of 2004 as they were able to sense the tsunami's approach using traditional knowledge, and were thus able to head to the highlands in time to survive.

Standard anthropometric measurements of height, weight, **mid-arm circumference** (MAC), and **fat-fold thickness** (FFT) at the triceps (on the other side of the humerus from the biceps) were used to study the nutrition of the people over a period of time from 1969 to 1989, and from 1989 to 1997. This was done along with analysis of the diet of the people. The results are summarized in Table 10.5, looking at the average numbers:

The date given for Y-Chromosome Adam is usually 60,000 to 70,000 years ago, around 100,000 years later than the date for Mitochondrial Eve (see Table 10.4). Out of Africa Adam (or 'OAA') has the mutation M168 ('M' stands for 'mutation'). He had three son groups. One carries the mutation M216 and is called 'C' (Oppenheimer's Cain). This haplogroup, said to originate almost 60,000 years ago, has descendants in New Guinea, Australia, Indonesia, and Japan, as well as among Aboriginal groups in the Americas, as we will see. A second son group, M145 or Abel, is sometimes called 'D' or 'DE' and is estimated by some to have lived roughly 55,000 years ago. A third group, associated with mutation M89 and sometimes called 'F' (or Seth by Oppenheimer), is said to have been initiated

Table 10.5	Comparison of mean anthropometric measurements of the Onges, 1969, 1989, and 1997				
Survey year	Number studied	Average height (cm)	Average weight (kg)	Average MAC (cm)	Average FFT (mm)
Females					
1969	13	140.4	43.0	24.9	11.7
1989	31	140.4	43.3	25.8	17.8
1997	23	140.5	41.4	24.4	6.0
Males					
1969	29	149.8	43.2	24.0	5.9
1989	36	151.8	47.7	26.1	7.7
1997	30	151.8	45.4	26.0	6.0

WHO, n.d.

What the data in Table 10.5 show is that the average weight of adult Onge males has increased by 2.2 kilograms (almost 5 lbs) over the 32-year period of the study, while their height has increased by 2 centimetres (just under an inch). The average adult female height has remained the same at about 55 inches (roughly 4′ 7″), a height that is not likely to see any decrease, while their average weight has decreased by 1.6 kilograms (roughly 3.5 lbs). These differences show up in the MAC and FTT, where males have seen slight increases, and females, slight decreases. The situation for the women of the tribe looks bad in that during this period the average intake of nutrients actually increased, yet only the males have exhibited the expected gain in anthropometric figures.

All of this shows that taking the measurement of a particular group can provide us with important information, in this case, that there is a need for women to receive better nutrition.

Race and Disease

Similar to forensic anthropologists, whose work we'll examine in Chapter 12, biomedical researchers (a group that includes physical anthropologists) often argue for the usefulness or 'necessary evil' of race in their work. There are clear links between particular geographic populations and certain diseases. Categorizing a geographic population by race may be an important first step towards isolating the hard-to-get factors that could explain why certain people are particularly susceptible to a given disease or condition. Take Ireland for example: the Irish have the highest rate of cystic fibrosis in the world. It has been estimated that one out of every seven people in the country is a carrier. Now the country has in place a thoroughgoing genetic test for 11 particular mutations indicating that

someone is a carrier. This screening has helped to reduce rates of the disease.

Here is another example. The Ashkenazi Jews, or *Ashkenazim*, are those Jewish people with ancestors who at one time lived along the Rhine River, in what is now Germany, before spreading out, in entire communities, into regions of present-day Poland and Russia. As a relatively endogamous group (meaning that they typically marry within their own racial community), they share a significant genetic heritage. Recent cancer research has uncovered a link between mutations of two cancer-suppressing genes (BRCA1 and BRCA2) and increased rates of breast cancer; these mutations have been shown to be especially prevalent among the Ashkenazim. A 2009 study of 2,080 women of Ashkenazi ancestry living in Ontario found that 22 of the research participants—slightly more than 1 per cent—had the BRCA1 or BRCA2 genetic mutations. Currently in Ontario, women are tested for these mutations only if they have been referred by their doctor because of a strong family or personal history with breast or ovarian cancer. Most of the women who tested positive for the mutations in the Ontario study would not have been eligible for the test, and so would not have been forewarned of their relatively high risk ('Breast Cancer Gene Test', 2009). Hopefully this study will demonstrate the benefits of extending the screening program to all Ashkenazim rather than just to families of women who have experienced breast cancer.

Burchard and associates, in an article published in the very prestigious *New England Journal of Medicine*, make the case for the necessary involvement of race in their work:

> We believe that ignoring race and ethnic background would be detrimental to the very populations and persons that this approach allegedly

seeks to protect. Information about patients' ethnic or racial group is imperative for the *identification, tracking, and investigation of the reasons for racial and ethnic differences in the prevalence and severity of disease and in responses to treatment.* (Burchard, et al., 2003; emphasis mine)

Of course, it is important to qualify this: people who are minoritized by readily perceived 'racial' characteristics are more likely to become sick. They also tend to have slower and lower recovery rates, because they tend to receive poorer-quality medical care, live in unhealthier (i.e. more polluted and unsafe) areas, and labour in unhealthier workplaces. In the late nineteenth and early twentieth centuries, when Aboriginal people—particularly those living in the Prairies—were dying of tuberculosis in rates that were disproportionately high when compared with those of the non-Aboriginal population, many government officials and doctors concluded that Aboriginal people had some built-in susceptibility to the disease. As it turned out the high rates of disease and death owed much less to 'race' than to the horrible overcrowded living conditions that the Prairie Aboriginal peoples were experiencing on reserves, a product of the enforced settlement of the people on those relatively small pieces of land, compounded by conditions of overcrowding, abuse, and poor diet at the residential schools.

While it's wrong to jump to conclusions about the susceptibility of particular geographic or racial groups to certain diseases, it is clear that there are many cases where the biologically influenced health effects of genetic differences cannot be ignored. As we will see in the sections that follow, there are many links between diseases and groups that can be categorized at least geographically, if not by race.

Blood Disorder and Race

1. Sickle-Cell Anemia

A classic case for a connection between disease and race involves **sickle-cell anemia** and black people. It is generally estimated that the condition affects 1 out of every 500 African Americans. As with the 'thrifty gene' (discussed below), this is a situation where evolutionary advantage and disadvantage are linked. Being **heterozygous** for sickle-cell anemia (i.e. receiving it from only one parent) offers a slight advantage in surviving the condition. Unfortunately, it is too often taught in high school science classes that being heterozygous for sickle-cell anemia gives automatic immunity to **malaria**, which it does not, and doctors sometimes misleadingly advise patients not to take malaria shots. These people can still get malaria, and be seriously harmed by it; it's just that they are likely to have fewer complications and are more likely to survive. 'Textbook examples' of particular phenomena—medical and otherwise—are often misleading (see Steckley, 2008,

ch. 3, on the popular textbook myth of the many words for snow in the language of the Inuit). Sickle-cell anemia can be fatal to a person who is **homozygous** for sickle-cell anemia (i.e. an individual who receives the gene from both parents). The life expectancy of someone who is homozygous for sickle-cell anemia is generally somewhere between 42 and 48 years.

The argument against branding sickle-cell anemia a 'black disease' is two-fold. One is that people whose genetic heritage is associated with other areas in which malaria is prevalent—people from parts of the Mediterranean (e.g. Greece and Sicily), the Middle East (e.g. Saudi Arabia and Turkey), and South Asia—also suffer disproportionately from sickle-cell anemia. As well, there are Africans from areas where malaria is not prevalent who do not get the disease. Nevertheless, as there are preventative treatments that can be used (e.g. a daily dose of folic acid or, with children aged from 1 to 5, penicillin), the racial identification of the disease is medically useful.

2. Thalassemia

Another blood disorder that is linked both with particular populations and with malaria is **thalassemia**, which has been found in archaeological evidence at Egyptian sites of more than 5,000 years old (Rabino Massa, 1977). The disease, which means 'sea blood' in Greek, was first associated primarily with Mediterranean peoples (i.e. people of Greek, Italian, Portuguese, Spanish, and Egyptian heritage); however, it now has been identified among South and Southeast Asians, among others. The highest rate in the world is in the Maldives (an island chain in the Indian Ocean), where 18 per cent of the population is affected, and on the Mediterranean island Cyprus, where 16 per cent of the people are affected. It is another 'good news/ bad news' disorder, as it makes one less susceptible to coronary heart disease and malaria.

Screening people likely to be carriers has significantly reduced the rate of thalassemia in Cyprus since the 1970s through early detection and abortion. The most heartening case happened in India in 2008, when the stem cells taken from a thalassemia-free embryo that had come from an egg that had been fertilized *in vitro* (i.e., 'in glass') were used to cure the disorder in his nine-year-old sister.

3. Tay–Sachs

The disorder whose connection to race has received perhaps the greatest amount of scholarly attention is the genetically programmed brain and nerve disorder **Tay–Sachs** disease. It is named after a British ophthalmologist, Warren Tay, who in 1881 was the first to notice the red spot on the retina of the eye that marks the disorder, and the American neurologist Bernard Sachs, who noted six years later that the disorder was primarily associated with the Ashkenazi Jews (those of central and eastern European

Science Gone Wrong

Eugenics and Racialized Sexual Sterilization in Canada

During the history of Alberta's Sexual Sterilization Act, which lasted from 1928 to 1972, some 2,832 people were sterilized, most of them women. Sterilizations of Métis and First Nations people account for a disproportionately high number of the total, an estimated 25 per cent (roughly 10 times their percentage of the total population).

The province of British Columbia passed an act similar to Alberta's Sexual Sterilization Act in 1933, the same year that the notorious Law for the Prevention of Genetically Diseased Offspring was passed in Nazi Germany. Recently there have been accusations that, following both racial and religious prejudice, hundreds of non-Christian Aboriginal people were sterilized by a United Church missionary doctor in a church-run hospital in the BC coastal community of Bella Bella, and that a good number of young Aboriginal women made pregnant by residential school staff, clergy, and visiting officials were coerced into having abortions. While numbers are being contested in the courts, there seems little doubt that such practices did take place.

heritage; see page 243). The condition is autosomal, which means that it is located on a chromosome other than the X and Y sex chromosomes; it is recessive, meaning that if you get it from one parent and not the other, then you don't exhibit the symptoms. It usually strikes in infancy (hitting at six months and killing before the age of 4), but can also strike in early childhood (stretching life expectancy to between 5 and 15 years). The rarer adult or 'late onset' Tay–Sachs disease (LOTS) causes disability, though it is seldom fatal.

Why does Tay–Sachs hit this group? Some scientists played with the idea that, like sickle-cell anemia, it might convey some evolutionary advantage, with some going as far as suggesting that being heterozygous for this brain-related disorder might translate into higher intelligence. The general view today is that it is much more likely that it is an example of the **founder effect**, which you may recall occurs with traits that arise when a relatively small group of people experience a **genetic bottleneck**, and breed within the group rather than outside of it. As the Ashkenazi Jews have long been a population that married and bred endogamously (that is within the group and not outside it), it is susceptible to the founder effect.

Two other groups affected by Tay–Sachs disease in disproportionately high numbers are also candidates for the founder effect. The first are the Cajuns, the Louisiana descendants of the French-speaking Acadians who were kicked out of the Maritime colonies during the eighteenth century. Linguistically and culturally separate from their southern American neighbours, the group for quite some time kept to itself. Researchers have traced the disorder back through several Cajun families to a single founding couple who developed the same mutation, and who came from France to North America during the eighteenth century. They are not known to have been Jewish.

French Canadians of southeastern Quebec and New England (the New Englanders being descendants of those who migrated from their home province during the late nineteenth and early twentieth centuries) also experience Tay–Sachs disease in disproportionately high numbers. Before it was discovered that the mutation causing the disorder among French Canadians was different from the one that causes Tay–Sachs among the Ashkenazi Jews and the Cajuns, it was believed that perhaps the first instance of the disorder could be traced to a very sexually prolific Jewish ancestor, in what has been called the Jewish Fur Trader Hypothesis. Again we have a population in which the founder effect seems most likely.

The association of some medical conditions with certain group or races has had effects both good and bad. During its peak in popularity from the 1910s through the 1930s, the policy of **eugenics** (lit. 'good gene') led many to believe that 'good' races and individuals should be encouraged to breed, while 'bad' races and individuals should be sexually sterilized. Among those who were targeted by the racist eugenics movement were Jews, who, based on the prevalence among them of Tay–Sachs, were identified as being an 'inferior race'. Another disadvantage of identifying Jewish people with the disorder is that non-Jewish people who may be carriers are much less likely to be identified; in fact, some people suffering from Tay–Sachs, particularly the adult form, might be misdiagnosed.

The positive side of group identification with the disorder is that a huge number of Jewish people have been tested for the mutation (estimates are that some 1.3 million Jews were screened between 1969 and 1998). This has resulted in a drastic reduction in the number of children born with Tay–Sachs in North America, Europe, and Israel.

The Thrifty Gene

When reviewing the way that certain racial groups have been targeted because of their high rate of a particular disease, it is hard to overlook the connection among

Aboriginal people between what is called the **thrifty gene** and type II diabetes. The idea of the thrifty gene (or, more accurately, the thrifty allele) was first hypothesized by geneticist James Neel in 1962. His theory was that individuals who can store calories in fatty tissues are better able to survive and reproduce in environments in which most people experience frequent famines or chronic food shortages. 'Thrifty' here refers to the body's ability to hold on to its sugars longer. The drawbacks of the thrifty gene become apparent when people with the hypothesized thrifty gene make high-calorie junk foods and even 'regular' twenty-first-century Western foods part of their diet, and subsequently become obese, experience hypertension, and contract type II diabetes. The primary focus of research on the subject has been the diabetes. Type I diabetes occurs when the pancreas does not produce enough insulin, so that a person suffering this form of the condition must take insulin shots to survive. Type I accounts for between 5 and 10 per cent of all diabetes cases, and usually develops in children and young adults. With type II diabetes, sometimes referred to as 'noninsulin-dependent diabetes', the pancreas produces sufficient quantities of insulin, but the cells are unable to use it effectively. It most often develops in adults over 40 but can occur at any age, and represents 90–95 per cent of all diabetes cases.

With type II diabetes, particular groups are more susceptible than others. In Canada, one of these groups, studied by Dr Robert Hegele, is the Sandy Lake Oji-Cree of northwest Ontario, a community with an astonishingly high rate of the disease. The people, traditionally, lived by hunting and gathering in an environment rich in resources. However, this lifestyle would occasionally present the community with short-term famine caused by food shortages, particularly during the winter. After World War II, the lives of the people became more connected with the lifestyle of the mainstream populace of southern Canada: the Oji-Cree were suddenly supplied with common foods of the south, and fewer people engaged in the highly active lifestyle of a hunter/trapper. During the time of these changes the rate of type II diabetes has gone from virtually nothing to one of the highest rates in the world. Dr. Hegele has found that the HNF1A G319S mutation was present in 40 per cent of the studied members of this community.

Lactose Intolerance

Lactose is milk sugar that requires an enzyme, **lactase**, to break it down into its constituent sugars (glucose and galactose) so that they can be absorbed. (To keep lactose and lactase straight, think of –o– as in *obese* (too much sugar) and –a– as in *attack*.) Mammal children typically produce lactase so that they can drink, process, and enjoy the nutritional benefits of the breast milk from their mothers. After a certain period they are weaned: they stop drinking breast milk, and, as lactase is no longer necessary, the ability to produce it is turned off. This is another case of the economics of genetics: why pay for something that you don't use? Mammal adults become lactose intolerant, unable to process lactose, with negative effects on digestion.

Lactase persistence (LP) occurs when a mammal's system continues to produce lactase into adulthood. That has an evolutionary advantage for humans living in cultures that have domesticated milk-producing animals. Unless those societies are able to develop cultural ways of

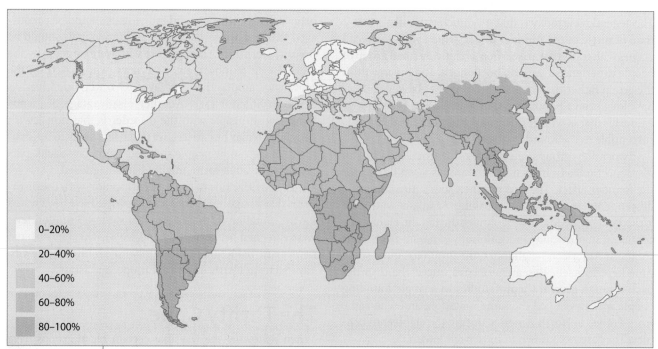

0–20%
20–40%
40–60%
60–80%
80–100%

Figure 10.3 | Frequency of lactose intolerance worldwide

enabling adults to digest milk (as with the addition of bacteria to yogurt), evolving into lactase persistence is ideal.

Most adults in the world are lactose intolerant rather than lactase persistent. LP is most common among people of European ancestry, although it is also found among other peoples who have domesticated cows, including some African groups (e.g. the Fulani of Nigeria and the Tutsi of Rwanda and Burundi) and certain Middle Eastern and South Asian groups (especially those of northern Indian and Pakistan). Among Europeans, there appears to be a relationship between LP and a substitution of C (cytosine) with T (thymine) in position 13910 of the lactase gene (Mineri, et al., 2009, p. 181). There may be other factors involved as well. A recent study of 3,344 women aged 60–79 years from 23 towns across Britain showed that the C variant (the non–lactase-persistent variant) decreased significantly from the south to the north, and from the sunnier east to the more rain-soaked west, in both ways showing that the greater the amount of sunlight, the higher the chance of the C variant (see Smith, et al., 2008).

LP was very rare or absent among European populations prior to 10,000 years ago, and yet in modern populations, frequency of the lactase-friendly allele ranges from 6 to 36 per cent in eastern and southern Europe, from 56 to 67 per cent in central and western Europe, and from 73 to 95 per cent in the British Isles and Scandinavia (Itan, Powell, Beaumont, Burger, & Thomas, 2009). LP itself ranges in frequency from 15 to 54 per cent in eastern and southern Europe, from 62 to 86 per cent in central and western Europe, and from 89 to 96 per cent in the British Isles and Scandinavia. The absence of the allele among other populations in which LP has, nevertheless, evolved is likely a case of **convergent evolution**, where two populations—often of different species and genera, or even different taxons—have developed the same phenotype (i.e. physical structure/characteristics) but by following somewhat different genetic paths.

HIV and CCR5–Delta 32

Not all genetic mutations are associated with greater susceptibility to disease. The following is an example of a positive association. CCR5 is a gene that encodes a protein whose (very) long name is chemokine (C-C motif) receptor 5. The human immunodeficiency virus (HIV) uses this protein as a co-receptor to enter human cells. A flawed CCR5 fails as a co-receptor. Drugs have been produced to achieve the same results by interfering with the co-reception process; however, they have not been as effective as CCR5–Δ32, which involves the deletion (hence the delta sign, Δ) of 32 bases. If you receive this mutation from both parents (i.e. if you are homozygous for it) you will be highly resistant to HIV; if you receive this mutation from one parent (making you heterozygous for it), HIV will proceed slowly, delaying the process of its becoming full-blown AIDS. This makes CCR5–Δ32 a **restriction gene**, the first one found (in 1996) to inhibit HIV.

There are two interesting aspects of this gene from a physical anthropologist's point of view. First, the deletion is unevenly distributed throughout different populations. It is found in highest numbers in northwestern Europe (Sweden, Norway, Finland, and northern Russia), with estimates that it occurs in about 16 per cent of the population (Itan, et al., 2009); the frequency decreases as you move south through Europe. Second, the deletion may have played a vital role in helping people survive the smallpox virus over the last few hundred years: it has been suggested that survivors of smallpox epidemics in northern Europe may have been helped by the mutation. These survivors, in turn, would have produced an unusually high number of descendants who would benefit in the future by being more resistant to HIV.

It's worth noting that this is not the only HIV restriction gene that has been found. There is evidence suggesting that sex-trade workers in certain parts of Africa who are also are resistant to HIV share an HIV-restricting genetic mutation.

Summary

We have seen in this chapter first that race takes a decidedly distant third place to traits and geography in terms of its bearing on human variation. As we started this chapter with a quotation from the American Association of Physical Anthropology's 'Statement on the Biological Aspects of Race', it seems fitting that we add another quotation from the AAPA for the summary, this time from the Preamble to the statement:

Popular conceptualizations of race are derived from 19th and early 20th century scientific formulations. These old racial categories were based on externally visible traits, primarily skin color, features of the face, and the shape and size of the head and body, and the underlying skeleton. They were often imbued with nonbiological attributes, based on social constructions of race. These categories of race are rooted in the scientific traditions of the 19th century, and in even earlier philosophical traditions which presumed that immutable visible traits can predict the measure of all other traits in an individual or a population. Such notions have often been used to support racist doctrines. Yet old racial concepts persist as social conventions that foster institutional discrimination. The expression of prejudice may or may not undermine material well-being, but it does involve the mistreatment of people and thus it often is psychologically distressing and socially damaging. Scientists should try to keep the results of their research from being used in a biased way that would serve discriminatory ends. (AAPA, 1996)

Science Gone Wrong

Taking Blood and Giving Back Nothing

While studying the blood of a people and its connection to disease is generally a good practice, there has to be, built into the process, a respect for the people being studied. In 1981, genetics professor Richard Ward, then affiliated with the University of British Columbia (UBC), sent a letter to the Nuu-chah-nulth (Nootka) people of Vancouver Island, inviting them to volunteer to have their blood taken as part of a study into the genetic links between the people and their high rate of arthritis. In part the letter said:

> We feel that if a proper study is carried out it will identify all people who have a problem with their joints, and a physiotherapy treatment can be started as a way of helping them. (as cited in Wiwchar, 2004)

Between 1982 and 1985, Dr Ward, who received a $330,000 Health Canada grant for the study, obtained 883 thirty-millilitre vials of Nuu-chah-nulth blood. The donors understood that they would receive information concerning the results of the blood tests and their implications within a year, and that this study would benefit them. No such information was ever distributed, and no benefits were received. Ward had not found the genetic link he was looking for. In 1986, he switched jobs to work at the University of Utah, and received a $172,000 grant to further study the blood, but again he found nothing of significance. In 1991, he studied the blood a third time (without obtaining permission), this time concluding, in a published article, that the Nuu-chah-nulth people, as a genetic group, had a distinct mtDNA 'lineage cluster' dating from between 41,000 and 78,000 years ago, before they migrated to the Americas (Ward, Frazier, Dew-Jager, & Pääbo, 1991).

In 1996, Ward accepted a position as the head of the newly formed Institute of Biological Anthropology at Oxford University. Once there, he treated the Nuu-chah-nulth blood as his personal possession, studying it for a variety of purposes and loaning it out to other researchers, all without asking the people who had donated the blood under a particular set of conditions that no longer applied to Ward's extended investigation.

Once the Nootka found out that their blood was still being examined—and without receipt of any of the information they had been promised initially—they felt used. Larry Baird, who had donated his blood and had given permission for his children's blood to be taken, spoke for his people when he said:

> We were of the understanding that we would have the results of the study within a year, but he [Ward] never told us anything after. He disappeared. . . . He published more than 200 papers and became the top guru in his field because he was carrying our blood around with him. He used us like cheap guinea pigs, and that incenses me. (as cited in Wiwchar, 2004)

In July 2003, the Nuu-chah-nulth Tribal Council established a Research Ethics Committee to monitor the research done on their people. They have evaluated requests to conduct research among the Nuu-chah-nulth, and have both granted and denied permission, depending on their careful assessment circumstances. In 2004, the blood collected by Ward's team was returned to Vancouver and stored in a freezer at UBC.

This case has become well known among academic researchers, and has been used a bad-case scenario to be avoided, a reason to pay greater attention to teaching ethics to students in the sciences. We will be looking generally at ethics of physical anthropological research in the next chapter. There is a general ethical rule in anthropology that when you take something from the people—their stories, their histories, their blood, or their artifacts—you should give something back, at the very least the results of your study.

Typical Student Questions

▶ *Why do we call ourselves 'the human race' when we are, in fact, a species?*

I think it is because 'the humans species' would make us sound too much like other animals, an idea that is unpalatable to many people (though we are, in fact, animals, and 'the human species' would be more apt).

▶ *Why are different breeds of dog that are so unalike—Great Danes and Shih Tzus, for instance—not considered separate species, or at least subspecies?*

Remember the species test mentioned in Chapter 4: if two individuals can mate and produce fertile young, then they belong to the same species. (This is known as the biological species concept.) Crossing a Great Dane with a Shih Tzu seems implausible, but it can happen.

Review Questions

1. Discuss the relationship between skin colour and racial categorization. In what way is it or is it not valid?
2. What are the mtDNA and Y-chromosome models of human migration?
3. Explain the relationship between earwax type and geographic population.
4. Describe the role of dental anthropology in categorizing human variation.

Discussion Questions

1. South Asians seem to be left out or underrepresented in both historical and contemporary systems of racial categorization.
   a) Why do you think this is so?
   b) If South Asian anthropologists of the nineteenth and early twentieth centuries had set up race categories, how might they have done so? Keep in mind that there is a clinal distinction of fundamentally lighter populations to the north and darker ones to the south in the Indian subcontinent.
2. Do you believe that biological race exists? What is the evidence for your position?
3. How is making the connection between disease or disorder and specific populations a useful form of study?

Key Terms

Ainu
allele
anthropometry
  (anthropometric)
Australoid
Austronesians
brachycephalic
breed
Capoid (Khoisan)
Caucasian
Caucasoid (or Caucasian)
cephalic (or cranial) index
convergent evolution
dental anthropology
derived

dolichocephalic
epicanthic fold
eugenics
facial angle
fat-fold thickness (FFT)
founder effect
generalized
genetic bottleneck
geographic population
Gloger's rule
haplogroups
heterozygous
holotype
homozygous
lactase

lactose
malaria
melanin
mesocephalic
mid-arm circumference
  (MAC)
Mongoloid (or Mongolian)
mutation
Negritos
Negroid
non-metric variation
Onge
phase
phylogeography
plasticity

population
race
racial time
racialization
red ochre
restriction gene
rickets
shovel-shaped incisors
sickle-cell anemia
sinodont
sundadont
Tay-Sachs disease
thalessemia
thrifty gene
type specimen

Key Individuals

Franz Boas
Johann Friedrich
  Blumenbach

Carleton Coon
Constantin Wilhelm
  Lambert Gloger

Aleš Hrdlicka
Richard Lewontin
Carolus Linnaeus

Ernest W. Mayr
Samuel George Morton
Christy Turner

Recommended Print and Online Resources

Adcock, G. J., Dennis, E. S., Easteal, S., Huttley, G. A., Jeremiin, L. S., Peacock, W. J., & Thorne, A. (2001, January 16). Mitochondrial DNA sequences in ancient Australians: Implications for modern human origins. *PNAS, 98*(2), 537–42.

American Association of Physical Anthropology. (Website). Welcome to AAPA—American Association of Physical Anthropology. Available at http://physanth.org/

American Association of Physical Anthropology (AAPA). (1996). AAPA statement on biological aspects of race. *American Journal of Physical Anthropology, 101*, 569–70. Retrieved from http://physanth.org/positions/race.html

Bass, E. J., & Jackson, J. F. (1977, September). Cerumen types in Eskimos. *American Journal of Physical Anthropology, 47*(2), 209–10.

Boas, F. (1912). *Changes in the bodily form of the descendants of immigrants.* New York: Columbia University Press.

Brace, C. L. (2000). Does race exist? An antagonist's perspective. *NOVA Online.* Retrieved from www.pbs.org/wgbh/nova/first/brace.html

Breast cancer gene test eligibility in Ont. misses some. (2009, December 14). *CBCnews* (online). Retrieved from www.cbc.ca/health/story/2009/12/14/breast-cancer-mutations-jewish-ontario.html

Burchard, E. G., Ziv, E., Coyle, N., Gomez, S. L., Tang, P. H., Karter, A. J., . . . Risch, N. (2003, March 20). The importance of race and ethnic background in biomedical research and clinical practice. *New England Journal of Medicine, 348*(12), 1170–5.

Canadian Association for Physical Anthropology. (Website). Welcome to the CAPA web site! Available at http://capa.fenali.net/

Cann, R. L., Stoneking, M., & Wilson, A. C. (1987, January 1). Mitochondrial DNA and human evolution. *Nature, 125*, 31–6.

Gill, G. W. (2000). Does race exist? A proponent's perspective. *NOVA Online.* Retrieved from www.pbs.org/wgbh/nova/first/gill.html

Hegele, R. A., Cao, H., Harris, S. B., Hanley, A. J., & Zinman, B. (1999). The hepatic nuclear factor-1 G319S variant is associated with early-onset type 2 diabetes in Canadian Oji-Cree. *The Journal of Clinical Endocrinology and Metabolism, 84*(3), 1077–82.

Itan, Y., Powell, A., Beaumont, M. A., Burger, J., & Thomas, M. G. (2009, August 28). The origins of lactase persistence in Europe. *PLOS Computational Biology, 5*(8). doi:10.1371/journal.pcbi.1000491

Krulwich, R. (2009, February 2). Your family may once have been a different color [Web log message]. Retrieved from www.npr.org/templates/story/story.php?storyId=100057939

Larsen, C. S. (2000). *Skeletons in our closet: Revealing our past through bioarchaeology*. Princeton: Princeton University Press.

Lewontin, R. C. (1972). The apportionment of human diversity. *Evolutionary Biology, 6*, 391–8.

Morton, S. G. (1839). *Crania Americana; or, A comparative view of the skulls of various Aboriginal nations of North and South America: To which is prefixed an essay on the varieties of the human species*. Philadelphia, PA: J. Dobson.

Nakbunlung, S., & Wathanawareekool, S. (2004). Talking teeth: Dental non-metric traits of wooden-coffin people in the Pang Ma Pha Cave sites, northwestern Thailand. *Indo-Pacific Prehistory Association Bulletin, 24*, 139–42. Retrieved from www.ejournal.anu.edu.au/index.php/bippa/article/view/119/109

Neel, J. V. (1962, December 14). Diabetes mellitus: A 'thrifty' genotype rendered detrimental by 'progress'? *American Journal of Human Genetics, 14*, 353–62.

Oppenheimer, S. (2004). *The real Eve: Modern Man's journey out of Africa*. New York: Carroll & Graf.

Perez, I., Bernal, V., Gonzalez, P., Sardi, M., & Politis, G. (2009). Discrepancy between cranial and DNA data of early Americans: Implications for American Peopling. *PLoS ONE, 4*(5), e5746. doi:10.1371/journal.pone.0005746

Rabino M. E. (1977). Presence of thalassemia in Egyptian mummies. *Journal of Human Evolution, 6*, 223–5.

Smith, G. D., Lawlor, D. A., Timpson, N. J., Baban, J., Kiessling, M., Day, I. N. M., & Ebrahim, S. (2008). Lactase persistence–related genetic variant: Population substructure and health outcomes. *European Journal of Human Genetics, 17*, 357–67. doi:10.1038/ejhg.2008.156

Steckley, J. L. (2008). *White lies about the Inuit*. Peterborough, ON: Broadview.

Turner, C. G. (1989). Major features of Sundadonty and Sinodonty, including suggestions about East Asian microevolution, population history, and late Pleistocene relationships with Australian Aboriginals. *American Journal of Physical Anthropology, 82*(3), 295–317.

Underhill, P., Shen, P., Lin, A., Jin, L., Passarino, G., Yang, W., . . . Oefner, P. (2000, November). Y chromosome sequence variation and the history of human populations. *Nature Genetics, 26*, 358–61.

Wade, N. (2006, January 29). Japanese scientists identify earwax gene. *The New York Times*. Retrieved from www.nytimes.com/2006/01/029/science/29cnd-ear.html

Ward, L. M., Gaboury, I., Ladhani, M., & Zlotkin, S. (2007, July 17). Vitamin D-deficiency rickets among children in Canada. *Canadian Medical Association Journal, 177*(2), 161–6.

Ward, R. H., Frazier, B. L., Dew-Jager, K., & Pääbo, S. (1991, October 1). Extensive mitochondrial diversity within a single Amerindian tribe. *Proceedings of the National Academy of Sciences of the United States of America, 88*(19), 8720-4.

Wells, Spencer. (2006). *Deep ancestry: Inside the Genographic Project*. Washington: National Geographic Society

WHO Regional Medical Research Centre (ICMR). (n.d.). Tribal health: Health, nutritional, and demographic study among the Onges. Retrieved from www.rmrc.res.in/projects/tribes/complete/Onges.htm

Wiwchar, D. (2004, December 16). Nuu-Chah-Nulth blood returns to west coast. *Ha-Shilth-Sa, 31*(25). Retrieved from http://caj.ca/wp-content/uploads/2010/mediamag/awards2005/(David%20Wiwchar,%20Sept.%2012,%202005)Blood2.pdf

11

Recent History
Ethics, Migrations, and the Physical Cost of Agriculture

Totem pole, Stanley Park, Vancouver
Miranda McMurray/iStockphoto

After reading this chapter, you should be able to:

- Outline the arguments both in favour of and against the rights of science in dealing with 'other people's bones'.

- Explain what can be learned, using physical anthropology, about the 'Iceman'.

- Discuss the effects on the bones and teeth of a people as they become agricultural.

- Assess the evidence against the Clovis First hypothesis.

- Critique the Solutrean hypothesis.

Introduction

In this chapter we'll look at how physical anthropologists study 'recent history'—the last 10,000 years—in terms of a number of areas of study. One of these is the ethics of digging up and studying the bones of the relatively recent dead. Excavating remains is not a problem with the early hominids, but it is for their more recent descendants. Another theme is the changes undergone by the human body, over time, with the shift to agriculture. The news is not good—for now, let's just say that the hunter–gatherer basketball team would probably beat the farmer squad.

Readers will also learn that the way we live our lives—from the good or bad eating habits we have as children to the repetitive actions (like keyboarding) we perform at work—shows up in our bones and teeth. A word to the wise: take good care of your teeth if you want to make a good impression on the physical anthropologist who may be examining your remains thousands of years from now.

Ethics

Ethics is an unavoidable part of all science. And ethics is problematic. After presenting several definitions of ethics, including 'the rules or standards governing the conduct of the members of a profession', noted biologist and talented science writer William Stansfield sagely remarked:

> One of the greatest problems with these definitions is that many of our most important ethical or moral problems do not have either all 'good' (right) or all 'bad' (wrong) solutions. (Stansfield, 2000, p. 12)

This is what we will be dealing with in this chapter. When it comes to ethics, there are only 'better' and 'worse' solutions. Sticking to one side exclusively is narrow-minded, and compromise is wise.

Ethics here relates primarily to the treatment of the remains of the dead. This was not and is not a problem with the ancient dead who go back hundreds of thousands of years and more; they have no readily identifiable descendants who care. However, as we get closer to and into our own time, the bones increasingly connect with living people, and there have to be ethical practices in place to satisfy the living that their dead are being treated with proper care and respect. But to what country, to what people, do the bones belong? What rights does science have as a spokesperson for humankind in general?

The Case for Science

Overcoming the Bias of the Book

So: what rights *does* science have as a spokesperson for humankind? I believe that we owe it to the people of the past to tell their story. This is especially true when a colonial relationship has been established, and the colonial power has been the only one to tell the story of a colonized people. This is known as the **bias of the book**. Archaeology and physical anthropology can work to overcome this bias.

Take the example of the **Great Zimbabwe** 'house of stone', an amazing structure in the country now named after it. Early European writers felt that it had to have been constructed by an ancient tribe of white people (a claim that may have helped to justify their dividing the African continent into a collection of European theme parks). The most striking feature of the Great Zimbabwe is known as the Great Enclosure, a double wall surrounding several edifices. It was constructed in almost a century during the 1300s, by the Shona people and consists of around a million granite blocks cut to fit without the use of mortar. It is about 250 metres in circumference and up to 11 metres high in places. The interior walls range in height from a little over a metre to 5 metres.

The Portuguese explorer João de Barros wrote in 1552 that the Great Zimbabwe must have been around 3,000 years old, and constructed by Europeans. This combination of facts would serve in later years as strong evidence

Ruins of the Great Zimbabwe house of stone, including the Great Enclosure

to back the claim that Europeans were entitled to conquer and colonize Africa; it was, after all, part of the ancient white world. In 1871, German writer Karl Mauch added to de Barros's speculation by proposing that the structure was built by the Phoenicians, a Mediterranean people who peaked culturally around 3,000 years ago. A few years later, in 1890, the English-born traveller, writer, and amateur archaeologist James Theodore Bent, who had been hired by the powerful diamond magnate and colonial administrator Cecil Rhodes (after whom the region, 'Rhodesia', was once named) wrote that he believed the Great Zimbabwe had been created by a 'civilized' nation—meaning, of course, an ancient nation of white people.

The story did not change until a pair of British archaeologists got to it. **David Randall-MacIver** (1873–1945) recognized the African characteristics of the construction in 1906, and **Gertrude Caton-Thompson** (1888–1985) supported his views in 1932, adding that it was definitely African, being an extreme example of early stone work in the area.

Archaeology is a great tool for overcoming the colonialist bias of the book, and physical anthropologists have a great contribution to make in that regard. American physical anthropologist **Clark Spencer Larsen** expresses this neatly:

Heretofore, the impact of colonization on native peoples was known mostly from written sources, written by Europeans. Native voices were unheard. Skeletons offer a new and unique source of information that is only now being tapped, offering new perspectives about this important period of history. (Larsen, 2000, p. 8)

The bones can tell us detailed stories about the original inhabitants of the land. Some of the themes of these stories are good (e.g. the remarkable longevity of some Aboriginal peoples), others are unsettling (e.g. cannibalism). Either way, they deserve to be studied. The study of ancient remains—biological as well as architectural—brings better balance to the story than that provided by written sources alone.

The Case against Science

The People's Rightful Connection to the Bones

While science, through the study of human remains, can work to correct biased accounts of a people's history, archaeologists and physical anthropologists must operate with a firm understanding of the strong emotional connection between people and the bones of their ancestors. I felt some of this in 1999, the day after I was adopted by the Wyandot nation of Kansas. That day we reburied the bones of some 500 Huron (of whom the Wyandot are a descendant group). These bones had originally been buried in 1636 at the site of the large Huron community of Ossossane. Most of the dead were victims of the first European diseases to strike the Aboriginal inhabitants of North America; Father Jean de Brébeuf provides an account of the burial of these individuals, written with a respect unusual in a European observer.

During the 1950s, the bones were dug up so that they could be examined by students and other scholars who, collectively, produced a great body of work, published and unpublished. In 1999, surviving members of the Huron and related Petun nations, having travelled from the Quebec community of Wendake as well as from First Nations communities in Michigan, Kansas, and Oklahoma, reburied the 500, with the aid of the Royal Ontario Museum, the institution that had once held these remains. Taking part in an event of such historical and emotional significance was one of the most profoundly spiritual experiences of my life.

You can see, then, that I am a little conflicted on the issue. On the one hand, I value the scientific opportunities the remains of our ancestors provide; on the other, I respect the wishes of surviving descendants to see any remains treated with the same respect we would show a recently deceased relative. I believe that scientists and the living representatives of bygone civilizations must work together to ensure that ancestral remains are treated with care, and with caring.

Stories of Trophies of Science

While respect for human remains should be the watchword of every anthropologist, there has been a double standard in Canada and the United States concerning the bones of the dead. The remains of non-Aboriginal people are nearly always treated with care and respect, yet Aboriginal bones have been viewed as what can be called 'trophies of science', more as objects than as the remains of people who were loved and buried respectfully and religiously.

Shanawdithit was the last of the Beothuk, an Aboriginal people who lived on the island of Newfoundland for almost 2,000 years. Shanawdithit died on 6 June 1829 and was buried in a Church of England cemetery in St John's. She did not stay there long. Her body was exhumed so that her skull could be taken for 'scientific purposes'. There is an irony here in that the Beothuk had been condemned by some as 'headhunters' for keeping the skulls of people who were presumably their enemies. Scientists kept the skulls of their presumed friends.

You will have seen by now several references to the excellent work of **Franz Boas** (1858–1942), a much respected anthropologist and a foundational figure in the history of anthropology, one who fought racism in many of its malignant forms. He will not shine so brightly in this chapter. There are two stories here that reflect badly both on Boas and on the scientific culture of which he was a part. They clearly show the colonial approach that was associated with early anthropology of all types. It is an approach that has led Aboriginal people critical of these practices to call Western anthropologists agents of 'vulture culture'.

The first incident took place in 1888. Boas was working for the Canadian government with the Kwakiutl people of Vancouver Island. His writings about the people became a tremendous source of information for all time (although some of it has been recently challenged as flawed). Boas at the time, then 30 years old, was having difficulty finding a full-time job. Museums were not hiring him, and anthropology was not yet a recognized discipline in North American universities. Boas decided to try to make some money by collecting the skulls of the people that he was studying, hoping to receive the then going rate of five dollars per skull and twenty dollars for a complete skeleton. By amassing his own collection of human relics, one that a museum might be interested in housing, he could also improve his odds of gaining a curator's position.

Eight years later, Boas was working as assistant curator of the American Museum of Natural History in New York, in addition to holding a professorial position at Columbia University, where he would train a generation of the leading lights in American anthropology. However, the respectability he gained by achieving these two prestigious positions was tarnished by the unprincipled nature of his research practices. Boas put in a request to American polar explorer Robert Peary, who was then planning an expedition to Greenland, to bring back an Inuk (singular for Inuit) for anthropologists to study at the museum. Peary brought back six: an adult man, Nuktaq; his wife, Atangana; their 12-year-old adopted daughter, Aviaq; another man, Qisuk; his 5-year-old son, Minik; and a young man named Uisaakassak. They arrived in New York on 30 September 1897. Sickness soon followed. Qisuk died on 17 February 1898; three others died within the year. The vultures struck. In 1901, physical anthropologist Aleš Hrdlička published 'An Eskimo Brain' in *American Anthropologist*. It was Qisuk's. This wasn't all. After Qisuk died, Boas and his colleagues staged a fake burial to fool Minik, then 6 years old, using a covered log in place of his father's body. According to Canadian historian Kenn Harper, Boas claimed that the purpose of the burial was to:

'. . . appease the boy, and keep him from discovering that his father's body had been chopped up and the bones placed in the collection of the institution.' Not only did Boas confirm the story of the fake burial, though, he defended it. He said that he saw 'nothing particularly deserved severe criticism' in the act. . . .

The reporter questioned the right of the museum to claim the body of a man whose relatives were still alive, but Boas responded, 'Oh that was perfectly legitimate. There was no one to bury the body, and the museum had as good a right to it as any other institution authorised to claim bodies.' But, protested the reporter, did not that body belong rightfully to Minik, the son of the deceased? 'Well,' was Boas's reply, 'Minik was just a little boy, and he did not ask for the body. If he had, he might have got it.' (Harper, 1986, pp. 93–4)

The museum and Bellevue Hospital, where Qisuk had died, disputed over who should have the body. In compromise, the hospital allowed their medical students to dissect and examine the body, while the skeleton was preserved for display in the museum. Qisuk's bones were finally returned to Greenland in 1993.

George A. Dorsey and the Haida

George A. Dorsey (1868–1931) was an early American physical anthropologist who collected many human remains and cultural artifacts from Aboriginal peoples of North and South America. Typical of the time, he felt that he was recording the data of a dying race and wanted to take note of their typical traits before they disappeared. In 1898, he wrote:

> The Haidas numbered seven thousand in 1840, and counted over thirty villages. To-day there are two inhabited villages and less than one thousand Haidas. They are a doomed race. Wars, small-pox, gross immorality, a change from old ways to new ways—their fate is the common fate of the American, whether he sails the sea in the North, gallops over the plain in the West, or sleeps in his hammock in the forests of Brazil. (Dorsey, 1898, pp. 164–5)

Adding together the two contemporary Haida bands of Old Massey Village and Skidegate, you get a combined population of 4,154. They haven't disappeared.

In any case, Dorsey and his staff transported a number of human remains from the Haida to the Field Museum in Chicago. It wasn't until October 2003 that the museum returned the remains of about 150 Haida to the people of Haida Gwai, British Columbia. Forty Haida travelled to Chicago to prepare the remains of their ancestors for their

final trip, wrapping them carefully in white muslin before placing them in shipping crates.

What About Now? Science and Ethics Today

NAGPRA and the Kennewick Man Dispute

In 1990, the US federal government passed the Native American Graves Protection and Repatriation Act (NAGPRA). The law requires federal agencies and any institutions that have received federal funding (including museums and universities) to return human remains and cultural items to the people of Aboriginal communities that can demonstrate a clear connection with the remains or items.

In 2000, when it appeared that, owing to the NAGPRA legislation, the Kennewick Man remains were going to be returned to Aboriginal peoples in the Washington state area where they were discovered, a dark shadow of pessimism fell over a number of white physical anthropologists who did their primary work in the North American context, engaging in the **osteology** (lit. 'bone study') of Aboriginal people. Here is a sample:

> The future of osteology, methodologically, is bright. The future of osteology as applied to questions about the prehistory of our continent is bleak, and while I understand the arguments and the wrongs of the past, I think this Dark Age is a loss to everyone. (Katzenberg, 2001)

The case of Kennewick Man (whose discovery is discussed later on in this chapter) brought greater public attention to the implications of the US law. Five Aboriginal groups in the area applied to have his remains repatriated under the provisions of the NAGPRA legislation. It turned out to be an especially problematic case. One complicating factor was that the land where the remains were found belongs to the US Army Corps of Engineers, making it property of the federal government; this opened the door to Aboriginal groups seeking to lay claim to the remains. It also left the door open for bureaucratic bungling. Another problem lay in the age of Kennewick Man, which was determined to be roughly 9,500 calendar years old, and the fact that his cranium had the long and narrow look of a Caucasoid. He looked different from contemporary Aboriginal people, and because of his great age, it would be scientifically virtually impossible to establish his connection to any one cultural group existing today. In February 2004, the United States Court of Appeals ruled that none of the five Aboriginal groups had managed to demonstrate Kennewick Man's cultural affiliation with any group surviving today. The next year, scientists began extensive research.

Canadian Cases

In a thoughtful article on the challenges to anthropology's use of osteology, **Jerome S. Cybulski**, curator of Physical Anthropology at the Canadian Museum of Civilization in Gatineau, argues that the only osteology that matters in the Canadian context is osteology applied to the anthropological study of Aboriginal peoples. 'This', he explains, 'is the anthropology that early in its history and for well over a century thereafter was the *raison d'être* for the interest in, practice, and evolution of human osteology in North America' (Cybulski, 2001). His use of the phrase 'current challenges' is, he states, 'another way of saying that biological anthropology is now having a difficult time and may, in fact, be in serious trouble as a socially viable academic discipline'.

Cybulski's concerns surround the unethical treatment of Aboriginal remains by some North American anthropologists (he cites as just one example the case discussed in Chapter 10, involving Richard Ward's use of Nuu-chah-nulth blood samples). Such incidents, he fears, damage the credibility of the discipline and threaten to make these communities resistant to research efforts in the future. Yet his conclusions show that he is hopeful:

> I should think some compromise for our work could be arranged if only we respect the beliefs of others—which is what we as anthropologists ought to know in the first place. We cannot act as though our research, ideas, and approaches take precedence simply because *science* is at work. We do not, in fact, have all of life's answers.
>
> . . . I do not agree, however, with those of my colleagues who have simply thrown up their hands and gone on to do osteology other than the 'Indian' kind, because they can't be bothered with the apparent obstacles. . . . Perhaps, it is time to look at bones as the stuff of once living human beings—and all that entails—rather than scientific objects, artifacts or curiosities. Our treating remains in that manner—which is in no way at odds with science—may just help to stabilize human osteology as a socially valued scientific pursuit. (Cybulski, 2001)

Kwäday Dän Ts'ìnchi

There is no NAGPRA-like legislation in Canada. Instead, it's up to physical anthropologists and archaeologists to govern themselves in accordance with certain 'best practices', such as those evident in the following example.

On 14 August 1999, a party of three hunters found human remains, associated with artifacts, at the foot of a glacier in northwestern British Columbia, close to the Yukon border. They contacted a local museum, whose staff notified officials within the Yukon territorial government, who in turn informed the local Champagne and Aishihik First Nations, speakers of Southern Tutchone, an **Athabaskan** (or Dene) **language**. Government archaeologists of both Yukon and British Columbia become involved. The people named the remains 'Kwäday Dän Ts'ìnchi', meaning 'Long-ago Person Found'.

Scientists together with members of the Champagne and Aishihik First Nations presented the results of their scientific and cultural research projects at the Kwäday Dän Ts'ìnchi Symposium and Public Lecture, held 25–27 April 2008 in Victoria. The research team, featuring the University of Alberta's Owen Beattie, has established, via carbon-14 dating, that Kwäday Dän Ts'ìnchi died sometime around AD 1670–1850. Researchers have also determined his mtDNA haplogroup, and have established, based on samples taken from local Aboriginal people who volunteered to have their mtDNA profiles done, a pool of living relatives. These surviving descendants will decide what is to be done with Kwäday Dän Ts'ìnchi's remains once samples of his remains have been collected for future study. The story offers an excellent example of what can happen when scientists, government officials, and Aboriginal communities work together to establish common goals.

Ethical Considerations in South and Central American Cases

Latin America: A Different Attitude toward the Dead

In his incredibly thorough article on ethics and physical anthropology, Phillip Walker discusses the different attitudes that Latin Americans exhibit when compared to people in Canada and the United States concerning ancestral remains:

> In Latin American countries where a strong sense of 'Indianness' has been integrated into the national identity, human remains are excavated and displayed without opposition in museums. In this context, they serve as symbols of a national past that is shared by and important to all citizens. The government of the United States [and that of Canada also], in contrast, has historically considered Native Americans as outsiders to be dealt with by isolating them on reservations ['reserves' in Canada] and suppressing their indigenous languages and beliefs to facilitate converting them into functioning members of the dominant Euro-American culture. (Walker, 2000, pp. 23–4)

Case Study Controversies

The Dead on Display #1: The Accidental Mummies of Guanajuato, Mexico

The tenth of October 2009 marked the start of a seven-city US tour of the Mexican museum exhibit called 'The Accidental Mummies of Guanajuato'. Originally at rest in a cemetery in the city of Guanajuato, northwest of Mexico City, the mummies now normally reside in El Museo de las Momias, the second-most popular museum in Mexico and an important institution that generates much needed income for the area. Tour buses can take you there from Mexico City.

They are called 'accidental' mummies because they were mummified by chance, thanks to a unique set of conditions, including the chemistry of the soil they were buried in and the dryness of the local climate, which effectively dried the corpses out before they could decompose. The mummies are also fairly recent, having been dug up during a period from 1896 and 1958, when

relatives of the dead were required to pay a 'grave tax'; the bodies of those who failed or were unable to pay were exhumed and put on display until the law changed in 1958.

The more than 100 mummies at El Museo de las Momias are distinctive. Some are wearing only socks or shoes, some are fully clothed. There is a baby labelled 'La momia más pequeño del mundo' (lit. 'the smallest mummy in the world').

What Do You Think?

1. From an ethical standpoint, what do you think should be done with the mummies?
2. Do you think that the descendants of these mummies are being victimized because their families are poor?
3. Who 'owns' these mummies?

This different attitude may have a lot to do with the fact that in Latin American countries, a greater percentage of people than in Canada and the United States have some Aboriginal genetic heritage. Hanis and colleagues (1991) assert that 31 per cent of the contemporary Latin American gene pool can be traced to Native American ancestors (the figure drops to 18 per cent for Puerto Ricans and for Cubans). French-Canadian biologist Jacques Rousseau stated in 1970 that he believed that over 40 per cent of all French Canadians had at least one Aboriginal ancestor (quoted in Smith, 1974, p. 88). But Aboriginality does not play a significant role in French-Canadian identity, although theoretically it could. Think of the Métis, who were for the most part the children of the French and the Cree, and whose culture has traces of both (including their language, 'Michif', which combines Cree verbs with French nouns). Quebecois in this way are not like Paraguayans, for whom the Aboriginal language Guarani is part of the national identity.

Inca Mummies: Whose Are They?

The Inca Empire was one of the great high civilizations of all time, with incredible stone architecture (including pyramids) and a population of over six million, occupying a region that extended more than 2,400 kilometres from one end to the other (covering parts or all of present-day Chile, Peru, Ecuador, and Argentina). The Inca capital of Cuzco had a population of between 50,000 and 100,000. Yet the empire lasted the length of just one lifespan— from 1476 until 1532—before it was brutally ended by

the Spanish. It was able to grow and achieve such sophistication so quickly because it built upon a long series of Andean civilizations (i.e. situated throughout the Andes mountain chain, on the western coast of South America) that had preceded it from roughly 3,000 BP to just before the Incas' rise.

Between 1995 and 1999, a series of expeditions sponsored in large part by the National Geographic Society uncovered a number of naturally frozen (i.e. accidental) mummies on or near the tops of some of the tallest mountains of the Andes chain. They were the victims of Inca sacrifices. The first one found, nicknamed 'Juanita' (also 'the Ice Maiden'), was discovered on Mount Ampato, southern Peru, at an elevation of 6,309 metres (20,700'). She was initially moved to a custom-designed freezer unit at the National Geographic Society's headquarters in Washington, DC; she now spends part of the year in a similar unit at the Catholic University in the Arequipa area of Peru in which she was found.

Juanita and the subsequent finds were discovered by anthropologist/mountaineer and National Geographic 'explorer-in-residence' **Johan Reinhard**. Already an award-winning field worker and well-published author, Reinhard gained considerable notoriety—while generating an equal measure of controversy—at the time of his discoveries. Two of his finds—that of Juanita in 1995 and the discovery in 1999 of six more Inca sacrifices—were honoured by *Time* magazine as being among the world's ten most important scientific discoveries for their respective years. Yet a review of his book *The Ice Maiden: Inca Mummies, Mountain Gods, and Sacred Sites in the Andes*

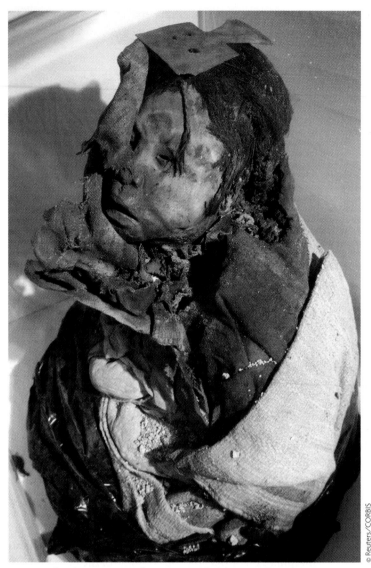

© Reuters/CORBIS

Mummified remains of an Inca child, sacrificed 500 years ago at the top of a North Argentine mountain

(2005), published in *The Wall Street Journal* (15 June 2005), touches on the charges of dishonourable motives and not-so-ethical conduct that have been levied against him:

> Mr Reinhard is sensitive to the charge that he is nothing more than a glorified grave-robber. He notes that if archaeologists don't recover mummies and the artifacts of distant cultures, looters will. Indeed, Mr Reinhard often finds that plunderers have made it to the mountaintops before he has. 'In my mind there is no conflict between having a scientific interest in the mummies and appreciating them as the dead deserving of respect,' he writes. (Ybarra, 2005)

While Inca high culture does not exist today (just as the ancient civilizations of Greece, Egypt, and Rome no longer exist), their language, Quechua, has more than

six million speakers. An mtDNA analysis of one of the six mummies discovered in 1999 shows that 1 out of 19 Peruvians tested belongs to the same haplogroup.

Trying to Resolve the Issue of Aboriginal Osteology

The issue of Aboriginal osteology is a complex one, with several aspects to consider carefully. One is the colonial history of archaeology and physical anthropology, which has often been carried out with a lack of respect for both the remains of the dead and the wishes of the living representatives of colonized peoples. This is not a situation unique to North America: a similar one exists in New Zealand and Australia. Anthropologists today need to change their mindset and their approach to studying the bones of indigenous peoples, and indeed, many have begun to distance themselves from their predecessors in this way. By and large, necessary changes are being made by contemporary physical anthropologists and archaeologists.

Another complication is that studying the dead, and even displaying the dead in museums, is not intrinsically wrong or disrespectful. People in certain parts of the world—especially in Latin America (as we have seen) and in Europe (as we will see)—are comfortable with it. Of course, the remains, the diggers, and the physical anthropological researchers are all linked by being 'from the same group.'

In the article mentioned earlier, Philip Walker states that 'The only justification for the study of skeletal remains from earlier human populations is that such research yields information that is useful to modern people' (Walker, 2000, p. 13). Information that helps us overcome the bias of the book is useful. Information concerning tribes and nations who have long ago disappeared also qualifies. In fact, any evidence that makes us more aware of Aboriginal history through as many means possible is of tremendous benefit. This includes knowledge of the oral traditions of Aboriginal peoples, which must be respected. The narratives of the people should never be completely dismissed; they have real value to offer in the interpretation of the remains and the artifacts of the dead. And of course, the people's wishes concerning the treatment of the dead must be heard with open ears and the utmost respect, not to mention a spirit of compromise.

At the same time, however, statements such as 'We have always been here' (i.e. in the Americas, or in a particular part of the Americas) do not contribute to a narrative that most anthropologists can respect as producing valid knowledge. The concept of a people's creation at the hands of a divine being or beings, whether it's a Creationist doctrine or an Aboriginal one, should never be accepted unquestioningly. All human life is connected. All humans have a history in Africa. DNA is shared between Aboriginal people, Asians, and others. These facts cannot be denied. Deep, connecting-with-others knowledge of one's history is ultimately more liberating than maintaining a deliberate ignorance that isolates people from their distant cousins.

Learning About the Culture of a People

The Iceman

In September 1991, the frozen body of a man was discovered near the top of a mountain on the Austria–Italy border. He had been stuck inside a glacier for between 5,300 and 5,200 years, so his body and clothing were well preserved. Archaeological research carried out on the man, nicknamed 'the Iceman', has revealed a great deal about his life, his culture, and his death. The following is a short summary of what we have learned. As you read on, consider what an amazing amount of knowledge Europeans must have had during his time.

The Iceman's clothing was different from today's fashions. He did not wear pants, just leggings and a loincloth. His shoes were made out of cowhide with a cord netting on top. Grass was stuffed into the shoes to keep his feet warm. He had a metre-long cloak made out of grass tied together in plaits. It would have been useful as a rain covering, as a sheet or blanket, and possibly as camouflage.

The Iceman carried a lot of useful things with him. Unusual is that many of these things were not completely made. The list of these items is included an unfinished bow, a broken quiver, an unattached cord, and a deer antler not yet carved into arrowheads. Why? We will see later.

In a birchbark container the Iceman had material for making and carrying fire: flints and iron pyrite to create a spark, and charcoal surrounded by green maple leaves. He also carried medicine with him: birch fungus as an antibiotic to fight intestinal worms and a sloe (a fruit similar to a plum) with a high vitamin and mineral content. Also intended as medicine were his tattoos, found on his lower spine, ankles, and knees. X-rays show that he had arthritis in these places. The historical record tells of people using hot brands with herbs for the purpose of easing pain.

We can do a little forensic anthropology concerning the Iceman. He had four cracked ribs on his right side, and died lying on his left side. X-rays show a normal calcium buildup in his left arm but not his right. This suggests that he hadn't used his right arm for 2 or 3 weeks. Add to this the evidence of his unfinished equipment, and it looks like he lost a fight. This is corroborated by recent evidence that his right hand had a stab wound. Injured, without proper equipment, and travelling in cold conditions, he died.

Mummified remains of a 5,000-year-old man found on the Similaun mountain on the Austria-Italy border

Case Study Controversies

The Dead on Display #2: The Bog Mummies of Europe

In northwestern Europe, including Britain and Ireland, Germany, and the Netherlands, hundreds of bodies have been discovered in peat bogs. They generally date from the final centuries BC to the first centuries AD. The people found were Iron Age farmers. Much of their iron came from boiling iron-rich water taken from these bogs, hence, I believe, that spiritual connection people felt towards the bogs. They were mystical places where water could be turned into the technological marvel of the age. As we have seen, bogs are also ideal for preserving once-living organic matter, including bodies. There are signs that the majority of individuals found in the bogs were murdered, often by garrotting (i.e. strangulation by a cord) and other ways that suggest the individuals may have been sacrificed.

There appear to be no reservations on the part of Europeans who may be descendants of the individuals to display their well-preserved bodies in museums, and in travelling displays. Certain ones have become particularly well known, not only as stars of travelling and permanent museum shows but also as central figures in novels, such as Erin Hart's beautiful *Lake of Sorrows* (2004). Some of the more celebrated bog people are the young 'Druid prince' found in northern England, called 'Lindow Man', and the 16-year-old 'Yde Girl', found in 1897 in the Netherlands, whose haunting face and blonde hair were constructed by medical artist Richard Neave in 1992. Looking at her reconstruction, you can feel empathy for her tragic life and death. Early reconstructions of bog bodies and their stories need to be checked and rechecked. 'Windeby I', found in a small German bog by peat cutters in 1952, was

long believed to be the body of a 14-year-old girl. Danish archaeologist P.V. Glob introduced her to the world in his classic work *The Bog People* (1965). Recently, though, Canadian anthropologist Heather Gill-Robinson has shown, via DNA testing, that the body is probably that of a boy (Lange, 2007).

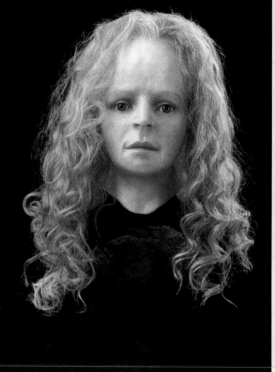

Mummified head of 'Windeby I', whose blindfolded body was discovered in 1952 in a bog in Germany

The mummified remains (top) and a reconstruction (above) of Yde Girl, a 16-year-old found in 1897 in a bog near the village of Yde in the Netherlands

The Shift to Agriculture: Effects on the Body

The case of the Iceman shows the abundance of evidence that a single body, preserved by chance, can produce for the observant physical anthropologist. The value of this kind of investigation lies in what it can tell us about the changing lifestyle of our human ancestors. One of the major changes, which took place in several places and different times from roughly 10,000 years ago, occurred when humans ceased to be nomadic hunter–gatherers and began setting down roots—literally and figuratively—as farm-tending agriculturalists. In this section, we'll look at what the physical evidence tells about life on the farm for the early stubble-jumpers.

DENTAL EFFECTS

In earlier chapters we've seen the usefulness of studying teeth to determine the age or geographic origin of a specimen. There are, in addition, a number of dental features that can tell us about the nutrition and lifeways of a people. Three of these features are **hypoplasia** (plural *hypoplasias*), **caries** (i.e. cavities), and tooth size.

Hypoplasias are:

> grooves, lines or pits in the teeth that reflect periods of time when, due to either poor nutrition or disease (or some combination thereof), the other covering of the tooth—the enamel—stops or slows in its growth. The cells that normally create the enamel . . . fail, and the hypoplasia caused by arrested growth results. (Larsen, 2000, p. 16)

Caries, or cavities due to tooth decay, relates to certain types of foods and how they are prepared. The main factor causing tooth decay is the high sugar content of some foods—sugar turns into a penetrating acid in our mouths. Prior to contact with Europeans, the Aboriginal people of Canada had a low rate of tooth decay, although those who ate and grew corn had a higher rate than others. Once the Aboriginal peoples became consumers of European food, their cavity rates shot up.

The increasing incidence of caries among North American people, then, reflects two factors: contact with Europeans and the production of corn. In a book chapter called 'From Foraging to Farming: A Regional Perspective', Larsen documents the effects of corn agriculture on the now extinct Guale Aboriginal people of the Georgia coast in the southeastern United States. The Guale become a chiefdom of agriculturalists in the pre-contact period, only to be devastated by contact first with the Spanish, then with the British. By the seventeenth and eighteenth centuries, the people had disbanded into scattered members left to join other peoples, independent no more. Clearly, contact with Europeans was responsible for effects worse than tooth decay.

Larsen's study of the Guale compares the health of hunters–gatherers of the early prehistoric period (400 BC–AD 1000) with that of these corn-growing agriculturalists of the late prehistoric period (1000–1450). Corn, he explains, is potentially a bad nutritional choice for a people to depend on as their primary food:

> Corn is deficient in calcium; it is deficient or lacking in the three (of nine) essential amino acids, lysine, tryptophan, and isoleucine; niacin (vitamin B3) is chemically bound, which prevents its bioavailability; it contains the chemical phytate, which prohibits iron absorption; and it has a large amount of sugar. These negative attributes should adversely affect the health and well-being of people whose dietary base includes a great deal of corn. (Larsen, 2000, pp. 78–9)

Of course, too much of anything can be a bad thing. The hunter–gatherers likely also benefited from their dependence on a greater diversity of foods—especially plants—than agriculturalists had access to.

Concerning dental health specifically, Larsen noticed, in the Guale's transition from hunting–gathering to corn production, a decrease in the overall number of hypoplasial lines but a definite increase in their size. This can be interpreted to mean that the Guale experienced fewer times of nutritional stress, but more serious stress when that stress occurred. Hunting-and-gathering societies are subject to periodic short periods with little or no food, which could be called involuntary fasting. When agriculturalists don't have food, it's called famine.

The farming Guale also had more cavities, showing a rate of incidence that increased over several hundred years, from 1.2 per cent of the foraging population of what Larsen termed the 'early prehistoric' (400 BC–AD 1000) to 9.6 per cent of the farming population of the 'late prehistoric (AD 1000–1450) (Larsen, 2000, p. 83). Of particular interest here is that the farming Guale women had a higher incidence of cavities (13 per cent) than did the men (9 per cent). It suggests that women ate more corn, while men ate more meat. Over time, the teeth of the people also became smaller, reflecting, according to Larsen's conjecture, the poorer nutritional state of mothers during pregnancy.

The link between nutrition and tooth size was the subject of a recent (2006) study of four British groups, historical and modern. Brook, Underhill, Foo, and Hector (2006) concluded that the individuals with the smallest tooth size, the Romano-Britons (the people who lived in Britain during the time of Roman dominance), had such small teeth because of recurrent illnesses, poor nutrition, and excessive lead ingestion (drinking vessels were often made of lead, water-bearing pipes were lead-based, and lead was added to drinks as a sweetener—talk about heavy drinkers!).

BONES

The effect on the bones of the shift to agriculture for the Guale was mixed. On the downside, there was an increase in periosteal reactions to infection (the **periosteum** is the outer surface of bones). The **tibia** (shinbone) seems to be generally the infection depository: Larsen calculated that the rate of detectable periosteal reaction in the tibias of the Guale increased from 9.5 per cent to 20.0 per cent. This would have been caused not so by poorer nutrition as by increased settlement size and closeness and frequency of human contact, which would have enabled infections to spread quickly. There was also a decrease in average height, thanks to a decrease in the length of the long bones. Adult males lost close to three centimetres, from 1.68 metres to 1.65 metres (5′ 6″ to 5′ 5″), while the average height of an adult woman fell from 1.57 metres to 1.52 metres (5′ 2″ to 5′ even). Bone strength diminished as well.

On the upside, Larsen noted that the early farmers had a significantly reduced rate of osteoarthritis, particularly of the neck and back. For the cervical vertebrae (the top seven vertebrae, found in the neck region), the rate of osteoarthritis dropped from 26.0 per cent to 16.4 per cent. The rate fell again, though not as far, for the thoracic vertebrae (the next 12), from 15.6 per cent to 11.4 per cent. The greatest decline in osteoarthritis was in the lumbar region (the next five vertebrae, just above the belt), where the rate fell from 41.9 per cent to 24.5 per cent. Consistent with this, osteoarthritis in the hip bones also dropped, from 12.0 per cent to 6.8 per cent. All of this would seem to relate to the large amount of bending and heavy lifting required in the collection of food (i.e. hunting and gathering) versus the production of food (agriculture).

CRIBRA ORBITALIA

'Cribra orbitalia' are '[p]orous, expanded lesions of the orbital roof of the frontal bone'(Pfeiffer, 2003, p. 195). They look rather like fungus growing out of the top of the orbital bone (i.e. above the eye). The condition has been blamed on various causes, but it is now generally believed that cribra orbitalia are produced in situations of iron-deficiency anemia, which may be caused by a deficiency of iron in the diet, chronic diarrhea, significant blood loss, or parasitic infection (Larsen, 2000, p. 17).

Cribra orbitalia has been studied in populations all over the world, but much of the focus has been on the condition among indigenous populations, such as those of North America and of Australia. The following discussion centres on two studies done in Canada, both of which identified a relationship between the disorder and a particular age group.

The first study examined the Moatfield site, an ossuary (i.e. a burial site) in present-day North Toronto, on land that was occupied by ancestors of the Huron from AD 1280 to 1320. Researchers found that 28.0 per cent of the individuals they studied (i.e. 23 out of 82) showed clear signs of cribra orbitalia. However, it was unevenly distributed among age groups: among adults, the rate was 17 per cent (10 out of 59); interestingly, 7 out of the 10 were male (although the sample size was, unfortunately, too small to draw conclusions about any significant gender difference). Among juveniles, a group composed of infants, young children, and adolescents, the rate was much higher at 56 per cent (13 out of 23); young children (aged 2–5) had the highest percentage at 73 per cent (8 out of 11) (Pfeiffer, 2003, p. 196).

While it would be easy to blame corn for this, van der Merwe and associates, in an article published in the same collection, assure us that the diet of the people was high in fish, and that '[t]he health problems that could have resulted from this foreign domestic crop, maize, were countered by the most complete food of all, fish from the streams and from Lake Ontario'(van der Merwe, Williamson, Pfeiffer, & Thomas, 2003, p. 220). In fact, the age contrast in cribra orbitalia is consistent with the general mortality rate. The 'young children' age group had the second highest mortality rate: specifically, there were 69.0 infant deaths per 1,000; 54.6 deaths per 1,000 of children aged 2–3 (the period of weaning stress); 28.7 deaths per 1,000 for children aged 4–6; and a much lower 2.3 deaths per 1,000 of children aged 7–11. The figure then increases gradually with age up to a rate of 15.8 deaths per 1,000 for adults over 50 (Merrett, 2003, pp. 176–7). Thus, while the incidence of cribra orbitalia among the Moatfield population is noteworthy, we probably shouldn't read too much into its high rate among children.

The uneven age distribution of cribra orbitalia is not an unusual finding in studies among the Aboriginal populations of Canada. Importantly, though, the condition is not always connected with agriculture. Jerome Cybulski, in a study of four BC First Nations (the Haida, Kwakiutl, Nootka, and Coast Salish), looked at 454 pre-contact skulls and found evidence of cribra orbitalia among 32.7 per cent of growing children and adolescents, 19 per cent of infants and toddlers, but only 13.3 per cent of adult females and 4.8 per cent of adult males. These Aboriginal peoples did not have agriculture, nor does diet appear to have been a factor, as there was ready access to iron-rich marine resources such as fish. The real cause may have been the long-term settlements, characteristic of the Northwest Coast, which would have produced increasing crowding and, consequently, infection. Increasingly, physical anthropologists who study cribra orbitalia are looking to the settlement conditions of farmers, not their diet, as the source of the problem.

Peopling of the Americas

They Came from Asia

When Europeans came to the Americas, it didn't take them long to conclude that that Aboriginal people they encountered had originally come from Asia. Why did the newcomers think this? The indigenous peoples bore a number of physical traits that the Europeans would have believed fit the Asian profile: long, straight, dark hair; brown eyes; skin of various shades of brown; and eventually, as noted by **Aleš Hrdlicka** (1869–1943) in the early twentieth century, **shovel-shaped incisors**. Europeans made a number of faulty assumptions about the Aboriginal people they encountered; in this case, though, they were right.

If the Aboriginal people came from Asia, how did they get here? The growing consensus is that they walked. All of that ice that covered continents during the Ice Age came from the oceans, which were, at the peak of the Ice Age, 120 metres lower than current levels. The Bering Sea, separating Alaskan North America from Siberian Asia, isn't very deep to begin with, so when the water level dropped it became **Beringia**, a land mass about 1,000 kilometres wide from north to south, connecting Siberia and Alaska.

The Archaeological Evidence

CLOVIS FIRST VERSUS PRE-CLOVIS

'Clovis' is the name given to a Paleo-Indian culture of North and Central America, dating between 12,700 and 13,200 calendar years ago. Clovis sites are characterized by the presence of finely worked stone projectile points of about 10–13 centimetres in length and 4 centimetres in width. Known as **Clovis points**, they are recognizable by a groove or flute (the points are described as 'fluted') along both sides or faces. This fluting would have allowed people to attach (the technical term is 'haft') the points to wooden spear shafts. Clovis points are widely distributed throughout the US and a few Canadian locations. The most famous Canadian Clovis site is Debert in Nova Scotia.

The points and related artifacts (including mammoth bones) first started appearing in the 1920s, when they were identified at a site near Clovis, New Mexico. When the material was first dated, using radiocarbon techniques, it revolutionized the way people viewed the history of the Americas by establishing that people had been in this part of the world at least 10,000 years ago—far longer than anthropologists had previously thought.

Acceptance of Clovis culture took a while, but as evidence of Clovis sites became widespread and fairly uniform in their artifacts, anthropologists warmed to the idea that the Clovis people were the first people to enter

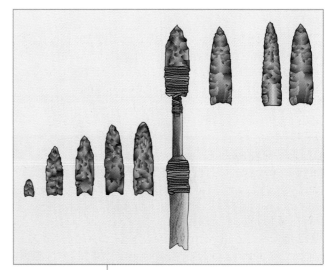

Figure 11.1 | A selection of sophisticated Clovis points, with a reconstruction of how they might have been attached to hunting spears by the Clovis people

the Americas. When American archaeologist **C. Vance Haynes** (b. 1928) published his classic work on the subject in 1964, a new orthodoxy set in: his intellectual lock sealed the vault of debate, with first settlement occurring around 11,000–11,500 radiocarbon years ago (roughly 13,000–13,500 calendar years). The religion of **Clovis First** had begun, and heaven help those who challenged the doctrine.

MONTE VERDE

Haynes's dates were widely accepted, but it didn't stop a few brave anthropologists and archaeologists from putting forward claims of human settlement predating Clovis. The most seriously considered pre-Clovis site is Monte Verde in southern Chile, which archaeologist Tom Dillehay has been working since 1977. The strongest evidence from the multi-layered site comes from a peat bog, which preserves organic materials incredibly well. This process is beautifully described in the words of Irish–English novelist Brian McGilloway, as spoken by a fictional archaeologist in *Bleed a River Deep*:

> The water in bogland contains a lot of organic acids and aldehydes [a chemical group that includes formaldehyde] which act almost like embalming fluid. More importantly, it stops bacteria growing. The body ages without ever decomposing; the skin toughens so much it becomes like tanned leather. (McGilloway, 2009, p. 32)

Figure 11.2 | Important early human sites, South America

The well-preserved organic materials at Monte Verde have been subjected to carbon-14 dating and fall within a range from 11,790 to 13,565 radiocarbon years ago; the average of 12,500 ya is now the standard figure given. When the radiocarbon years are converted to calendar years, we get an average date of 14,500 ya. Older artifacts have been found at the site, but they are more controversial. Nevertheless, the widely accepted date for Monte Verde is much earlier than the standard Clovis dates would allow, especially given that it is some 12,000 kilometres south of Alaska, where the people who made up Clovis culture are believed to have entered the American continents.

In 1997, Dillehay and his financial sponsor invited some of the foremost specialists in the study of Paleo-Indians, including the skeptical C. Vance Haynes—so see the Monte Verde site up close and to review copies of an extensive site report. Following the visit, a consensus report was issued, giving 14,500 ya as the agreed-upon date. The back of Clovis First was officially broken, or so it seemed.

Two years later, Stuart Fiedel, a defender of Clovis First, savaged the site report, with bitter sarcasm, in the popular magazine *Discovering Archaeology* (1999). Dillehay had written his own response in a proper scientific way, indicating where he was sure, where he was unsure, and where he had to change his mind. Fiedel's attack was more characteristic of a prosecuting attorney demanding acceptance of his views in court. The fight continued, with Haynes backing down from his compromise in accepting Dillehay's work. A balanced account of this issue is David Thomas's thoughtful piece 'One Archaeologist's Perspective on the Monte Verde Controversy' (1999). For my money, Haynes and Fiedel are fighting a battle after the war is lost.

MEADOWCROFT ROCKSHELTER

Meadowcroft Rockshelter is an archaeological site in southwestern Pennsylvania. **James Adovasio** (b. 1944) excavated the locality from 1973 to 1977, and then again from 1994 on to the present. A rock shelter is a cave-like cavity in a hillside or cliff where a protruding or overhanging slab or rock forms a natural cover; it makes a good dwelling. Meadowcroft Rockshelter is the longest continuously used site in eastern North America, being part of some people's lives from at least 14,000 year ago to 700 ya. It boasts the largest collection of plant and animal remains at a single site in North America. But it is a site contested by the defenders of Clovis First, much to the frustration of Adovasio, who has claimed—with good evidence—that a tool-bearing layer dates to roughly 14,800 BP.

According to the account promoted by Clovis defender C. Vance Haynes, and parroted in many of the standard textbooks, there is a very real possibility that the Meadowcroft Rockshelter site has been contaminated by coal in the groundwater, which could distort the radiocarbon dates. Certainly, this part of Pennsylvania is rich in coal. However, if you read Adovasio (2002)—something that I suspect many textbook writers have *not* done, preferring instead to repeat criticisms from earlier books—you are left with a compelling sense that groundwater contamination is not possible here; the nearest coal seam is around a kilometre away.

PEDRA FURADA

Perhaps the most controversial pre-Clovis claim comes from the Pedra Furada site in Serra da Capivara National Park in northeastern Brazil. Even the *un*controversial part is fascinating: the site features cave paintings of humans

Case Study Controversies

What Was Found at Monte Verde

Preserved on the [14,500]-year-old surface were 38 chunks of animal meat and hide, 11 specimens of wild potato, 4 kinds of exotic seaweed, and 180 architectural elements, wooden lances, and mortars. The archaeologists mapped at least a dozen huts on this surface, including a 60-foot-long tent-like structure that could house 20 to 30 people—maybe year-round. Framed with logs and planks, the superstructure was covered by wet mastodon and paleo-llamal hide and anchored by wooden stakes. Skins also covered the floor inside, where families maintained some degree of privacy by defining their living spaces with hide-covered walls. To the west stood a wishbone-shaped structure, made of wooden uprights set into sand with a gravel foundation stabilized with hardened animal fat. Mastodons had been butchered inside, and the inedible hide and bones fashioned into tools and building materials. Incredibly, inside the west building were some 23 kinds of medicinal plants. . . . The wet clay floor of Monte Verde preserved three human footprints—from a bulky teenager or . . . a smallish adult who once walked the banks of Chinchihualpi Creek.

Thomas, 2000, p. 159.

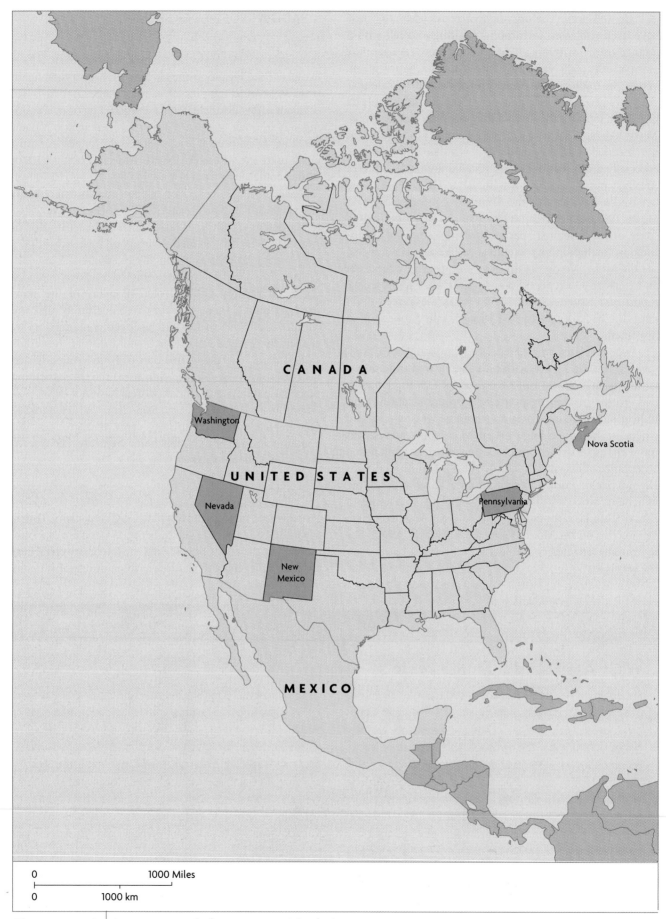

Labels on map: CANADA, UNITED STATES, MEXICO, Washington, Nevada, New Mexico, Pennsylvania, Nova Scotia

0   1000 Miles

0   1000 km

Figure 11.3 | Important early human sites, North America

and animals from roughly 6,000 to 10,400 years ago. The controversial layers lie underneath, and the problem they pose is a perennial difficulty for archaeologists—that of **geofacts** versus **artifacts**. Geofacts are items created by nature, not by humans. Typically, though, the term is applied to natural items that *look* like they were created by humans. In the case of the Pedra Furada site we have evidence of fire and what look like flaked pebble tools. The problem is that the dates for the layers or strata in which these 'tools' are found have been dated as far back as 48,000 to 56,000 years ago. So, if you accept that the fires are evidence of hearths made by humans and that the pebbles are tools, you are saying that anatomically modern human beings were in Brazil before AMH first moved into Europe.

In a strangely civil act of open scholarship, Niede Guidon in 1993 invited James Adovasio, Tom Dillehay, and David Meltzer to come to the site to study the evidence. The visiting trio were not convinced, particularly as the quartzite pebbles bore no discernible wear patterns consistent with human use, although they left the matter open to further research (see Meltzer, Adovasio, and Dillehay, 2004). That measured conclusion seems to have brought the civility to an end, and Guidon's response—published together with Meltzer, Adovasio, and Dillehay in the same issue of *Antiquity*—seems to me to be mean-spirited and defensive, and lacking the scientific tone of the trio's piece (see Guidon & Pessis, 2004).

In sum, the all-important combination of reliable dating and credible proof of human occupation has yet to be achieved. This would seem to reflect the geofact status of the early material.

The Linguistic Evidence

When considering the peopling of the Americas, there are two key linguistic facts to consider. The first is that the farther south you go in the Americas, the greater the diversity of languages. In Canada, for example, we have 11 language families and isolates (although two of these isolates, Haida and Tlingit, are often lumped together with the Athabaskan family to make up the Na-Dene grouping); overall, there are about 50 languages still spoken. Algonquian, the language family with the greatest number of languages in the Americas, has speakers in every province. In North America altogether, there are about 55 language groupings. In South America, there are 143, although as these languages have been little studied, this number is quite rough. Sometimes the borderline between language and dialect is as shifting as the one between species and subspecies.

The second key fact is that there is far greater diversity of languages along the west coast of North America than there is inland, to the east. Of the 11 groupings in Canada, 8 are in British Columbia, and 6 of those are found in no province other than BC.

Meadowcroft Rockshelter archaeological site, Pennsylvania

© Tom Uhlman/Alamy

Why mention this? Well, generally speaking, linguistic diversity reflects two factors. One is age of settlement. The longer people have been in an area, the greater the diversity of languages. Africa has greater language diversity than Europe or Asia. The Austronesian languages (the group that includes the Polynesian languages, such as Maori in New Zealand and Hawaiian) show the greatest diversity on the island of Taiwan, where they have been spoken for longer than they have been around in the Pacific Islands. The second factor is accessibility. Where there is little accessibility of communities to outsiders or to one another, language diversity is relatively large. The languages of the jungles of New Guinea reflect this, as, to a certain extent, do the languages of the jungles of South America.

The point here is that language diversity in the Americas suggests two things. One is that South America may have been broadly settled before North America (see the discussion below of the Beringia Refuge hypothesis). The other is that the first settlers took a western coastal route in their migration; this is known as the **Coastal Migration hypothesis** (see the discussion in the feature box on page 271). This latter idea is gaining popularity, at the expense of the old idea of an ice-free corridor through Alberta, a hypothesis first proposed by Canadian geologist W.A. Johnston in 1933.

The challenge to the Alberta corridor idea comes from the work of archaeologists such as Alan Bryan, emeritus professor at the University of Alberta, who point out that such a route would have been blocked during the early period that people were moving south. In addition, as **Knut Fladmark** and others have demonstrated, the Ice Age coastline of western North America (most of it now under water), including British Columbia's Haida Gwai (formerly the Queen Charlotte Islands), was ice-free. Read Canadian writer Tom Koppel's *Lost World: Rewriting Prehistory—How New Science Is Tracing America's Ice Age Mariners* (2003) for a good account of how the facts related to the Coastal Migration hypothesis are beginning to appear.

Joseph Greenberg, the ultimate lumper when it comes to languages, claimed that there are three linguistic groups native to the Americas: Amerind (which includes just about everybody), Na-Dene (including the Athabaskan languages, along with Haida and Tlingit), and Eskimo–Aleut. Archaeologists, physical anthropologists, and geneticists seem to like this distinction more than linguists do; then again, we linguists seem always to be more critical of people in our own discipline. While there is no good reason to lump most Aboriginal languages into one big pile, there are good reasons to look at Na-Dene and Eskimo–Aleut as distinctive. In the case of the former, some of the languages are characterized by subtle variations in tone or pitch, in a way that is similar to (but simpler than) features associated with East Asian languages such as Mandarin and Vietnamese; no other Aboriginal languages exhibit tones in this way. Also, there is evidence suggesting that these languages are related to Ket, a Siberian language. Eskimo–Aleut has speakers in Siberia, indicating a greater connection to that area than is found with other Aboriginal languages. Otherwise Greenberg's statement is rather like my saying that there are three kinds of birds: parrots, ducks and others. We can distinguish who are and who are not parrots and ducks, but the 'others' grouping is too diverse.

The Beringia Refuge Hypothesis

The story gaining momentum among geneticists using mtDNA analysis to interpret the peopling of the Americas is the **Beringia Refuge hypothesis**. It goes like this. First, sometime long before Clovis points—perhaps 21,000 to 30,000 years ago, before the **Last Glacial Maximum** (**LGM**)—there was a series of migrations from Asia that reached the grassy, ice-free, newborn semi-continent of Beringia. There, the people experienced a moderate **genetic bottleneck** that limited the number of mtDNA haplogroups they possessed. Some of them, including representatives of A1, B2, C1, D1, and X2a, were able to travel south, probably taking a western coastal route that has now mostly flooded since the end of the LGM. When the LGM struck with its powerful, chilly fists, perhaps 20,000 to 21,000 years ago, a number of people were 'trapped' within the ice-covered perimeters of Asia and North America in the refuge of Beringia. In a more serious genetic bottleneck, the number of haplogroups in Beringia was reduced to two basic mtDNA lines: A2 and D2. Once the temperature began to warm up, they, too, expanded, although in a more limited way. Today, A2 is found among the Siberian Inuit and the Chukchi of eastern Siberia, among the Inuit and Aleut, and among speakers of Athabaskan languages (e.g. Dogrib and Navajo), and Tlingit. D2 is found among the Inuit and Aleut.

The Genetic Evidence: Y-Chromosomes and the Americas

You'll recall from the discussion of Y-chromosomes in Chapter 10, where we introduced and described the various groups, that genetic biologists have only recently reached consensus on the haplogroups A–R. The first comprehensive Y-chromosome studies of Aboriginal people were not carried out until the mid-1990s. A study published in 2004 by Zegura and colleagues used 63 single-nucleotide polymorphisms (SNPs) and 10 short-tandem repeats (STRs) on a sample of 2,344 Y-chromosomes from 51 populations representing Aboriginal peoples (a total of 588 individuals from 18 populations, including the three proposed linguistic groups, Eskimo–Aleut, Na-Dene, and Amerind) and peoples from Asia and European groups. The researchers followed the 2002 consensus naming of the Y-chromosome Consortium (YCC) to determine which letter labels to attach to the haplogroups to (before then there was taxonomic chaos). They found that 76.4 per

Knut Fladmark: Sticking to the Coast

Knut Fladmark is easy to admire. He developed an idea that was very unpopular for a long time, he got shot down, often, but he stuck to his belief and, at last, has been proven right. He is Simon Fraser University's answer to Raymond Dart.

The idea that Fladmark is most associated with is the **Coastal Migration hypothesis**, the idea that the first Aboriginal people to travel down from the northwest corner of North America to what is now the United States did so by following an ice-free path down the West Coast. Fladmark's hypothesis ran counter to the traditional (and increasingly discredited) belief that the path taken by the migrants was in an ice-free corridor through Alberta. A key aspect of this theory was that the coast was covered with ice, making it an impossible route. The problem with both ideas is that neither had any solid evidence to support it.

Fladmark first latched onto the idea of the Coastal Migration hypothesis from the dissertation of a palynologist (one who studies historic pollen), Calvin Heusser. Heusser wrote about early pollen sites along the coast, which he claimed were between 11,000 and 12,000 years old. Where there is pollen, there are plants, and, therefore, no large sheets of ice. Heusser suggested that the sites of these plants could have been used by the first Aboriginal people to move south. Fladmark writes:

> [A]s a graduate student I invested a week's meal money xeroxing a copy of Heusser's dissertation and I just built on his ideas. So I certainly did not invent the idea of the coastal route—it was Heusser's idea. (retrieved from SFU Museum of Archaeology and Ethnology)

The humility is refreshing.

Part of Fladmark's research has involved looking at an early location in an area of the old ice-free corridor (the Peace River District), specifically the Charlie Lake Cave site

Knut Fladmark at the historic signing of the repatriation agreement with the Haida Nation

Simon Fraser University

in northern British Columbia, which he excavated in 1983, and again with Jon Driver in 1991. This site, which dates to about 10,500 years ago, is too recent for a north-to-south migration but:

> sites like Charlie Lake Cave . . . indicate that people were moving from south to north up the corridor, after they had already moved southwards along the coast and then begun to spread inland, perhaps along the Columbia River [where Kennewick Man was found] and developed terrestrial adaptations in that process. (retrieved from SFU Museum of Archaeology and Ethnology)

cent of the Aboriginal individuals studied belonged to haplogroup 'Q', while only 5.8 per cent belonged to 'C'; the 17.3 per cent belonging to 'R' they dismissed as the product of admixture with Europeans.

The distribution of haplogroup 'C' was particularly interesting, having a concentration in the three Na-Dene populations studied; the researchers didn't find any trace of C among the Eskimo–Aleut. This was consistent with previous research, or so the team thought. But as they were writing the article, the researchers came across a 2003 study that found 6 Greenlandic Inuit (in a study of

69) with haplogroup C. They believe that this presents evidence of a single migration, as proposed in the Beringia Refuge hypothesis, and that the C does not represent a separate migration. It could fit into the second expansion theorized by the mtDNA researchers.

Zegura's description of the single migration as 'recent', reflected in the title of the team's paper (Zegura, et al., 2004), seems to me to be a little weak. As Y-chromosome material generally shows a more recent founder, it doesn't lend itself to creating an accurate timeframe. It is important to be critical of these studies: the lab work may look impressive, but oftentimes the discussion or conclusions are not as strong. After taking many careful small steps in the lab, the writers may leap to conclusions that are sometimes not warranted by their data.

While the initial work concerning Aboriginal people and the sinodont/sundadont distinction emphasized that the people were sinodont (see Chapter 10), more recent research has suggested that there are some sundadont Paleo-Indians and populations that do not fit neatly into that pattern. Rebecca Haydenblit's 1996 study of four pre-contact Mexican populations living from 1300 BC to AD 750 reveals variation, with only the most recent group showing a clear sinodont pattern. Taken as a whole, there were more sundadont examples found than sinodont. The presence of sundadont teeth among such pre-contact Aboriginal populations is used to argue that Southeast Asian populations (especially the ancestors of the Austronesians) migrated northward along the Pacific Coast, and leaving behind the sundadont Ainu (indigenous people of Japan).

The Cranial Evidence: Skulls versus Genes

The remains of **Kennewick Man** were discovered on 28 July 1996 along the banks of the Columbia River, near Kennewick in the state of Washington. The discoverers, two students, brought the skull they found to the local police, who turned it over to James Chatters, an archaeologist who did contract work for the county coroner. Chatters looked at the long, narrow skull, with its narrow chin and high-bridged nose, and concluded that the cranium had belonged to a European. However, when other bones of the individual were retrieved and dated as being between 9,330 and 9,580 calendar years old, and then when an ancient-style spear point was found embedded in the hip of the skeleton, a change of view was required: the body must be Aboriginal.

Chatters' mistake was understandable. The skull did not conform to the usual rounded, wider face with non-protruding jaw that was considered characteristic of Aboriginal people and East Asians. The skull not only

Science Gone Wrong

The Mysterious Haplogroup X and the Migrating Europeans

During the 1990s, genetic researchers established that there were four major founding mtDNA lineages, or haplogroups, conveniently labelled A–D and representing some 97 per cent of the Aboriginal haplogroups of the Americas. But from the exceptions emerged a clear minor founding group. What was especially intriguing about this group (known as group 'X') was that there were also Europeans who belonged to it—4 per cent of those Europeans studied. There are three possible interpretations of this:

1. Indigenous people of North or South America travelled to Europe in ancient times.

2. Europeans travelled to the Americas in ancient times (well before the eleventh-century Viking visits).

3. Both groups share a common founder.

A number of scientists unthinkingly seized upon the second named interpretation and took it to heart. Part of the blame goes to Michael D. Brown and a team of colleagues, who studied 22 Aboriginal people (Ojibwa or Anishinabe from northern Ontario and Wisconsin) and 14 Europeans who belonged to group X. There were 11 mutations shared by almost all of the individuals studied—Aboriginal and European—on a strand of 1,349 nucleotide bases. In addition, there were two mutations that were shared by almost all of the Aboriginal individuals studied.

Among Aboriginal peoples of the Americas, group X seemed to show a pretty exclusive membership. In the peoples studied by that time it had been found only among the Ojibwa or Anishinabe (25.0 per cent and 25.7 per cent in two different studies), the Sioux (14.6 per cent) the Nuu-Chah-Nulth or Nootka (13.1 per cent and 11.1 per cent), the Navajo (6.5 per cent) and Yakima (4.8 per cent). The last three of those groups share a solid long-term connection to the west, the Nuu-Chah-Nulth being from Vancouver Island, the Yakima from the state of Washington, and the Navajo having lived in Alaska, Yukon, and Northwest Territories up until a thousand years ago.

The writers dated the break between the Aboriginal and European groups to between 23,000 and 30,000 years ago—too long ago for the sharing to relate to a recent European presence in the Americas. However, they played with the idea that maybe there were two founder groups of X in the Americas, lowering the possible date to 12,000–17,000 years ago. The problem with this more recent date, and also with the claims made in the abstract generally, was that the writers did not limit themselves to scientifically demonstrated conclusions, but went out on a limb to say:

> To date, haplogroup X has not been unambiguously identified in Asia, raising the possibility that some Native founders were of Caucasian ancestry. (Brown, et al., 1998, p. 1852)

resembled the cranium of a European but was very much like that of an Ainu or Austronesian. Dental analysis suggested it had sundadont teeth.

Generally speaking, when it comes to the peopling of the Americas, the story told by the skulls seems different to that told by the genes. As we have just seen, the genes say, in the Beringia Refuge hypothesis, that four major mtDNA lines (A–D) and one minor line (X) entered Beringia at roughly the same period, and expanded into the Americas twice, once before the Last Glacial Maximum, and then again after the big melt began. The skulls, however, seem to be telling us that there was an earlier group, the **Paleo-Indians**, who came from Southeast Asia, migrating north along the Asian coastline, and then crossed over before any other group did, possibly by boat, riding the west-to-east Japan Sea current to the coastline of the Americas. From there, they went along the western coastline all the way south to the tip of South America and then up the Atlantic coastline to Brazil.

The **craniometrics**—the cranial measurements used to try to differentiate between human populations—suggest that the crania of the vast majority of the earliest skeletons found are different from those that came later. For example, Kennewick Man had a long, narrow skull, similar enough to 'non-Mongoloid' Asians such as the Austronesians (Polynesians and their linguistic relatives) and the Ainu to lend some support to this theory. The following are other examples.

THE SPIRIT CAVE MUMMY

The **Spirit Cave** mummy, discovered in Nevada by S.M. and Georgia Wheeler in the 1940s, is carbon-dated at 9,430 ± 60 years ago. A craniometric study by Jantz and Owsley in 1997 concluded that it was different from modern Aboriginal people but similar to other groups, such as the Ainu (as reported in Barker, Ellis, & Damadio, 2000, p. 39). The teeth show signs of being sinodont.

This opened the door to several Eurocentric interpretations. One of these was the **Solutrean hypothesis**, put forward by Dennis Stanford of the Smithsonian Institution and Bruce Bradley of the University of Exeter. Solutrean material culture, which existed from roughly 21,000 to 17,000 years ago in present-day France and Spain, is defined by bifacial, pressure-flaked points. Stanford and Bradley claim that the European carriers of this culture travelled west across the icefields of the northern Atlantic into North America, where the spear points characteristic of the Solutrean tool tradition sparked the Clovis or pre-Clovis tradition.

Two subsequent studies of haplogroup X help to disprove this hypothesis, without even addressing the supposed tool similarities (which may simply reflect similarity of purpose and of tool-making resources). First, Derenko and colleagues, in 2001, sampled 10 different indigenous Siberian populations in Altai, the name of both an administrative district and a republic of Siberia. They discovered haplogroup X in 7 individuals, or roughly 3.5 per cent of the 202 Altaians studied; other Altaians exhibited haplogroups A–D. It would appear that the ancestors of every Aboriginal haplogroup lived or at least went through the Altai.

Secondly, Malhi, Schultz, and Smith (2001) demonstrated that the Aboriginal people of North America need to be studied more as well. Algonquian is the language family to which the Ojibwa belong; in fact, it is the most widespread Aboriginal language family in North America. Linguists have long suggested that the Pacific Northwest was an early location of the Algonquian language family (the evidence comes from a few languages still located there, which are distantly related to the Algonquian family). Now, there are three branches of the Algonquian language family: west, central, and east. If all three groups showed haplogroup X, that would suggest that the Algonquian connection to the haplogroup had a great time depth. It had already been established that the Ojibwa were heavily represented in group X; Ojibwa is a central Algonquian language. Malhi's team took mtDNA samples from a representative group of western Algonquian speakers (the Cheyenne/Arapaho) and from a group of eastern Algonquian speakers (the Mi'kmaq); they found haplogroup X in both, demonstrating the long history of the Algonquian connection to haplogroup X. Based on the timing the study established and the fact that the earliest speakers of the Algonquian language family were then in the west, Europeans, if they had come across the Atlantic in Solutrean times, would not have encountered the Algonquians. The more eastern-associated Iroquoian peoples do not exhibit haplogroup X.

LUZIA

Luzia (Brazilian Portuguese for 'Lucy'), also known as Lapa Vermelha IV Hominid 1, has, at roughly 11,500 years old, been dubbed the oldest skeleton in the Americas. She was discovered in Lapa Vermelha, Brazil, in 1975 by Annette Laming-Emperaire, put into storage, and then retrieved by Brazilian anthropologist **Walter Neves** in the mid-1990s. Her cranium is narrow and oval, and she has the projecting face and pronounced chin typical of non-Mongoloid Southeast Asians.

THE LAGOA SANTA SKULLS

The strongest case in support of the hypothesis that early crania were different from later ones was made in a study published in 2005 by Neves and Mark Hubbe of 81 early crania (42 males and 39 females) from the Lagoa Santa site in central Brazil. The skulls varied in age from 7,500 to 11,000 years old. Neves and Hubbe found that the skulls

> exhibit a cranial morphology that is very different from late and modern Northeastern Asians and Amerindians (short and wide neurocrania; high orthognathic [i.e. non-protruding] faces; and relatively high and narrow orbits and noses) but very similar to present Australians/Melanesians and Africans, especially with the former (narrow and long neurocrania, prognatic [i.e. protruding], low faces; and relatively low and broad orbits and noses) (Neves & Hubbe, 2005, p. 313).

Neves and Hubbe put forward two possible hypotheses that they believed could explain this craniometric difference. The first, which they dismiss, that can be called the *local diversification hypothesis*; it proposes that a local microevolutionary process took place, in which the crania changed from the morphology described above to one that is more typical of Aboriginal people. The second, which they support, is called the *migratory hypothesis*; it suggests that there were two different populations that moved into the area, the first and by far the earliest being different from the second and larger migration. Two reasons Neves and Hubbe give for favouring the second hypothesis are:

- that the change between the crania they studied and the more typical 'Mongoloid' skulls seems to have been too abrupt, occurring over too short a time period to represent microevolution; and

- that cranial morphology typically changes very slowly and only to extreme environmental conditions.

Perez, Bernal, Gonzalez, Sardi, and Politis (2009) have challenged this conclusion, in part arguing for a relatively high plasticity of the skulls. They compared the craniometric data and mtDNA of individuals found in east-central Argentina at an early site (Arroyo Seco 2, with carbon-14 dates ranging from about 7,800 to 4,500 years ago) with corresponding measures taken from individuals found at younger sites and from contemporary populations. They found that the individuals studied at the Arroyo Seco 2 site had skulls that were 'more elongated' than the more recent skulls. There wasn't that kind of variance with the mtDNA: the mtDNA of 8 individuals found at the older site showed haplogroups 'B' (3 individuals), 'C' (4), and 'D' (1), while the 15 individuals found at the more recent Laguna del Juncal site also had 'C' (4), and 'D' (11); the contemporary population, the Mapuche, showed haplogroups 'A' (6 per cent), 'B' (36 per cent), 'C' (24 per cent), and 'D' (34 per cent), all very typical of modern South American populations.

Perez and his colleagues gave several theories for why the skulls would be so different while the mtDNA would not. One of them relates to the plasticity of the skulls, which mostly has to do with jaw structure and how it reflects the primary food eaten. Perez in another study had noted some differences of this kind that were based on whether a population lived by hunting and gathering (with more dolichocephalic, or long-headed, skulls) or by agriculture (more brachycephalic, or broad-headed, skulls). The Arroyo Seco 2 population lived by hunting and gathering, while populations from the later sites were agricultural.

Summary

We have seen in this chapter that ethical concerns surrounding what to do with the recent (relatively speaking) dead is a complex issue. In the past, anthropologists of a colonizing culture would often treat the bones of other peoples (the colonized) with too little respect. In the present, anthropologists and indigenous groups are attaining a greater degree of compromise and co-operation, whether it is enforced through legislation like NAGPRA in the United States or achieved through more voluntary means as here in Canada.

We have also seen that in recent times (i.e. the last 10,000 years), as people moved from hunting-and-gathering societies to agricultural ones, the effects on bones and teeth have been significant, and often damaging. The evidence has provided anthropologists with fairly reliable means of charting the transformation of early human societies in the Americas.

Finally, it seems fair to state, on the strength of recent research, that Aboriginal people have been in the Americas significantly longer than anthropologists traditionally believed. The Clovis First hypothesis is an untenable intellectual position. On the other side of things, there is no good reason to believe that so-called Caucasians arrived in the Americas any earlier than the Vikings, who stayed for a short period on the cold northeastern shores of the island of Newfoundland, about a thousand years ago. Be very suspicious if you hear anything that suggests otherwise.

Typical Student Questions

▶ *So how long have Aboriginal people been in the Americas?*

Aboriginal people have been here somewhere between 20,000 and 15,000 years. The more recent dates associated with the Clovis First hypothesis are no longer considered valid. Clovis First is dead; it has left the anthropology building.

Review Questions

1. What were the effects on the Guale of shifting from hunting and gathering to agriculture?
2. Discuss the arguments used in the Clovis First versus Pre-Clovis First debate.
3. Why is it difficult to prove the Coastal Migration hypothesis?
4. Identify and state the problems with the Solutrean hypothesis.

Discussion Questions

1. How do you feel about displaying the dead in museums or travelling exhibits? How do you think your position might change if you were of (a) North American Aboriginal descent, (b) South American Aboriginal descent, (c) European descent?
2. Is it fair to judge Franz Boas by the ethical standards of our times? Why or why not?
3. Why do you think that 'white' scholars (most of them outside the field of physical anthropology) keep coming up with hypotheses about early (i.e. pre-Norse) migrations of Europeans to North America?

Key Terms

Amerind
artifact
Athabaskan language
Beringia
Beringia Refuge
    hypothesis
bias of the book

caries
Clovis First hypothesis
Clovis points
Coastal Migration
    hypothesis
craniometrics
genetic bottleneck

Geofact
Great Zimbabwe
hypoplasia
Kennewick Man
Last Glacial Maximum
    (LGM)
Luzia

osteology
Paleo-Indians
periosteum
shovel-shaped incisors
Solutrean hypothesis
Spirit Cave
tibia

Key Individuals

James Adovasio	Jerome S. Cybulski	C. Vance Haynes	David Randall-MacIver
Franz Boas	George A. Dorsey	Aleš Hrdlicka	Walter Neves
Gertrude Caton-Thompson	Knut Fladmark	Clark Spencer Larsen	Johan Reinhard

Recommended Print and Online Resources

Adovasio, J., & Page, J. (2002). *The first Americans: In pursuit of archaeology's greatest mystery*. New York: Random House.

Barker, P., Ellis, C., & Damadio, S. (2000). *Determination of cultural affiliation of ancient human remains from Spirit Cave, Nevada*. Bureau of Land Management, Nevada State Office.

Brook, A., Underhill, C., Foo, L., & Hector, M. (2006). Aprroximal attrition and permanent tooth crown size in a Romano-British population. *Dental Anthropology, 19*, 23–8.

Brown, M., Hosseini, S., Torroni, A., Bandelt, H-J., Allen, J., Schurr, T., . . . Wallace, D. (1998). mtDNA haplogroup X: An ancient link between Europe/western Asia and North America? *American Journal of Human Genetics, 63*, 1852–61.

Cybulski, J. (2001). Current challenges to traditional anthropological applications of human osteology in Canada. In L. Sawchuk & Susan Pfeiffer (Eds), *Out of the past: The history of osteology at the University of Toronto*. Toronto, ON: CITD Press.

Derenko, M., Grzybowski, T., Malyarchuk, B., Czarny, J., Miścicka-Śliwka, D., & Zakharov, I. (2001, July). The presence of mitochondrial haplogroup X in Altaians from South Siberia. *American Journal of Human Genetics, 69*(1), 237–41.

Dorsey, George. (1898). A cruise among the Haida and Tlingit villages. *Appleton's Popular Science Monthly*.

Guidon, N., & Pessis, A.M. (1996). Falsehood or untruth? *Antiquity, 70*(268), 408–16.

Hanis, C., Hewett-Emmett, D., Bertin, T. & Schull, W. (1991, July). Origins of US Hispanics: Implications for diabetes. *Diabetes Care, 14*(7), 618–27.

Harper, K. (1996). *Give me my father's body: The life of Minik, the New York Eskimo*. Iqaluit, NU: Blacklead Books.

Haydenblit, R. (1996). Dental variation among four prehispanic Mexican populations. *American Journal of Physical Anthropology, 100*, 225–46.

Koppel, T. (2003). *Lost world: Rewriting prehistory—How new science is tracing America's Ice Age mariners*. New York: Atria Books.

Lange, K. E. (2007, September). Tales from the bog. *National Geographic* (online). Retrieved from http://ngm.nationalgeographic.com/2007/09/bog-bodies/bog-bodies-text

Larsen, C. S. (2000), *Skeletons in our Closet: Revealing Our Past through Bioarchaeology*. Princeton: Princeton University Press

Malhi, R. S., Schultz, B. A., & Smith, D. G. (2001). Distribution of mitochondrial DNA lineages among Native American tribes of Northeastern North America. *Human Biology 73*, 23–51.

Meltzer, D., Adovasio, J., & Dillehay, T. (1994). On a Pleistocene human occupation at Pedra Furada, Brazil. *Antiquity, 68*(261), 695–714.

Merrett, D. (2003). Moatfield demography. In R. F. Williamson & S. Pfeiffer (Eds), *Bones of the ancestors: The archaeology and osteobiography of the Moatfield Ossuary* (pp.171–88). Gatineau, QC: Canadian Museum of Civilization.

Mountain Institute. (Website). Welcome to the Mountain Institute. Available at www.mountain.org/

Neves, W., & Hubbe, M. (2005, December 20). Cranial morphology of early Americans from Lagoa Santa, Brazil: Implications for the settlement of the New World. *PNAS, 102*(51), 18309–14. doi: 10.1073/pnas.0507185102

SFU Museum of Archaeology and Ethnology. (Website). A journey to a new land. Available at www.sfu .museum/journey/home1.php

Thomas, D. H.. (1999). One archaeologist's perspective on the Monte Verde controversy. *Archaeological Institute of America*. Retrieved from www .archaeology.org/online/features/clovis/ thomas.html

———. (2000). *The skull wars: Kennewick Man, archeology, and the battle for Native American identity*. New York: Basic Books.

van der Merwe, N., Pfeiffer, S., Williamson, R., Thomas, S. (2003). The diet of the Moatfield people. In R. F. Williamson & S. Pfeiffer (Eds), *Bones of the ancestors: The archaeology and osteobiography of the Moatfield Ossuary* (pp. 205–22). Gatineau, QC: Canadian Museum of Civilization.

Walker, P. (2000). Bioarchaeological ethics: A historical perspective on the value of human remains. In M. A. Katzenberg & Shelley R. Saunders (Eds), *Biological anthropology of the human skeleton* (pp. 3–39). New York: Wiley-Liss.

Weiss, E. (2009). Sex differences in humeral bilateral asymmetry in two hunter–gatherer populations: California Amerinds and British Columbia Amerinds. *American Journal of Physical Anthropology, 140*, 19–24.

Zegura, S., Karafet, T., Zhivotovosky, L. & Hammer, M. (2004). High-resolution SNPs and microsatellite haplotypes point to a single, recent entry of Native American Y chromosomes into the Americas. *Molecular Biology and Evolution, 21*(1), 164–75.

12

Forensic Anthropology

CHAPTER OUTLINE

Remains of a human skull are reconstructed
on a frame of wire and wax
Michael Donne, University of Manchester/
Science Photo Library

After reading this chapter, you should be able to:

- Describe the work that a forensic anthropologist does and distinguish it from the work of specialists in similar fields.

- Outline the history of forensic anthropology as a field of study.

- Discuss the role of the Body Farm in forensic anthropology.

- Explain how to determine the age, sex, and race of a skeleton.

- Outline the skeletal signs that a child has been abused.

Introduction: Fact Meets Fiction

The scene is a familiar one on TV. A couple walking their dog stumble upon the gruesome remains of a body in the bushes just off the trail. The police are notified, and they call upon (among other forensic experts) the **forensic anthropologist**, who quickly identifies the age, sex, race, and cause of death of the deceased; in the process, she or he also discovers and reveals some impossibly detailed facts that cause the audience and the fictional police to be in awe of the breadth and depth of knowledge of the great anthropologist.

It doesn't work quite like that in real life. Physical anthropologists who do forensic work are not—unlike characters in the TV series *Bones* (which features the fictional Temperance Brennan, who was dreamed up by real-life physical anthropologist **Kathy Reichs**) or *CSI*—knowledgeable in all areas of science. They know bones and teeth very well, and tend to leave the soft tissue stuff (skin, flesh, internal organs, and so on) to the **forensic pathologist**. They also are not involved in studying insects such as blowflies, whose precisely datable life stages (eggs, three stages of maggots, and pupae) greatly assist in determining how long a body has been dead—that is the work of a **forensic entomologist** (an entomologist being one who studies insects). Unlike the *CSI* specialists, but like other scientists, the forensic anthropologist may speculate about the weapons used to kill and injure the diseased and the possibility of animal interference with the body. He or she may provide educated guesses (much like meteorologists discussing the likelihood of rain in their long-range forecast) as to the body's age range, the gender and race, and the height.

Yet people are so intrigued by forensic anthropology that it has become an immensely popular subject of fiction—movies, TV, novels—and in fact it is not unusual for a forensic anthropologist to turn his or her hand to writing murder mysteries. One of my favourites is Kathy Reichs (b. 1950), professor of anthropology at the University of North Carolina (previously at Concordia and McGill in Montreal), author of 12 books of fiction, and, as noted above, producer of the TV series *Bones*. Another is **Aaron Elkins** (b. 1935), creator of fictional forensic anthropologist Gideon Oliver, 'the Skeleton Detective' (go to www.aaronelkins.com and check out the link to 'forensic tidbits' for fascinating facts from the world of forensic science). All of this has helped make forensic anthropology a popular area of study right now, the main growth area for new jobs in physical anthropology and in anthropology generally.

As we will see, the work of forensic anthropologists is varied, and not all of it is made for television. They do assist police with their inquiries, helping to identify the dead and providing evidence that can lead to a conviction. They also are found at the scene of mass casualties, helping to determine the number of dead from such events as airplane crashes, acts of genocide, or brutal mass murders, such as 9/11.

How Do You Know You're Going to Be a Forensic Anthropologist?

In *The Bone Woman*, **Clea Koff** (b. 1972), describes the early moments that reveal her inclination to devote her life to forensic anthropology, before she even knew that it existed:

> When I was seven, I was collecting the dead birds I found outside our Los Angeles house and burying them in a little graveyard. A few years later, in Kenya with my family, I collected bleached animal bones in Amboseli National Park while monkeys yelled in the trees above me. I put all the bones back the next day, after I had cleaned them gently. By the time I was thirteen, . . . I was burying dead birds in plastic bags so I could dig them up later—I was curious how long it took them to 'turn into' skeletons. I took the stinking bags to my (somewhat horrified) science teacher

© Reuters./Corbis

Forensic anthropologists in Cordoba, Argentina, work at the site of a mass grave of 40 suspected victims of the 1976–83 military junta

for extracurricular and wholly self-motivated death investigations. (Koff, 2005, p. 8)

Forensic anthropology was clearly 'in her bones': after graduating with degrees in anthropology and forensic anthropology, Koff took a job with the United Nations, where she worked on two major UN tribunals investigating cases of genocide, first in Rwanda, then in the former Yugoslavia. The moral is that if you're the sort of person who looks with scientific interest at the dead mouse the neighbour's cat leaves on your step or the squirrel that has fallen from the hydro lines, forensic anthropology just might be for you. We'll read more about Koff's fascinating life and work later in the chapter.

A Brief History of Forensic Anthropology

Thomas Dwight, Father of American Forensic Anthropology

It's safe to say that the field of forensic anthropology first developed in the United States. **Thomas Dwight** (1843–1911) is often referred to as the 'father of American forensic anthropology', although no famous cases are associated with his pioneering anatomy work at Harvard Medical School. Instead, the epithet comes from the osteology collection Dwight put together for the Warren

Profiles in Canadian Anthropology

Moira McLaughlin, the 'Bone Lady'

Physical anthropologist **Moira McLaughlin** (a.k.a. the 'Bone Lady') teaches at St Thomas University in Fredericton, New Brunswick. After she graduated from the University of Toronto with a degree in physical anthropology, specializing in osteology and archaeology, she was first introduced to the gruesome joys of forensic anthropology at the University of Tennessee at Knoxville, where she did her graduate studies with the famous (see below) and intellectually influential William Bass. In an interview with Ellen Creighton, published in the Moncton *Times & Transcript* (3 December 2005), McLaughlin talks about the place of forensic anthropology within the broader discipline of physical anthropology, and how her real work differs from what you might see on TV:

> There's really no such thing as a forensic anthropologist—it would be impossible to do full-time. 'I only get to do forensics when bones are found, and how often is that? First and foremost, I'm a physical anthropologist and within that, I do a variety of things.' When she does find bones, however, McLaughlin says there's more to the job than just piecing together recovered materials. 'I draw on a full set of skills—I bring that person back to life. I think of who they were, their communities. You're figuring out uncertainty.'

Being an anthropologist, according to McLaughlin, can be emotionally demanding and is not always the exciting job that is portrayed on crime-scene television shows. 'Luckily most of the bones that I identify turn out to be animal bones.'

McLaughlin says popular television shows glamorizing her line of work are hindering, not helping, young people to gain interest in her field. '*CSI* is horrible. It's crazy. We spend a lot of time on that. Some students watch it for fun, but you have

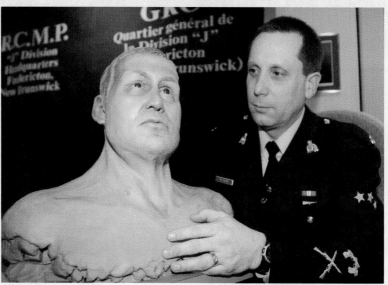

An RCMP constable examines a facial reconstruction of the unidentified man found in the Gulf of St Lawrence, 29 January 2004; the clay model produced a number of tips from across Canada, though the case remains unsolved.

Museum of Anatomy at Harvard. Collections of this kind are an important part of forensic anthropology: they are a source of many useful statistics that can help us calculate stature from a single bone or find out what vital data might be considered 'typical' for someone of a certain gender, age, or ethnic background. His first book, available online, is *Frozen Sections of a Child* (1881), a 69-page work describing a frozen 3–year-old girl. The original work features 15 beautiful drawings of microscopic plates 'drawn from nature' by Dr Henry P. Quincy. People using the book for teaching or learning were

encouraged 'To study the plates, turn the book so that the vertebra is nearest to you and imagine that you are looking down into your own body' (Dwight 1881, p. v). That would give me nightmares.

George A. Dorsey and the Case of the Sausage Maker's Wife

Adolph Louis Luetgert was born in Germany in 1845 and sailed to the United States at the age of around 20 to seek his fortune and fame. He became a successful sausage

to realize that this is a Hollywood version of what we do, and it's very different.' (Creighton, 2005)

My favourite Canadian 'cold case' for teaching purposes comes from McLaughlin's work. I first read about it in the New Brunswick *Telegraph-Journal*, in an article by Shawn Berry (2003): the headline read 'Investigators hoping to put a name to body found in ocean 23 years ago,' I've since read the official report on the investigation from the RCMP's cold case file. Here are the facts (see Table 12.1). On 3 May 1980, a naked male body was found floating in the Gulf of St Lawrence by crab fishermen from the northeastern New Brunswick community of Le Goulet. After three weeks of lying in the morgue, unclaimed, the body was buried with no name. Fortunately, the funeral director put extra embalming fluid into the body, and placed it in a coffin made properly airtight; it was ideal for exhuming. Twenty-three years later, a forensic team made up of Moira McLaughlin, a forensic artist, and a detective constable was assembled to try to find out the dead man's story. In McLaughlin's words,

> Each [team member] has their own area of expertise . . . and when you put us together, it seems to be the right combination to solve these cases. I think this kind of work is important because out there is a family somewhere who has no idea where their loved one went. This is what this is all about, returning these people to their families and giving them back their dignities. (as cited in Berry, 2003)

Unfortunately, in this case, the team has been unsuccessful: the case remains unsolved.

What do *you* think?

1. What other information (e.g. regarding his teeth and his bone structure) might help you determine the body's identity (e.g., his country of origin, and possible occupation)?
2. What would you look for in the bones that might give you some additional clues?
3. How could you use the body's DNA in your search for his identity?
4. What social characteristics would you attribute to the person, given the physical description in Table 12.1?
5. What would your strategy be for discovering his identity?

Table 12.1	Dead in the water: Forensic details of the body found in the Gulf of St Lawrence
Sex	male
Race/ethnicity	white
Age	30–50
Stature (height)	~ 1.68 m (5' 6")
Weight	~ 63.5–72.5 kg (140–160 lbs)
Build	lean and fit
Cause of death	unknown; no sign of foul play
Distinguishing features	4 gold teeth; porcelain bridge

maker in late nineteenth-century Chicago. In 1897, he reported to the police that his second wife, Louisa, had disappeared. Her brother was convinced that Adolph had murdered her, but without a body, there was no evidence to make a case. When Louisa's brother finally persuaded the police to search the five-storey sausage factory, they found a ring with the initials 'L.L.' inscribed on it, and four small pieces of bone embedded in a congealed substance that had leaked out of a vat. The vat was filled with potash, a strong caustic agent used in the manufacture of soap and, more important, capable of breaking down bone.

The case became famous. Rumours abounded that Luetgert had rendered his wife into sausages sold to his customers, but these stories proved to be untrue. **George A. Dorsey** (1868–1931), the first person to receive a PhD in anthropology from Harvard, and possibly a student of Thomas Dwight, appeared in court as an expert witness for the prosecution, the first anthropologist to play this role. He first identified the bones as human. The following statements come from articles in the New York Times:

LUETGERT MURDER TRIAL

Two Experts for the Prosecution Identify Bones Exhibited as Those of a Woman.

EFFECT OF CAUSTIC POTASH

Juror J. E. Flowler, Jr., Recovering His Health—Judge Tuthill Exasperated—Prosecution Expects to be Through with Its Witnesses by Saturday.

CHICAGO, Sept. 15 [1897]—The prosecution in the Luetgert trial is nearing its end, and the announcement was made today that it would probably have all of its witnesses on the stand by Saturday noon. . . .

In the afternoon session Prof Dorsey positively identified a bone handed to him by Assistant State's Attorney McEwen as the upper portion of the left femur. 'The bone is undoubtedly that of a female,' said Prof. Dorsey. 'I base my opinion upon the formation of the socket at the hip or thigh joint. This bone has the appearance of having been somewhat burned. It has lost all animal matter.'

'Could a solution of caustic potash have produced the effect you speak of?' queried Assistant State's Attorney McEwen.

'In my judgment it could,' replied the witness.

Prof. Dorsey also identified another piece of bone as the left temporal bone of a female. This bone also bore evidences of having been seared by fire.

LUETGERT MURDER TRIAL

Attempt of the Defense to Shake Expert Testimony of Prof. Dorsey on Femurs.

LONG WRANGLE OVER BONES

Luetgert Hobbles into Court Supported by Crutches—Had Been Engaged in a Friendly Boxing Bout with One of His Guards.

CHICAGO, Sept. 16 [1897].—The day was given over to expert testimony in the Luetgert trial, and the attorneys for the defense and the witnesses for the State wrangled vigorously regarding femurs of human beings, femurs of sheep, and of hogs. . . . Attorney Vincent, for the defense, made it his business to show the jury that Prof. Dorsey did not know anything about bones, . . . and that he was densely ignorant on the subject of femurs. . . .

Attorney Vincent tried in vain for over an hour to break the force of the professor's identification of the femur, questioning him sharply as to the difference between the femur of a male and a female. Prof. Dorsey calmly explained with the aid of the bone in evidence.

The closing hour of the afternoon sessions of the Luetgert trial was lively. Attorney Vincent . . . made a supreme effort to break down the evidence of Prof. Dorsey. After a number of questions touching upon the qualifications of the witness of testify as an expert, Attorney Vincent picked up a handful of bones, and, holding them out to the witness, asked him to identify each of them. Prof. Dorsey refused to attempt to do so without first having an opportunity to examine them.

'Can you do it!' shouted the attorney for the defense. 'If you are the expert you claim to be it should not be difficult for you to name these bones.'

This exasperated the witness, and Assistant State's Attorney McEwan came to his aid with an objection, which was sustained by the Court. But Attorney Vincent was not to be stopped by objections. 'Did you ever dissect a hog?' he inquired.

'No, Sir,' answered Prof. Dorsey. There was an objection to this line of cross-examination, and Attorney Vincent arose to explain that he wished to show that the femur of a hog was identical with the femur the State had introduced, and which Prof. Dorsey had identified as the right femur of a human being, a female. After further questioning with reference to whether or not Prof. Dorsey had every dissected a calf, or a sheep, and receiving a negative reply as to the former and an affirmative answer as to the latter animal, Attorney Vincent suddenly asked: 'How do you know, then, that this is not a calf's femur?'

'Because, from a careful examination of it, I believe it to be a human femur,' replied the witness.

'Why do you consider it the femur of a woman?'

'Because it is smooth, and has the shape and other distinguishing characteristics of the female femur,' replied the witness.

Dorsey's later testimony, which related to the four small bones, held up, and Luetgert was convicted of murder.

Clyde Snow and the Angel of Death

Clyde Snow (b. 1928) is another pioneer of forensic anthropology, having done much work to develop and advance the discipline. One of his more innovative studies involved the Civil Aeromedical Institute, for whom he was engaged in measuring pilots to provide soundly engineered seats and safety equipment, beginning in 1960. He then spent several months measuring 423 airline stewardesses, as part of a study into a spate of injuries among flight attendants owing to badly designed protection, sideways seating, and inadequately researched survival equipment, which was markedly inferior to the better engineered systems made to fit the male pilots.

In 1973, Snow published a famous paper detailing who walked away from 'survivable accidents' in airplanes:

> Males get out. Businessmen who fly a lot, they get out. The odds are roughly two to one in favor of adult males. There's a lot of physical stuff going on in there, and they certainly do not behave like gentlemen. Another thing is travelling companions. That is significant in that if you are traveling in a group, family groups tend to do a little better. Blood relatives are more likely to help each other. Age-wise, it amounts to the fact that if you are under 12 years old, forget about it, only about 10 percent of them get out, at least in the 1960s. And if you are male over 55, you survive about the same as females do. As the toxic environment becomes extreme . . . age, sex, and experience become less important in survival. (as cited in Joyce & Stover, 1991, p. 27)

Snow's paper got him into some trouble, especially over his attitude (it didn't help that he called crashes 'natural experiments'). However, it did help him get work with a medical examiner in Oklahoma.

One of Snow's most famous cases was the identification of Dr Joseph Mengele, Hitler's notorious 'Angel of Death'. Mengele was medical officer of the concentration camp at Auschwitz, where, during his 21-month tenure, there were over 400,000 deaths, many coming as a result of hideous experiments he oversaw.

Following the war, Mengele fled Europe and disappeared to avoid standing trial for war crimes. He was never found, despite several alleged sightings of him. In 1979, he was rumoured to have drowned in Brazil, where he was buried, it was believed, under the name of Wolfgang Gerhard. When the body was exhumed, Snow was called in to verify the identity.

The bones had been disturbed and fractured during the exhumation, so Snow had a complicated case on his hands. Early in his investigation he determined that the individual was right-handed (typically confirmed by thicker, stronger bone growth on the preferred side), Caucasian, and between the ages of 60 and 70 (Mengele supposedly died at age 67). Estimates of the body's stature, based on key bone measurements, came within half a centimetre of Mengele's known height. Snow pieced together the shattered skull and marked it with pins at 30 points of comparison. He marked a photograph of Mengele in the same places, and performed one of the earliest video skull–face superimposition studies. In 1985, Snow, along with a team of scientists from Brazil and West Germany, declared that the exhumed skull was indeed that of Joseph Mengele. Later studies performed using dental X-rays and DNA evidence proved that Snow's identification was correct.

William Bass and The Body Farm

In *Death's Acre* (2003), renowned forensic anthropologist **William Bass** introduced readers both to his own extensive background experience—consisting largely of his work on thousands of skeletons of the ancestors of the Arikara (an American Aboriginal group) that were going to be flooded from their earthly home by the creation of a new dam—and to his scientific research facility, best known as the Body Farm. Bass had moved to Knoxville in 1971 to head up the anthropology department of the University of Tennessee. In 1980, he got a great idea for further research:

> In 1980, on a couple of acres of junk land behind the University of Tennessee Medical Center—a patch of scraggly woods with a charred open area at the center, where the hospital had burned trash for years—I poured a concrete pad measuring sixteen feet square; atop this pad I built a chain-link enclosure, complete with a chain-link 'roof'. I planned to place, inside the fence and safe from predators (except for predators small enough to slip through the weave of the wire), human bodies, which my graduate students and I would observe closely, recording the sequence and the timing of human decomposition during the extended post-mortem interval. (Bass & Jefferson, 2007, pp. xv–xvi)

The first experiments were basic: How long does it take a body in particular circumstances to become skeletonized? How can you tell if a body has been moved since it died? Gradually, the complexity and importance of the experiments grew. The first donated body arrived in May of 1981. Donations were slow at first, but now, due to the great fame of the research facility, hundreds are offered

The 'Body Farm', in Knoxville, Tennessee, where crime scenes are staged using donated cadavers for students of forensic science to excavate

every year. The fame is due in part to publicity generated by the bestselling novel *The Body Farm*, by celebrated crime writer Patricia Cornwell (1994). Cornwell had first contacted Bass while conducting research for one of her mysteries: she wanted to know whether a coin underneath a body would leave a mark. Bass invited her to set up an experiment to find out, and Cornwell went on to make Bass's research facility a central part of her 1994 book. Bass's own crime fiction series The Body Farm Novels, co-authored by Bass and Jon Jefferson under the pen name of Jefferson Bass, together with his equally popular works of non-fiction *Death's Acre* and *Beyond the Body Farm*, have also added to the facility's reputation. The three case studies that follow shed a little light on the kind of work Bass does when he is not teaching or tending his 'body farm'.

BILL BASS CASE STUDY #1

It was March 1974. William Bass was three years into his job at the University of Tennessee; it would be seven years before the Body Farm was established.

Bass was asked to have a look at a skull found by a boy along the shoreline of a lake whose water level had recently fallen. Bass analyzed the skull in the following way:

From the heavy, prominent brow ridge over the eyes, this was clearly a male skull. The narrow nasal opening and the vertical structure of the teeth and jaws told me that it was Caucasoid. . . . And the fact that the incisors were basically flat across the back, rather than shovel-shaped, ruled out the possibility that his was a Native American skull. Besides, . . . the skull was in remarkably good shape—far too pristine to have been stuck in the mud for decades or centuries. Amalgam fillings—done by a good dentist, judging from the handiwork—confirmed the skull's recent vintage. The skull was unblemished; if it belonged to a murder victim, the murder was definitely not committed by a blow or gunshot to the head. From the wear on the teeth and the prominence of the cranial sutures—the joints between the bones of the cranial vault—I estimated the age to be somewhere between 30 and 34. (Bass & Jefferson, 2007, pp. 21–2)

Table 12.2 summarizes the features Bass identified and the characteristics he inferred from them.

Table 12.2	Findings of Bass's Case Study #1
Skull feature	Characteristic inferred
prominent brow ridge	gender = male
narrow nose	race = Caucasoid (vs Negroid)
vertical jaw and teeth	race = Caucasoid (vs Negroid)
flat-backed incisors	race = not Aboriginal (or East Asian)
amalgam fillings	date = recent
no trauma marks	cause of death = undetermined
moderate tooth wear	age = early 30s
prominent cranial sutures	age = ‹35

Bass, 2007, pp. 31–2.

BILL BASS CASE STUDY #2

Table 12.3 summarizes the findings of an analysis Bass performed on a skeleton found by the local police. From these data Bass determined that the skeleton was that of a 13-year-old girl. In the second row of the table, 'medial' means 'middle', and **condyle** (derived from the Greek and Latin terms for 'knuckle') refers to the knuckle-like part of long bones such as the humerus. The **greater sciatic notch** refers to a groove on both hip bones that widens to at least the width of a thumb when females reach puberty.

Table 12.3	Findings of Bass's Case Study #2
Skeletal feature	Characteristic inferred
fused distal articular humerus	age = ›12
unfused humeral medial condyle	age = ‹18–20
fully developed second maxillary molars ('12-year molars')	age = ›12
absence of third molars (wisdom teeth)	age = ‹18
greater sciatic notch beginning to widen	age = early puberty sex = female
hips not yet widened	age/sex = not complete puberty (if female)

Bass, 2007, pp. 31–2.

BILL BASS CASE #3

In this third case, which occurred in 1991, Bass was presented with a burned skeleton. From the skeletal evidence, Bass determined that the skeleton belonged to a young man between the ages of 25 and 30, with a stature a little shorter than average (Bass & Jefferson, 2007, pp. 115–18).

Table 12.4	Findings of Bass's Case Study #3
Skeletal feature	Characteristic inferred
skinny hips	sex = male
external occipital protuberance	sex = male
48 mm proximal humeral head	sex = male
well-developed nasal spine (i.e. narrow nose)	race = Caucasoid
nasal sill	race = Caucasoid
two fused maxilliary sutures	age = early 30s
one unfused maxilliary suture	age = early 30s
state of pubic symphysis	age = 22–28 years old
breadth of proximal femur	stature = 1.68 m ± 5.0–7.5 cm (5' 6" ± 2–3")

Bass, 2007, pp. 31–2.

Forensic Anthropology Step by Step

1. Determining that the Body Is Human

The first step when bones are found is to determine whether the skeleton is human. Obviously, this is fairly easy when there is a skull; it is not so easy when the remains consist of small bones.

I have an experience that relates to this. When I was about 12 years old, my cousin Gordon and I were digging in the tidal pools of the North River in Charlottetown, Prince Edward Island. We discovered a bone, and our curiosity got the better of us. With visions of pirates and pirate treasure spurring our imaginations, we dug on to discover large ribs, and we felt sure in what we'd found. We ran up the stairs from the beach to Inkerman, a large house that had belonged to one of the Fathers of Confederation but that now was owned by our grandparents. Breathless, we told our grandmother what we had found. But instead of being pleased with our sensational discovery, she was angry with us: it turned out to have been the body of a horse that had been buried in the 1930s. We had to return the bones to the beach, our dreams of pirate treasure dashed.

The moral of the story is that is not always easy for the non-scientist (and that includes many police officers) to tell whether bones are human or non-human. This is a crucial first step for the forensic anthropologist called in to investigate possible human remains.

2. Numbering the Skeletons

The next step, determining how many individual bodies are represented by the bones that have been found, requires the help of an archaeologist, or a physical anthropologist trained in archaeology. The digging must be precise and methodical. This kind of expertise is in demand in southern Ontario because of the historical presence in the area of some Aboriginal groups that practised mass burials. The ancestors of the Huron/Wendat and related peoples had a ceremony called the **Feast of the Dead**. During this important funerary event, the bodies of those who had died over the previous 10–15 years (typically the period since the village was last closed and moved) would be buried together in one pit. It has several times occurred in southern Ontario that one of these burial sites has been uncovered, with bones found in great number. Recent cases include a discovery in Milton Heights, west of Toronto, made by a family digging in their backyard, and one in Midland, where excavations were underway for the expansion of a sports complex near the Huronia Museum by Little Lake. Residential development is often the occasion of such discoveries—one occurred quite recently near Canada's Wonderland north of Toronto.

Determining the exact number of skeletons located at a grave site is work more commonly associated with airline crashes, terror bombings, building fires and explosions, and cases of genocide. It often requires painstaking lab work. However, looking for 'repeats' of certain bones is a relatively easy first step in determining the ultimate number of bodies; imagine determining how many identical jigsaw puzzles you have in a pile of pieces by looking for how many duplicates you have. More complicated methods include sorting specimens by age. This is the kind of work that can test the forensic anthropologist's skills as a lab technician, but it is quite successful if carried out with due care. In a case in 1979, Clyde Snow, then working for the US Federal Aviation Administration, had to sort through over 10,000 body parts from a single airline disaster; he was able to identify an impressive 234 of the 273 victims.

3. Sexing the Skeleton

Once the human nature of the remains has been verified and the number of bodies established, the forensic anthropologist begins to determine specific traits of the individual(s) found. One of the tasks involved is sexing the skeleton (which sounds kinkier than it really is). There are a number of fairly clear bodily landmarks that help the anthropologist do this (see Table 12.5). With skulls, a lot has to do with size and robustness, though there are other features to look for, too. For example, there is the external occipital protuberance or 'inion'. In more accessible English, that's a bump on the middle of the bottom back line of the skull, just above the neck. Men usually have a rather clearly defined one; women do not. Difficulty in distinguishing male from female skeletons arises when the individual is a child; in fact, it is much more difficult—almost impossible in some cases—to sex the skeleton of a child. DNA testing is frequently required.

4. Racing the Skeleton

The fictional forensic anthropologist considers the skull before him, giving particular attention to the teeth:

> The mouth structure is orthagnic—strongly vertical—so it doesn't appear Negroid. No appreciable occlusal wear, so she didn't have an edge-to-edge bite, and the incisors are definitely not shovel-shaped. That probably rules out Native American or Asian, though to be sure, we should put the skull measurements into ForDisc. (Bass, 2006, p. 45)

Very controversial among anthropologists, as we have seen, is the idea of determining the 'race' or ethnicity of a skeleton. That being said, there are certain points that can be used as racial clues or indicators, certainly among North American populations of blacks, whites, Chinese, and Aboriginal people (see Table 12.6). It should be noted that there is significant variation across African populations, and that most African Americans have ancestors originally from West Africa. The same can be said for North American Aboriginal peoples (who differ in some ways from the indigenous peoples of Central and South America) and for people of Chinese descent, as China is a country with more ethnic diversity (and, thus, more physical diversity) than many North Americans are aware of. Table 12.6, adapted from Jeffrey Schwartz's rendition of St Hoyme and İşcan's work of 1989, summarizes the cranial features that give clues to a skeleton's ethnicity. *Prognathism* (adj. **prognathic**) is a facial profile characterized by a protruding lower face (i.e. below the nose); the opposite is *orthognathism* (adjective **orthognathic**). The nasal spine referred to in the table is the bone that forms the bridge of the nose. The alveolar ridge refers to the upper inside of the mouth.

As Table 12.6 illustrates, the main indicators of race, traditionally, have been found in the cranium. Bass reinforces this view in the following:

> For virtually my entire half-century career, and for many decades before I began my studies in

Table 12.5	Sexing the skeleton		
Feature		In females	In males
Skull			
size/weight		smaller/lighter (average 595 g)	larger/heavier (average 795 g)
build		gracile	robust
supraorbital ridge		less pronounced	more pronounced
orbits		large relative to face rounded to circular	small relative to face square to rectangular
external occipital protuberance		smooth to weak	marked to massive (stud-like)
nasal bones		smaller, less protrusive	larger, more protrusive
palate		shorter, rounder, flatter	broader, longer, more vaulted
mandible		short, narrow, low, gracile	long, broad, high, robust
chin		v-shaped, pointed	u-shaped, square
mouth		narrower	wider
cranial base		flatter, less marked	rounder, more marked
Postcranials			
pelvis (after puberty)		wider	narrow
ilium	blade of ilium	more flared, wide, low	more vertical, narrow, high
	greater sciatic notch	u-shaped, broader, shallower	v-shaped, narrower, deeper
femur		inclines inward	goes straight down
	femoral head	‹ 81 mm	› 81 mm
	femoral shaft	‹ 47 mm (diameter)	› 47 mm (diameter)
humerus	humeral head	‹ 43 mm (diameter)	› 43 mm (diameter)

Adapted with additions from Schwartz, 2007, pp. 293–4.

osteology (bones) and anthropology, the skull was considered the only reliable basis for determining the race of an unknown victim's skeleton. In fact, as recently as 2006, I retained this passage in a new, updated edition of my guide to bones, *Human Osteology: A Laboratory and Field Manual*: 'The skull is the only area of the skeleton from which an accurate estimation of racial origin may be obtained.' (Bass & Jefferson, 2007, p. 127)

In some ways, the work of identifying race intersects with the work of identifying sex. Among East Asians and, to a lesser extent, Aboriginal peoples, there is less sexual dimorphism than among white and black people. (In this regard, think of East Asians as being like gibbons, and white and black people being like gorillas.) So, if your skeleton turns out to be that of an East Asian or Aboriginal person, sexing the skeleton will be difficult, especially if you do not have the pelvis to analyze.

RACE AND DENTAL CHARACTERISTICS

With teeth as with other diagnostic racial traits, it is important to bear in mind that the data we are dealing with are variable. For example, some Caucasoid people have **sinodont** teeth (the shovel-shaped incisors and three-rooted molars more characteristic of Chinese, Mongolian, and Native North American people). My wife, Angie, though she is white, has sinodont teeth, probably relating to the Mongolian strains in her eastern European heritage. Some black people from the Caribbean also have sinodont teeth, likely stemming from the Aboriginal heritage that many Caribbean people share (see Chapter 13 for more on this).

FORDISC'S FIRST CASE

ForDisc (short for forensic discrimination function analysis) is a software program developed at the University of Tennessee by Richard Jantz and Steve Ousley. It's designed to create a profile of an unidentified deceased person

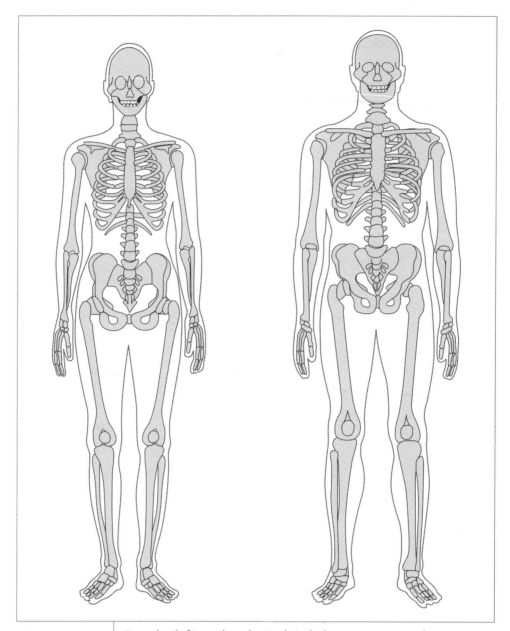

Figure 12.1 | Female (left) and male (right) skeletons compared

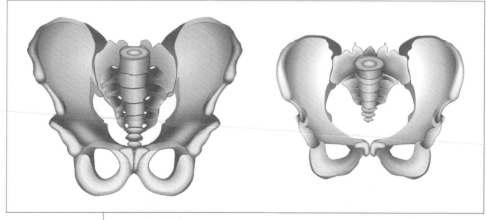

Figure 12.2 | Female (left) and male (right) pelvises compared

Table 12.6	Distribution of cranial features among various humans						
Cranial feature	Population						
	Inuit	Aboriginal	European	African	Melanesian	North Asian	South Asian
chin	projecting	projecting	projecting	intermediate	rounded	projecting	rounded
anterior nasal spine	prominent	prominent	prominent	short/blunt	prominent	prominent	prominent
nasal bridge projection	least	least	most	medium	least	least	least
prognathism	face vertical/ very flat	face vertical	upper face/ mid-face	alveolar/ dental	face vertical	face vertical	face vertical

Table 12.7	Distribution of dental features among various humans			
Dental feature	Population			
	Black	White	East Asian	Aboriginal
sinodonty/sundadonty	sundadont	sundadont	sinodont	sinodont
overbite	no	yes	no	no
occlusional wear (clacking)	no	no	yes	yes
crenulated (rugged/bumpy) molars	yes	no	no	no

by comparing the remains with skeletal data held in the **Forensic Anthropology Data Bank** and producing a set of likely traits based on the results. Established in 1986, the data bank has developed over the years to include statistics on skeletal material from all over the world.

ForDisc's first test case came in 1991. It was fall, and Bill Bass and his graduate students received a call from the police of a local county saying that a bundle of plastic containing human skeletal remains (it would prove to be an almost complete skeleton) had been found in the interestingly named Polecat Creek. As it was his forty-second forensic case of 1991, it received the label 91-42.

The police had received an anonymous tip that a pregnant white female had been shot and dumped into the river. Could this be the body? Perhaps—yet when the University of Tennessee team first observed the bones, they were unanimous in their assessment that the skeleton was male. The police were reluctant to believe them because of the tip they had received.

It would take the computer voice of science to convince the doubting police. The other reason for trying ForDisc out on this case was that the individual had skull features that Bass referred to as 'racially ambiguous'. Here's how he described the skeleton's mixed features:

He possessed some Negroid features, including a long, low skull, wide-set eyes, and a short nasal opening. But he also exhibited Caucasoid features, including a 'sill' at the base of the nasal opening, a sharp nasal spine (the bone that forms the bridge of the nose), and a hooked inion (the point on the end of the external occipital protuberance, which is the bump men have at the base of their skull). (Bass & Jefferson, 2007, p. 137)

Bass's best educated guess was that the individual was white (Caucasoid) and male. He also estimated that the individual stood about 1.8 meters (5′ 11″) tall, based on the length of his femur. He called in Richard Jantz, his former student, to do the type of measuring necessary to conduct an analysis using ForDisc. After the measurements were entered into the computer, ForDisc produced a profile that contradicted Bass's, although the computer 'spoke' with all the assurance and cover-your-backside qualifications of a TV meteorologist discussing the possibility of snow for Christmas:

Like me, the software found ambiguities in the skull—initially ForDisc gave a 55 percent probability that 91-42 was black. Looking just at the measurements of the postcranial skeleton though . . . ForDisc was far more confident in its prediction, assigning a probability of 95 percent that our victim was black. Performing two different calculations based on measurements from four bones (the clavicle, the radius, the humerus, and the tibia), ForDisc estimated his stature to be six feet—an inch taller than I'd estimated. (Bass & Jefferson, 2007, p. 141)

This did the trick. It wasn't long before the police matched this profile to a black man who had gone missing in 1980; three years later, in 1991, his killer was convicted. A computer detective was born.

5. Aging the Skeleton

We've seen in past chapters the many ways that skeletal remains can be dated to see how long ago an individual lived. For the forensic anthropologist tasked with aging a skeleton—of 'guesstimating' how old an individual was at the time he or she died—the techniques involved are quite different. Indeed, there are various means by which a skeleton can be aged, and it's important to keep in mind that it is through a combination of these means, not through any one means alone, that the highest degree of accuracy can be attained. Even so, aging methods typically produce a range of ages rather than a specific figure.

The importance to this kind of work of achieving **consilience**—broad agreement by different measurement means—was established in one of the pioneering studies on aging skeletons, engaged in by T.D. Stewart of the Smithsonian Institution and Thomas W. McKern of the Anthropology Branch of the Environmental Protection Research Division, an American governmental body, in 1957 (McKern & Stewart, 1957). They studied the skeletonized remains of 450 identified soldiers who had fought in the Korean War, and who had been repatriated in 1954. In 10 different chapters, they looked at 10 features that could be examined for signs of aging. A strength of the study was that the ages were already known, so the accuracy of the methodology could be tested. Of course, there were some weaknesses, too. Their subjects were male, and that weakened somewhat the ability to generalize to females, certainly when looking at the pubic symphysis (see below). Also, the vast majority were white. Still, this does not take away from the importance of their concluding statement:

One of the principal conclusions emerging from the foregoing study is that individual maturation features or events are highly variable in the chronological sense. An epiphysis, for instance, which is completely closed in 100% of the population only after the 24th year, can have reached

this stage for the majority of individuals several years earlier. It is incorrect and misleading, therefore, to state categorically, as has been done all too often in the past, that this epiphysis closes, let us say, at 21 years, or even at 20.5–21.5 years. We feel that the present documentation of full variability and the emphasis we have placed thereon is a step forward in identification procedure. (McKern & Stewart 1957, p. 172)

The authors here identify the closing of the epiphysis, otherwise known as epiphyseal union, as key to aging a skeleton. In the sections that follow, we'll take a closer look at this process and other changes that occur during the development of a skeleton and that can give an indication of its age.

EPIPHYSEAL UNION

One of the primary biological processes that is examined in assessing the ages of skeletons belonging to adolescents or to young adults is **epiphyseal union**. This concept is introduced well by Bass and Jefferson in the following passage:

During our youth, the ends of our long bones called the epiphyses—are not yet fused to the shafts; instead they're connected to the shafts by cartilage, allowing the shafts to grow throughout childhood. Beginning in early adolescence, the epiphyses begin to fuse, gradually halting the growth of the bone shafts and bringing an end to that remarkable growth spurt occurring around puberty. (Bass & Jefferson, 2007, pp. 31–2)

A forensic anthropologist can look at the degree of epiphyseal union in a skeleton (particularly that of a teenage or twenty-something person) in order to gauge its age. The bones involved in this kind of examination are those of the arm (radius, ulna, humerus) and leg (tibia, fibula, femur), the innominate (one of the hip bones), the scapula, and the clavicle.

There is typically a difference between the timing of the **proximal** (closer to the trunk of the body) and the **distal** (farther from the trunk of the body) epiphyseal union. With the tibia and fibula, for example, the distal ends fuse about two years ahead of the proximal ends: epiphyseal union of the distal ends of these bones occurs around the age of 15.5–16.5, while union of the proximal ends occurs at 17.5–18.5 years of age. But the distal ends aren't always the first to fuse: the proximal radius and ulna (the bones that are part of the elbow) are about four years ahead of their distal equivalents (14.5–15.5 years of age, as opposed to 18–19). The humerus undergoes three different unions: proximal and distal union (which occur at about the same time, in early puberty) plus union of the medial condyle, which occurs about two or more years later.

Table 12.8	Age of epiphyseal union of the distal femur and proximal tibia of 19 Irish males		
Degree of epiphyseal union (score)	Age of youngest	Age of oldest	Total number
0	10	13	7
1	—	—	0
2	15	19	3
3	19	24	7
4	21	24	2

O'Connor, Lewis, Last, & Spence, 2004, 543.

Table 12.9	Age of epiphyseal union of the proximal fibula of 19 Irish males		
Degree of epiphyseal union (score)	Age of youngest	Age of oldest	Total number
0	10	13	7
1	15	17	2
2	19	—	1
3	19	24	4
4	20	24	5

As with most other bone-and-tooth–based means of determining age, forensic anthropologists use a scale of 0–4 to measure the completeness of epiphyseal union (with 0 representing non-union and 4 being complete union). Absolute precision is impossible to achieve; a narrow range is the best you can hope for. This is apparent in tables 12.8 and 12.9, which summarize the results of a study done among 19 Irish males aged 10–24 who had their knees X-rayed. The thing to look at is the range for each stage, which extends from a minimum of three years to a maximum of five years.

PUBIC SYMPHYSIS MORPHOLOGY

Another way of determining the age of a skeleton is by checking the **pubic symphysis**. This is the front (or 'anterior') of the pubic bone. A symphysis is a kind of false joint in that it isn't a true joint but has some characteristics of one. This one is located at the junction of the left and right pubic bones. Over time, the surface of the pubic symphysis changes from rough to smooth, and a physical anthropologist well schooled in the sequence of changes can examine the symphysis and make a good educated guess at how old the person was when she or he died.

This method was first developed by the master anatomist of the early twentieth century, **T. Wingate Todd** (1885–1938), who studied the bones of 306 individuals collected between 1912 and 1920 in Cleveland, Ohio; these were the first of a now extensive collection of specimens called the **Hamman–Todd Collection**. The next researchers to study pubic symphysis morphology and age were the earlier mentioned McKern and Stewart, who studied the male pubic symphyses of 349 of the Korean War dead that they worked with. The system of calculating age was later refined by Alison S. Brooks and Judy Suchey, who studied skeletons whose ages were known from the Los Angeles County Morgue. Their scale, known as the **Suchey–Brooks system**, is the standard used today. One part of its improvement over previous forms of assessment is that it accounted for differences in men and women—Brooks and Suchey demonstrated that it's impossible to generalize well from males (the first ones assessed) to females in this form of analysis. Their system comprises 6 stages, each divided into early and late sections. That means there are 12 sections for each sex. Physical anthropology students are taught the system using plaster casts of 24 different profiles.

There is some criticism of the Suchey–Brooks system. Some critics contend that the generalization does not always work well across ethnicities and races (particularly in the later stages), nor is it always completely accurate when applied to historic populations. That being said, the system is, on the whole, quite reliable, and it's improving: recently there have been attempts to move the

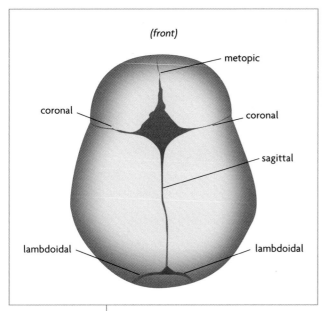

(front)

metopic

coronal

coronal

sagittal

lambdoidal

lambdoidal

Figure 12.3 | Cranial sutures (top view)

Cranial sutures have been extensively studied since the nineteenth century, when it was noted that there is a general order to the closure. For example, anatomists noted that of the so-called **vault sutures**, the **sagittal suture** (which joins the left and right **parietal** bones) closes first, then the **coronal suture** (which joins the **frontal** and the parietal bones), and then the **lambdoidal suture** (which joins the **occipital** with the parietal and the **temporal** bones). They also observed that complete closure generally happens sooner in the endocranial side than in the ectocranial side. When T. Wingate Todd (pioneer in developing methods of aging skeletons) and D.W. Lyon engaged in their comprehensive study published in a series of articles in 1924 and 1925 in the *American Journal of Physical Anthropology*, they arrived at the age sequence shown in Table 12.10. The standard deviation from actual age (when such was known through other means) was ±6 years, for a total range of 12 years. With that kind of broad variance, and with the idiosyncratic timing of closure for some individuals (some sutures never completely close), this means of aging the skull is used more to support findings already made by other means, than to determine on its own, with *CSI*-style accuracy, the actual age of the individual.

Suchey–Brooks system out of the hands of the subjective (albeit well-trained) anthropologist and into a specialized form of software capable of generating a 3-D assessment. Still the sense of what is significant will be down to subjective decision-making. Machines can measure very, very carefully, but they can't tell you what is significant. That is a human choice.

CRANIAL SUTURES

Joining the bones of the cranium are fibrous joints known as **sutures**. The bundled fibres (known as Sharpey's fibres) that make up the sutures allow for some elasticity in the skull, particularly useful during the first year of life, when the brain is still growing substantially. Typically these joints ossify or fuse over time, so that they are virtually invisible (even on a balding head like mine, you cannot see them); this process is often called 'closure'. Human biologists and physical anthropologists traditionally talk about two different types of closure: **endocranial** (occurring on the inside of the skull) and **ectocranial** (on the outside of the skull).

DENTAL AGING

A great way to determine the age of children is to look at their teeth. Developmental changes in the teeth occur along a fairly tight schedule, although, as with other aspects of aging, there is some variance.

First we have our baby teeth, otherwise known as our **deciduous teeth** (as they fall out, like the leaves of deciduous trees fall). These emerge at fairly predictable times, as shown in Table 12.11. The emergence of **permanent teeth** differs slightly for females and males: as Table 12.12 shows, girls usually have their teeth come in somewhat earlier than boys do.

6. Estimating Skeletal Stature

In 1952, physical anthropologist and anatomist **Mildred Trotter** (1899–1991), together with statistician Goldine C. Gleser, developed a formula for using the length of

Table 12.10	Todd and Lyon's estimates of age based on degree of suture closure					
Suture	**Endocranial**			**Ectocranial**		
	open	partial	closed	open	partial	closed
sagittal	‹23	20–37	›34	‹24	23–30	›29
coronal	‹25	23–38	›36	‹26	23–84	›26
lambdoidal	‹26	24–43	›41	‹26	25–84	›30

Adapted from Todd & Lyon, 1924, 1925a, 1925b, 1925c.

Table 12.11	Emergence of deciduous teeth
Tooth	Predicted time of emergence
lower first incisor	6 months
lower second incisor	7 months
upper first incisor	7.5 months
upper second incisor	9 months
lower first molar	12 months
upper first molar	14 months
lower second molar	20 months
upper second molar	24 months

Adapted from Schwartz, 2007, p. 222.

Table 12.12	Emergence of permanent teeth	
	Predicted time of emergence	
Tooth	In girls	In boys
First molars		
upper first molar	7.2 years	7.8 years
lower first molar	7.2 years	7.8 years
Incisors		
lower first incisor	7.3 years	7.3 years
lower second incisor	7.3 years	8.1 years
upper first incisor	7.4 years	8.3 years
upper second incisor	8.1 years	9.1 years
Canines		
lower canine	9.2 years	10.9 years
upper canine	9.4 years	11.0 years
Pre-molars		
upper first pre-molar	9.7 years	11.1 years
lower first pre-molar	9.9 years	11.2 years
upper second pre-molar	10.6 years	11.6 years
lower second pre-molar	10.6 years	11.9 years
Second molars		
upper second molar ('12-year molar')	11.8 years	12.4 years
lower second molar	11.8 years	12.5 years
Third molars		
lower third molar ('wisdom tooth'*)	17.7 years	17.4 years
upper third molar	17.8 years	17.4 years

Remember that in this system of tooth labelling, there are two of every kind of tooth, for a total of 32 in a full mouth.

* Remember: this does not signify that people become wise at 17.

Adapted from Schwartz, 2007, pp. 222–3.

a long bone to estimate the height of the body from which the bone came. From her work comes the **femur length formula**:

$$\text{height} = \text{femur length} \times 2.11 + 70.35$$

Trotter arrived at the formula after studying American fatalities from the Korean War for 14 months in 1948 and 1949. She was first involved in identifying the war dead. She then obtained permission to do scientific research on the identified bodies, and this work, along with research she carried out on specimens from the **Terry Collection** (named after Robert J. Terry, who had built up the collection at Washington University between 1910 and 1941), led her to develop the long-bone formulae that formed the basis for much of the work done in the field today.

It's worth noting the pioneering aspect of Trotter's work with the war dead, who had never been studied in this way before. She paved the way for McKern and Stewart's work with Korean War fatalities. Consistent with her lifelong dedication to science, Trotter arranged to have her body donated for medical research to the Washington University Medical School, where she had worked for over 40 years.

7. Establishing Cause of Death

Every fracture tells a story, and if you study it closely, you can often piece together a surprisingly detailed narrative.

Bass & Jefferson, 2007, p. 232

As every fan of murder mysteries knows, the cause of death often shows up in the bones. Nicks in the ribs can tell you that the person was stabbed in the heart or other internal organs. There are some forensic anthropological specialists who can tell you what kind and size of knife was used, and even how sharp it was when it was used as a murder weapon. Bullets leave a smear called a **lead wipe** as they enter, hit, or merely graze a bone; these marks show up as bright white streaks on X-ray film. Bullets that hit the skull leave different marks on entry and exit (bevelled inward for the former, bevelled out for the latter). One bone that often figures prominently in a forensic post-mortem is the **hyoid**, which, because it is situated deep in the throat, is often examined for signs of death by strangulation. In the

Skeletal Evidence for Child Abuse

The suspicious death of a young child or infant often sparks accusations of child abuse, and this inevitably becomes a matter for the forensic anthropologist to determine. So, when you have the skeleton of a small child, how do you know that it was abused over time before it died? There are several telltale signs, as Walker, Cook, and Lambert explain in their provocative paper 'Skeletal Evidence for Child Abuse: A Physical Anthropological Perspective' (1997). Unfortunately, many of the signs they discuss cannot be detected when the child is alive and prevention or intervention can happen.

There are particular kinds of skeletal injuries that speak to a child's having been abused over an extended period of time. Walker, et al., describe one type as follows:

> Rib fractures without major chest trauma . . . are a strong indication of abuse. . . . Child abuse tends to result in bilateral [on both sides of the body], multiple rib fractures close to the spine. . . . In infants, this kind of fracture typically results when an adult assailant grasps the victim's chest with both hands and shakes or squeezes it. Compression of the chest in this way fractures ribs near their vertebral ends. . . . (Walker, et al., 1997, p. 203)

The **metaphysis** is the part of a growing long bone where it begins to widen before it reaches the epiphyseal plate. It grows during childhood until the age of 18–25, when it ossifies or turns to bone. Metaphyseal injuries are common in child abuse. According to Walker, et al., such injuries often are multiple and:

> often occur when an infant is held by the trunk or extremities and violently shaken. High-risk metaphyses for this kind of injury include the proximal [near the shoulder] humerus, knee, and distal [near the foot] tibia. . . . (Walker, et al., 1997, p. 203)

Another sign of child abuse is the presence on the skeleton of a significant number of **Harris lines**. Harris lines, like growth rings on a tree and hypoplasias on teeth, are produced when growth is disrupted, or stops and then resumes. Injury is a major cause of Harris lines.

Frequent injury is also indicated by changes in the **periosteum**, the dense connective tissue that forms

---

Body Farm novel *Carved in Bone*, by Jefferson Bass (i.e. Jon Jefferson and William Bass), anthropologist Dr Bill Brockton gives the following account of the hyoid following his examination of a young female murder victim:

> The hyoid is an arch measuring an inch to an inch-and-a-half high, and about the same in width. Under the magnifying glass, it looked five times that size. Once upon a time this hyoid had supported the dead woman's tongue and the other muscles she used to talk. Now I hoped the bone itself could tell us how she died.
>
> Attached to the central arch, or 'body', of the hyoid are two thinner arches, called the 'horns'. Normally, the arch's height is roughly the same as its span, or the distance between the tips of the horns. In this case, though, the horns were much closer together. It was easy to see why: where the horns joined the central body, the cartilage looked ripped from the bone, and the body itself was cracked at the midline. I had seen dozens of damaged hyoids in the time, but none so mangled as the one I held now. This young woman had been strangled with crushing force. The story of her death was written in bone. (Bass, 2006, p. 47)

The passage may come from a work of fiction, but Bass's science, as we have seen, is sound and authoritative.

Forensic Anthropology in Action: Two Canadian Cases

The Case of the Swansea Skull

In August 2004, near the Swansea Town Hall, south of Bloor in west-end Toronto (closer, really, to Etobicoke), a skull was found some 20 centimetres below the ground. It was discovered beside a drainpipe that had been installed when the town hall was built in 1958.

One of the local papers, the *Toronto Star*, looked into missing persons reports dating back to the 1950s in an attempt to identify the remains. It was a sensational story

a membrane at the outer surface of bones. (The periosteum that covers the cranial bones is known as the **pericranium**.) When a child is injured, a new bone formation develops under the periosteum; this is called a subperiosteal formation. When forensic anthropologists detect several areas of subperiosteal new bone formation, with some in different stages of healing, this is a classic sign of child abuse. According to Walker, et al.:

'Much of the evidence of chronic abuse in our cases comes from the identification of localized, asymmetrical distributed subperiosteal lesions in different stages of healing. These are thin layers of new bone that form beneath the periosteum in response to trauma and subperiosteal bleeding.

The subperiosteal lesions we have encountered fall into two groups. Some of them are the result of the stripping of the periosteum from the bone that occurs when an arm or leg is used as a 'handle' to punish the child violently. . . . The forces applied by the adult hand in such circumstances are predominantly traction and torsion.

They are likely to result in epiphyseal separations and periosteal shearing, especially in the metaphyseal area. . . . Infants are especially susceptible to this kind of injury because their periosteum is loosely attached to the cortex. . . .

A different type of subperiosteal lesion results when the child is hit with a hand or hand-held object, or when the child is lifted and beaten against an object. Lesions produced in this way are likely to be in areas with little overlying soft tissue, such as the cranial vault, where the bone is especially vulnerable to this kind of trauma. . . . (Walker, et al., 1997, p. 204)

The details are upsetting, almost too painful to read. Performing forensic work on the bodies of children and infants requires an uncommon measure of equanimity, a requirement that makes this one of the most challenging areas of specialization within forensic anthropology. Given, though, what's at stake—the health and safety of children and the opportunity to charge and convict those who would threaten that safety—it is undeniably one of the most important.

that produced a bevy of rumours about who the victim (if it *was* a victim) was.

When consulting archaeologist Ron Williamson, of Archaeological Services Inc. (that's ASI, not CSI), arrived on the scene, his task was to determine (a) whether there were other bones belonging to the skeleton, and (b) whether there was any evidence that this was the site of an ancient mass burial ground—a distinct possibility in territories once occupied by Iroquoian peoples such as the Huron, Petun, and Neutral who celebrated the Feast of the Dead (see page 288). Williamson found that the skull was isolated.

The next scientist-specialist to enter the scene was **Kathy Gruspier**, who teaches law and forensic science at the University of Toronto. Since 1992 she has been the consulting forensic anthropologist to the chief coroner of Ontario. She determined, through careful skull measurement (which involved feeding over 20 measurements into the ForDisc program), that the cranium was relatively robust, suggesting that it belonged to a male. She also noted that the jaw bore shovel-shaped incisors, a trait shared, as we know, by Aboriginal people and those of East Asian genetic heritage. Given the apparent age of

the skull (i.e. that it looked as though it had been in the ground for a long time), the informed guess was that this was the skull of an Aboriginal person who lived sometime after AD 750. The age at death was determined by the fact that the maxilla had recently erupted **wisdom teeth**, suggesting that the person had been in his twenties when he died. The teeth also gave other clues: they were cavity-free, suggesting that the person died before coming into contact with sugar-bearing European foods, and they were worn down, consistent with the person's having lived on a diet of non-refined flour. Given the context, that would be cornmeal, since corn had been in the area for over a thousand years.

Finally, the cause of death was determined. There was a golfball-sized hole in the top of the skull, but there was no sign of an entry wound (no, this does not mean death by golf). The cause of death was probably an infectious disease of the brain, probably tubercular meningitis, which can build up sufficient pressure in the brain to blow a hole in the skull.

Thus was the case of the Swansea skull solved—or nearly solved. While it was possible, through forensic anthropology, to compile a fairly detailed profile of the bearer of the skull, it was impossible to say how the skull arrived at the site of Swansea Town Hall, or why it was placed—deliberately, it would seem—at the base of the drainpipe.

The Robert Pickton Case

Probably the most dramatic use of forensic anthropology in Canada occurred in compiling a case against then-accused serial killer Robert Pickton, who was arrested for murder in 2002. Pickton, a pig farmer living in Port Coquitlam, British Columbia, was convicted in 2007 of the second-degree murders of six women, and has been charged in the deaths of 20 more. Most of the victims were prostitutes or drug users, typically Aboriginal women, the most vulnerable people living on the streets of Vancouver's Downtown Eastside. The early forensic anthropology work in this case involved the use of heavy equipment for excavation and a 15 metre–long conveyor belt and soil sifters to take the bones from the soil. The work was difficult in a number of ways. Human bones had to be identified as separate from pig bones. Multiple individuals had to be identified, some of whom were known to have disappeared, some of whom had disappeared long before their death. And then, there was the powerful smell, as can be readily imagined. The emotional difficulties were also intense; psychological counsellors had to be kept onsite.

By the time the work was done, over 100 forensic anthropology students (archaeologists and physical anthropologists) had been employed in the project. Across Canadian universities there are professors, young and old, who have the Pickton Farm case on their resumés and can recount certain gruesome aspects of the case to their students (there is still a publication ban on some of the information). Forensic anthropologists also appeared in the case to testify as expert witnesses. There is always something conflicting about being an expert witness: the law demands certainty, but good science demands its own measure of tentativeness (requiring ample use of expressions such as 'is consistent with', 'to the best of my knowledge', and 'suggests'). Further, expert witnesses appear on both sides of the case, occasionally pitting colleague against colleague. When the series *Bones* still

A forensic investigator rakes through a pile of sifted dirt for evidence at a Port Coquitlam pig farm during the investigation of then-accused serial killer Robert Pickton

CP Photo/Richard Lam

focused more on science and plot than on cutie-pie story-lines, there was an interesting story in which Temperance Brennan had to be the expert witness for the prosecution in a case in which a good friend was her counterpart for the defence. In a similar way, there were forensic anthropologists on both side of the Pickton case, prosecution and defence. To a certain extent I can see why an anthropologist would stand as expert witness for the defence in such an apparently one-sided case; after all, without such an expert, the case cannot proceed, justice cannot be served. But I just can't see testifying on the side of a serial killer.

Unidentified Persons: A Sample from the RCMP Cold Case Files

Location: Near Rosthern, Saskatchewan

Details: On 26 May 1982, the body of a badly decomposed male was discovered floating in the South Saskatchewan River. The male was wearing Howick brand denim jeans, knitted polyester fabric trousers, Size 9½ Texas brand beige cowboy boots, size 34 grey belt, and a size 34 Caswell brand black money belt. The male also wore a complete set of dentures, and his appendix had been removed.

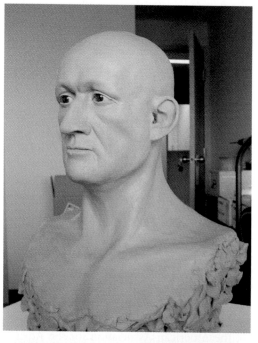 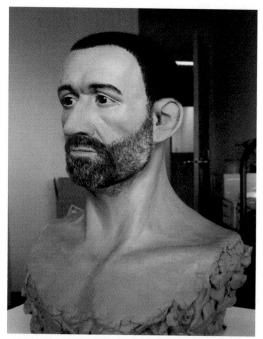

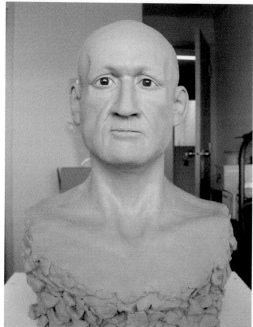 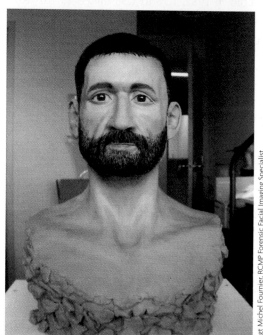

The RCMP's facial reconstruction of John Doe. The four images show the clay model with hair and without hair, from two different angles.

The *CSI* Effect

It is called the **CSI** (*Crime Scene Investigation*) **effect**, and it refers to the impact, primarily on jurors, of forensic science as it is presented on the popular franchise of shows that begin with those three initials: *CSI*, *CSI: Miami*, and *CSI: NY*.

Some early studies suggest that the *CSI* effect has led jurors in high-profile court cases to be skeptical of evidence that does not measure up to the unrealistically high standards of fictional forensic science. The alleged effect (see: I am writing now like a lawyer, not a scientist) causes viewers of the show to develop unreasonable expectations concerning the degree of accuracy and reliability of evidence provided by forensic scientists, including forensic anthropologists. These expectations extend to the experts themselves: TV viewers believe that a specialist in one area of forensic science must, to be considered credible, be an expert in all areas of forensics (I am tempted to refer to this as the 'Bones effect'). DNA tests must be performed quickly and on demand, regardless of the cost, as they are shown on television, which tends not to develop story lines around overworked lab staff and the backlog of cases.

It is not just jurors and the general public who may be susceptible to the *CSI* effect. College and university students may be moved by it to choose a career path that they hope will lead them to the front lines of the fast-paced, never-dull world of forensic science. In a way, I don't want to complain about this too much, as it has doubtless increased enrolment in postsecondary courses and programs related to forensic anthropology. I've had to accept that the *CSI* effect shares credit for the popularity of my physical anthropology course with the comedic genius and academic brilliance of its instructor. But students moved by what they see on television should keep in mind that while forensic anthropology may be—at times—exciting, it can also be painstakingly slow and methodical work, not to mention gut-wrenchingly painful when it deals with the heartbreaking cases of children whose untimely deaths capped brief lifetimes of physical abuse. Forensic anthropology may be many things, but if anything, it is far more complex and multifaceted a science than the impression TV accounts could ever give.

In July 2004, the body of John Doe was exhumed and re-examined by a forensic anthropologist. The forensic anthropologist determined that the male was Caucasian, 40 to 50 years old in 1982, 1.78 to 1.38 meters (5′ 10″ to 6′ 0″) tall, and with a stocky build.

The skull of John Doe was sent to an RCMP forensic artist to have a facial reconstruction performed. The reconstruction was completed in June 2005.

At the time of writing (October 2010), the file was still open. This is not unusual.

Human Rights Work: Witnesses to World Oppression

The *Equipo Argentino de Antropología*, or Argentine Forensic Anthropology Team: From Disappeared to Witness for the Prosecution

From 1976 to 1983, the military regime, or junta, that ruled Argentina committed crimes ranging from abduction and torture to rape and murder against nearly 20,000 of the country's citizens. The victims just disappeared, and were hence called '*los desaparecidos*'—literally, 'the disappeared'.

In 1984, when the junta was finally toppled, the civilian government that replaced it invited forensic scientists from the American Association for the Advancement of Science to help them excavate believed sites of mass graves, return the remains of the dead to their relatives, and prosecute the criminals in charge. It should be pointed out that it was not that Argentina did not have qualified scientific professionals, but that many were implicated in the crimes as having colluded with the military regime.

Clyde Snow rode to the rescue. He recruited a team of Argentinean graduate students to help with the investigation. Though they were not specifically trained in forensic anthropology (the regime had not been keen on having too many scientists knowing how to investigate murders), they were committed to the hard work the situation required.

The forensic evidence collected through archaeological and lab identification methods helped to convict six of the nine junta leaders in 1985. The legal strategy was to take a few well-researched cases that would stand for all of the victims. The best-known case was that of young Liliana Pereyra, a woman who disappeared in October 1976, and was kidnapped, tortured, raped, and murdered. The forensic team was able to identify her body and compile evidence of the crimes that had been committed against her, evidence that was crucial in the 1985 convictions. Among other things, the team detected pitting marks on the preauricular sulcus (a groove on the sacroiliac, a joint at the back of the pelvis), which demonstrated that shortly before she was killed, Liliana had given birth to a baby, who was never found.

Snow's team of grad students became the founding members of *El Equipo Argentino de Antropología Forense* (EAAF), the Argentine Forensic Anthropology Team, a non-governmental organization dedicated to forensic investigation. They have since engaged in work in Bolivia, El Salvador, and Guatemala, and outside of Latin America in Ethiopia and Iraqi Kurdistan. In addition, they have trained many doing work elsewhere. According to their website, they have 13 part-time and full-time staff, specializing in such areas as archaeology, physical and social anthropology, information technology, and law (see http://eaaf.typepad.com/founding_of_eaaf/).

Clea Koff: The Bone Woman

Clea Koff (b. 1972) was 23 when she first witnessed the effects of genocide. It was 1996, and she was in Rwanda. Two years before, there had been a mass killing of over 800,000 people, most of them of the Tutsi tribe. Western media tend to buy the old excuse of 'intertribal violence', but the conflict in Rwanda was the direct result of colonialism. The Belgians had been running Rwanda and neighbouring Burundi as a colony from just after World War I (when the Germans had been forced to give it up), and their administration had taught tribal hatred, playing one side off against the other. Employing a system known in the political science literature as dual colonialism, the Belgian overlords had set the Tutsi up as rulers over the majority Hutu, racializing the small differences between them so that the former (as taller, whiter, and 'more civilized') were given educational advantages, privileges in political position, and praise, while the latter went from independence to being doubly ruled and exploited. When the Belgians left, their legacy was a series of mass killings that led to the slaughter of 1994.

Koff, as we've read, was radicalized from a young age by her parents, her Tanzanian mother, Msindo Mwinyipembe, and her white American father, David Koff, both of them documentary filmmakers. Shortly after receiving degrees in anthropology and forensic anthropology, Koff took a job with the United Nations, where she worked on two international criminal tribunals.

The first site that Koff visited in Rwanda was the *préfecture* (or county) of Kibuye, in the western part of the country, where she worked for two months beginning in January 1996. The *préfet*—the governor—of Kibuye was rumoured to have organized troops and police to direct victims-to-be to assemble in churches and the region's major stadium,

where the innocent were eventually executed. In Kibuye, almost 250,000 people died or 'disappeared'. Koff's first work involved retrieving and analyzing unburied skeletons from the hillsides surrounding a Catholic church. The work generated a mixture of strong negative emotions (sorrow and disgust) and exhilaration in being able to 'make a difference':

> As we began clearing vegetation wherever we found bones on the hillside, it became clear that these were the remains of people who had tried to escape from the killings in the church and were murdered as they either hid in the banana groves or brought water at night to wounded survivors in the church. Remarkably, they still lay where they had fallen, all over the hillside, all exhibiting sharp force trauma to their heads. As I worked with Roxana on one skeleton whose postcranial elements put its age at under twenty-five years, I felt sad in the way that one does at the loss of a young life, but my sorrow was tempered by a thrilling

Clea Koff

Possessions collected from victims of the Rwandan genocide

In the course of their investigation, Koff and her team were plagued by fundamental questions, arising largely because the work was new both to the team and to forensic sciences generally:

> In the end, we asked ourselves how much forensic evidence is enough. Is it necessary to dig every grave in Rwanda to prove the genocide happened? We didn't know. It was up to the Tribunal judges, and *they* didn't know because this was the first time such forensic evidence would be used in an international court. (Koff, 2005, p. 75)

In the end, the evidence they compiled was enough to convict Clement Kayishema, the *préfet* of Kibuye and a medical doctor, of four counts of genocide; he was sentenced to prison for the rest of his life. Also convicted was Obed Ruzindana, a Kibuye businessman, who was found guilty of one count of genocide and was sentenced to 25 years' imprisonment.

© Physicians for Human Rights

> awareness that I was really there, doing what I had longed to do, and, dammit, we had a chance to identify these people and help make meaning out of all this death. (Koff, 2005, pp. 34–5)

The 'sharp force trauma' came from the mode of execution. Koff and her co-investigators found other signs of the state-sanctioned killers' horrific practices, including stabbing gouges to the back of the distal tibia, which, the investigators theorized, were designed to sever the Achilles tendon of fleeing victims so that they could not run away but could be executed at the leisure of their pursuers.

Koff's next assignment in Rwanda involved excavating the bodies that had been buried in a mass grave. With the rest of her team members, she would separate the bodies (in various states of decomposition, depending on how deep they were buried). This work sometimes required her to remove flesh with a high-pressure water hose. Her unit within the team autopsied about 30 bodies a day.

READ BEFORE YOU JUDGE

It is easy, after reading most books about Rwanda (including the work of Clea Koff, although it is certainly not her intent), to judge the Hutu negatively. I would be remiss as an anthropologist if I did not discourage a hasty judgement. Before you judge, read any one (and preferably more) of the following books, some of which predicted the massacre long before it happened:

- Liisa Maalki, *Purity and Exile: Violence, Memory, and National Cosmology among the Hutu Refugees in Tanzania* (1995; Chicago, University of Chicago Press)

- Catharine Newbury, *The Cohesion of Oppression: Clientship and Ethnicity in Rwanda, 1860–1960* (1988; New York, Columbia University Press)

- René Lemarchand, *Burundi: Ethnocide as Discourse and Practice* (1995; Cambridge, Cambridge University Press)

- Gérard Prunier, *The Rwanda Crisis: History of a Genocide* (1996; New York, Columbia University Press)

- Paul Rusesabagina, *An Ordinary Man* (2007; New York, Penguin)

- Marie-Béatrice Umutesi, *Surviving the Slaughter: The Ordeal of a Rwandan Refugee in Zaire* (2004; Madison: University of Wisconsin Press).

The last two are especially insightful as they are written by Rwandans. The first was made into the movie *Hotel Rwanda*; the last was written by a Hutu refugee and sociologist to document her experiences after the genocide.

AFTER KIBUYE

Koff's later work took her to Kigali, Rwanda (June 1996), and then to Bosnia (July–August 1996) and Croatia (September 1996), where she worked on the UN's Criminal Tribunal for the Former Yugoslavia. After completing her master's degree in 1999, she returned to the former Yugoslavia, performing a UN mission in Kosovo (April–July 2000).

What is she doing now? Koff is the founder of the Missing Persons Identification Center (MPID), a non-profit organization that tries to link families of missing people with the coroner's offices in a number of American jurisdictions, which together have thousands of unidentified bodies in their files. In this work, she continues to try to bring closure to those who are missing loved ones.

I want to add an extra personal note here. In the fall of 2008, I was teaching my course in physical anthropology, and, as usual, I included Koff's *The Bone Woman* in the list of required readings. One of my students, in a paper discussing the book, noted the impact that it had on her life. She was a student in the Fashion Arts program, a business program geared to those interested in working in the fashion industry. After reading about Koff's life, she decided that she needed to do something that—in her words—might have a more meaningful impact on the world:

> As soon as I started reading *The Bone Woman* I felt some sort of connection with the work that Clea Koff was doing. I felt like she had inspired me to go after what I really wanted in life, no matter how challenging the path is to get there. How a college fashion student connects with this dramatic turn in career direction, I'm not sure; I just know that since I read this book my life is different. (2008)

I don't know what this student is doing now, but I suspect that whatever career path she has chosen to follow, she will be one of those people whose work gives more than it takes from the world.

At a December 2010 news conference, MPID founder Clea Koff points to a memorial site in California where the remains of a young woman—the daughter of the woman in the photo—were found. Koff, on behalf of the family, has asked the Coroner's Office to exhume the body for further testing to determine a definitive cause of death.

Mel Mecon/Los Angeles Times

Typical Student Questions

▶ *How long does it take for a body to become a skeleton?*

There is no fixed time for a body to skeletonize. It depends on the conditions. You know from an earlier chapter that 'accidental' mummies do not become completely skeletonized because they exist in conditions of low oxygen where the bacteria and other small critters that help in the process of decomposition cannot function. Warm weather speeds up the process; cold weather slows it down. Exposure to the elements quickens the rate of decomposition; being covered by earth can reduce it.

▶ *Can physical anthropologists determine to the year how old a skeleton is?*

No, not unless the person's birth and death dates are tattooed on their bones (although this could become a fad, the next popular area of body mutilation). In all seriousness, there is always a range—narrower for younger people, broader for older people—that a forensic anthropologist presents.

Review Questions

1. How does the work of forensic anthropologists and forensic pathologists differ?
2. For each of the following, name and discuss at least three features of a skeleton that can help a forensic anthropologist determine:
   a) the age of the individual at the time of death
   b) the sex of the individual
   c) the race or ethnicity of the individual
   d) the stature of the individual
   e) the cause of death.
3. Why is the determination of 'race' controversial in the work of forensic anthropologists? How can the American origins of many of the statistics and statistical formulae used by forensic anthropologists be problematic?
4. What is the international role of forensic anthropologists in human rights cases?

Discussion Questions

1. What is the *CSI* effect? Do you think that prior to reading this chapter you were afflicted with this effect?
2. Discuss the challenges a forensic anthropologist might face in acting as an 'expert witness' in a courtroom. Why might a forensic anthropologist be hesitant to appear as an expert witness?
3. If you were to design an experiment to be undertaken at the Body Farm, what would it be? What would your hypothesis be?

Key Terms

condyle
consilience
coronal suture
CSI effect
deciduous teeth
distal
ectocranial
*El Equipo Argentino
de Antropología
Forense* (EAAF)
endocranial

epiphyseal union
Feast of the Dead
femur length formula
ForDisc
forensic anthropology
Forensic Anthropology
Data Bank
forensic entomology
forensic pathology
frontal
greater sciatic notch

Hamman-Todd Collection
Harris lines
hyoid
lambdoidal suture
lead wipe
metaphysis
occipital
orthognathic
parietal
pericranium
periosteum

permanent teeth
prognathic
proximal
pubic symphysis
sagittal suture
Suchey–Brooks system
sutures
temporal
Terry Collection
vault sutures
wisdom tooth

Key Individuals

William ('Bill') Bass
George A. Dorsey
Thomas Dwight

Aaron Elkins
Clea Koff
Kathy Gruspier

Moira McLaughlin
Kathy Reichs
Clyde Snow

T. Wingate Todd
Mildred Trotter

Recommended Print and Online Resources

American Academy of Forensic Sciences. (Website). American Academy of Forensic Sciences: A professional society dedicated to the application of science and law (Home page). Available at www.aafs.org/

Argentine Forensic Anthropology Team. (Website). Argentine Forensic Anthropology Team/*Equipo Argentino de Antropologia Forense* (EAAF). Available at www.eaaf.org/

Bass, Bill, & Jefferson, Jon. (2003). *Death's acre: Inside the legendary forensic lab, the Body Farm, where the dead do tell tales.* New York: G.P. Putnam's Sons.

———. (2007). *Beyond the Body Farm: A legendary bone detective explores murders, mysteries, and the revolution in forensic science.* New York: HarperCollins.

Berry, Shawn. (2003, November 19). Investigators hoping to put a name to body found in ocean 23 years ago. *New Brunswick Telegraph-Journal*, p. A3.

Canadian Society of Forensic Science. (Website). Canadian Society of Forensic Science/*La Société Canadienne des Sciences Judiciaires* (Home page). Available at: www.csfs.ca/

Creighton, Ellen. (2005, December 3). Anthropologists see life beyond bones. *The Times & Transcript* (Moncton), pp. D1, D3.

Ferllini, Roxana. (2002). *Silent witness: How forensic anthropology is used to solve the world's toughest crimes.* Toronto: Firefly.

Joyce, Christopher, & Stover, Eric. (1991). *Witnesses from the grave: The stories bones tell.* New York: Ballantine.

Koff, Clea. (2005). *The Bone Woman: A forensic anthropologist's search for truth in Rwanda, Bosnia, Croatia, and Kosovo.* Toronto: Random House.

McKern, T.W., & Stewart, T.D. (1957). *Skeletal age changes in young American males: Analysed from the standpoint of age identification.* Natick, MA: Quartermaster Research & Development Center, US Army, Environmental Protection Research Division.

Bass, Jefferson. (2006). *Carved in bone: A Body Farm novel.* New York: William Morrow.

Steadman, Dawnie Wolfe. (2003). *Hard evidence: Case studies in forensic anthropology.* Upper Saddle River, NJ: Prentice-Hall.

Walker, P.L., Cook, D.C., & Lambert, P.M. (1997, March). Skeletal evidence for child abuse: A physical anthropological perspective. *Journal of Forensic Sciences, 42*(2), 196–207. doi: 10.1520/JFS14098J

13

Conclusions

CHAPTER OUTLINE

The 'invisible ape': a gibbon departs
© Johner Images/Alamy

- Outline the arguments for and against the existence of species as a scientifically viable concept.

- Describe the major changes that have taken place in physical anthropology over the last 20 years.

Introduction: A Changing Field

Physical anthropology is changing, with several large and recent developments worth noting. Spectacularly new finds, many of them complete, have been made in many parts of the world and covering a broad spectrum of time. The splitters in the discipline have added 12 species names over the last 20 years—slightly more than a species every two years (see Table 13.1). And physical anthropologists today are more than just entertaining the idea that our species was recently not alone as a hominid, thanks to the revelation of the Hobbit-like *Homo floresiensis*. We have known about it for several years now, and no one yet has mounted a serious claim that it is nothing more than a hoax; indeed, the voices suggesting it is a one-off freak are becoming quieter and quieter.

New Thoughts on the Species Concept

The **biological species concept** (BSC), or species reproductive model, theorized by **Ernst Mayr** (1904–2005), has taken a battering in recent years, as molecular biology has put a new priority on strings of DNA and individual traits, challenging the notion that a species comprises those members that can breed together and produce fertile young. In writing this textbook I've had the opportunity to question seriously this concept. My thoughts? I cannot be sure that *species* is much more than a conceptual necessity with a weak definition. The BSC may be a useful guideline, but it cannot be applied too rigidly. After all, once you have members of supposedly separate species interbreeding, how much should the fertility or infertility of their descendants really count in delimiting the species? Is it really accurate, or useful, to say that wolves and dogs belong to the same species, just to keep an old definition of *species* intact? Again, to recall an example from close to home (well, my home), two of my parrots, Stanee and Louie, are a mated pair, although they belong to different species. My money says that they will have viable young someday. Does this mean that for conures (the non-scientific term for certain small parrots and large parakeets), the species divisions are wrong? Or is it that the concept of species altogether is wrong?

As I have been saying over the last few months to anyone who will listen (and to a few who won't or who really don't want to), the species concept is an artificial framework

Table 13.1	Hominid species named in the last 20 years	
Species name	Year of naming	Place of discovery
Australopithecus sediba	2010	Malapa Fossil Site, South Africa
Homo floresiensis	2004	Flores, Indonesia
Ardipithecus kadabba	2004	Afar region, Ethiopia
Homo cepranensis	2003	Ceprano, Italy
Homo georgicus	2002	Dmanisi, Republic of Georgia
Sahelanthropus tchadensis	2002	Djurab desert, Chad
Kenyanthropus platyops	2001	Lake Turkana, Kenya
Orrorin tugenensis	2001	Tugen Hills, Kenya
Australopithecus garhi	1997	Afar region, Ethiopia
Homo antecessor	1997	Gran Dolina, Spain
Australopithecus anamensis	1995	Lake Turkana, Kenya
Australopithecus bahrelghazali	1995	Bahr el Ghazal valley, Chad

©Joseph White./iStockphoto

The species concept: not all animals—parrots included—can be easily pigeon-holed

imposed on nature; it is not, itself, natural. That might be a bit of an overstatement, but it is true that science, with ample data at hand, has been unable to come up with a sound set of guidelines for defining what makes a species. We have a long heritage of binominal classification, of sorting living things into categories according to genus and species. The tradition dates at least as far back as the eighteenth-century biologist **Carolus Linnaeus** (1707–1778). As we are rethinking the criteria for defining the upper levels of taxonomy, we should also do so for the lower ones.

Of course, something does not need to be defined well in order for it to exist, or even be understood. I have been known to say in class that a definition is something that instructors sometimes equate with knowledge, and yet knowing how to define something (on an exam, say, especially with multiple-guess or short-answer questions) is not the same as understanding it. On the other hand, it's impossible to deny the usefulness of separating living things, past and present, into different groups according to their shared characteristics, and scientists need to strive for consensus on naming in order to share and expand their knowledge of things. But all that epistemological stuff being said, without clear definition, species as a concept is kind of up in the air right now.

How does this relate to physical anthropology? Be prepared for a lumper manifesto. If we keep the species

concept—and I think we probably should—then we have to make the scientific court in which it is declared a very rigourous one. Evolutionary biologist **Clive Finlayson** and other critical anthropologists have challenged current species designations and the process of designation. (Mind you, Finlayson still accepts *Homo heidelbergensis*, and there is no solid definition for that possible species.) When it comes to the transition from *Homo erectus* to *Homo sapiens*, Tattersall and Schwartz (1999) argue, with good reason, that we lack sufficiently precise definitions of *H. sapiens* characteristics. They contrast the trait 'robust brow ridge', which is highly variable and cannot be neatly set in an either/or distinction with, say, 'gracile brow ridge', with **bipartite brow ridge**, which a skull either has or does not have. But here, perhaps, we can take a page out of the forensic anthropologist's book and develop a measurable scale (e.g. 0–4) with agreed-upon criteria for each score so that we can assess a specimen, and compare it with others, with greater accuracy. Perhaps it would become part of a species version of the **ForDisc** program. Then we might arrive at a score that says 'You have found a *Homo sapiens*.'

As well as having measurable criteria, I think we have to establish a principle that nothing should be officially named until there are enough fossil specimens available to speak intelligently about it. The history of physical anthropology is littered with hasty judgements based on too little evidence. Basic rule number one should be (and my students get tired of hearing this in my class) *One fossil specimen does not establish a species*. Perhaps we should call this the **Kenyanthropus platyops** rule: no more single-specimen species. In addition, it is not a sign of scientific weakness to say that we have transitionals, or to say that this is a *Homo erectus* that is approaching *Homo sapiens* and is closer to *Homo sapiens* than a certain other example. If you want to declare that a specimen represents a distinct transitional species—if you want to say there is a **Homo heidelbergensis**—then you must present a comprehensive set of criteria that define it, as well as a solid argument for why we need, taxonomically speaking, to recognize it. In the case of *H. heidelbergensis*, I don't think the evidence supports recognizing it as a unique species; however, as a reasonable scientist I am open to hearing arguments.

New Thoughts on Genetic Dating: The Penguins Speak

Although they have triumphed in making the rest of us look more at traits and less at species, the molecular biologists have not had everything go their way. Their dating methods—the trick of their trade and a skill they once displayed proudly in the face of the stones-and-bones anthropologists—have come under serious questioning.

The stones-and-bones people—and I count myself in their camp—take an unseemly measure of delight in watching the smug scientific certainty of the lab-coated genetics set rocked to the core. We do need one another. However, a recent article in *Trends in Genetics* warmed the hearts of at least a few physical anthropologists when it raised doubts about the accuracy of molecular biology's cherished genetic dating methods (see Subramanian, et al., 2009). A team led by **Dee Denver**, an evolutionary biologist working with the Center for Genome Research and Biocomputing at Oregon State University, reported that they were able to obtain frozen DNA samples of penguins that lived in Antarctica from 250 to 44,000 years ago. The consistently cold temperatures found there preserved the DNA in the penguin bones much better than in warmer areas, where DNA deteriorates much more quickly. With this data, the team were able to seriously challenge the 'traditional' (i.e. 20- to 30-year-old) methods of genetic aging and evolutionary rates based on DNA analysis. According to the researchers, the traditional methods were, with some regularity, underestimating the ages of specimens by a measure of 200 to 600 per cent—an unacceptable error level. This means that if a specimen had been determined, using the old methods, to be some 100,000 years old, it might actually be from 200,000 to 600,000 years old!

Now, before the gloating begins on the part of stones-and-bones scientists (and, unfortunately, the science-skeptical Creationists), it must be remembered that this is the way that science advances: it self-corrects. That is its very nature. Raising doubts about the earliest methods used is not to criticize the principle of genetic dating (although the researchers suggested that different types of DNA sequences change at different rates, so this too must be allowed for). These researchers learned from their predecessors about genetic dating and its great potential, and, admirably, they've striven to perfect the science. Unfortunately, when a generation of scientists stand on the shoulders of their pioneering predecessors, they kind of kick the older ones in the face sometimes.

New Thoughts on Race: The Place of Race in Physical Anthropology Today

And what do we do with **race**? **William Bass**, author of fascinating accounts of forensic anthropology, non-fiction and fiction, would simply say, keep it. For as a forensic tool used to deal with the limited set of subspecies characteristics found in the United States (and in Canada, one could argue), he is right: it makes sense. He is not a racist; he is just seeing patterns, and is able to use those patterns to make predictions—the true quality of a scientist. In fact, he is not unlike those social scientists engaged in anti-racism research, who observe and comment on the patterning of systemic or institutional racism. They, too, are just commenting on patterns. Human interbreeding and our increased *globality* (that is a word, no matter what my

spell check says) create such a mix that any classification based on simple racial categories will become decreasingly useful. I would like to think that with an increased emphasis on genetic traits and concrete characteristics, not abstract and sometimes arbitrary constructs like genus and species, the somewhat arbitrary concept 'race' will disappear. Our role, as physical anthropologists, can be to continue to stress the ultimate genetic sameness of humans, the independence and social neutrality of characteristic difference.

New Thoughts on Physical Anthropologists

Two related trends are affecting physical anthropologists themselves, as well as the work they do. Both have to do with the discipline's colonialist past, a history shared with all branches of anthropology.

In physical anthropology's deep, dark, distant past, as we have seen, white men collected trophies of science—primarily skulls—from non-white people in order to tell the story of these people (and to publish, and to get jobs in the academy). In this way, anthropology both mirrored and perpetuated the colonial practices of the nations these white European explorers came from.

There have been two dramatic changes that have altered this practice in a very positive way. As discussed in Chapter 11, new ethical principles—demanded by the colonized people, supported by physical anthropologists everywhere, and enforced by government legislation—have become an inescapable consideration in the work done by physical anthropologists when the dead they study lived within the last 15,000 years or so. Perhaps more important, the face of physical anthropology itself is changing. We no longer read about the largely anonymous local assistants who find the treasured specimens but whose names are missing from the journal articles, news items, and popular books in which the finds are documented. African names (most of them Ethiopian) such as **Awoke Amzaye**, **Berhane Asfaw**, and **Yohannes Haile-Selassie** now appear in the scientific record, typically with 'Dr' before them or 'PhD' after them. They have achieved a prominent place in the discipline. We remember that Toronto-born **Davidson Black** was an early leader in promoting non-white scientists, through the respect and support he gave his Chinese colleagues, notably **Zhongjian Yang** (1897–1979) and **Wenzhong Pei** (1904–1982).

New Thoughts on Bipedalism

Sometimes you come across two or more bits of information that, put together, suggest a new idea. Call it a type of consilience, to use a term introduced earlier in the text. Here is an example of a new (to me) idea that presented itself over the last few weeks. I have been reading Clive Finlayson's *The Humans Who Went Extinct: Why Neanderthals Died Out and We Survived* (2009). Finlayson is very good at suggesting alternatives to traditional anthropological interpretations, although it seems to me that he does this by replacing them with new ideas that require a little bit more thinking through and evidence before they can be stated as surely as he seems to state things. Anyhow, Finlayson borrowed an idea presented by

Thorpe, Holder, and Crompton (2007) about orangutans walking on branches high up in trees:

> Orang-utans share something with humans that gorillas and chimpanzees do not. All of them stand upright but when chimpanzees and gorillas do so, the hind limbs are flexed. Orangutans and humans on the other hand, stand on straight hind limbs. (Finlayson, 2009, p. 35)

Orangutan knuckle-walking; compare the gait with that of the chimp (next) and then gibbon (p. 312)

Chimpanzee knuckle-walking

After stating the potential benefits to orangutans of walking in this way, Finlayson goes on, extending this idea and applying it to the ancestry of hominoids, or at least the great apes in bipedalism:

> It seems that this is an ancestral form of loco-motion present in the ancestor of all the great apes. It was retained by the orang-utans who kept to a similar lifestyle [i.e., being primarily arbor-eal] in the south-east Asian rainforests. . . . The orang-utan was probably the only conservative ape to have made it to today; all others either went extinct or changed their ways. (Finlayson, 2009, p. 35)

What he is saying here is that the changing of the ways for chimps (and bonobos) and gorillas was to innovate by developing knuckle walking in a life that is spent more significantly on the ground. He argues that the ances-tor both of chimps and bonobos (he neglects the latter, as many anthropological writers do) and of humans never made that shift, but kept the ancestral tree bipedalism, and shifted it to the ground. After the break from humans, the other two became knuckle walkers.

This is quite a leap into the trees, so to speak, a huge shift from the traditional position that bipedalism came from a shift from the trees to the infamous savannah, a position that has been criticized much of late. Here, thanks to Finlayson, we have the novel idea that bipedal-ism developed in the trees and was maintained when our ancestors 'hit the ground'.

If this is true—and, unfortunately, it would take a lot to prove its truth or lack of truth, like many speculations about bipedalism—why stop at the orangutan? Why not include as well that invisible ape, the gibbon (the so-called 'lesser ape' or 'small ape', the latter being, I think, a far more appropriate appellation)? This is where my personal experience over the last few months comes in. I have seen gibbons walk. I saw them do so before I read the passage in Finlayson that I have just quoted and referred to. I was struck, in my admittedly short period of observation, by how 'natural' this arboreal ape species looked walking on two limbs. I don't know precisely what the muscles were doing, but they did not appear to be in a non-flexed pos-ition to me: they appeared to be moving naturally. So perhaps we have another thing to share with the gibbon, another reason why studying the gibbon might be import-ant in learning about our own history.

New Thoughts on Human Evolution

Students often ask whether we will evolve—for instance, will our thumbs get larger the more we become emotionally/

Gibbon walking upright

socially dependent on text messaging? Will we lose our pinkie toes, appendices, or third molars (a.k.a. wisdom teeth)? Will our bodies adapt ways to make better use of the fat-laden fast foods that are becoming the staple of the Western diet? The standard anthropological answer to this is that culture works faster than nature, and that humans now have reached a level of sophistication that makes us able to invent ways to adapt to changing circumstances—or even to change the circumstances themselves, to adapt them to our needs. Moreover, evolutionary changes hap-pen slowly over the course of many generations, and we humans are slow breeders (no matter how many pregnant teens there are out there); our evolution won't happen in time to save us from the challenges we face today.

But I have been thinking about that standard anthropo-logical answer, and I'm not convinced. For one, I do not believe that technology will solve all of our problems; it is a far greater cause of our problems, and I believe that changes in social practices are more important than innovations in technology. Based on this belief and influenced by the likely connection between resistance to the ancient mass killer smallpox and resistance to the current epidemic AIDS, I have come up with a potential scenario for future human evolution. At least, it would make a good movie premise for one of those disaster thrillers informed by the American-style apocalyptic worldview.

In 2009–10, public health officials had us fearing the H1N1 virus, otherwise known as swine flu. I didn't really think that it would be the nasty pandemic that some people thought it would be, and it looks as though I was right. Much overlooked amid the fear-mongering and disaster planning has been the cause of this flu. It comes down to an American company, Smithfield Farms, the world's largest producer of pork. Their pig-crowding, environmentally unsafe practices earned them a huge fine in their home state of Virginia, so they set up shop in Mexico, a country known for its lax environmental standards and enforcement. Not only are Smithfield's practices bad for pigs and the environment: they also create great conditions for flu evolution. And yet, industrial farms such as this are quickly seizing market share from the much romanticized but increasingly scarce family-owned farm.

If the Smithfield model is typical of the growing industrialization of farming practices, then we may be in for more and more types of flu, including some very serious strains that could decimate the human population. If this happens—and particularly if there are more cases of bungling large-scale immunization programs the way the H1N1 vaccinations were mismanaged in Canada in the latter half of 2009—then there will be an evolutionary advantage to being 'naturally' resistant to flu, an advantage as great to humans as long necks were to the treetop-grazing pre-giraffes. That is one way in which humans could evolve, thanks to fortunate mutations brought about in response to the continuation of unthinking social and industrial practices. Getting a big corporate sponsor for the movie might be a little difficult.

New Thoughts on Nations: Seeking the Taino

In the fall of 2009, I gave a lecture on the nonrecombinant genes: **mitochondrial DNA** (or **mtDNA**) and the **nonrecombinant Y-chromosome** (or **NRY**). **Nonrecombinant**, as you may remember, means that there is no combining of the genetic input of both parents, mother (for mtDNA) and father (for NRY). When you study both of these genetic markers, you can learn a great deal about history: about human history generally and—to my mind just as important—the history of particular regions and countries.

My lecture was informed by some recent reading I had done on the history of Jamaica. A number of my students are of Jamaican and other Caribbean background, and I'd

wanted to know more than the little that was in my head (most of that stemming from my study of reggae legend Bob Marley, a longtime favourite singer–songwriter). It had been very, very difficult to find decent books—it was a lot like finding good books about gibbons. The economically and politically powerful countries research and write a lot about themselves, so they are over-represented in print. At last, though, I'd found some decent books on Jamaican history, all of them published by the University of the West Indies Press.

During my fall 2009 lecture, I spoke about a study entitled 'Genetic Origin, Admixture, and Asymmetry in Maternal and Paternal Human Lineages in Cuba', written by Isabel Mendizabal and associates, and published in 2008. The authors of the study, having analyzed the mtDNA of 245 Cuban individuals and the NRY of 132 Cuban males, had found that the genetic contributions broke down as follows:

Ancestral ethnic group	Genetic contribution (%)	
	mtDNA	NRY
African	45	20
Aboriginal	33	0
European	22	80

There is such a history lesson in this study. The lasting effects of conquering, colonialism, and slavery and the surprising (to most historians) persistence of the Aboriginal genetic component are in evidence in all six numbers. The main Aboriginal group in Cuba was the **Taino**. And while European and African males had significant status and opportunity to be able to add to the Cuban population, Taino males seemed to have been shut out—a predictable outcome, based on their low status. Taino females, also because of their low status, would have been considered 'fair game' by European and African men.

A similar study, published just a few years earlier (2005) by J.C. Martínez-Cruzado and colleagues, recorded the following findings on the mtDNA of 800 people in Puerto Rico:

Ancestral ethnic group	Genetic contribution to mtDNA (%)
Aboriginal	61.3
African	27.2
European	11.5

As an 'Aboriginal specialist', I was very surprised by the high percentage of the study group tracing its origins to Aboriginal roots. You can imagine how the NRY would differ along the same lines as the Cuba study. My point with this is that it was the Taino, again, who were the main

Aboriginal group contributing genetically to this mix. They have made a significant contribution to the genome of this (and no doubt many other) Caribbean people.

I presented these statistics to two classes. In both classes, I had Jamaican students come up to me afterwards to talk and ask about the Taino, the museums the students had visited, and the stories they had heard. Now, from my admittedly slight reading knowledge of the area, I have come to understand that the terms Taino and **Arawak** are pretty much interchangeable. On the Jamaican coat of arms are two Arawak, one female and the other male. Beneath them is the motto, 'Out of Many, One People'. One of those many, the Arawak/Taino are a source of great interest to the people of the island nation, but they need to be studied genetically so that Jamaican history can be written more completely, and Jamaican heritage strengthened. In earlier classes, I'd noted how sometimes my Caribbean (a term derived from the Caribs, another Aboriginal group whose complete identity is still somewhat mysterious) students exhibited Aboriginal features, most particularly the **sinodont** teeth and famous **shovel-shaped incisors**. Even a more in-depth study of that feature would help Caribbean people embrace their Aboriginal heritage. It could make for some interesting conversations between Aboriginal people born in Canada and their Caribbean 'cousins'. Physical anthropologists can help the people claim their Aboriginal heritage. I am tempted now to do a study of my Caribbean students, especially the ones who show Aboriginal features—it could yield tremendous insights that might cause us to reconsider the way we view and define 'nations'.

Closing Remarks

At the same time that all of these changes are taking place in the discipline, some things remain the same. We still don't know whether Neandertal is part of our modern human gene pool in any significant way. We still cannot trace our ancestry along a straight line, drawn over millions of years, from *them* and *that time and place* to *us, now*. Science's certainty about evolution is still being challenged by the narrow religious interpretation of the world's history. Creationists may cloak their beliefs in the disguise of Intelligent Design theory, but it's the same old present of ignorance repackaged to look like a new gift, and then re-gifted to the world.

Finally, it has long been said that teaching a subject is a very good way to learn about it. It is even truer of writing a textbook. I've rewritten all of my lecture notes in physical anthropology because of the writing, reading, and research that I have done over the last year or so. I could not have predicted what I would learn, or how much! It is an experience that has set me on a path of learning that will continue for years.

Discussion Questions

1. In what ways has physical anthropology changed over the past 20 years? What changes do you foresee over the next 20?

2. Discuss the pros and cons of the biological species concept.

3. What are some of the benefits Caribbean people might gain from learning that they have an Aboriginal heritage?

Key Terms

Arawak

biological species concept (BSC)

bipartite brow ridge

ForDisc

mitochondrial DNA (mtDNA)

nonrecombinant

nonrecombinant Y-chromosome (NRY)

race

shovel-shaped incisors

sinodont

Taino

Key Species

Homo heidelbergensis   *Kenyanthropus platyops*

Key Individuals

Awoke Amzaye

Berhane Asfaw

William Bass

Davidson Black

Dee Denver

Clive Finlayson

Yohannes Haile-Selassie

Carolus Linnaeus

Ernst Mayr

Wenzhong Pei

Zhongjian Yang

If You Were to Read 10 Books . . .

As this is a textbook designed for introductory courses, it provides only an overview of the ideas and thinkers that make up the fascinating field of physical, or biological, anthropology. When I am interested in an area of study (say, gibbons), I like to read on about the subject. If you want to read on in physical anthropology, here are 10 books that I highly recommend. Take note that my list is going to be different from your instructor's list, and from any other anthropologist's list. On another day, I might make up another list myself. . . .

William Bass & Jon Jefferson, *Beyond the Body Farm: A Legendary Bone Detective Explores Murders, Mysteries, and the Revolution in Forensic Science* (2007, New York: HarperCollins)

Noel T. Boaz & Russell L. Ciochon, *Dragon Bone Hill: An Ice Age Saga of* Homo erectus (2004; Oxford, UK: Oxford University Press)

Dorothy L. Cheney & Robert M. Seyfarth, *Baboon Metaphysics: The Evolution of a Social Mind* (2007; Chicago: University of Chicago Press)

Richard Dawkins, *The Ancestor's Tale: A Pilgrimage to the Dawn of Life* (2004; Boston: Houghton Mifflin)

Clive Finlayson, *The Humans Who Went Extinct: Why Neanderthals Died Out and We Survived* (2009; New York: Oxford University Press)

Donald C. Johanson & Kate Wong, *Lucy's Legacy: The Quest for Human Origins* (2009; New York: Random House)

Clea Koff, *The Bone Woman: A Forensic Anthropologist's Search for Truth in Rwanda, Bosnia, Croatia, and Kosovo* (2005; Toronto: Random House)

Neil Shubin, *Your Inner Fish: A Journey Into the 3.5 Billion Year History of the Human Body* (2005; New York: Vintage Books)

Mike Morwood & Penny van Oosterzee, *A New Human: The Startling Discovery and Strange Story of the 'Hobbits' of Flores, Indonesia* (2009; Walnut Creek, CA: Left Coast Press)

Clark Spencer Larsen, *Skeletons in Our Closet: Revealing Our Past through Bioarchaeology* (2000; Princeton: Princeton University Press).

Other Recommended Resources

Martinez, J.C., Toro-Labrador, G., Viera-Viera, J., Rivera-Vega, M.Y., Startek, J., Latorre-Esteves, M., Roman-Colon, A., . . . & Valenci-Rivera, P. (2005). Reconstructing the population history of Puerto Rico by means of mtDNA phylogeographics analysis. *American Journal of Physical Anthropology, 128*, 131–55.

Mendizabal, I., Sandoval, K., Berniell-Lee, G., Calafell, F., Salas, A., Martínez-Fuentes, A., & Comas, D. (2008). Genetic origin, admixture, and asymmetry in maternal and paternal human lineages in Cuba. *Biomedical Evolutionary Biology, 8*(213), 1–10. doi:10.1186/1471-2148-8-213

Subramanian, S., Denver, D.R., Millar, C.D., Heupink, T., Aschrafi, A., Emslie, S.D., Baroni, C., & Lambert, D.M. (2009, October 15). High mitogenomic evolutionary rates and time dependency. *Trends in Genetics, 25*(11), 482–6.

Glossary

The following definitions correspond to the words set in bold in chapters 1–13 of the textbook. These are intended to help you recall concepts and ideas explained elsewhere in the book. Remember that the ability to repeat even a detailed definition does not substitute for a solid understanding of what the term means and how it can be applied.

GENERAL TERMS

absolute dating dating techniques that supply an age, whether in years or (more typically) ranges of years, for artifacts, features, or archaeological sites (see Feder, 2004, p. 314). *Also called* **chronometric dating**.

accelerator mass spectrometry (AMS) a method of determining the composition of material by bombarding it with ions—this breaks the material down into constituent atoms, allowing them to be counted. AMS is frequently used to enhance the accuracy of carbon-14 dating.

acculturation a social or cultural process through which members of one society pick up the cultural traits of another; example: during the 2004 summer Olympics in Athens, people in bars across Canada acquired the habit of saying 'Opa!' whenever someone broke a glass.

Acheulean denoting a mode-2 tool tradition associated with *Homo erectus*, beginning about 1.5 million years ago. The diagnostic trait is the hand axe.

adaptive radiation a relatively rapid evolution into a broad diversity of species or sub-species, occurring through adaptation to a diversity of environmental circumstances. The classic example is Darwin's finches, which acquired a variety of beak sizes and structures as adaptations to the different micro-environments (e.g. featuring seeds of different sizes and shell thicknesses) of the various Galapagos Islands.

adenine (A) a compound that is one of the four constituent bases of nucleic acids; a purine derivative, it is paired with thymine in double-stranded DNA.

Ainu the indigenous people of Japan, whose language, traditional culture, and appearance differ from those of the people who later came to and settled in the country.

alarm calls calls made by primates to warn of the presence of dangerous predators; example: vervet monkeys have specific learned calls to signal the presence of eagles, leopards, and snakes.

Algonquian a language family that in Canada includes the languages of the Mi'kmaq, Malecite, Innu, Cree, Ojibwa, Algonquin, Delaware, Blackfoot, Peigan, and Blood. *Do not confuse with* **Algonquin**, *a dialect of the Algonquian language family and a name for the people who speak it.*

allele one of a series of different pairs of nucleotide bases at any one site or location (short for **allelomorph**, lit. 'other form').

Allen's rule the principle that states that body appendages (beaks, ears, noses, tails, and especially arms and legs) of a particular widespread endothermic (warm-blooded) genus or species are relatively shorter in cooler than in warmer areas.

alternative splicing a form of mutation in which the exon from a gene that had communicated with the RNA is spliced out, and a different exon within a gene or even an intron is left to do the communication.

Amerind the Aboriginal languages of the Americas, considered together as a hypothetical language family, as suggested by linguist Joseph Greenberg; it includes nearly all of the indigenous languages except for the Eskimo–Aleut and Athabaskan languages.

amniotes animals that lay eggs that are adapted to land, including mammals, reptiles, and birds. The embryos are protected and fed by membranes inside the egg. As these eggs do not have to be laid in water (unlike those of amphibians), amniotes do not have to go through the metamorphosis from water-dwelling to (at least partial) land-dwelling animal.

anagenesis the evolution of the entire population of a species into another species; the first, or ancestral, species goes extinct. *Also known as* **phylogenetic change**. ▪ **anagenetic** (*adjective*).

analogy *See* **homology**.

anatomically modern human (AMH) an individual *Homo sapiens* that has all the main physical characteristics of humans living today. There is no strict definition or listing of what this entails. *Compare* **behaviourally modern human**.

ancestral traits *See* **generalized traits**.

Animalia the kingdom of species that includes animals but excludes plants and bacteria.

anthropodenial the act or attitude of denying any connection between animal and human behaviours, emotional states, etc.

anthropoids a group of higher primates that includes apes, monkeys, and humans.

anthropological linguistics *Another name for* **linguistic anthropology**.

anthropology the holistic study of humans, particularly those living deep in time and within smaller groups than those usually studied by other social sciences.

anthropometry the comparative measurement, particularly of bones, of different human groups.

anthropomorphism the tendency to project onto non-humans human emotions, actions, and goals that cannot be demonstrated to be there.

Arawak name used to refer to one or several of the Aboriginal peoples encountered by Christopher Columbus and the early Spanish visitors to the Caribbean and northern South America; these include the people now known as the Taino.

archaeology the excavation and study of the material remains of human actions and presence.

archaic *Homo sapiens* a term used to refer to *Homo sapiens* who bear some *Homo erectus* traits. This is not a strictly defined concept.

arthropod any member of the phylum Arthropoda, which includes invertebrate animals such as insects, spiders, and crustaceans.

artifacts (*or* **artefacts**) typically, portable objects manufactured or altered by humans. Some archaeologists reserve the term for objects that have been made intentionally by humans, distinguishing them from objects that are the products of human actions but not intent (e.g. charcoal, nutshells, crustacean shells), which they call **ecofacts** (see Feder, 2004, p. 323).

Athabaskan languages a family of North American Aboriginal languages that includes Chipewyan, Carrier, Hare, Apache, and Navaho, mostly spoken in Alaska, Yukon, Northwest Territories, and the northern parts of Canada's western provinces.

atheistic evolution a view that combines an acceptance of evolution with the belief that there is no God or divine creator.

Aurignacian denoting or having to do with the material culture of the Upper Paleolithic in Europe and southwestern Asia, frequently associated with Cro-Magnon and thought to date to between 34,000 and 29,000 years ago.

Australoid a racial term used by Carleton Coon (among others) to refer to the Aborigines of Australia.

Austronesians denoting or having to do with the peoples who speak languages in the Austronesian language family, including Hawaiian, Maori (of New Zealand), Amis (of Taiwan), and Samoan (among many others).

baboon a relatively large member of the parvorder Catarrhini (Old World Monkeys), living in Africa and belonging to the same subfamily (Cercopithecinae) as macaques and vervet monkeys.

bacteria (*sing.* **bacterium**) a large group of single-celled microorganisms that represent one of the three domains of living entities; they have thick walls but no nucleus or organelles.

behaviourally modern human (BMH) an individual of the *Homo* genus typically considered to be *Homo sapiens* because of associated behaviour seen as 'modern' (examples: the practice of deliberate burial; the tendencies to show compassion and express ideas artistically). *Compare* **anatomically modern human**.

Bergmann's rule the principle that among endotherms with a long north–south geographic range, the subspecies or variants with the larger/stockier size are to be found in the colder parts of the species' range, while those of smaller body size and greater slenderness are found in the warmer parts.

Beringia the large landmass formed during the last glacial period, when the waters of the relatively shallow Bering Sea that separates northeastern Asia (i.e. Siberia) from northwestern North America (i.e. Alaska) dropped to sufficiently expose the seafloor. It was roughly 1,000 kilometres wide, from north to south.

Beringia Refuge hypothesis a theory that immigration to and settlement in the Americas occurred in two stages. According to the hypothesis, one very large group reached Beringia, experienced a small reduction in genetic variation, and then travelled down the western coast of the Americas. During the Last Glacial Maximum, a smaller number of people were 'trapped' between the ice-covered perimeters of Asia and North America in the refuge of Beringia; these people, the ancestors of the Athabaskans and the Inuit, experienced a more serious genetic bottleneck before they were able to move into North America in a limited way.

bias of the book the biased interpretation of history presented in (and adopted from) the writings of literate peoples when writing about non-literate peoples.

binominal nomenclature the system of naming animals and plants, initiated by eighteenth-century biologist Carolus Linnaeus, in which each species is assigned a two-part name consisting of genus (e.g. *Homo*) and species (e.g. *sapiens*)

biological anthropology *Another name for* **physical anthropology**.

biological species concept (BSC) the notion that a species may be defined as any population or group of populations capable of interbreeding and producing fertile offspring.

bioturbation the disturbance of silt layers or sedimentary deposits by various life forms.

bipedalism the ability to walk using just two legs or limbs. **Obligate bipedalism** is the ability to walk on two limbs and having no choice but to do so (as with an ostrich or baby parrot, which cannot fly); **optional bipedalism** is the ability to walk on two limbs but not as a full-time mode of locomotion (as with an adult parrot, which can fly). ▪ **bipedal** (*adjective*)

blade a thin strip of stone made and used in the making of Mode-4 tools.

brachial index a score used to determine whether a specimen is more apelike or human, achieved by dividing the length of the ulna by the length of the humerus and multiplying by 100; a higher score suggests the specimen is more apelike.

brachiation movement through the trees by swinging, using arms as primary mode of locomotion. ▪ **brachiate** (*verb*)

brachycephalic (according to an outdated system of classification) broad-headed, esp. with reference to east Asians and the Aboriginal people of the Americas. *Compare* **dolichocephalic, mesocephalic**.

breed a subspecies, esp. with reference to domesticated animals, such as dogs.

bush model of evolution a conceptual model of evolution that allows for broad diversification of many different species. *Compare* **tree model of evolution**.

calibration a way of calculating a score by taking raw figures and refining or adjusting them to fit another scale. For example, results of radiocarbon dating are reported in radiocarbon years; calibration must be used to convert the results to calendar years.

Cambridge Reference Sequence (CRS) the first human mitochondrial DNA ever sequenced, as published in 1981 by a Cambridge University research team. The work contained some minor errors that were corrected in 1999.

Capoid a term used by Carleton Coon to refer to the Khoisan, or click-speaking, Bushmen of southern Africa.

Caucasoid (or **Caucasian**) denoting or having to do with white- or light-skinned people of European (and sometimes North African and Indian) origin. The term, which is part of an outdated racial classification developed in the nineteenth century, derives from the idea that the holotype was found in the Caucasus mountains. Although the classification is now considered unscientific, the term Caucasoid (or Caucasian) has acquired the more restricted meaning 'white, of European origin'.

centromere a constricted area in the centre of a chromosome. It helps direct the chromosome to its proper position when cells divide.

Châtelperronian a Mode-4 tool tradition associated with Neandertal, in evidence especially in central and southwestern France and adjacent parts of northern Spain.

chronometric dating *See* **absolute dating**.

clade a taxonomic unit consisting of an ancestor and all of its descendants.

cladistics a taxonomic system that classifies living beings according to their evolutionary history. *Also called* **phylogenetics**.

class a cladistic taxon between order and phylum.

Clinal Replacement hypothesis a theory of human population that combines human origins in African with continuity of characteristics. It proposes that there were waves of cline movements over time, with repeated instances in which a cline with a 'newer' evolved trait gradually replaced an older, more 'primitive' cline, yet with remnant features of the former remaining.

clines geographically separate subspecies that differ from one another in one or more characteristics; the differences are graded, so that they may be subtle for neighbouring clines but will become more significant and apparent the greater the distance of separation.

Clovis points long, thin, bifacial (i.e. worked on both sides), fluted (i.e. with grooves cut out of each side) projectile points (in this case, spear tips) that were widespread in North America from 13,200 to 12,700 years ago.

Clovis First hypothesis the theory, long held but now discredited, that the makers of Clovis points were the first peoples to inhabit the Americas.

Coastal Migration hypothesis the theory that the first people to settle in the Americas took a route following the West Coast along shores now under water.

coding the process whereby genes in the DNA send genetic messages to the RNA to initiate the building process.

comparative anatomy a field of study that involves investigating the similarities and differences in physical structures (especially bones) among animals.

compassion high degree of caring for other humans. Compassion is viewed as a trait of behaviourally modern humans.

concestor a common ancestor. This useful term was coined by Richard Dawkins.

cones light-sensitive cells in the retina of the eye, responsible for sharpness of vision and colour perception.

consilience convergence or agreement among results (particularly a date) achieved by several methods.

contingency (in evolutionary theory) the view that small, isolated incidents greatly affected the course(s) of evolution, and that other, equally small incidents, had they occurred, could have dramatically altered the course of evolution. *Compare* **convergent evolution**.

convergent evolution the development of very similar physical structures by species of very different backgrounds. In terms of evolutionary theory, it suggests that certain evolutionary developments are inevitable. *Compare* **contingency**.

core tool a tool made from a stone from which flakes or pieces have been knocked off. Core tools are associated with Mode-2 toolmaking.

C/P3 honing complex a dental process in which the canines are sharpened by rubbing against the nearest premolar.

cranial having to do with the skull.

craniometry the scientific measurement of the skull or cranium.

creation science the reinterpretation of scientific knowledge by fundamentalist Christians to reconcile this knowledge with belief in the literal truth of the Bible.

Cro-Magnon the earliest anatomically modern *Homo sapiens* in Europe, associated with the Aurignacian tool culture and dating back to at least 35,000 years ago.

CSI effect the impact, primarily on jury members, of the way forensic science is portrayed on TV shows. It is seen especially in the way it influences people into thinking that the evidence presented in cases involving serious crimes must attain unrealistically high standards of certainty.

cultural primatology the study of the behaviour of primates. One area of focus is behaviour that might be learned in one group and not in others.

culture (in primatology) a set of learned behaviours shared by one group of a particular primate but not by others.

cytosine (C) a compound—specifically a single-ringed or pyrimidine nucleotide base—found in living tissue as a constituent base of DNA; it is paired with guanine in double-stranded DNA.

deep time the 4.5-billion-year span of the earth's history, esp. as viewed as the backdrop to human history and long-term evolutionary changes.

deletion (in genetics) a mutation involving the loss of a section from a nucleic acid molecule or chromosome.

dendrochronology a method of assigning dates to trees or wooden objects based on the principle that every year, a tree forms a growth ring.

dental anthropology the study of the varying dental characteristics of historic and current human populations.

derived traits traits that are newer and less widespread than those referred to as ancient or generalized. Examples include dry, flaky earwax and sinodont teeth. *Also called* **specialized traits**.

diabetes type II *See* **type II diabetes**.

direct observation of species change the ability to perceive a change in species characteristics over a short period of time. This typically occurs with fast-breeding organisms such as bacteria, viruses, and fruit flies.

distal denoting or having to do with the end of the bone farthest away from the centre of the body. *Compare* **proximal**.

DNA (deoxyribonucleic acid) a long strand of genetic material that sends instructions to RNA that ultimately result in the construction of the living entity's body.

dolichocephalic (according to an outdated system of classification) long-headed, esp. with reference to people of European origin. *Compare* **brachycephalic**, **mesocephalic**.

domain the highest taxonomic rank, one level above kingdom. The three biological domains are bacteria, archaea, and eukarya.

double-blind test (in primatology) a test of communication skills in which a primate (for example, a gorilla) conveys the identity of an object to a person who must interpret what is being represented without seeing the original object.

dura mater the outer of the three layers of the meningeal membranes surrounding the brain. Literally, the term means 'hard mother', as it is the toughest of the three layers.

ecofacts *See* **artifacts**.

ectocranial having to do with the outside of the cranium, esp. in terms of the closure of cranial sutures. *Compare* **endocranial**.

electron spin resonance (ESR) a method of absolute dating that measures the number of electrons trapped in an artifact of bone or shell.

encephalization the ratio of brain size to body size, taken as an indication of an animal's intelligence.

endemics species that are native to a particular place and are not found elsewhere

endocranial having to do with the inside of the cranium, esp. in terms of the closure of cranial sutures. *Compare* **ectocranial**.

endogamous retrovirus a virus that has left a genetic 'footprint' in the DNA of a species.

endotherms animals such as mammals and birds whose body temperatures are maintained at a relatively constant level and do not vary significantly with the temperature surrounding the body. These animals are better known (if inaccurately) as 'warm-blooded'.

epicanthic fold a fold of skin at the inner corner of the eye that gives the eye a more almond-like shape. It is common in East Asian populations.

epiphyseal union the fusion and ossification of the epiphyses with the rest of the long bone.

epiphyses the end of a long bone. It is initially separated from the main bone by a layer of cartilage that eventually ossifies so that the parts become fused or united.

Eskimo–Aleut a language family whose members in the extreme north extend from Greenland to Siberia.

ethology the study of the behaviour of primates and other animals.

eugenics the science of improving a population by controlled breeding to increase the occurrence of desirable heritable characteristics.

Eurocentrism a biased view that privileges the perspectives of Europe or the West. Examples include claims that Christopher Columbus 'discovered' America, when in fact many inhabitants of the Americas were already very much aware of the continent's existence. ■ **Eurocentric** (*adjective*)

evolution the adaptation and change of species through mutation and natural selection of variants.

evolvability the capacity of a species to evolve or change over time.

exaption the adaptation of a trait to a function other than the one it was originally adapted for. Example: feathered wings may originally have been adaptations useful for cooling or heating the body.

exon a segment of a gene in the DNA that typically communicates with the RNA. This is the part that can code.

facial angle the degree of protrusion of the lower face (i.e. below the nose). If the facial angle protrudes, then the facial angle is **prognathic**; if it is relatively flat, then the facial angle is **orthognathic**.

facilitated variation processes that can increase the likelihood of mutated variation.

fact a scientific observation that is true as far as we can determine it to be.

family (in taxonomy) a rank between order and genus. There is an –id marker in the family name that differs from the –oid of a super-family and the –in of a subfamily (example: hominoid (superfamily), hominid (family), hominin (subfamily)).

fat-fold thickness (**FFT**) a measurement of the skinfold thickness, taken from the right side of the body, used to gauge body fat as a way of judging the overall health of an individual or group.

faunal having to do with the animals of a particular region, habitat, or period. In anthropology the term is used esp. of bones.

Feast of the Dead a ceremony performed mainly by Iroquoian speakers (such as the Huron) during the pre-contact and early contact periods. During the ceremony, those who have died and been buried in temporary graves over the previous 10–15 years are re-buried on one site, creating what archaeologists refer to as an **ossuary**.

fission track dating an absolute dating technique performed by analyzing the tracks or damage trails left by fragments from nuclear fission, in particular uranium-bearing minerals and glasses subject to the decay of uranium-238.

flake tool a tool made from the stone flakes that have been chipped off of a rock.

fluorine analysis a relative dating technique used to sort fossils found at a single site by measuring the amount of fluorine each one contains (older fossils are associated with greater quantities of fluorine).

foramen magnum the hole in the skull where the bones of the neck and the various nerve strands approach and enter the brain (from Latin, lit. 'big hole').

ForDisc a computer software package, developed at the University of Tennessee by Stephen Ousley and Richard Jantz, designed to create a reliable profile of a skeleton, including gender, age, and race or ethnicity.

forensic anthropology the use of methods of physical anthropology—especially osteology—in the analysis of physical remains for legal purposes, such as a criminal trial. It should not be confused with **forensic pathology**, which involves soft tissues, or **forensic entomology**, which involves the study of insects that infest the body and aid in its decomposition.

fossils the products of a slow decay-and-replacement process of the parts of once-living beings replaced by stone.

founder effect the loss of genetic variation that occurs when a new population is established by a very small number of individuals from a larger population; it can lead to the development of a new species or at least of new traits not possessed by the larger population.

Galagidae a family belonging to the primate suborder Strepsirrhini, made up of African bush babies (also called galagos).

gene a distinct sequence of nucleotides forming part of a chromosome.

generalized traits characteristics that are old and shared widely among divergent species, and are considered to have been shared at one time by a common ancestor. Examples: (1) the two bones forming the lower part of the limb and one bone forming the upper part of the limb found among amphibians, reptiles, birds, and mammals; (2) wet and gooey ear wax among humans. *Also called* **ancestral traits**, **primitive traits**.

genetic bottleneck a situation in which either only a few genetic variants are able to pass through some kind of natural geographic barrier or (more generally) a significant percentage of a geographic population or species is prevented from reproducing (e.g. by being killed off by a weather event or predators). *Also called* **population bottleneck**.

genetic cost the amount of genetic material that it takes to produce and maintain a particular body part or process.

genetic drift variation in the relative frequency of different genotypes in a small, relatively isolated population, owing the chance disappearance of particular genes as individuals die or do not reproduce.

genetic mutation naturally occurring changes (substitutions in nucleotide bases, insertions, deletions, and alternative splicing) in the genetic structure of a newly formed biological entity.

genetics the branch of biology concerned with the study of nucleotide bases, genes, and DNA to better understand heredity and population variation.

genus the principal taxonomic category ranking above species and below family; it is denoted by a capitalized and italicized Latin name (example: *Homo* in the species name *Homo erectus*).

geofacts objects that have been altered by natural as opposed to human forces. *Compare* **artifacts**.

geographic population a term used to refer to a relatively distinct people who have an association with a particular area or areas

Gloger's rule the principle that endotherms (i.e. 'warm-blooded' animals, such as mammals and birds) living in warm, humid climates have darker feathers, fur, or skin than closely related endotherms living in cooler and drier areas.

gorilla model a model of species variation that, owing to the tremendous sexual dimorphism that separates the size of adult male and female gorillas, allows for considerable variation within a species. *Compare* **lemur model**.

Great American Interchange the intercontinental movement of animals (primarily mammals) that occurred when the Isthmus of Panama rose, volcanically, to join the Americas together.

Great Zimbabwe a series of huge granite edifices, including a large wall known as the Great Enclosure, built by the ancestors of the Shona people of southern Africa (specifically Zimbabwe) between the thirteenth and fifteenth centuries.

grooming claws claws possessed by prosimians (with the exception of aye-ayes), used for grooming the fur.

growth arrest lines *See* **Harris lines**.

guanine (**G**) a double-ringed, or purine, nucleotide base, one of the four constituent bases of nucleic acid; it is paired with cytosine in double-stranded DNA.

half-life the time taken for the radioactivity of a specified isotope to decay and fall to half its original value.

Hamman–Todd Collection a collection of over 3,000 humans started by and housed in the Cleveland Museum of Natural History.

haplogroups groups of genetic markers or mutations, or one genetic marker or mutation, inherited together as a unit.

Haplorrhini a primate suborder that includes tarsiers, monkeys, apes, and us (from Greek, lit. 'simple noses'; compare Strepsirrhini, lit. 'twisted noses').

Harris lines quoting Feder: 'Horizontal cracks at the end of the diaphysis of a long bone that persist through adulthood, resulting from a bout of malnutrition [or physical or emotional trauma] during the developing years' (Feder, 2004, p. 330).

heterozygote an individual possessing two different alleles of a particular gene or genes, and so giving rise to varying offspring. ▪ **heterozygous** (*adjective*). *Compare* **homozygote**.

history traditionally, the period of time covered by the written record; use of the term can reflect a bias against non-literate peoples.

holotype an individual of a species chosen to be the model for a description of what the species is like. *Also called* **type specimen**.

hominids members of a family (**Hominidae**) that includes orangutans, gorillas, chimpanzees, bonobos, and humans. Traditionally, and still often today, it has been defined as the taxonomic family that includes *Homo sapiens* and its ancestors.

hominins (or **hominines**) members of a subfamily (**Homininae**) that includes African apes, humans, and their immediate ancestors; in earlier use, the term referred only to humans and their direct ancestors.

hominoids members of a superfamily (**Hominoidea**) that includes gibbons, apes, humans, and their immediate ancestors.

homology similarities between characteristics of living entities that are due to their shared ancestry; it is distinguished from **analogy**, which refers to characteristics in common between different species that may have developed by performing the same or similar functions but that evolved along different paths. ▪ **homologous** (*adjective*); **analogous** (*adjective*). *See also* **convergent evolution**.

homozygote an individual having two identical alleles of a particular gene or genes and so breeding true for the corresponding characteristic. ▪ **homozygous** (*adjective*). *Compare* **heterozygote**.

humerofemoral index a score used to determine whether a specimen is more apelike or human, achieved by taking the length of the humerus and dividing it by the length of the femur. Generally speaking, the lower the score, the more like *Homo sapiens* the primate or fossil specimen is.

Huron an Iroquoian-language-speaking people who, at the time of early European contact, were living in southern Ontario. Descendant groups currently live near the city of Quebec and in the states of Kansas and Oklahoma. *Also called* **Wendat**.

hypothesis a tentative scientific explanation for the relationship between observed phenomena. Ideally (in scientific terms) it is testable.

industrial melanism a theoretical explanation for the darkening of a particular type of moth, supposedly so that it could match better with the industrial soot-covered bark of trees.

infanticide the killing of a child less than a year old.

insertion (in genetics) a mutation that involves the addition or repetition of a sequence or strand of nucleotide bases.

intelligent design *See* **creation science**.

interrupted gene a gene that contains at least one intron.

intron a part of a gene that is usually spliced out so that it cannot communicate with the RNA.

Iroquoian denoting or having to do with a family of languages that includes the living languages of Mohawk, Oneida, Onondaga, Cayuga, Seneca, Tuscarora, and Cherokee, as well as Huron (Wendat), which is being brought back to life.

irreducible complexity the argument, put forward in support of intelligent design theory, that certain biological systems are too sophisticated and complex to have evolved from simpler forms.

island dwarfism a phenomenon that occurs when mid-sized or large species become smaller owing to the limited resources of islands.

isotope any one of two or more forms of a chemical element that have the same number of protons but a different number of neutrons in their atomic structure; isotopes of the same element will have different physical properties but the same chemical ones.

kingdom the second highest taxonomic rank. The kingdoms include Plantae, Animalia, Protista, Fungi, Eubacteria, and Archaebacteria.

kin-mediated reconciliation a practice observed among female baboons, in which a relative of two disputing female baboons will grunt to signify to both combatants that the dispute is over.

lactase an enzyme needed to break down the sugar lactose in digestion. Deficiency in this enzyme causes lactose intolerance.

lactose a sugar found in milk.

lactose intolerance the inability to break down lactose in digestion.

Lamarckism a theory of organic evolution, not generally considered scientifically viable today, according to which acquired characteristics—physical traits that individuals have picked up during their lifetimes—could be passed on to offspring.

lanugo a downy coat of fur-like hair, resembling hair grown by early human ancestors, that fetuses grow and then usually reabsorb prior to birth.

Last Glacial Maximum (**LGM**) the time of maximum extent of the glacial ice sheets during the last glacial period, roughly 20,000 years ago.

late-onset diabetes *See* **type II diabetes**.

Later Stone Age (**LSA**) the Upper Palaeolithic as it occurred in Africa from around 30,000 years ago.

lateral line system a sensory organ developed by the blind Mexican tetra (*Astyanax mexicanus*), which helps the fish detect, by pressure and vibration, objects that it cannot see.

law of superposition the principle stating that lower or deeper strata, or layers, are older than those that are higher up or shallower.

lead wipe a smear left by a bullet as it enters or grazes a bone.

lemur model a model of species differentiation based on the lemur, whose bones are poor indicators of the species to which they belong. *Compare* **gorilla model**.

Levallois technique a Mode-3 tool technique in which the core is carefully prepared in a kind of tortoiseshell shape, with well-struck blows that pop the flakes tool out.

linguistic anthropology the study of languages; it differs from regular linguistics in putting greater emphasis on the culture, history, and society of the speakers. *Also called* **anthropological linguistics**.

living fossil an animal that has changed little over perhaps millions of years.

Lorisidae a family belonging to the primate suborder Strepsirrhini, which includes the lorises of the Philippines, Indonesia, Thailand, Vietnam, India, and Bangladesh and the pottos and golden pottos of Africa.

lumpers biologists and physical anthropologists who feel that within each species there is usually a significant amount of variation; they are not inclined to declare distinct species and subspecies based on what they perceive to be small differences. *Compare* **splitters**.

malaria a mosquito-borne infectious and often lethal disease that affects millions of people in tropical and subtropical countries of Africa, the Americas, and Asia.

marsupial any individual of the order Marsupialia, whose members are characterized by a distinctive pouch (the *marsupium*), in which females carry their young through early infancy. Marsupials include kangaroos, wallabies, koalas, and (in Canada) possums.

matrilines   family lineages based on kinship with the mother or the female line, common among primates, including humans. ▪ **matrilineal** (*adjective*).

matrix   a mass of rock in which fossils are found.

melanin   a dark brown to black pigment occurring in the hair, skin, and iris of the eye in humans and animals; it is responsible for tanning of skin exposed to sunlight.

members   layers or strata of rock in which specimens are found.

meromictic   denoting a lake in which water below a certain depth does not circulate with the upper layers, often as a result of high salt concentration.

mesocephalic   (according to an outdated system of classification) having a medium-sized head (intermediate between broad-headed and long-headed). *Compare* **brachycephalic**, **dolichocephalic**.

Mesolithic Period   *Another term for* the **Middle Stone Age**.

metabolic budget   the amount of metabolic energy needed or available for a particular function; the term is associated with the concept that the development of any organ or other part of the body takes up metabolic energy, and that sometimes some parts become too 'expensive' for the benefits they offer.

microwear   small use patterns on tools or bones.

mid-arm circumference (MAC)   a measure of the body fat content of an individual, used to produce statistics relating to the individual's particular group.

Mid-Atlantic Ridge   an S-shaped underwater ridge system 16,000 kilometres long running the length of the Atlantic Ocean east of North, Central, and South America, from the Arctic to the Antarctic. It is relatively young in geological terms, with molten rock pushing up and then flowing both east and west.

Middle Stone Age (MSA)   a period of stone toolmaking in Africa from about 200,000 to 30,000 years ago. *Also called* **Mesolithic Period**.

missing link   the species that forms the latest common link between humans and non-human primates.

mitochondrion (*pl.* **mitochondria**)   is a membrane-enclosed organelle found in large numbers outside the nucleus in most cells; it acts as a kind of cellular power plant converting sugar into chemical energy.

mitochondrial DNA (mtDNA)   a relatively short (15,000–17,000 base pairs), round strand of DNA found outside of the nucleus of the cell; it is passed down along the mother's line.

mitochondrial Eve   the first woman who bears the mtDNA patterns shared by all women living on earth today.

Mode-1   denoting the earliest toolmaking tradition, represented by the Oldowan tools.

Mode-2   denoting the earliest toolmaking tradition, represented by Acheulean tools—bifacial, symmetrical tools made by pressure flaking rather than other methods of percussion.

Mode-3   a toolmaking tradition, represented by Mousterian tools, characterized by flake tools produced by the Levallois technique.

Mode-4   denoting a toolmaking tradition, represented by Aurignacian tools, characterized by the making of blade tools.

molecular clock   a genetic approach to separating living entities based on mutational differences and the dates at which they occurred.

Mongoloid   belonging or relating to a division of humankind represented by the indigenous peoples of East Asia, Southeast Asia, and the Arctic region of North America. The term is part of an outdated system of classification introduced in the nineteenth century to categorize human races.

monogamous   having only one spouse at any one time.

monophyletic   (of a clade) descended from a common evolutionary ancestor or ancestral group, esp. one not shared with any other group.

mosaic evolution   the concept that major evolutionary changes tend to take place in stages, not all at once, with some parts evolving more quickly than others.

Mousterian   a Mode-3 toolmaking tradition associated with the Neandertal Levallois technique, which involves a careful preparation of the core into a kind of tortoiseshell shape, with well-struck blows used to pop the flakes tool out.

MRCA (**most recent common ancestor**; also mt-mrca: **matrilineal most recent common ancestor**)   mitochondrial Eve, viewed as having the mtDNA mutational pattern shared by all women.

multilevel inheritance   the inheritance of features passed down by means other than the usual genetic route.

Multiregionalism   a hypothesis proposing that anatomically modern *Homo sapiens* emerged in different areas (possibly including Africa, Australia, China, Europe, and Indonesia) at around the same time.

mutation   a naturally occurring process wherein the DNA of a newly formed individual has had nucleotide bases replaced, and strands of bases repeated and inserted, deleted, or spliced differently, for communication with the RNA. This forms the metaphorical engine for evolution.

natural selection   the process, first identified and described by Charles Darwin in *Origin of Species* (1859), in which mutated variants of a species have an advantage over others, either by being more attractive to the opposite sex ('sexual selection') or by some enhanced opportunity for survival. Natural selection enables those with the advantage in the given circumstances to produce more young than those without that advantage.

Negrito   a member of any one of several peoples located in isolated parts of Southeast Asia, including the Philippines, the Malay peninsula, Thailand, and the Andaman Islands. They are characterized as short, with tightly curled hair and dark skin, but in other ways they are genetically very distant from Africans.

Negroid   belonging or relating to a division of humankind represented by the indigenous peoples of central and southern Africa. The term is part of an outdated system of classification introduced in the nineteenth century to categorize human races.

neutralism   the belief that many mutations that are perpetuated are adaptively neutral, meaning that they do not confer any advantage but are more or less fortuitous in their time and place.

non-insulin diabetes   *See* **type II diabetes**.

non-metric variation   variation that is not a matter of larger or smaller but of distinctly different traits; differences in hair colour are an example of non-metric variation.

nonrecombinant   denoting genetic elements that are passed down by only one parent, not both.

nonrecombinant Y-chromosome (NRY)   genetic elements of the Y-chromosome (the male sex chromosome) that are passed down solely by the father.

notochord   a skeletal rod supporting the body in all embryonic and some adult chordate animals.

nuclear fission   a nuclear reaction in which a heavy nucleus is torn apart by the explosion of an atom, with the release of energy.

nucleotide base   a paired structure that represents the lowest or most basic unit of nucleic acids such as DNA.

objective   an attitude, striven for by traditionalist scientists, that requires detachment from one's personality, beliefs, aspirations, feelings, and prejudices so that these do not become involved in the interpretation of data. ▪ **objectivity** (*noun*)

obligate bipedalism *See* **bipedalism**.

occipital bun an extension of the back bone of the skull, associated with the crania of Neandertal.

Ojibwa an Aboriginal people whose communities (including the Anishinabe, Mississauga, Saulteaux, Chippewa, Algonquin, and Odawa) stretch from western Quebec to British Columbia. The Ojibwa speak a central Algonquian language and were one of the first groups to exhibit the X haplogroup mtDNA.

Oldowan a Mode-1 toolmaking tradition dating to about 2.0–1.5 million years ago (think 'old one'). It is characterized by primitive stone tools that are associated chiefly with *Homo habilis*. It is named after the Olduvai Gorge in Tanzania.

olfactory having to do with the sense of smell.

Onge a population or community of people living in the Andaman Islands in the Indian Ocean.

ontogenesis the development of an individual organism from the earliest stage to maturity. ▪ **ontogenetic** (*adjective*)

ontogeny the branch of biology concerned with ontogenesis.

opposable thumb a relatively large first digit that opposes or presses in the opposite direction of the other digits, making it useful for grasping.

opsin a protein that forms part of the visual pigment rhodospin and that is released by the action of light.

optional bipedalism *See* **bipedalism**.

orangutan paradox (in primatology) the once-common view that orangutans, as well as chimps, bonobos, and gorillas, demonstrate intelligence in performing tests for humans that does not seem to be matched in their natural lives. It has recently been demonstrated that both activities demonstrate significant intelligence, and there is no paradox.

order (in taxonomy) a rank occurring one level below class and one level above family.

orthogenesis evolution in which variations follow a particular direction as if by some kind of driving force that overrides natural selection; changes are not sporadic or merely fortuitous.

orthognathic having a relatively flat face beneath the nose. *Compare* **prognathic**.

ossuary *See* **Feast of the Dead**.

osteology the study of bones.

Out of Africa a hypothesis—the prevailing theory in human evolution today—suggesting that anatomically and behaviourally modern humans first developed in Africa and then spread to other parts of the world without significant genetic input from hominids already living there.

paleoanthropology the study of fossil remains and artifacts, usually with the goal of making statements about human biological evolution.

paleography the study of old writing, often with a focus on means of dating a text.

palynology the study of pollen grains and other spores, particularly as found in archaeological deposits. Pollen extracted from such deposits may be used for radiocarbon dating and for studying past climates and environments by identifying plants then growing. ▪ **palynogist** (*noun*).

panins chimps, bonobos

paraphyletic (as used by cladists) denoting or having to do with traditional biological, morphology-based taxonomies.

percussion flaking *See* **pressure flaking**.

phase a genetic or seasonal variety of an animal's coloration; example: (1) the blue phase of the Quaker parrot; (2) the black phase of the eastern grey squirrel.

phylogeography the science of situating mutations that characterize certain haplogroups in specific times and places.

phylogenetic change *Another term for* **anagenesis**.

phylogenetics *Another term for* **cladistics**.

phylum (*pl.* **phyla**) a taxonomic rank between kingdom and class.

physical anthropology the study of humans as biological organisms, dealing with their evolution, their primate relatives, their variation, and their identification for forensic purposes. *Also called* **biological anthropology**.

pithecophobia the fear of a genetic or ancestral connection between human and non-human primates.

placentals mammals that develop a placenta during the birth process; humans and dogs are examples.

plasticity the capacity of the cranium to change in size and shape over generations.

plate tectonics a geological theory describing the structure of the earth's crust and many associated phenomena, based on the interaction of seven or eight major lithospheric plates and many more minor plates moving slowly over the underlying mantle.

platyrrhine any member of a parvorder (**Platyrrhini**) that includes monkeys, apes, and humans (from Greek, lit. 'flat nose').

playbacks taped primate communications used to elicit behavioural and vocal responses.

point mutation a mutation affecting only one or very few nucleotides in a gene sequence.

polygeny the theory (not generally held today) that humans evolved from several independent pairs of ancestors.

pongid a member of a subfamily (**Pongidae**) within hominids that includes orangutans.

population either (a) a community of animals, plants, or humans among whose members interbreeding occurs, or (b) a finite or infinite collection of individuals under consideration or study.

population bottleneck *Another term for* **genetic bottleneck**.

population genetics the branch of science that deals statistically with the genetics of biological populations.

potsherd *Another term for* **sherd**.

prehensile tail a tail that has the capacity to grab and hold onto objects.

prehistory a term traditionally used to refer to a time before writing, or before a particular people wrote. Use of this term can reflect a prejudice in favour of literate over non-literate peoples.

pressure flaking a toolmaking technique that typically involves making flakes from larger ones by pressing with a tool such as an antler. This method is in opposition to **percussion flaking**, where striking is involved.

primate a member of the order of mammals including lemurs, tarsiers, monkeys, baboons, humans, gorillas, chimps, and (of course) gibbons.

primatology the study of mammals of the primate order, including human beings.

primitive traits *Another term for* **generalized traits**.

principle of cross-cutting relationships the idea that any rock that cuts through a given stratum was formed after that stratum was.

principle of lateral continuity the idea that strata originally extend unbroken over a great distance, only to have this lateral continuity broken by intruding rocks.

principle of original horizontality the idea that rock, typically, is first formed in a horizon position, like a calm surface of water.

prognathic (of the part of the face below the nose) sticking out or protruding. *Compare* **orthognathic**.

prosimians (in older primate taxonomy) a primate of the suborder Prosimii, which includes the strepsirrhines and the tarsiers.

proximal denoting or having to do with the end of the bone that is closer to the centre of the body. *Compare* **distal**.

pseudogenes genes that do not code (i.e. send messages to the RNA leading to the creation of some part of the body).

purine a double-ringed nucleotide base, esp. either adenine (A) or guanine (G).

pyrimidine a single-ringed nucleotide base, esp. either cystosine (C) or thymine (T).

race a category used to distinguish subspecies within a species. Humans lack clear biologically defined sets of features that allow for the identification of separate races.

racial time the last four to five centuries, when humans across the world have been described, categorized, and treated in racialized terms.

racialization the social process of using a relatively small set of physical characteristics to distinguish between populations in the world.

reconstruction the process of piecing together a fossil specimen. Reconstruction can be a source of controversy, as it was, famously, with *Kenyanthropus platyops*.

red ochre an earth dye long used by humans for artistic and symbolic purposes.

relative dating dating techniques that involve determining whether something is 'older than' or 'younger than' something else.

restriction gene a gene whose mutation blocks or restricts the entry or effectiveness of an attacking virus or disease.

retroelements products of a type of genetic copy-and-paste, in which the DNA makes an RNA copy of itself, which in turn makes a DNA copy of itself.

rhinarium the moist patch on the nose that facilitates the smelling process in dogs, cats, and lemurs.

rickets a softening of bones in children, caused esp. by a deficiency of vitamin D and/or calcium; the condition leads to deformity, particularly a bending of the weight-bearing long bones, and increases the chances of fractures (from old English *wrickken*, lit. 'to twist').

ring species any one of a set of subspecies situated along a continuum, in which nearby subspecies can interbreed but species at both ends of the continuum cannot.

RNA (**ribonucleic acid**) a nucleic acid present in all living cells, which acts as a messenger, carrying genetic instructions from the DNA to produce amino acids for protein and, ultimately, the production of body structures.

savannah hypothesis the theory, based on the work of Raymond Dart but not widely accepted today, that a drying climate change beginning about 10 million years ago caused the jungle or forest lands of Africa to give way to tall grasses and occasional bushes, forcing hominid species to adapt; the ancestors of modern humans adapted by becoming bipedal to see over tall grasses, according to the theory.

Schöningen spears four ancient (380,000–400,000 years old) javelin-like wooden spears found in Germany and formed with great technological sophistication.

secular trends differences in human growth and development measures over time; example: people in Canada are taller than they were 200 years ago, and reach puberty at an earlier age.

selective pressure the extent of advantage of a particular mutation or adaptation.

seriation a relative dating technique that is based on changes in the patterns or styles of particular items of material culture.

sexual dimorphism the differences in size and body structure between females and males.

sherd a broken piece of ceramic material found at an archaeological site (*short for* **potsherd**).

sickle-cell anemia an inherited blood disorder caused when a mutated form of hemoglobin distorts the red blood cells into a sickle or crescent shape.

silverback a mature male mountain gorilla, distinguished by an area of white or silvery hair across the back; it is the dominant member of its social group.

Simians (or **Simiformes**) an infraorder within the primate suborder Haplorrhini that includes monkeys, apes, and humans.

single-crystal laser fusion a superior method of radio-potassium dating that permits scientists to take samples from small, single crystals instead of from larger (i.e. several grams) specimens of rock.

sociocultural anthropology the branch of anthropology that studies (through fieldwork, traditionally) the societies or cultures of people not usually studied by other social scientists.

Solutrean hypothesis a flawed hypothesis of settlement in the Americas, advanced by Dennis Stanford and Bruce Bradley, according to which carriers of the European Solutrean material culture, roughly 21,000–17,000 years ago, travelled across the Atlantic and established the Clovis Point tradition, which is superficially similar.

specialized traits *Another term for* **derived traits**.

species a group of living organisms consisting of similar individuals capable of exchanging genes or interbreeding.

species concept *Another term for* **biological species concept**.

splitters biologists and physical anthropologists who are quick to declare a separate species based on relatively small differences among specimens. *Compare* **lumpers**.

St Lawrence Iroquoians a group of Iroquoian language speakers who lived along the St Lawrence River up until the late sixteenth century, after which they were dispersed and forced to join other peoples (e.g. the Huron); these people are *not* the Iroquois.

stereoscopic vision the capacity to see most objects in three dimensions, or to have depth perception.

stratigraphy the branch of geology concerned with analyzing the order and relative position of strata of stone or soil and their relationship to the geological timescale.

strepsirrhines members of a primate suborder (**Strepsirrhini**), distant from human beings, that includes lemurs, bushbabies, and lorises.

subjective, subjectivity (of information) produced by or coming from the subject, or the person who is reporting the information and thereby constructing the knowledge. ▪ **subjectivity** (*noun*).

Suchey-Brooks system a method of calculating age from pubic symphysis morphology, developed by Alison Brooks and Judy Suchey.

supplanting a behaviour noted in baboons in which a senior female approaches a junior-ranked female who automatically gives up her position to the more senior female.

swim bladder an internal gas-filled organ of a fish that enables it to control its buoyancy and stabilize itself in an upright position; it can sometimes act as a resonating chamber to produce or receive sound.

Taino an Aboriginal people of the Caribbean and northern South America, whose genetic heritage has recently been identified among supposedly non-Aboriginal people of the area.

taphonomy the study of the changes that take place over time in objects, primarily the bones and general remains of the bodies of animals (including humans), after the animals die or have their remains deposited somewhere.

Tarsiidae (or **Tarsiiformes**) an infraorder within the primate suborder Haplorrhini that includes tarsiers.

taxonomy a system of classifying organisms into different ranks, or taxons.

telomere a compound structure at the end of a chromosome, which keeps the chromosome from coming apart.

Terry Collection a collection of some 1,728 human skeletons, established by Robert J. Terry (1871–1966) and currently held by the Department of Anthropology of the National Museum of Natural History of the Smithsonian Institution.

tetrapod an animal with four limbs, including amphibians, reptiles, birds, and mammals.

thalessemia a blood disorder linked with particular populations and with malaria.

theistic evolution a belief that reconciles acceptance of evolution with a belief in a divine Creator or God.

theory a scientific explanation for a fact or set of facts.

thermoluminescence a method of dating ancient pottery, brick, or tile, in which the researcher reheats the material and counts the number of trapped electrons to find out how old the manufactured product is.

thrifty gene a genetic pattern that involves a high degree of maintaining blood sugars.

Thymine (T) a single-ringed or pyrimidine nucleotide base.

transposons (in genetics) a chromosomal segment that can undergo transposition, esp. a segment of bacterial DNA that can be translocated from one area to another.

transitions and consistency in the fossil record a proof of evolution based on evidence from the fossil record, which consistently shows living entities gradually becoming closer to their current forms over time.

tree model of evolution a conceptual model of evolution that involves only a few species deriving from each ancestor's species. *Compare* **bush model of evolution**.

tuff light, porous rock that consists of consolidated volcanic ash from a volcanic eruption.

type specimen *See* **holotype**.

type II diabetes a type of diabetes that typically strikes adults, not children, and which is not affected by insulin shots. *Also called* **late-onset diabetes**, **non-insulin diabetes**.

tyranny of the discontinuous mind a phrase coined by Richard Dawkins to represent his belief that by imposing 'species' on living beings, we overemphasize the discontinuity or separation between beings when the continuity or linking is much greater.

Upper Palaeolithic the last part of the 'Old Stone Age', associated with modern humans (typically European) of 40,000–10,000 years ago.

varve a pair of thin layers of clay and silt of contrasting colour and texture, which represent the deposit of a single year (summer and winter) in a lake or along a melting glacier. Varves can be measured to determine the chronology of glacial sediments.

wahoo contests duels between adult male baboons as to who is the loudest and can produce the longest call.

wear patterns microscopic signs that an object has been scratched or worn by human use.

Wendat *Another name for* **Huron**.

Y-chromosome the male sex chromosome, which is paired with but not matched by an X-chromosome

BONES

anvil *Another term for* **incus**.

baculum a bone in the penis of carnivores and some other mammals.

bipartite brow ridge (among modern *Homo sapiens*) a two-part brow ridge, which marks a departure from a supraorbital torus.

brow ridge *Another name for* **supraorbital torus**.

calcaneus the large bone forming the heel.

carpals (*sing.* **carpus**) the cluster of bones found in the wrist in tetrapods (including humans), which separate the radius and ulna from the metacarpus.

cephalic (*or* **cranial**) **index** the maximum width of the head multiplied by 100 and divided by the maximum length of the head, front to back.

condyle a large, knuckle-like projection often jutting out at the point of articulation between one bone and another; example: the lateral and medial condyles of the femur.

coronal suture the wavy lines separating the frontal bone from the left and right parietals of the skull.

cranium (*pl.* **crania**) the skull.

crista sagittalis a sagittal keeling of the parietal bones of *Homo erectus*, distinct from the sagittal ridge of the robust Australopithecines.

femur the single large bone of the upper leg; the thighbone. ▪ **femoral** (*adjective*).

femur length formula a formula (length × 2.11 + 70.35) developed by Mildred Trotter and Goldine Gleser in 1952 for using long bone lengths to estimate the height of the body from which the bone came.

fibula the thinner of the two lower leg long bones; the calf bone.

frontal the major bone of cranium or skull, located (unsurprisingly) at the front.

glenoid fossa a slight depression in the scapula or shoulder blade, which articulates with the head of the humerus to make the shoulder joint.

glenoid process the head of the scapula.

greater sciatic notch a groove occurring on each of the hip bones; it develops to become wider than a thumb in adult females but remains quite narrow in males.

hammer *Another name for* **malleus**.

humerus the single long bone of the upper arm. (Note: it is strictly coincidence that the elbow is called the 'funny bone'.)

hyoid a short, arch-shaped bone situated deep in the throat, often checked for signs of death by strangulation.

incus the small, anvil-shaped bone in the middle ear. *Also called* **anvil**.

keeling the curving or rounding of the *Homo erectus* skull, which ends with an edge resembling the keel of a boat.

lambdoidal suture the wavy lines that separate the parietal and temporal bones from the occipital bone.

malleus a small, hammer-shaped bone in the middle ear. *Also called* **hammer**.

mandible the lower jaw.

mandibular tori *Plural of* **torus mandibularis**.

maxilla the upper jaw.

metacarpals (*sing.* **metacarpus**) the group of five bones of each hand between the wrist and fingers.

metaphysis the part of a long bone where it begins to widen before it reaches the epiphyseal plate.

metatarsals (*sing.* **metatarsus**) the bones of the foot.

occipital (*or* **occiput**) a major cranial bone that forms the back of the skull.

parietals the two major cranial bones (left and right) on the top of the head.

patella the kneebone.

pericranium a layer of dense connective tissue (periosteum) that covers the cranial bones.

periosteum dense connective tissue that forms a membrane at the outer surface of bones.

phalanx (*pl.* **phalanges**) the set of three bones in each finger and toe (except for the big toe and the thumb): the distal, at the tip; the middle, or intermediate; and the proximal, closest to the hand or foot and articulating with the metacarpals of the hand or the metatarsals of the foot.

postcranial denoting the bones other than or below the cranium.

postorbital bar the bone that surrounds or encircles the orbit or eye socket of strepsirrhine primates (e.g. lemurs, bushbabies, and lorises).

pubic symphysis the false joint or symphysis at the front or anterior of the left and right pubic bones.

radius the forearm bone that extends from the thumb side of the wrist and, when the palm is extended upward, is positioned laterally, which means that it is the bone farthest away from the body; it is slightly shorter than the ulna.

sagittal crest a bony ridge along the sagittal suture; it is a diagnostic trait of *Australopithecus boisei*.

sagittal suture the wavy line down the centre of the top of the skull, separating the left and right parietals

scapula the shoulder blade.

stapes (*or* **stirrup**) the small, stirrup-shaped bone in the middle ear.

supraorbital torus a ridge or lump of bone above the orbits of the eyes. *Also called* **brow ridge**.

sutures wavy lines formed of dense, fibrous connective tissue that separate the major bones of the skull or cranium and the craniofacial area. *See* **coronal suture**; **lambdoidal suture**; **sagittal suture**; **vault sutures**.

tarsals (*sing.* **tarsus**) the bones that link the leg (through the tibia and fibula) to the foot (through the metatarsus) (from Greek, lit. 'flat basket').

temporals the cranial bones on the left and right side of the skull.

torus angularis a thickening or bulging of the top line (superior edge) of the temporal bone as it approaches the occipital bone in the back; it is a diagnostic trait of *Homo erectus*.

torus mandibularis (*pl.* **mandibular tori**) a bony growth in the mandible, typically one on the left and one on the right of the tongue, along the surface nearest to the tongue, near the premolars; it is a common trait among Asian and Inuit populations.

torus occipitalis a ridge at the back of the occipital bone.

tibia the frontmost and larger of the two bones of the lower leg; the shinbone.

ulna the forearm bone that begins on the 'bumpy' side of the wrist and that is positioned medially, which means that when the palm is extended upward, it is the bone closest to the body; it is slightly longer than the radius.

vault sutures the sutures associated with the cranium.

zygomatic arch the cheekbone.

TEETH

baby teeth *Another name for* **deciduous teeth**.

canines the pointed teeth that make up the second tooth of four types; there is one in each quadrant of the mouth.

caries decay and crumbling of a tooth or bone.

clacking *Another term for* **occlusional wear**.

deciduous teeth the first, temporary teeth that an individual has; like the leaves of deciduous trees, they eventually fall (out). *Also called* **baby teeth**, **milk teeth**.

dental arcade the shape of the arrangement of teeth in the upper jaw.

dental enamel hypoplasia a condition characterized by striations and grooves in the dental enamel, indicative of periods in which an individual experienced trauma, disease, or a scarcity of food.

diastema a space separating teeth of different functions, esp. that between the biting teeth (incisors and canines) and grinding teeth (premolars and molars).

hypoplasia *See* **dental enamel hypoplasia**.

incisors the cutting teeth at the front of the mouth; each quadrant has two.

milk teeth *Another name for* **deciduous teeth**.

molars the teeth situated the furthest back in the mouth; the usual human pattern is to have three in each quadrant.

occlusion the manner in which the upper and lower teeth come together when the mouth closes. ▪ **occlude** (*verb*).

occlusional wear the wear that comes from occlusions, statistically more common among East Asians. *Also called* **clacking**.

overbite the extent of vertical overlap of the maxillary central incisors ahead of the maxilla (upper jaw) over those of the mandible (lower jaw).

permanent teeth the second set of teeth that grow in, succeeding the deciduous teeth and lasting through adulthood.

premolars the third of the four types of teeth characteristic of mammals. Humans and other Old World monkeys have two of these in each quadrant of the mouth.

shovel-shaped incisors incisors that have a shovel shape or (more accurately) a scoop shape on the side that faces the back of the mouth.

sinodont denoting a derived set of dental traits that include shovel-shaped maxilla or upper incisors, single-rooted maxilla premolars, and three-rooted lower first molars (lit. 'China teeth').

sundadont denoting a set of dental traits that are more generalized that those in sinodont teeth, and are characteristic of the Ainu, the indigenous people of Taiwan, Southeast Asians, Filipinos, Indonesians, and Malaysians.

toothcomb a comb-like structure, useful for grooming, consisting of the incisors and canines of strepsirrhines (e.g. lemurs).

twelve-year molars a popular term for the second molars of a human being; they generally have erupted by age 12.

wisdom teeth the third molars of humans, which typically erupt sometime around the age of 18.

References

Adcock, G.J., Dennis, E.S., Easteal, S., Huttley, G.A., Jeremiin, L.S., Peacock, W.J., & Thorne, A. (2001). Mitochondrial DNA sequences in ancient Australians: Implications for modern human origins. *PNAS, 98*(2), 537–42.

Adovasio, J.M., & Page, J. (2002). *The first Americans: In pursuit of archaeology's greatest mystery*. New York, NY: Random House, Inc.

Agnew, N., & Demas, M. (1998). Preserving the Laetoli footprints. *Scientific American,* 44–55.

Alemseged, Z., Spoor, F., Kimbel, W., Bobe, R., Geraads, D., Reed, D., & Wynn, J. (2006). A juvenile early hominin skeleton from Dikika, Ethiopia. *Nature, 443,* 296–301.

Alexeev, V.P. (1986). *The origin of the human race*. Moscow: Progress Publishers.

American Association of Physical Anthropology. (1996). AAPA statement on biological aspects of race. *American Journal of Physical Anthropology, 101,* 569–70. Retrieved from http://physanth.org/association/position-statements/biological-aspects-of-race

American Museum of Natural History. (2003). The humans of Sima de los Huesos. Retrieved from www.amnh.org/exhibitions/atapuerca/simahumans/index.php

Amundson, R. (1987). Watson and the hundredth monkey phenomenon. *Skeptical Inquirer,* 303–4.

Anikovich, M.V., Sinitsyn, A.A., Hoffecker, J.F., Holliday, V.T., Popov, V.V., Lisitsyn, S.N., . . . , & Praslov, N.D. (2007). Early upper Paleolithic in Eastern Europe and implications for the dispersal of modern humans. *Science, 314*(5809), 223–26.

Arnold, J.R., & Libby, W.F. (1949). Age determinations by radiocarbon content: Checks with samples of known age. *Science, 110.* Retrieved from http://hbar.phys.msu.ru/gorm/fomenko/libby.htm

Arsuaga, J.L. (2001). *The Neanderthal's necklace: In search of the first thinkers*. New York, NY: Four Walls Eight Windows.

ASU News. (2007, October 17). Researchers find earliest evidence for modern human behaviour in South Africa. *Science Centric News.*

Bahn, P. (Ed.). (2003). *Written in bones: How human remains unlock the secrets of the dead*. Toronto, ON: Firefly Books.

Barker, P., Ellis, C., & Damadio, S. (2000). *Determination of cultural affiliation of ancient human remains from Spirit Cave, Nevada*. Reno, NV: Bureau of Land Management Nevada State Office.

Bass, E.J., & Jackson, J.F. (1977). Cerumen types in Eskimos. *American Journal of Physical Anthropology, 47*(2), 209–10.

Bass, Jefferson. (2006). *Carved in bone: A Body Farm novel*. New York: William Morrow.

Bass, W., & Jefferson, J. (2003). *Death's acre: Inside the legendary forensic lab, the Body Farm, where the dead do tell tales*. New York, NY: G. P. Putnam's Sons,

———, & ———. (2007). *Beyond the Body Farm: A legendary bone detective explores murders, mysteries, and the revolution in forensic science*. New York, NY: HarperCollins.

Beattie, O., & Geiger, J. (1987). *Frozen in time: Unlocking the secrets of the Franklin expedition*. Saskatoon, SK: Western Producer Prairie Books.

Berger, L.R., Churchill, S.E., De Klerk, B., Quinn, R.L., & Hawks, J. (2008). Small-bodied humans from Palau, Micronesia. *PLoS ONE, 3*(3), e1780. doi:10.1371/journal.pone.0001780

Berry, Shawn. (2003, November 19). Investigators hoping to put a name to body found in ocean 23 years ago. *New Brunswick Telegraph-Journal,* p. A3.

Bischoff, J.L., Shamp, D.D., Aramburu, A., Arsuaga, J.L., Carbonell, E., & Bermudez de Castro, J.M. (2003). The Sima de los Huesos hominids date to beyond U/Th equilibrium (>350 kyr) and perhaps to 400–500 kyr: New radiometric dates. *Journal of Archaeological Science, 30*(3), 275–80.

Bischoff, J.L., Williams, R.W., Rosenbauer, R.J., Aramburu, A., Arsuaga, J.L., Garcia, N., & Cuenca-Bescós, G. (2007). High-resolution U-series dates from the Sima de los Huesos hominids yields 600 kyrs: Implications for the evolution of the early Neanderthal lineage. *Journal of Archaeological Science, 34*(5), 763–70.

Blackwell, Bonnie A.B., Gong, Jane J.J., Blickstein, Joel I.B., Blais-Stevens, Andree, Skinner, Anne R., & Nelson, Robert E. (2008). ESR analyses with barnacles: A new method for dating shell middens and sealevel changes. Joint Meeting of the Geological Society of America, Soil Science Society of America, American Society of Agronomy, Crop Science Society of America, Gulf Coast Association of Geological Societies with the Gulf Coast Section of SEPM.

Boas, F. (1912). *Changes in the bodily form of the descendants of immigrants*. New York, NY: Columbia University Press.

Boaz, Noel T., & Almquist, Alan J. (1997). *Biological anthropology: A synthetic approach to human evolution*. Upper Saddle River, NJ: Prentice Hall.

Boaz, Noel T., & Ciochon, R.L. (2004). *Dragon Bone Hill: An Ice Age saga of* Homo erectus. Oxford, England: Oxford University Press.

Bocquet-Appel, J.-P., & Arsuaga, J.-L. (1999). Age distributions of hominid samples at Atapuerca (SH) and Krapina could indicate accumulation by catastrophe. *Journal of Archaeological Science, 26*(3), 327–38.

Brace, C. Loring. (2000). Does race exist? An antagonists's perspective. *NOVA Online.* Retrieved from www.pbs.org/wgbh/nova/first/brace.html

Breast cancer gene test eligibility in Ont. misses some. (2009, December 14). *CBCnews* (online). Retrieved from www.ca/health/story/2009/12/14/breast-cancer-mutations-jewish-ontario.html

Breuer, T., Ndoundou-Hockemba, M., & Fishlock, V. (2005). First observation of tool use in wild gorillas. *PLoS Biology, 3*(11). doi:10.1371/journal.pbio.0030380

Bromage, T., McMahon, J., Thackeray, J., Kullmer, O., Hogg, R., Rosenberger, A., Schrenk, F., & Enlow, D. (2008). Craniofacial architectural constraints and their importance for reconstructing the early Homo skull KNM-ER 1470. *Journal of Clinical Pediatric Dentistry, 33*(1), 43–54.

Brook, A.H., Underhill, C., Foo, L.K., & Hector, M. (2006). Approximal attrition and permanent tooth crown size in a Romano-British population. *Dental Anthropology, 19,* 23–8.

Brown, M., Hosseini, S., Torroni, A., Bandelt, H.-J., Allen, J., Schurr, T., Scozzari, R., Cruciani, F., & Wallace, D. (1998). mtDNA haplogroup X: An ancient link between Europe/Western Asia and North America? *American Journal of Human Genetics, 63*(6), 1852–61.

Brown, P. & Maeda, T. (2009). Liang Bua *Homo floresiensis* mandibles and mandibular teeth: A contribution to the comparative morphology of a new hominin species. *Journal of Human Evolution, 57,* 571–96.

Brunet, M., Beauvilain, A., Coppens, Y., Heintz, E., Moutaye, A., & Pilbeam, D. (1995). The first australopithecine 2,500 kilometres west of the Rift Valley (Chad). *Nature, 378*(6554), 273–75.

Brunet, M., Guy, F., Pilbeam, D., Mackaye, H., Likius, A., Ahounta, D., . . . , & Zollikofer, C. (2002). A new hominid from the Upper Miocene of Chad, Central Africa. *Nature, 418,* 145–51.

Burchard, Esteban González, Ziv, Elad, Coyle, Natasha, Gomez, Scarlett Lin, Tang, P. Hua, Karter, Andrew J., Mountain, Joanna L., Pérez-Stable, Eliseo J., Sheppard, Dean, & Risch, Neil. (2003, March 20). The importance of race and ethnic background in biomedical research and clinical practice. *New England Journal of Medicine, 348*(12), 1170–5.

Buss, D., Haselton, M., Shakelford, T., Bleske, A., & Wakefield, J. (1998). Adaptations, exaptations, and spandrels. *American Psychologist, 53*(5), 533–48.

Cann, R.L., Stoneking, M., & Wilson, A.C. (1987). Mitochondrial DNA and human evolution. *Nature, 325,* 31–6. doi:10.1038/325031a0

Caramelli, D., Lalueza-Fox, C., Vernesi, C., Lari, M., Casoli, A., Mallegni, F., Chiarelli, B., Dupanloup, I., Bertranpetit, J., Barbujani, G., & Bertorelle, G. (2003). Evidence for a genetic discontinuity between Neandertals and 24,000-year-old anatomically modern Europeans. *PNAS, 100*(11), 6593–97.

Caramelli, D., Milani, L., Vai, S., Modi, A., Pecchioli, E., Girardi, M., Pilli, E., Lari, M., Lippi, B., Ronchitelli, A., Mallegni, F., Casoli, A., Bertorelle, G., & Barbujani, G. (2008). A 28,000 years old Cro-Magnon mtDNA sequence differs from all potentially contaminating modern sequences. *PLoS One, 3*(7). doi:10.1371/journal.pone.0002700

Cartmill, M., Smith, F.H., & Brown, K.B. (2009). *The human lineage.* Hoboken, NJ: John Wiley and Sons.

Chen, T., Yang, Q., & Wu, E. (1994). Antiquity of *Homo sapiens* in China. *Nature, 368,* 55–6.

Cheney, D.L., & Seyfarth, R.M. (1990). *How monkeys see the world: Inside the mind of another species.* Chicago, IL: University of Chicago Press.

———, & ———. (2007). *Baboon metaphysics: The evolution of a social mind.* Chicago, IL: University of Chicago Press.

Ciochon, R., Olsen, J., & James, J. (1990). *Other origins: The search for the giant ape in human prehistory.* New York, NY: Bantam Books.

Coqueugniot, H., Hublin, J., Veillon, F., Houët, F., & Jacob, T. (2004). Early brain growth in *Homo erectus* and implications for cognitive ability. *Nature, 431,* 299–302.

Coyne, J. (2009). *Why evolution is true.* New York, NY: Penguin Books.

Creighton, Ellen. (2005, December 3). Anthropologists see life beyond bones. *The Times & Transcript* (Moncton, NB), pp. D1, D3.

Cybulski, J. (2001). Current challenges to traditional anthropological applications of human osteology in Canada. In L. Sawchuk & Susan Pfeiffer (Eds), *Out of the past: The history of osteology at the University of Toronto.* Toronto, ON: CITD Press.

Damon, P.E., Donahue, D.J., Gore, B.H., Hatheway, A.L., Jull, A.J.T., Linick, T.W., . . . , & Tite, M.S. (1989). Radiocarbon dating of the Shroud of Turin. *Nature, 337*(6208), 611–15.

Dart, R.A., & Craig, D. (1959). *Adventures with the missing link.* New York, NY: The Viking Press.

Darwin, Charles. (1859). *On the origin of the species by means of natural selection, or the preservation of favoured races in the struggle for life.* London: John Murray.

Dawkins, R. (2004). *The ancestor's tale: A pilgrimage to the dawn of life.* Boston, MA: Houghton Mifflin.

———. (2006). *The God delusion.* Boston, MA: Houghton Mifflin.

———. (2009). *The greatest show on earth: The evidence for evolution.* New York, NY: Free Press.

Dawson, William. (1901). *Fifty years of work in Canada: Scientific and educational. Being autobiographical notes by Sir William Dawson.* Ed. Rankine Dawson. London and Edinburgh: Ballantyne, Hanson, & Co.

Deacon, H.J. (2001). *Guide to Klasies River.* Stellenbosch, South Africa. Retrieved from http://academic.sun.ac.za/archaeology/KRguide2001.PDF

DeLaurier, A., & Spence, M.W. (2003). Cranial genetic markers: Implications for postmarital residence patterns. In R.F. Williamson & S. Pfeiffer (Eds), *Bones of the ancestors: The archaeology and osteobiography of the Moatfield Ossuary* (Mercury Series, Archaeology Paper 163). Gatineau, QC: Canadian Museum of Civilization.

Dennell, R. (1997). The world's oldest spears. *Nature, 385,* 767–68. doi:10.1038/385767a0

Derenko, M., Grzybowski, T., Malyarchuk, B., Czarny, J., Miścicka-Śliwka, D., & Zakharov, I. (2001). The presence of mitochondrial haplogroup X in Altaians from South Siberia. *American Journal of Human Genetics, 69*(1), 237–41.

d'Errico, F., Zilhao, J., Julien, M., Baffler, M., & Pelegrin, J. (1998). Neanderthal acculturation in Western Europe? A critical review of the evidence and its interpretation. *Current Anthropology, 39*(2), S1–S44. doi:10.1086/204689

de Waal, F. (1995). Bonobo sex and society: The behaviour of a close relative challenges assumption about male supremacy in human evolution. *Scientific American, 272*(3), 82–8.

———. (1997a). Are we in anthropodenial? *Discover, 18*(7), 50–3.

———. (1997b). *Bonobo: The forgotten ape.* Berkeley, CA: University of California Press.

DiCenzo, David. (2000, April 28). When did history begin? An anthropology professor searches for 'the origin of everyone'—and turns a popular theory on its head. *Folio, 37*(16). Retrieved from www.folio.ualberta.ca/37/17/back.html

Dienhart, Charlotte M. (1979). *Basic human anatomy and physiology.* Philadelphia: W.B. Saunders.

Dobzhansky, T. (1973). Nothing in biology makes sense except in the light of evolution. *The American Biology Teacher, 35,* 125–29.

Dorsey, G. (1898). A cruise among the Haida and Tlingit villages about Dixon's entrance. *Popular Science Monthly.*

Duke University. (2009, July 22). Human spear likely cause of death of Neandertal. *ScienceDaily.* Retrieved from www.sciencedaily.com/releases/2009/07/090720163729.htm

Ember, Carol R., et al. (2009). *Physical anthropology and archaeology* (3rd Cdn edn). Toronto: Pearson.

Fairbanks, D.J. (2007). *Relics of Eden: The powerful evidence of evolution in human DNA.* Amherst, NY: Prometheus Books.

Feder, K.L. (1990). *Frauds, myths, and mysteries: Science and pseudoscience in archaeolgy.* Palo Alto, CA: Mayfield Publishing Company.

Ferllini, R. (2002). *Silent witness: How forensic anthropology is used to solve the world's toughest crimes.* Toronto, ON: Firefly Books.

Finlayson, C. (2009). *The humans who went extinct: Why Neanderthals died out and we survived.* New York, NY: Oxford University Press.

Finlayson, C., Pacheco, F., Rodríguez-Vidal, J., Fa, D., López, J., Pérez, A., . . . , & Sakamoto, T. (2006). Late survival of Neanderthals at the southernmost extreme of Europe. *Nature, 443,* 850–53. doi:10.1038/nature05195.

Fossil Man 1. (1913, August 28). *Nature, 91,* 662–4. doi:10.1038/091662b0

Foster, J. Bristol. (1964, April 18). Evolution of mammals on islands. *Nature, 202,* 234–5. doi:10.1038/202234a0

Foster, M.W., Gilby, I.C., Murray, C.M., Johnson, A., Wroblewski, E.E., Pusey, A.E. (2009). Alpha male chimpanzee grooming patterns: implications for dominance 'style'. *American Journal of Primatology, 71*(2), 136–44. doi:10.1002/ajp.20632.

Galdikas, B.M.F. (1995). *Reflections of Eden: My years with the orangutans of Borneo.* New York, NY: Little, Brown and Company.

Galik, K., Senut, B., Pickford, M., Gommery, D., Treil, J., .Kuperavage, A.J., & Eckhardt, R.B. (2004). External and internal morphology of the BAR 1002'00 *Orrorin tugenensis* femur. *Science, 305*(6589), 1450–53. doi:10.1126/science.1098807

Gibbons, A. (2006). *The first human: The race to discover our earliest ancestors.* New York, NY: Doubleday.

Giberson, K.W. (2008). *Saving Darwin: How to be a Christian and believe in evolution.* New York, NY: HarperOne.

Gill, George W. (2000). Does race exist? A proponent's perspective. *NOVA Online.* Retrieved from www.pbs.org/wgbh/nova/first/gill.html

Glob, P.V. (2004). *The bog people: Iron Age man preserved.* New York, NY: New York Review Books Classics.

González Burchard, E., Ziv, E., Coyle, N., Gomez, S.L., Tang, H.P., Karter, A.J., Mountain, J.L., Pérez-Stable, E.J., Sheppard, D., & Risch, N. (2003). The importance of race and ethnic background in biomedical research and clinical practice. *New England Journal of Medicine, 348*(12), 1170–75.

Goodall, J. (1971). *In the shadow of man.* Boston, MA: Houghton Mifflin.

———. (1990). *Through a window: 30 years observing the Gombe chimpanzees.* Boston, MA: Houghton Mifflin.

Gould, S.J. (1977). The misnamed, mistreated, and misunderstood Irish elk. In *Ever since Darwin*, (pp. 79–90). New York, NY: W.W. Norton.

———. (1981a). Evolution as fact and theory. *Discover, 2,* 34–7. Reprinted in S.J.

Gould, (1994), *Hen's teeth and horses' toes* (pp. 253–62). New York, NY: W.W. Norton.

———. (1981b). *The mismeasure of man*. New York, NY: W.W. Norton.

———. (1987). *Time's arrow, time's cycle*. Cambridge, MA: Harvard.

———. (1989). *A wonderful life: The Burgess Shale and the nature of history*. New York, NY: W.W. Norton.

Gould, S.J., & Vrba, E.S. (1982). Exaptation—a missing term in the science of form. *Paleobiology, 8*(1), 4–15.

Green, R., Krause, J., Briggs, A., Maricic, T., Stenzel, U., Kircher, M., . . . , & Pääbo, S. (2010, May 7). A draft sequence of the Neandertal genome. *Science, 328*(5979), 710–22.

Groves, C. (1999). *Australopithecus garhi*: A new found link? *Reports of the National Center for Science Education, 19*(3), 10–13. Retrieved from www.noanswersingenesis.org.au/cg_australopithecus_garhi.htm

Guidon, N., & Pessis, A.-M. (1996). Falsehood or untruth? *Antiquity, 70*(268), 408–16.

Haile-Selassie, Y. (2001). Late Miocene hominids from the Middle Awash, Ethiopia. *Nature, 412*, 178–81.

Hanis, C., Hewett-Emmett, D., Bertin, T., & Schull, W. (1991). Origins of U.S. Hispanics. Implications for diabetes, *Diabetes Care, 14*(7), 618–27.

Hardy, J., Pittman, A., Myers, A., Gwinn-Hardy, K., Fung, H.C., de Silva, R., Hutton, M., &. Duckworth, J. (2005). Evidence suggesting that *Homo neandertalensis* contributed to the H2 MAPT haplotype of *Homo sapiens*. *Biochemical Society Transactions, 33*(4), 582–85.

Harper, K. (1996). *Give me my father's body: The life of Minik, the New York Eskimo*. Iqaluit, NU: Blacklead Books.

Harvard University. (2007, October 25). At least one percent of Neanderthals were redheads. *Science Centric News*.

Haydenblit, R. (1996). Dental variation among four prehispanic Mexican populations, *American Journal of Physical Anthropology, 100*(2), 225–46.

Hegele, R.A., Cao, H., Harris, S.B., Hanley, A.J.G., & Zinman, B. (1999). The hepatic nuclear factor-1 G319S variant is associated with early-onset type 2 diabetes in Canadian Oji-Cree. *The Journal of Clinical Endocrinology and Metabolism, 84*(3), 1077–82.

Hirst, K.K. (n.d.(a)) Stratigraphy and seriation. Timing is everything: A short course in archaeological dating. *About.com*. Retrieved from http://archaeology.about.com/cs/datingtechniques/a/timing.htm

———. (n.d.(b)) Seriation: A step by step discussion of the dating technique. *About.com*. Retreived from www.archaeology.about.com/od/dating/ss/seriation.htm

Hooper, J. (2002). *Of moths and men: An evolutionary tale: The untold story of science and the peppered moth*. New York, NY: W.W. Norton.

Howells, W.W. (1973). *Cranial variation in man: A study by multivariate analysis of patterns of difference among recent human populations*. Cambridge, MA: Harvard University, Peabody Museum of Archaeology and Ethnology.

Ingman, M., Kaessmann, H., Pääbo, S., & Gyllensten, U. (2000). Mitochondrial genome variation and the origin of modern humans. *Nature, 408*, 708–13. doi:10.1038/35047064.

Itan, Y., Powell, A., Beaumont, M.A., Burger, J., & Thomas, M.G. (2009). The origins of lactase persistence in Europe. *PLoS Computational Biology, 5*(8). doi:10.1371/journal.pcbi.1000491

Johanson, D.C., & Wong, K. (2009). *Lucy's legacy: The quest for human origins*. New York, NY: Random House.

Joyce, C., & Stover, E. (1991). *Witnesses from the grave: The stories bones tell*. New York, NY: Ballantine Books.

Jull, A.J.T., Donahue, D.J., Broshi, M., & Tov, E. (1995). Radiocarbon dating of scrolls and linen fragments from the Judean desert. *Radiocarbon, 37*(1), 11–19.

Kaifu, Y., Aziz, F., & Babo, H. (2005). Hominid mandibular remains from Sangiran: 1952–1986 collection. *American Journal of Physical Anthropology, 128*(3), 497–519. doi:10.1002/ajpa.10427.

Katzenberg, M. Anne. (2001, May 2). Destructive analyses of human remains in the age of NAGPRA and related legislation. In Larry Sawchuk & Susan Pfeiffer (Eds), *Out of the past: The history of human osteology at the University of Toronto*. Scarborough, ON: CITDPress. Retrieved online at https://tspace.library.utoronto.ca/citd/Osteology/Katzenberg.html

Klaatsch, Hermann. (1923). *The evolution and progress of mankind* (trans. Joseph McCabe). London: T. Fischer Unwin.

Koff, C. (2005). *The bone woman: A forensic anthropologist's search for truth in Rwanda, Bosnia, Croatia, and Kosovo*. Toronto, ON: Random House.

Koppel , T. (2003). *Lost world: Rewriting prehistory—how new science is tracing America's Ice Age mariners*. New York, NY: Atria Books.

Kreger, C. David. (n.d.). *Homo habilis*. Archaeology Info. Retrieved from www.archaeologyinfo.com/Homohabilis.htm

Krulwich, R. (2009, February 2). Your family may once have been a different color. *NPR*. Retrieved from www.npr.org/templates/story/story.php?storyId=100057

Kuhn, S.L., & Stiner, M.C. (2006). What's a mother to do? The division of labor among Neandertals and modern humans in Eurasia. *Current Anthropology, 47*(6), 953–63.

Lange, Karen E. (2007, September). Tales from the bog. *National Geographic* (online). Retrieved from http://ngm.nationalgeographic.com/2007/09/bog-bodies/bog-bodies-text

Larsen, C.S. (2000). *Skeletons in our closet: Revealing our past through bioarchaeology*. Princeton, NJ: Princeton University Press.

Leakey, L.S.B., Tobias, P.V., & Napier, J.R. (1964). A new species of genus *Homo* from Olduvai Gorge. *Nature, 202*, 7–9.

Leakey, Mary D, & Hay, Richard L. (1979). Pliocene footprints in the Laetoli beds at Laetoli, northern Tanzania. *Nature, 278*, pp. 317–23.

Leakey, R. (1994). *The origin of humankind*. New York, NY: Basic Books.

Le Gros Clark, W.E. (1967). *Man-apes or ape-men? The story of discoveries in Africa*. Holt, Rinehart and Winston: New York.

Lewin, R. (1987). *Bones of contention: Controversies in the search for human origins*. New York, NY: Simon and Schuster.

Lewontin, R.C. (1972). The apportionment of human diversity. *Evolutionary Biology, 6*, 391–98.

Linville, S., Breitwisch, R., & Schilling, A. (1998). Plumage brightness as an indicator of parent care in northern cardinals. *Animal Behaviour, 55*(1), 119–27. doi:10.1006/anbe.1997.0595

Lovejoy, C.O. (2009). Reexamining human origins in light of *Ardipithecus ramidus*. [Special issue]. *Science, 326*(5949), 74, 74e1–74e8. doi:10.1126/science.1175834

Lovejoy, C.O., Latimer, B., Suwa, G., Asfaw, B., & White, T. (2009). Combining prehension and propulsion: The foot of *Ardipithecus ramidus.*, [Special issue]. *Science, 326*(5949), 72, 72e1–72e8. doi:10.1126/science.1175832

Lovejoy, C.O., Simpson, S., Matternes, J., & White, T. (2009). The great divides: *Ardipithecus ramidus* reveals the postcrania of our last common ancestors with African apes. [Special issue]. *Science, 326*(5949), 73, 100–6. doi:10.1126/science.1175833

Lovejoy, C.O., Simpson, S., White, T., Asfaw, B., Suwa, G. (2009). Careful climbing in the Miocene: The forelimbs of *Ardipithecus ramidus* and humans are primitive. [Special issue]. *Science, 326*(5949), 70, 70e1–70e8. doi:10.1126/science.1175827

Lovejoy, C.O., Suwa, G., Spurlock, L., Asfaw, B., & White, T. (2009). The pelvis and femur of *Ardipithecus ramidus*: The emergence of upright walking. [Special issue]. *Science, 326*(5949), 71, 71e1–71e6. doi:10.1126/science.1175831

Lu, Zun'e. (1989). Date of Jinniushan man and his position in human evolution. *Liaohai Wenwu Xuekan, 1*, 44–55.

McAndrews, J.H., & Turton, C.E. (2005). The goosing of Crawford Lake with prehistoric corn pollen. *CAP Newsletter, 28*(1), 5–7. Retrieved from www.scirpus.ca/cap/articles/paper035.htm

McGilloway, Brian. (2009). *Bleed a river deep*. London: Macmillan.

McKern, T.W., & Stewart, T.D. (1957). *Skeletal age changes in young American males: Analysed from the standpoint of age identifi-*

cation. Natick, MA: Quartermaster Research & Development Center, U.S. Army, Environmental Protection Research Division.

Malhi, R.S., Schultz, B.A., & Smith, D.G. (2001). Distribution of mitochondrial DNA lineages among Native American tribes of Northeastern North America. *Human Biology 73*, 23–51.

Manzi, G., Mallegni, F., & Ascenzi, A. (2001). A cranium for the earliest Europeans: Phylogenetic position of the hominid from Ceprano, Italy. *PNAS, 98*(17), 10011–16. doi:10.1073/pnas.151259998

Marean, C.W. (1998). A critique of the evidence for scavenging by Neandertals and early modern humans: New data from Kobeh Cave (Zagros Mountains, Iran) and Die Kelders Cave 1 layer 10 (South Africa). *Journal of Human Evolution, 35*(2), 111–36. doi:10.1006/jhev.1998.0224

Martinez, J.C., Toro-Labrador, G., Viera-Viera, J., Rivera-Vega, M.Y., Startek, J., Latorre-Esteves, M., Roman-Colon, A., Rivera-Torres, R., Navarro-Millan, I.Y., Gomez-Sanchez, E., Caro-Gonzalez, H.Y., & Valenci-Rivera, P. (2005). Reconstructing the population history of Puerto-Rico by means of mtDNA phylogeographics analysis. *American Journal of Physical Anthropology, 128*(1), 131–55. doi:10.1002/ajpa.20108

Mayr, Ernst. (1942). *Systematics and the origin of species, from the viewpoint of a zoologist.* Cambridge: Harvard University Press.

Mellars, P., Gravina, B., & Bronk Ramsey, C. (2006). Confirmation of Neanderthal/modern human interstratification at the Chatelperronian type-site. *PNAS, 104*(9), 3657–62. doi:10.1073/pnas.0608053104

Meltzer, D., Adovasio, J., & Dillehay, T. (1994). On a Pleistocene human occupation at Pedra Furada, Brazil. *Antiquity, 68*(261), 695–714.

Mendizabal, I., Sandoval, K., Berniell-Lee, G., Calafell, F., Salas, A., Martínez-Fuentes, A., & Comas, D. (2008). Genetic origin, admixture, and asymmetry in maternal and paternal human lineages in Cuba. *Biomedical Evolutionary Biology, 8*(213), 1–10. doi:10.1186/1471-2148-8-213

Merrett, D. (2003). Moatfield demography. In R.F. Williamson & S. Pfeiffer (Eds), *Bones of the ancestors: The archaeology and osteobiography of the Moatfield Ossuary* (pp.171–88). Gatineau, QC: Canadian Museum of Civilization.

Michael, J. (1988). A new look at Morton's craniological research. *Current Anthropology, 29*(2), 349–54.

Mineri, Rossana, Ghilardi, Giosuè, Brambilla, Simona, Sarno, Tiziana, Bologna, Muriel, Valaperta, Serenella, & Montanelli, Alessandro. (2009, January 27). The genetic variant of lactase persistence C/T (13910) and blood glucose control in type 1 diabetes mellitus patients. *Mediterranean Journal of Nutrition*

and Metabolism, 1(3): 181–5. doi:10.1007/s12349-008-0027-y

Morgan, J. (2009, February 12). Neanderthals 'distinct from us'. *BBC News*. Retrieved from http://news.bbc.co.uk/2/hi/science/nature/7886477.stm

Morin, E. (2008). Evidence for declines in human population densities during the early Upper Paleolithic in Western Europe. *PNAS, 105*(1), 48–53. doi:10.1073/pnas.0709372104

Morton, S.G. (1839). *Crania Americana; or, a comparative view of the skulls of various Aboriginal nations of North and South America: To which is prefixed an essay on the varieties of the human species.* Philadelphia, PA: J. Dobson.

Morwood, M., & van Oosterzee, P. (2009). *A new human: the startling discovery and strange story of the 'Hobbits' of Flores, Indonesia.* Walnut Creek, CA: Left Coast Press.

Mounier, A., Marchal, F., & Condemi, S. (2009). Is *Homo heidelbergensis* a distinct species? New insight on the Mauer mandible. *Journal of Human Evolution, 56*(3), 219–46. doi:10.1016/j.jhevol.2008.12.006

Myers, Elaine. (1985, Spring). The hundredth monkey revisited: Going back to the original sources puts a new light on this popular story. *In Context, 9*, 10. Retrieved from www.context.org/ICLIB/IC09/Myers.htm

Nakbunlung, Supaporn, & Wathanawareekool, Sukhontha. (2004). Talking teeth: Dental non-metric traits of wooden-coffin people in the Pang Ma Pha Cave sites, northwestern Thailand. *Indo-Pacific Prehistory Association Bulletin*, 24, 139–42. Retrieved from www.ejournal.anu.edu.au/index.php/bippa/article/view/119/109

Nash, S.E. (2000). Seven decades of archaeological tree-ring dating. In S. Nash (Ed.), *It's about time: A history of archaeological dating in North America* (pp. 60–82). Salt Lake City, UT: University of Utah Press.

Neel, J.V. (1962). Diabetes mellitus: A 'thrifty' genotype rendered detrimental by 'progress'? *American Journal of Human Genetics, 14*, 353–62.

Neves, W., & Hubbe, M. (2005). Cranial morphology of early Americans from Lagoa Santa, Brazil: Implications for the settlement of the New World. *PNAS, 102*(51), 18309–14. doi:10.1073/pnas.0507185102

O'Connor, J.E., Lewis, M.E., Last, J., & Spence, L. (2004, December). A method to establish chronological age from radiographic assessment of epiphyseal union at the knee. *Journal of Anatomy, 205*(6), 543.

Ogilive, M., Curran, B., & Trinkaus, E. (1988). Incidence and patterning of dental enamel hypoplasia among the Neandertals. *American Journal of Physical Anthropology, 79*(1), 25–41.doi:10.1002/ajpa.1330790104

Ohman, J.C., Lovejoy, C.O., White, T.D., Eckhardt, R.B., Galik, K., & Kuperavage,

A.J. (2005). Questions about *Orrorin* Femur. *Science, 307*(5711), 845–46. doi:10.1126/science.307.5711.845b

Oppenheimer, S. (2004). *The real Eve: Modern man's journey out of Africa.* New York, NY: Carroll & Graf Publishers.

Orlando, L., Darlu, P., Toussaint, M., Bonjean, D., Otte., M., & Hänni, C. (2006). Revisiting Neandertal diversity with a 100,000 year old mtDNA sequence. *Current Biology, 16*(11), R400–R402. doi:10.1016/j.cub.2006.05.019

Ovchinnikov, I., & Goodwin, W. (2001). The isolation and identification of Neanderthal mitochondrial DNA. *Profiles in DNA*, 7–9. Retrieved from www.promega.com/profiles/402/ProfilesinDNA_402_09.pdf

Palmer, D. (2006). *Seven million years: The story of human evolution.* London: Phoenix Paperbacks.

Palumbi, Stephen R. 2001. *The evolution explosion: How humans cause rapid evolutionary change.* New York: Norton

Perez, I., Bernal, V., Gonzalez, P., Sardi, M., & Politis, G. (2009). Discrepancy between cranial and DNA data of early Americans: Implications for American peopling. *PLos One, 4*(5). doi:10.1371/journal.pone.0005746

Pfeiffer, Susan. The health of the Moatfield People as reflected in palaeopathological features. In R.F. Williamson & S. Pfeiffer (Eds), *Bones of the ancestors: The archaeology and osteogeography of the Moatfield Site.* Archaeological Survey of Canada, Mercury Series Paper #163. Ottawa: Canadian Museum of Civilization.

Quintyn, C.B. (2006). *The morphometric affinities of the Qafzeh and Skhul hominans: Method and theory.* Bloomington, IN: AuthorHouse.

Rabino Massa, E. (1977). Presence of thalassemia in Egyptian mummies. *Journal of Human Evolution, 6*, 223–5.

Rahman, P., Jones, A., Curtis, J., Bartlett, S., Peddle, L., Fernandez, B.A., & Freimer, N.B. (2003). The Newfoundland population: a unique resource for genetic investigation of complex diseases. *Human Molecular Genetics, 12*(2), R167–R172. doi:10.1093/hmg/ddg257

Ramirez Rozzi, F.V., Bromage, T., & Schrenk, F. (1997). UR 501, the Plio-Pleistocene hominid from Malawi. Analysis of the microanatomy of the enamel. *Comptes Rendus de l'Académie des Sciences, 325*(3), 231–34. doi:10.1016/S1251-8050(97)88294-8

Reader, John. (1981). *Missing links: The hunt for earliest man.* Boston: Little Brown & Co.

Relethford, J.H. (2001). Ancient DNA and the origin of modern humans, *PNAS, 98*(2), 390–91.

Repcheck, J. (2003). *The man who found time: James Hutton and the discovery of the Earth's antiquity.* Cambridge, MA: Perseus Publishing.

Richmond, B., & Jungers, W. (2008). *Orrorin tugenensis* femoral morphology and the evolution of hominin bipedalism. *Science, 319*(5870), 1662–65. doi:10.1126/science.1154197

Rightmire, G.P. (1990). *The evolution of Homo erectus: Comparative anatomical studies of an extinct human species*. Cambridge: Cambridge University Press.

Ross, A., & Robins, D. (1990). *The life and death of a Druid prince: The story of Lindow man an archaeological sensation*. New York, NY: Touchstone.

Rougier, H., Milota, S., Rodrigo, R., Gherase, M., Sarcină, L., Moldovan, O., . . . , & Trinkaus, E. (2007). Peștera cu Oase 2 and the cranial morphology of early modern Europeans. *PNAS, 104*(4), 1165–70. doi:10.1073/pnas.0610538104

Russon, A.E. (1998). The nature and evolution of intelligence in orangutans (*Pongo pygmaeus*). *Primates, 39*(4), 485–503.

St Hoyme, Lucile B., & İşcan, Mehmet Yaşar. (1989). Determination of sex and race: Accuracy and assumptions. In Mehmet Yaşar İşcan & Kenneth A.R. Kennedy (Eds), *Reconstruction of life from the skeleton* (53–93). New York: Wiley-Liss.

Sarkar, S. (2007). *Doubting Darwin: Creationist designs on evolution*. Malden, MA: Blackwell Publishing.

Sawchuk, L., & Pfeiffer, S. (Eds), *Out of the past: The history of human osteology at the University of Toronto*. Toronto, ON: CITD Press.

School of Archaeology, University of Oxford. (2010). Luminescence dating: Age range and precision. Retrieved from www.arch.ox.ac .uk/L.html

Schwartz, Jeffrey H. (2007). *Skeleton keys* (2nd edn). New York: Oxford University Press.

Senut, B., Pickford, M., Wolpoff, M.H., Hawks, J., & Aherne, J. (2006). An ape or the ape: Is the Toumai cranium TM266 a hominid? *PaleoAnthropology*, 36–50.

Shermer, M. (2006). *Why Darwin matters: The case against intelligent design*. New York, NY: Henry Holt and Company.

Shipman, P. (1998). *Taking wing: Archaeopteryx and the evolution of bird flight*. New York: Simon & Schuster.

———. (2001). *The man who found the missing link: Eugène Dubois and his lifelong quest to prove Darwin right*. Cambridge, MA: Harvard University Press.

Shubin, N. (2009). *Your inner fish: A journey into the 3.5 billion year history of the human body*. New York, NY: Vintage Books.

Skelton, Randall R. (1996). *Skull graphics in jpg and bmp formats*. Missoula, MO: University of Montana.

Smith, Donald B. (1974). *Le sauvage: The Native people in Quebec historical writing of the Heroic Period (1534–1663) of New France*. Mercury Series, History Division Paper #6. Ottawa: National Museum of Man.

Smith, G.D., Lawlor, D.A., Timpson, N.J., Baban, J., Kiessling, M., Day, I.N.M., & Ebrahim, S. (2008). Lactase persistence-related genetic variant: Population substructure and health outcomes. *European Journal of Human Genetics, 17*, 357–67. doi:10.1038/ejhg.2008.156

Smith, William E., & Vollers, Maryanne. (1986, Sept. 1). Rwanda case of the Gorilla Lady murder. *Time*. Retrieved from www.time.com/time/magazine/article/0,9171,962196,00.html

Soficaru, A., Dobos, A., & Trinkaus, E. (2006). Early modern humans from the Peștera Muierii, Baia de Fier, Romania. *PNAS, 103*(46). 17196–201. doi:10.1073/pnas.0608443103

Spalding, L. (1999). *The follow*. Toronto, ON: Key Porter Books.

Spoor, F., Leakey, M., Gathogo, P., Brown, F., Antón, S., McDougall, I., . . . , & Leakey, L. (2007). Implications of new early *Homo* fossils from Ileret, east of Lake Turkana, Kenya. *Nature, 448*, 688–91. doi:10.1038/nature05986

Stanford, Craig, Allen, John S., Anton, Susan C., & Lovell, Nancy. (2008). *Biological Anthropolgy* (Cdn edn). Toronto: Pearson Education Canada.

Stansfield, William D. (2000). *Death of a rat: Understandings and appreciations of science*. Amherst, NY: Prometheus.

Steadman, D.W. (2003). *Hard evidence: Case studies in forensic anthropology*. Upper Saddle River, NJ: Prentice-Hall.

Steckley, John L. (2008). *White lies about the Inuit*. Peterborough, ON: Broadview.

Stieglitz, Tara. (2007, October 3). Tools may dig up historical clues. *The Gateway (The official student newspaper at the University of Alberta)*. Retrieved from http://thegatewayonline.ca/tools-may-dig-up-historical-clues-20071003-1005

Stiner, M. (1991). The faunal remains from Grotta Guattari: A taphonomic perspective. *Current Anthropology, 32*(2), 103–17.

Subramanian, S., Denver, D.R., Millar, C.D., Heupink, T., Aschrafi, A., Emslie, S.D., . . . , & Lambert, D.M. (2009). High mitogenomic evolutionary rates and time dependency. *Trends in Genetics, 25*(11), 482–86. doi:10.1016/j.tig.2009.09.005

Suwa, G., Asfaw, B., Kono, R., Kubo, D., Lovejoy, C.O., & White, T. (2009). The *Ardipithecus ramidus* skull and its implications for hominid origins. [Special issue]. *Science, 326*(5949),68,68e1–68e7.doi:10.1126/science .1175825

Swanscombe skull: Committee of Investigation [news story]. (1937, June 19). *Nature, 139*, 1048. doi:10.1038/1391048a0

Swisher III, C.C., Curtis, G.H., & Lewin, R. (2000). *Java Man: How two geologists' dramatic discoveries changed our understanding of the evolutionary path to modern humans*. New York, NY: Scribner.

Swisher III, C.C., Rink, W., Anton, S., Schwarcz, H., Curtis, G., Supryo, A., & Widiasmoro. (1996). Latest *Homo erectus* of Java: Potential contemporaneity with *Homo sapiens* in Southeast Asia. *Science, 274*(5294), 1870–4. doi:10.1126/science.274.5294.1870

Tattersall, I., & Schwartz, J.H. (1999). Hominids and hybrids: The place of Neanderthals in human evolution. *PNAS, 96*(13), 7117–19.

Taylor, R.E. (2000). The introduction of radiocarbon dating. In S. Nash (Ed.), *It's about time: A history of archaeological dating in North America* (pp. 84–104). Salt Lake City, UT: University of Utah Press.

Thomas, David Hurst. (1999). One archaeologist's perspective on the Monte Verde controversy. *Archaeological Institute of America*. Retrieved from www.archaeology .org/online/features/clovis/thomas.html

———. (2000). *The skull wars: Kennewick Man, archeology, and the battle for Native American identity*. New York: Basic Books.

Thorpe, S.K.S., Holder, R.L., & Crompton, R.H. (2007). Origin of human bipedalism as an adaptation for locomotion on flexible branches. *Science, 316*(5829), 1328–31. doi:10.1126/science.1140799

Tobias, P.V. (1992). J.C. Boileau Grant and the changing face of anatomy. *Clinical Anatomy, 5*(5), 409–16. doi:10.1002/ca.980050508

Todd, T. Wingate, & Lyon, D.W. (1924). Endocranial suture closure. Its progress and age relationship. Part I: Adult males of white stock. *American Journal of Physical Anthropology, 7*(3), 325–84.

———, & ———. (1925a), Endocranial suture closure. Its progress and age relationship. Part II: Ectocranial closure in adults males of white stock. *American Journal of Physical Anthropology, 8*(1), 23–45.

———, & ———. (1925b). Endocranial suture closure. Its progress and age relationship. Part III: Endocranial closure in adults males of negro stock. *American Journal of Physical Anthropology, 8*(1), 47–71.

———, & ———. (1925c). Endocranial suture closure. Its progress and age relationship. Part IV: Ectocranial closure in adults males of negro stock. *American Journal of Physical Anthropology, 8*(2), 149–68.

Tremblay, M.S., & Willms, J.B. (2000). Secular trends in the body mass index of Canadian children. *Canadian Medical Association Journal, 163*(11), 1429–33.

Trinkaus, E., Moldovan, O., Milota, S., Bîlgăr, A., Sarcina, L., Athreya, S., . . . , & van der Plicht, J. (2003). An early modern human from the Peștera cu Oase, Romania. *PNAS, 100*(20), 11231–36. doi:10.1073/pnas.2035108100

Trinkaus, E., & Shipman, P. (1994). *The Neandertals: Of skeletons, scientists and scandal*. New York, NY: Vintage Books.

Tudge, C., & Young, J. (2009). *The link: Uncovering our earliest ancestor*. New York, NY: Little, Brown and Company.

Turner, C.G. (1989). Major features of Sundadonty and Sinodonty, including suggestions about East Asian microevolution, population history, and late Pleistocene relationships with Australian aboriginals. *American Journal of Physical Anthropology, 82*(3), 295–317. doi:10.1002/ajpa.133082030Tyler-

Smith, C., Hurles, M., & Jobling, M. (2005). Don't mix radiocarbon and calendar years. *Nature, 434,* 697. doi:10.1038/434697c

Tyson, Edward. (1751). *Orang-outang, sive* Homo sylvestris: *Or, the anatomy of a pygmy compared with that of a monkey, an ape, and a man* (2nd edn). London: T. Osborne.

Underhill, P., Shen, P., Lin, A., Jin, L., Passarino, G., Yang, W., . . . , & Oefner, P. (2000). Y chromosome sequence variation and the history of human populations. *Nature Genetics, 26,* 358–61. doi:10.1038/81685

van der Merwe, N., Pfeiffer, S., Williamson, R., & Thomas, S. (2003). The diet of the Moatfield people. In R.F. Williamson & S. Pfeiffer (Eds), *Bones of the ancestors: The archaeology and osteobiography of the Moatfield Ossuary* (pp. 205–22). Gatineau, QC: Canadian Museum of Civilization.

van Oosterzee, P. (1999). *Dragon bones: The story of Peking Man.* St. Leonard's, Australia: Allen & Unwin.

van Schaik, C. (2004). *Among orangutans: Red apes and the rise of human culture.* Cambridge, MA: Belknap Press.

Vekua, A., Lordkipanidze, D., Rightmire, G.P., Agusti, J., Ferring, R., Maisuradze, . . . , & Zollikofer, C. (2002). A new skull of early *Homo* from Dmanisi, Georgia. *Science, 297*(5578), 85–9. doi:10.1126/science.1072953

Wade, N. (2006, January 29). Japanese scientists identify ear wax gene. *The New York Times.* Retrieved from www.nytimes.com/2006/01/29/science/29cnd-ear.html

Walker, A., & Shipman, P. (1996). *The wisdom of bones: In search of human origins.* London, England: Weidenfeld and Nicolson.

——, & ——. (2005). *The ape in the tree: An intellectual and natural history of Proconsul.* Cambridge, MA: Harvard University Press.

Walker, M., Gibert, J., Lopez, M., Lombardi, A.V., Perez-Perez, A., Zapata, J., . . . , & Trinkaus, E. (2008). Late Neandertals in southeastern Iberia: Sima de las Palomas del Cabezo Gordo, Murcia, Spain. *PNAS, 105*(52), 20631–36. doi:10.1073/pnas.0811213106

Walker, P. (2000). Bioarchaeological ethics: A historical perspective on the value of human remains. In M.A. Katzenberg & S.R. Saunders (Eds), *Biological anthropology of the human skeleton* (pp. 3–39). New York, NY: Wiley-Liss.

Walker, P.L., Cook, D.C., & Lambert, P.M. (1997). Skeletal evidence for child abuse: A physical anthropological perspective. *Journal of Forensic Sciences, 42*(2), 196–207. doi:10.1520/JFS14098J

Walker, A.C., Zimmerman, M.R., & Leakey, R.E. (1982). A possible case of hypervitaminosis A in *Homo erectus. Nature, 296,* 248–50.

Ward, L.M., Gaboury, I., Ladhani, M., & Zlotkin, S. (2007). Vitamin D-deficiency rickets among children in Canada. *Canadian Medical Association Journal, 177*(2), 161–66.

Ward, R.H., Frazier, B.L., Dew-Jager, K., & Pääbo, S. (1991, October 1). Extensive mitochondrial diversity within a single Amerindian tribe. *Proceedings of the National Academy of Sciences of the United States of America, 88*(19), 8720-4.

Weiner, J.S. (1980). *The Piltdown forgery.* New York, NY: Dover Publications Inc.

Weiss, E. (2009). Sex differences in humeral bilateral asymmetry in two hunter-gatherer populations: California Amerinds and British Columbia Amerinds. *American Journal of Physical Anthropology, 140*(1), 19–24. doi:10.1002/ajpa.21025

Wells, H.G. (1922). *A short history of the world.* London: Waterlow & Sons, Ltd.

Wells, S. (2006). *Deep ancestry: Inside the Genographic Project.* Washington, DC: National Geographic Society.

Wheatley, B.P. (1999). *The sacred monkeys of Bali.* Prospect Heights, IL: Waveland Press.

White, T., Asfaw, B., Beyene, Y., Haile-Selassie, Y., Lovejoy, C., Suwa, G., & WoldeGabriel, G. (2009). *Ardipithecus ramidus* and the paleobiology of early hominids. [Special issue]. *Science, 326*(5949), 64, 75–86. doi:10.1126/science.1175802

White, T.D., Asfaw, B., De Gusta, D., Gilbert, H., Richards, G., Suwa, G., & Howell, F.C. (2003). Pleistocene *Homo sapiens* from Middle Awash, Ethiopia. *Nature, 423*(6941), 742–47. doi:10.1038/nature01669

Whiten, A., Goodall, J., McGrew, W.C., Nishida, T., Reynolds, V., Sugiyama, Y., Tutin, C.E.G., Wrangham, R.W., & Boesch, C. (1999). Cultures in chimpanzees. *Nature, 399*(6737), 682–85. doi:10.1038/21415

WHO Regional Medical Research Centre (ICMR). (n.d.). Tribal health: Health, nutritional, and demographic study among the Onges. Retrieved from www.rmrc.res.in/projects/tribes/complete/Onges.htm

Wilford, John Noble. (1999, April 25). Discovery suggests Man is a bit Neanderthal. *The New York Times.* Retrieved from www.nytimes.com

Wiwchar, D. (2004). Nuu-Chah-Nulth blood returns to west coast. *Ha-Shilth-Sa Newspaper, 31*(25). Retrieved from http://caj.ca/wp-content/uploads/2010/mediamag/awards2005/(David%20Wiwchar,%20Sept.%2012,%202005)Blood2.pdf

WoldeGabriel, G., White, T.D., Suwa, G., Renne, P., de Heinzelin, J., Hart, W.K., & Heiken, G. (1994). Ecological and temporal placement of early Pliocene hominids at Aramis, Ethiopia. *Nature, 371*(6495), 330–33. doi:10.1038/371330a0

Wolpoff, M.H., & Caspari, R. (1997). *Race and human evolution.* New York: Simon and Schuster.

Wood, B., & Lonergan, N. (2008). The hominin fossil record: taxa, grades and clades. *Journal of Anatomy, 212,* 354–76. doi:10.1111/j.1469-7580.2008.00871

Wright, J.V., & Pilon, J.-L. (Eds). (2004). *A passion for the past: Papers in honour of James F. Pendergast.* (Mercury Series, Archaeology Paper 164). Gatineau, QC: Canadian Museum of Civilization.

Yamei, H., Potts, R., Baoyin, Y., Zhengtang, G., Deino, A., Wei, W., Clark, J., Guangmao, X., Weiwen, H. (2000). Mid-Pleistocene Acheulean-like stone technology of the Bose basin, South China. *Science, 287*(5458), 1622–26. doi:10.1126/science.287.5458.1622

Ybarra, Michael. (2005, June 15). Out of thin air. *The Wall Street Journal.*

Zegura, S., Karafet, T., Zhivotovosky, L., & Hammer, M. (2004). High-resolution SNPs and microsatellite haplotypes point to a single, recent entry of Native American Y chromosomes into the Americas. *Molecular Biology and Evolution, 21*(1), 164–75. doi:10.1093/molbev/msh009

Zilhão, J. (2001). The Lagar Velho child and the fate of the Neanderthals. *Athena Review, 2*(4), 33–9.

Credits

Grateful acknowledgement is made for permission to reproduce the following:

TEXT

Page 79, block quote: Stephen Jay Gould, *Ever Since Darwin*, pp79–90, New York: W. W. Norton (1977).

Page 114 and 115, block quotes: Corothy L. Cheney and Robert M. Seyfarth, *Baboon Metaphysics: The Evolution of a Social Mind*, University of Chicago Press, 2007.

Page 120 and 121, block quotes: Biruté M. F. Galdikas, *Reflections of Eden: My Years with the Orangutans of Borneo*, New York: Little, Brown and Company, 1995

Page 122–3, block quote: Thomas Breuer, Mireille Ndoundou-Hockemba, Vicki Fishlock, 2005, 'First Observation of Tool Use in Wild Gorillas' www.plosbiology. org/article/info:doi/10.1371/journal. pbio.0030380

Page 141, 143–4, block quotes: Donald C. Johanson and Kate Wong, *Lucy's Legacy: The Quest for Human Origins*, Random House, New York, 2009.

Page 145, block quote: Raymond A. Dart with Dennis Craig, *Adventures with the Missing Link*, Raymond A. Dart with Dennis Craig, The Viking Press, New York, 1959.

Page 166, block quote: Noel T. Boaz and Russell L. Ciochon, *Dragon Bone Hill: An Ice Age Saga of Homo erectus*, Oxford: Oxford University Press, 2004.

Page 170, block quote (in 'Profiles' box): Mike Morwood and Penny Van Oosterzee, *A New Human: the Startling Discovery and Strange Story of the 'Hobbits' of Flores, Indonesia*, Walnut Creek, CA: Left Coast Press, 2009

Page 184, block quote: Spencer Wells, 2007, *Deep Ancestry: Inside the Genographic Project*, Washington: National Geographic Society, page 38

Page 204, block quote: Erik Trinkaus and Pat Shipman, *The Neandertals: Of Skeletons, Scientists and Scandal*, Vintage Books, New York, 1994

Page 282, news clip: New York Times, September 17, 1897

Pages 286–7, block quote: Bill Bass & John Jefferson, *Beyond the Body Farm: A Legendary Bone Detective Explores Murders, Mysteries, and the Revolution in Forensic Science*, HarperCollins, New York, 2007

Pages 299–300, block quote: Clea Koff, *The Bone Woman: A Forensic Anthropologist's Search for Truth in Rwanda, Bosnia, Croatia, and Kosovo*, Random House, Toronto, 2005

FIGURES AND TABLES

Page 6, Figure 1.1: Angelika Steckley

Page 12, Figures 1.2, 1.3: Based on figures in: Dienhart, Charlotte M. 1979. *Basic Human Anatomy and Physiology*. Philadelphia: W.

B. Saunders Company; Howells, W. W. 1973. *Cranial Variation in Man: A Study by Multivariate Analysis of Patterns of Difference Among Recent human Populations*. Cambridge, Massachusetts: Harvard University, Peabody Museum of Archaeology and Ethnology; and Skelton, Randall R. 1996. *Skull Graphics in jpg and bmp formats*. Missoula, Montana: University of Montana.

Page 31, Table 2.1: 'Seriation: A Step by Step Discussion of the Dating Technique' by K. Kris Hirst. www.archaeology.about.com/od/dating/ss/seriation.htm

Page 65, Figure 3.4: Coyne, *Why Evolution is True* (Penguin, 2009), p. 41

Page 68, Figure 3.6: Darwin, Beagle Notebook B-1837, p. 36

Page 69, Figure 3.7: Modified from Encyclopaedia Britannica, Inc.; http://www.britannica.com/EBchecked/topic/270557/homology

Page 87, Figure 4.3: Shubin, Neil, 2009, *Your Inner Fish: A Journey Into the 3.5 Billion Year History of the Human Body*, New York: Vintage Books p. 183

Page 93, Figure 4.5: Adapted from Ember, Carol R. et al. (2009). *Physical Anthropology and Archaeology* (3rd Canadian ed.). Toronto: Pearson, p. 110

Page 94, Figure 4.6: *The Ape In the Tree: An Intellectual and Natural History of Proconsul* Alan Walker and Pat Shipman, Cambridge, Harvard University Press, 2005 p. 146

Page 97, Table 4.5: Alan Walker and Pat Shipman, 1996, *The Wisdom of Bones: In Search of Human Origins*, London: Weidenfeld and Nicolson

Page 106, Figure 5.2: Adapted from *Physical Anthropology and Archaeology*, 3/e, Ember et al (Pearson)

Page 107, Figure 5.3: Adapted from: Illustrator: Helen Beare; © Australian Museum

Page 108, Figure 5.4: Palmer, Doulas (2006). *Seven Million Years: The Story of Human Evolution*. London: Phoenix Press (p. 26)

Page 118, unlabelled table: Adapted from Ember, Carol R. et al. (2009). *Physical Anthropology and Archaeology* (3rd Canadian ed.). Toronto: Pearson, p. 115

Page 111, Figure 5.5: Noel T. Boaz and Alan J. Almquist , *Biological Anthropology: A Synthetic Approach to Human Evolution*, Upper Saddle River, NJ: Prentice Hall (1997), p. 203

Page 116, Table 5.2: Alan Walker and Pat Shipman, 1996, *The Wisdom of Bones: In Search of Human Origins*, London: Weidenfeld and Nicolson

Page 123, Figure 5.6: Anne E. Russon, 1998 'The Nature and Evolution of Intelligence in Orangutans (Pongo pygmaeus)', *Primates*, 39 (4), October. p. 497

Page 133, Figure 6.1: Jeffrey H. Schwartz, *Skeleton Keys*, 2/e (OUP US, 2007), p. 11

Page 137, Figure 6.4: Palmer, Doulas (2006). *Seven Million Years: The Story of Human Evolution*. London: Phoenix Paperbacks, p. 116

Page 159, Figure 7.2: Noel T. Boaz and Russell L. Ciochon, 2004, *Dragon Bone Hill: An Ice Age Saga of Homo erectus*, Oxford: Oxford University Press, p. 56.

Page 165, Table 7.3: G. Philip Rightmire, 1990, *The Evolution of Homo Erectus: Comparative anatomical studies of an extinct human species*, Cambridge: Cambridge University Press

Page 174, Figure 7.5: Douglas Palmer, *Seven Million Years: The Story of Human Evolution*, 2006, London: Orion Books, p. 208

Page 196, Figure 8.3: *Biological Anthropology*, Canadian Edition, by Craig Stanford et al., p. 381 (Pearson).

Page 207, Figure 9.1: Encyclopaedia Britannica, Inc. 2010.

Page 208, Figure 9.2: *Biological Anthropology*, Canadian Edition, by Craig Stanford et al., p. 301 (Pearson).

Page 211, Figure 9.2: Based on Stanford et al, *Biological Anthropology*, C/e, p. 298 (Pearson).

Page 234, Figure 10.1: Mayo Foundation for Medical Education and Research

Page 237, Table 10.2: Supaporn Nakbunlung, and Sukhontha Wathanawareekool, 2004 'Talking Teeth: Dental Non-Metric Traits of Wooden-Coffin People in the Pang Ma Pha Cave Sites, Northwestern Thailand', ejournal.anu.edu.au/index.php/bippa/article/view/119/109

Page 263, Figure 11.1: Palmer, Doulas (2006). *Seven Million Years: The Story of Human Evolution*. London: Phoenix Paperbacks, p. 197.

Page 285, Table 12.2: William Bass, *Beyond the Body Farm: A Legendary Bone Detective Explores Murders, Mysteries, and the Revolution in Forensic Science*, New York: HarperCollins, 2007

Page 285, Table 12.3: William Bass, *Beyond the Body Farm: A Legendary Bone Detective Explores Murders, Mysteries, and the Revolution in Forensic Science*, New York: HarperCollins, 2007

Page 285, Table 12.4: William Bass, *Beyond the Body Farm: A Legendary Bone Detective Explores Murders, Mysteries, and the Revolution in Forensic Science*, New York: HarperCollins, 2007

Page 287, Table 12.5: Adapted with additions from Schwartz 2007: 293–4.

Page 293, Table 12.11: Adapted from Schwartz 2007: 222

Page 293, Table 12.12: Adapted from Schwartz 2007: 222–3

Index